Proceedings of the NATO Advanced Study Institute on
Thermodynamics and the Optimization of Complex Energy Systems
Neptun, Romania
13-24 July 1998

A C.I.P. Catalogue record for this book is available from the Library of Congress.

ISBN 978-0-7923-5725-4 ISBN 978-94-011-4685-2 (eBook)
DOI 10.1007/978-94-011-4685-2

Printed on acid-free paper

Table of Contents

Preface

This book provides a comprehensive assessment of the methodologies of thermodynamic optimization, exergy analysis and thermoeconomics, and their application to the design of efficient and environmentally-sound energy systems. The contributors cover a wide spectrum of educational backgrounds and research orientations, from engineering to physics, and from fundamentals to actual industrial applications. Each chapter was presented to a select and highly-animated audience of 100, at the NATO Advanced Study Institute held on July 13-24 in Neptun, Romania.

The chapters are organized in a sequence that begins with pure thermodynamics and progresses toward the blending of thermodynamics with other disciplines (e.g., heat transfer and cost accounting) in the pursuit of more realistic models and more significant optimization results. This is also how this very active field has developed in time. In this sequence, three methods of analysis and optimization stand out, and are documented through specific examples and applications.

First, entropy generation minimization combines from the start the laws of classical thermodynamics with principles of transport phenomena (e.g., heat transfer and fluid mechanics) in the construction of simple models. Entropy generation is expressed as a function of the physical parameters (geometry, size, materials) of the device. Their minimization is then carried out subject to global constraints such as overall size and time of operation. The specific tutorial examples collected in this book show how the method became established in the decades of the '50s, '60s and '70s in engineering. The significant growth that has occurred in the '80s and '90s, and the new titles that have been attached recently to the method (endoreversible, exoirreversible, finite-time, finite-size, finite-resources) are also documented (see e.g., *Am. J. Phys.*, 1994, Vol. 62, pp. 11-12 and *J. Non-Equilibrium Thermodynamics*, 1996, Vol. 21, pp. 239-242). It is shown that system "structure" springs out of the optimization subject to global constraints.

Second, exergy (or availability) analysis shows how the destruction of "useful" energy can be traced at the component and sub-component level, by accounting for irreversible flows, interactions, and regions of nonequilibrium. The contributors stress the importance of describing the exergy-reference environment (the "dead" state) unambiguously and completely. They also document the merits of the method with respect to pinpointing system features that could be improved thermodynamically. Presented are also new techniques (computer-aided, modular, artificial intelligence), and several exergy-based efficiencies for evaluating thermodynamic performance and the improvements that result from optimization.

Third, thermoeconomics (or exergoeconomics) combines thermodynamics with principles of cost accounting, and the ultimate objective is cost minimization. The cost of each component and stream is related to its exergy content. Capital costs of the components and the environmental impact are taken into account systematically. The authors present several computer-based modular techniques for implementing the method. They also document the expansion of the method into life-cycle analysis, and its application to specific industrial examples of complex energy systems.

In summary, this book reviews the current directions in a field that is both extremely important in practice and alive intellectually. Additionally, new directions for research on thermodynamics and optimization are unveiled. For example, the occurrence of shape and structure in natural flow systems is now reasoned on the basis of geometry optimization subject to global constraints (constructal theory).

* * *

We close with a few thoughts directed to student readers of this book. The Neptun institute was truly a historic event in a field that is experiencing tremendous growth. The invited contributors to this volume are arguably the best known authors in the field. No less than six of the invited speakers have won the premier thermodynamics award bestowed by the American Society of Mechanical Engineers International—the James Harry Potter Gold Medal. They have written books and chronicled our discipline. They have come together to pass on to the next generation *the method*, with all its strengths, limitations and areas for real growth.

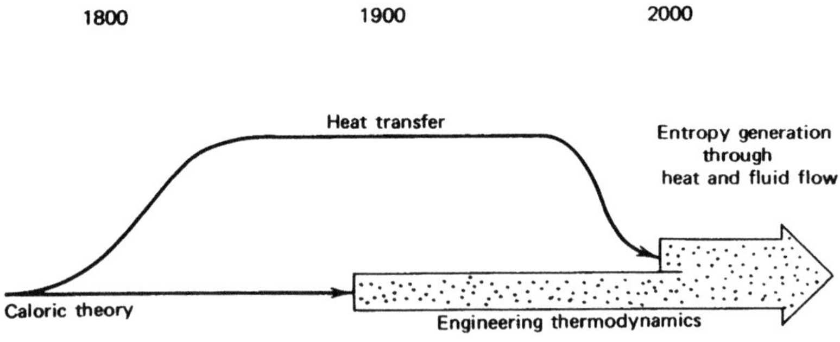

Figure 1. The predicted merger of classical thermodynamics with heat transfer and fluid mechanics (A. Bejan, *Entropy Generation through Heat and Fluid Flow*, Wiley, New York, 1982).

Creative individuals have a tendency to overstate the importance of their contributions, especially in the early stages when they must pursuade others that the contributions are both novel and valuable. Students are advised to read this book critically and to question authority. Conflicting claims and disagreements between the authors of the following chapters can be clarified by reading the original sources. By knowing the history of your discipline, your work can be more original and valuable to others.

Thermodynamic optimization has made spectacular progress. The prediction made graphically in 1982 in the first textbook dedicated the method (Fig. 1) has come true: classical thermodynamics has merged with heat transfer and fluid mechanics into a method of modeling and optimization subject to global constraints. This method unites education with research, engineering with physics, and fundamentals with applications. There is even a rush to teach thermodynamics with heat transfer in a single course, as was predicted in Fig. 1. Today, this broad field is being enlarged still more by additional considerations such as cost accounting, life-cycle analysis, computer-aided design, and natural self-organization and self-optimization (constructal theory). The path is open to new ways and new ideas.

July 1998

Adrian Bejan
Duke University
Durham, NC, USA

Eden Mamut
Ovidius University
Constanta, Romania

Acknowledgement

We wish to thank the following sponsors of the NATO Advanced Study Institute on *Thermodynamics and the Optimization of Complex Energy Systems*, Neptun, Romania, July 13-24, 1998:

NATO Scientific and Environmental Affairs Division
Ovidius University, Constanta
Duke University
Neptun - Olimp S.A.
Romanian Electric Power Authority (RENEL)
Tomis S.A.
Santierul Naval Constanta
Petromidia S.A.
Danubius Travel Agency
Adrian & Mary Bejan

We also thank the following colleagues for their personal contributions before and during the institute:

Adrian Bavaru, President, Ovidius University, Constanta
Aureliu Leca, President, Romanian Electric Power Authority
Constantin Bratianu, General Director, Ministry of National Education, Romania
Dumitru Filip, General Director, s.c. Neptun-Olimp s.a.
Florentin Filote, Technical Director, s.c. Neptun-Olimp s.a.
Yalcin Gögüs, International Center for Applied Thermodynamics
Adina Gogan, Assistant Professor, Ovidius University, Constanta
Teodor Popa, Assistant Professor, Ovidius University, Constanta
Ion Aldea, Assistant Professor, Navy Scientific Researcher Center, Constanta

The Scientific Commitee of the institute was:

Adrian Bejan, Duke University, USA
Elias P. Gyftopoulos, Massachusetts Institute of Technology, USA
Sadik Kakac, University of Miami, USA
Eden Mamut, Ovidius University, Romania
Michael J. Moran, The Ohio State University, USA
George Tsatsaronis, Technical University of Berlin, Germany

This book and the preliminary version used in Neptun were prepared by

Deborah Alford, Duke University
Madalina Ursu, Ovidius University
Rita Avram, Ovidius University

A.B. & E. M.

PRESENTATION OF THE FOUNDATIONS OF THERMODYNAMICS IN ABOUT TWELVE ONE-HOUR LECTURES

E.P. GYFTOPOULOS
Massachusetts Institute of Technology
Department of Nuclear Engineering, Room 24-111
Cambridge, MA 02139, USA

1. Introduction

In many expositions of thermodynamics, heat is introduced at the outset of the logical development as an intuitive and self-evident concept, independent of the laws of the subject [1, 2].

For example, Feynman [3] describes heat as one of several different forms of energy related to the jiggling motion of particles stuck together and tagging along with each other [3, pp. 1–3 and 4–2], a form of energy which really is just kinetic energy — internal motion [3, p. 4–6], and is measured by the random motions of the atoms [3, p. 10–8]. Tisza [4] argues that such slogans as "heat is motion," in spite of their fuzzy meaning, convey intuitive images of pedagogical and heuristic value. Brush [5] entitles his two-volume history of the kinetic theory of gases "The kind of motion we call heat."

There are at least two problems with these illustrations. First, work and heat are not stored in a system. Second, and perhaps more important, concepts of mechanics are used to justify and make plausible the notion of heat which is beyond the realm of mechanics. In spite of these logical drawbacks, the trick works because at first the student finds the idea of heat harmless, even natural. But the situation changes drastically as soon as heat is used to define a host of new concepts, less natural and less harmless. Heat is contrasted to work and used as an essential ingredient in the first law. The student begins to worry because heat is less definite than and not as operational as work.

An attempt to address the first problem is made in some expositions. Landau and Lifshitz [6] define heat as the part of an energy change of a body that is not due to work. Guggenheim [7] defines heat as an exchange of energy that differs from work and is determined by a temperature difference. Keenan [8] defines heat as that which transfers from one system to a second system at lower temperature, by virtue of the temperature difference, if the two are brought into communication. Similar definitions are adopted in other textbooks.

These definitions, however, are ambiguous, and none addresses the basic problem, that is, the existence of exchanges of energy that differ from work, and cannot be justified by mechanics. Such exchanges are one of the striking results of thermodynamics, and are due to entropy being a property of matter. Hatsopoulos and Keenan [9] have pointed out explicitly that without the second law heat and

A. Bejan and E. Mamut (eds.), Thermodynamic Optimization of Complex Energy Systems, 1–18.

work would be indistinguishable and, therefore, a satisfactory definition of heat is unlikely without a prior statement of the second law.

In our experience, whenever heat is introduced before the first law, and then used in the statement of the second law and in the definition of entropy, the student cannot avoid but sense ambiguity and lack of logical consistency. This results in the wrong but unfortunately widely spread conviction that thermodynamics is a confusing, ambiguous, and hand-waving subject.

During the past three decades of teaching thermodynamics to students from all over the globe, we have sensed a need to address these concerns. In response, we have composed an exposition [10] in which we strive to develop the basic concepts without ambiguities and logical inconsistencies, building upon the student's sophomore background in introductory physics.

The basic concepts and principles are introduced in a novel sequence that eliminates the problem of incomplete definitions, and that is valid for both macroscopic and microscopic systems, and for both stable or thermodynamic equilibrium states and for states that are not stable equilibrium. The laws of thermodynamics are presented as partial complements to the incomplete law of mechanical dynamics. Heat plays no role in the first law, the definition of energy, the second law, the definition of entropy, and the general concepts of energy and entropy exchanges between interacting systems. It emerges as a consequence of these concepts and laws. The discussion that follows is a summary of the key aspects of the novel logical sequence that we propose in our exposition. Though it makes no reference to the ideas of quantum thermodynamics, the exposition is construed so as to be entirely compatible with quantum theory. Such compatibility is absolutely essential to the resolution of the dilemma about the relation between mechanics and thermodynamics.

Experience at the University of Brescia, Italy, and MIT, Tufts University, Northeastern University, Washington University in St. Louis, and the University of Florida in the USA indicates that the fundamental ideas can be presented in about twelve one-hour lectures, with sufficient time for explanations and illustrations short of elaborate proofs. In what follows, we provide a syllabus of the twelve lectures without the necessary numerical and graphical illustrative examples given in [10]. However, the use of numerical and graphical illustrations is absolutely necessary and strongly recommended.

A variant of the sequence of the topics is to begin with the energy versus entropy graphs of Lecture 11 without defining any term. This approach gives the students a visual conception of the ideas that are presented in Lectures 1 to 10, and stimulates interest in the details of the course.

2. Lecture 1

2.1. GENERAL THERMODYNAMICS

We define *general thermodynamics* or simply *thermodynamics* as the study of motions of physical constituents (particles and radiations) resulting from externally applied forces, and from internal forces (the actions and reactions between constituents). This definition is identical to that given by Timoshenko and Young about mechanical dynamics [11]. However, because of the second law, the definition encompasses a much broader spectrum of phenomena than mechanical dynamics. We will see that thermodynamics accounts for phenomena with both zero and positive values of entropy, whereas classical mechanics and ordinary quantum mechanics account only for phenomena with zero values of entropy.

2.2. KINEMATICS: CONDITIONS AT AN INSTANT IN TIME

In kinematics we give verbal definitions of the terms system, environment, property, and state so that each definition is valid without change in any physical theory, and involves no statistics attributable to lack of information. The definitions include innovations.

2.2.1. *System*

A *system* is defined as a collection of constituents. For our purposes, we consider only one constituent which is determined by the following specifications: (a) The type of the constituent and the range of values of its amount; (b) The type and the range of values of the parameters that fully characterize the external forces exerted on the constituent by bodies other than the constituent itself, such as the volume of a container. The external forces do not depend on coordinates of bodies other than those of the constituent of the system. For our purposes, we consider only volume as a parameter; and (c) The internal forces between particles of the constituent, such as intermolecular forces. The internal forces depend on the coordinates of all the interacting particles.

Everything that is not included in the system is the *environment*. We denote the amount of the constituent by n, and the value of the volume by V. At any instant in time, the amount of the constituent and the volume have specific values. We denote these values by n and V with or without additional subscripts.

2.2.2. *Property*

By themselves, the values of the amount of the constituent and of the volume at an instant in time do not suffice to characterize completely the condition of the system at that time. We also need the values of properties at the same instant in time. A ***property*** is an attribute that can be evaluated at any given instant in time by means of a set of measurements and operations that are performed on the system and result in a numerical value — the ***value of the property***. This value is independent of the measuring devices, other systems in the environment, and other instants in time.

Two properties are ***independent*** if the value of one can be varied without affecting the value of the other.

2.2.3. *State*

For a given system, the values of the amount of the constituent, the value of the volume, and the values of a complete set of independent properties encompass all that can be said about the system at an instant in time and about the results of any measurements that may be performed on the system at that same instant. We call this complete characterization of the system at an instant in time the ***state*** of the system. Without change, this definition of state applies to any branch of physics.

3. Lecture 2

3.1. DYNAMICS: CHANGES OF STATE IN TIME

3.1.1. *Equation of Motion*

The state of a system may change in time either spontaneously as the constituent tries to conform to the external and internal forces or as a result of interactions with other systems, or both. A system that experiences only spontaneous changes of state is called *isolated*. A system that is not isolated interacts with the environment in a number of different ways, some of which may result in net flows of properties between the system and the environment. For example, collision between a system and the environment results in the flow or transfer of momentum to or from the system.

The relation that describes the evolution of the state of an isolated system — spontaneous changes of state — as a function of time is the *equation of motion*. Certain time evolutions obey *Newton's equation* which relates the total force F on each system particle to its inertial mass m and acceleration a so that $F = ma$. Other evolutions obey the *time-dependent Schroedinger equation*, that is, the quantum-mechanical equivalent of Newton's equation. Other experimentally observed time evolutions, however, do not obey either of these equations. So the equations of motion that we have are incomplete. The discovery of the complete equation of motion that describes all physical phenomena remains a subject of research at the frontier of science — one of the most intriguing and challenging problems in thermodynamics.

Many features of the equation of motion have already been discovered. These features provide not only guidance for the discovery of the complete equation but also a powerful alternative procedure for analyses of many time-dependent, practical problems. Two of the most general and well-established features are captured by the consequences of the first and second laws of thermodynamics presented in subsequent lectures.

3.1.2. *Interactions and Processes*

Rather than through the explicit time dependence which requires the complete equation of motion, a change of state can be described in terms of: (a) the *interactions* that are active during the change of state; (b) the *end states* of the system, that is, the initial and final states; and (c) *conditions on the values of properties* of the end states that are consequences of the laws of thermodynamics, that is, conditions that express not all, but the most general and well-established features of the complete equation of motion. Each interaction is characterized by means of well-specified net flows of properties across the boundary of the system. For example, after defining the properties energy and entropy, we will see that some interactions involve the flow of energy across the boundary of the system without any flow of entropy, whereas other interactions involve the flows of both energy and entropy. Among the conditions on the values of properties, we will see that the energy change of a system must equal the energy exchanged between the system and its environment, whereas the entropy change must not be less than the entropy exchanged between the system and its environment.

4. Lecture 3

4.1. ENERGY AND ENERGY BALANCE

Energy is a concept that underlies our understanding of all physical phenomena, yet its meaning is subtle and difficult to grasp. It emerges from a fundamental principle known as the first law of thermodynamics but is not a part of the statement of that law.

4.2. FIRST LAW OF THERMODYNAMICS

The *first law* asserts that any two states of a system may always be the initial and final states of a change (weight process) that involves no net effects external to the system except the change in elevation between z_1 and z_2 of a weight, that is, a mechanical effect. Moreover, for a given weight, the value of the expression $Mg(z_1 - z_2)$ is fixed only by the end states of the system, where M is the inertial mass of the weight, and g the gravitational acceleration.

4.2.1. *Definition of Energy*

One consequence (theorem) of the first law is that every system A in any state A_1 has a property called *energy*, with a value denoted by the symbol E_1. The energy E_1 can be evaluated by a weight process that connects A_1 and a reference state A_0 to which is assigned an arbitrary reference value E_0 so that

$$E_1 = E_0 - Mg(z_1 - z_0) \tag{1}$$

Energy is an *additive property*, that is, the energy of a composite system is the sum of the energies of its subsystems. Moreover, energy either has the same value at the final time as at the initial time if the process is a zero-net-effect weight process, or remains invariant in time if the process is spontaneous. In either of these two processes, $z_2 = z_1$ and $E(t_2) = E(t_1)$ for time t_2 greater than t_1, that is, energy is *conserved*. Energy conservation is a time-dependent result. Here it is obtained without use of the general equation of motion. It is noteworthy that energy is defined for any system (both macroscopic and microscopic) and for any state, and is not statistical.

4.2.2. *Energy Balance*

In the course of interactions, energy can be exchanged between a system and its environment. Denoting by $E^{A\leftarrow}$ the amount of energy exchanged between the environment and system A in a process that changes the state of A from A_1 to A_2, we can derive the *energy balance*. This balance is based on the additivity of energy and energy conservation, and reads

$$(E_2 - E_1)_{\text{system } A} = E^{A\leftarrow} \tag{2}$$

where $E^{A\leftarrow}$ is positive if energy flows into A. In words, the energy change of system A must be accounted for by the net energy crossing the boundary of the system.

4.3. MASS BALANCE

The values of energy disclosed by the first law are relative because the choice of the reference value E_0 is arbitrary. We can assign absolute values by using the theory of special relativity and the concept of (inertial) mass.

In general, and in contrast to energy, mass is neither additive nor conserved. However, in the absence of nuclear reactions, and creation and annihilation reactions, the mass changes caused within a system by energy exchanges between either constituents and the electromagnetic field or by chemical reactions, or both, are negligible with respect to the mass of the system. As a result, we establish a very useful tool, the *mass balance*, i.e.,

$$(m_2 - m_1)_{\text{system } A} = m^{A\leftarrow} \tag{3}$$

where $m^{A\leftarrow}$ is the net amount of mass exchanged with the environment of system A. The mass $m^{A\leftarrow}$ is positive if mass flows into A.

4.4. UNITS OF BOTH ENERGY AND MASS

Provide a table of the different units and the conversion factor from one unit to another for both energy and mass.

5. Lecture 4

5.1. TYPES OF STATES

Because the number of independent properties of a system is infinite even for a system consisting of a single particle with a single translational degree of freedom — a single variable that fixes the configuration of the system in space — and because most properties can vary over a range of values, the number of possible states of a system is infinite. To facilitate the discussion of these states, we classify them into different categories according to their time evolutions. This classification brings forth many important aspects of thermodynamics, and provides a readily understandable motivation for the the introduction of the second law of thermodynamics.

An *unsteady state* is one that changes as a function of time because of interactions of the system with its environment. A *steady state* is one that does not change as a function of time despite interactions of the system with its environment. A *nonequilibrium state* is one that changes spontaneously as a function of time, that is, a state that evolves in time without any effects on or interactions with the environment. An *equilibrium state* is one that does not change as a function of time while the system is isolated — a state that does not change spontaneously. An *unstable equilibrium state* is an equilibrium state that may be caused to proceed spontaneously to a sequence of entirely different states by means of a minute and short-lived interaction that has only an infinitesimal temporary effect on the state of the environment. A *stable equilibrium state* is an equilibrium state that can be altered to a different state only by interactions that leave net effects in the environment of the system. These definitions are identical to the corresponding definitions in mechanics but include a much broader spectrum of states than those encountered in mechanics.

Starting either from a nonequilibrium state or from an equilibrium state that is not stable, energy and only energy can be transferred out of a system and affect a mechanical effect without leaving any other net changes in the state of the environment. For example, a charged electricity storage battery can supply energy to an electrical appliance. In contrast, experience shows that, starting from a stable equilibrium state having any value of energy, the mechanical effect just cited is impossible. For example, an internally discharged battery can have the same amount of energy as a charged battery and yet the former cannot energize an electrical appliance.This impossibility is one of the most striking consequences of the first and second laws of thermodynamics. It is consistent with innumerable experiences. The second law will be introduced in the next section.

6. Lecture 5

6.1. STABILITY AND THE SECOND LAW

To a scientist or engineer familiar with mechanics, the existence of stable equilibrium states at any value of energy is not self-evident. It was first recognized by Hatsopoulos and Keenan [12] as the essence of all correct statements of the second law of thermodynamics. We concur with this recognition.

6.2. SECOND LAW OF THERMODYNAMICS

The *second law* asserts that, among all the states of a system with a given value of energy E, and given values of the amount n of the constituent and of the volume V, there exists one and only one stable equilibrium state.

The existence of a stable equilibrium state for each set of values E, n, V and, therefore, the second law of thermodynamics cannot be derived from the laws of mechanics. In mechanics among all the states of a system with a fixed value of the amount of the constituent and a fixed value of the volume, the only stable equilibrium state is that of lowest energy, that is, the state of least potential energy and zero kinetic energy. In contrast, the second law avers that a stable equilibrium state exists for each value of the energy. It follows that the class of states contemplated by thermodynamics is broader than the class of states contemplated by mechanics.

6.2.1. *Impossibility of Perpetual Motion Machine of the Second Kind*

The existence of stable equilibrium states for various conditions of a system has many theoretical and practical consequences. One consequence is that, starting from a stable equilibrium state of any system, no energy can be extracted to affect solely a mechanical effect while the values of the amount of the constituent and of the volume of the system experience no net changes. If the extraction were possible, the energy could be returned to the system, and accelerate it to a nonzero speed. Thus the stable equilibrium state would have been changed to another state without any effect on the environment. Such a change violates the definition of stable equilibrium. Because the concept of stable equilibrium is part of the second law, such a violation is also a violation of that law. This consequence is often referred to as the impossibility of the perpetual motion machine of the second kind (PMM2). In some expositions of thermodynamics it is taken as the statement of the second law. Here, it is only one aspect of both the first and the second laws.

7. Lecture 6

7.1. ADIABATIC AVAILABILITY

For given values of n and V, another consequence of the two laws is that not all states can be changed to the state of minimum energy of the system by means of interactions which result solely in the rise of a weight — interactions which have solely a mechanical effect on the environment. This is a generalization of the impossibility of a PMM2 because the initial state is not a stable equilibrium state. In essence we prove the existence of another property that we call adiabatic availability and denote by Ψ.

The *adiabatic availability* Ψ_1 of a system in state A_1 represents the largest amount of energy that can be transferred from the system to a weight in a reversible process, as A changes from initial state A_1 to a stable equilibrium state A_0. The value of Ψ is nonnegative.

7.1.1. *Reversible and Irreversible Processes*

Among other distinctions, a process may be either reversible or irreversible. A process is *reversible* if it can be performed in at least one way such that both the system and the environment can be restored to their respective initial states. A process is *irreversible* if it is impossible to perform it in such a way that both the system and its environment can be restored to their respective initial states.

7.2. GENERALIZED ADIABATIC AVAILABILITY

The concept of adiabatic availability can be generalized to a reversible process of system A in the course of which the only effect on the environment is the rise or fall of a weight as the system changes from state A_1 with values n_1 and V_1 to a stable equilibrium state A_0 with values n_0 and V_0. We define the energy exchanged between the system and the weight in the course of the process just cited as *generalized adiabatic availability*. The energy exchanged between the system and the weight is optimum. It is the largest if transferred from the system to the weight, and the smallest if transferred from the weight to the system. For simplicity, we denote generalized adiabatic availability by the same symbol Ψ as adiabatic availability. If energy is transferred out of the system, the adiabatic availability is denoted by a positive number, and if transferred into the system, by a negative number.

Like energy, Ψ is a well-defined property of any system in any state. Unlike energy, neither adiabatic availability nor generalized adiabatic availability is additive. If $n_1 = n_0$ and $V_1 = V_0$, it is noteworthy that the generalized adiabatic availability reduces to the adiabatic availability.

8. Lectures 7 and 8

8.1. ADIABATIC AVAILABILITIES OF A SYSTEM AND A RESERVOIR

Additive properties are very useful because they facilitate the discussion of properties of composite systems — systems that consist of two or more subsystems. In striving to define an additive availability, we investigate the adiabatic availabilities of a composite consisting of a system and of a special reference system called a reservoir. These availabilities are generalizations of the concept of motive power of fire first introduced by Carnot [13]. They are generalizations because he restricted the composite to consist of two reservoirs with fixed values of both the respective constituents and the respective volumes, and we do not use his restrictions.

8.2. MUTUAL STABLE EQUILIBRIUM

If a composite of two or more subsystems is in a stable equilibrium state, the subsystems are said to be in *mutual stable equilibrium*. For example, if system C is in a stable equilibrium state and is a composite of two systems A and B, we say that A is in mutual stable equilibrium with B.

In contrast, if each of two systems is in a stable equilibrium state, their composite is not necessarily in a stable equilibrium state and, therefore, the two systems are not necessarily in mutual stable equilibrium.

8.3. RESERVOIR

A *reservoir* is an idealized kind of system that provides useful reference states both in theory and in applications, and that behaves in a manner approaching the following limiting conditions: (a) it passes through stable equilibrium states only; (b) in the course of finite changes of state, it remains in mutual stable equilibrium with a duplicate of itself that experiences no such changes; and (c) at constant values of both the constituent and the volume of each of two reservoirs initially in mutual stable equilibrium, energy can be transferred reversibly from one reservoir to the other with no net effect whatsoever on the environments of the two reservoirs.

8.4. AVAILABLE ENERGY

We consider a composite of system A in state A_1, and reservoir R in state R_1. Without net changes of the values of both the amounts of constituents and the volumes of A and R, we raise the question: What is the largest energy that can be exchanged between the composite and its environment so that the only effect on the latter is the rise of a weight?

The answer is the adiabatic availability of the composite and, therefore, conforms to the following assertions: (a) it is limited and equal to the energy exchanged during a reversible weight process in which system A starts from state A_1, interacts with both the weight and the reservoir, and ends being in mutual stable equilibrium with the reservoir; (b) it is the same for all reversible weight processes such that system A begins from a given state A_1 and ends in a state in which A and R are in mutual stable equilibrium; (c) it is independent of the initial state of R; and

(d) it is the same for all reservoirs that are initially in mutual stable equilibrium with R.

We denote this largest value by Ω_1^R, and conclude that Ω^R is a property of the composite of A and R. We call Ω^R *available energy*.

8.5. GENERALIZED AVAILABLE ENERGY

We consider weight processes of a composite of system A and reservoir R in which the values of both the amount of the constituent and of the volume of A experience net changes. For example, system A is in state A_1 with values n_1 and V_1 and, at the end of a weight process of the composite that affects only the elevation of a weight in the environment of the composite, the values are n_0 and V_0. Again, system A is in state A_2 with values n_2 and V_2 and at the end of the weight process of the composite the values are the same n_0 and V_0 as in the first example. With these specifications, we raise the question: Under the conditions just cited, what is the optimum amount of energy that can be exchanged between the composite and the weight?

The answer is the generalized adiabatic availability of the composite. It is the largest if the energy transfer is from the composite to the weight, and the smallest if the energy transfer is from the weight to the composite. For simplicity, we denote the optimum by the same symbol Ω^R as for available energy, and call it the *generalized available energy* of the composite of A and R. For state A_1 with n_1 and V_1, and final values n_0 and V_0, the value of Ω^R is denoted as Ω_1^R. For state A_2 with n_2 and V_2, and final values again n_0 and V_0, the value of Ω^R is denoted Ω_2^R. For $n_1 = n_2 = n_0$ and $V_1 = V_2 = V_0$, and a given reservoir R, the generalized available energy is identical with the available energy.

It is noteworthy that generalized available energy is defined for any state of any system regardless of whether the state is steady, unsteady, nonequilibrium, equilibrium, or stable equilibrium, and regardless of whether the system has many degrees of freedom or one degree of freedom, or whether the size of the system is large or small. In addition, like energy, Ω^R is an additive and nonstatistical property of the composite system of A and R.

8.5.1. *Energy and Generalized Available Energy Relations*

In principle, given any two states A_1 and A_2 of system A and a reservoir R, we can tabulate the values of the energies E_1 and E_2, and the generalized available energies Ω_1^R and Ω_2^R. For a process of A alone from state A_1 to state A_2 that affects only the elevation of a weight in the environment of A, that is, a weight process of A and not of the composite of A and R, we can prove the following. *If the process of A alone is reversible, then*

$$E_1 - E_2 = \Omega_1^R - \Omega_2^R \; ; \tag{4}$$

and if the process of A alone is irreversible, then

$$E_1 - E_2 < \Omega_1^R - \Omega_2^R \; . \tag{5}$$

9. Lecture 9

9.1. ENTROPY AND ENTROPY BALANCE

Entropy is the concept that distinguishes thermodynamics from all other branches of science and engineering. It is an additional dimension of the space defined by the properties of a system, and has important, strange and unusual characteristics that have attracted the attention of thousands of scientists and engineers, and that have stimulated animated discussions.

9.1.1. *Definition of Entropy*

A system A in any state A_1 has the two properties: energy E_1, and generalized available energy Ω_1^R with respect to a given auxiliary reservoir R. These two properties determine a third one we call *entropy* and denote by the symbol S_1. It is a property in the same sense that energy is a property, or momentum is a property. It can be evaluated by means of the auxiliary reservoir R, a reference state A_0, with energy E_0 and generalized available energy Ω_0^R, to which is assigned a reference value S_0, and the expression

$$S_1 = S_0 + \frac{1}{c_R}\left[(E_1 - E_0) - (\Omega_1^R - \Omega_0^R)\right] \tag{6}$$

where c_R is a well-defined positive constant that depends on the auxiliary reservoir R only. Entropy S is shown to be independent of the reservoir, that is, indeed the reservoir is auxiliary and is used only because it facilitates the definition of S. It is also shown that S can be assigned absolute values that are nonnegative.

The concept of entropy introduced here differs from and is more general than that in practically all textbooks. It does not involve the concepts of heat and temperature which have not yet been defined; it is not restricted to large systems; it applies to both macroscopic and microscopic systems, including a system with one particle with only one (translational) degree of freedom, that is, even one particle has entropy; it is not restricted to stable or thermodynamic equilibrium states; it is defined for both stable equilibrium and not stable equilibrium states because both energy and generalized available energy are defined for all states; and most certainly, it is not statistical because both energy and generalized available energy are not statistical.

The dimensions of S depend on the dimensions of both energy and c_R. It turns out that the dimensions of c_R are independent of mechanical dimensions, and are the same as those of temperature (defined later). The unit of c_R chosen in the International System of units is the kelvin, denoted by K. So entropy can be expressed in J/K or other equivalent units.

Because both energy and generalized available energy are additive, entropy is also an additive property. In contrast, whereas energy remains constant in time if the system experiences either a spontaneous process or a zero-net-effect interaction only with a weight, using relations 4 to 6 we prove that entropy remains constant if either of these processes is reversible, and increases if either of these processes is irreversible. These two features are known as *the principle of nondecrease of entropy*.

The spontaneous entropy creation or increase during an irreversible process is called *entropy generated by irreversibility*. It is positive. Like energy conservation, entropy nondecrease is a time-dependent result which here is obtained without use of the unknown general equation of motion.

9.1.2. *Entropy Balance*
Like energy, entropy can be exchanged between systems by means of interactions. Denoting by $S^{A\leftarrow}$ the amount of entropy exchanged between system A and the environment as a result of all interactions involved in a process in which the state of A changes from A_1 to A_2, we derive a very important analytical tool, the *entropy balance*, that is,

$$(S_2 - S_1)_{\text{system } A} = S^{A\leftarrow} + S_{\text{irr}} \tag{7}$$

where S_{irr} is positive or at least zero and represents the entropy generated spontaneously within system A during the time interval from t_1 to t_2 required to affect the change from state A_1 to state A_2. Spontaneous entropy generation within a system occurs if the system is in a state that is not stable equilibrium as the constituent tries to conform to both the external and internal forces, and tends towards stable equilibrium.

10. Lecture 10

10.1. STABLE EQUILIBRIUM STATES AND THE STATE PRINCIPLE

An isolated system admits an infinite number of states that share the same values of energy E, amount of constituent n, and volume V. Some of these states are nonequilibrium, some are equilibrium and, according to the second law, one and only one is a stable equilibrium state. It follows that the value of any property of the system in a stable equilibrium state is uniquely determined by the values E, n, V, that is, it can be written as a function of E, n, V only. This result is known as the ***stable equilibrium state principle*** or simply the ***state principle***.

10.2. THE FUNDAMENTAL RELATION

A system in general has an infinite number of independent properties. If we focus on the special family of states that are stable equilibrium, however, the state principle asserts that the value of each of these properties is uniquely determined by the values E, n, V or the values of an equivalent set of three other independent properties. In contrast, for states that are not stable equilibrium, the values of E, n, V are not sufficient to specify the values of all independent properties.

In general, one property of great interest is the entropy of stable equilibrium states which, according to the state principle, must satisfy the relation of the form

$$S = S(E, n, V) \tag{8}$$

that is, S must be solely a function of E, n, V. The function $S(E, n, V)$ is called the *fundamental relation*, and is concave — has a negative second partial derivative — with respect to each of the independent variables E, n, V. The entropy of each unique stable equilibrium state is larger than that of any other state with the same values E, n, V. This result is known as the ***highest entropy principle***.

Upon solving the fundamental relation for E, we find the ***energy relation***, that is, we express the energy as a function of S, n, V

$$E = E(S, n, V) \;. \tag{9}$$

Traditionally, if the energy of a system in a stable equilibrium state depends only on S, n, V, it is called *internal energy*, and is denoted by U. However, this terminology must not be misconstrued as implying that some systems have both internal and external energy.

10.3. TEMPERATURE, CHEMICAL POTENTIAL, AND PRESSURE

Upon writing the differentials of the fundamental relation and the energy relation, we find

$$\mathrm{d}S = \left(\frac{\partial S}{\partial E}\right)_{n,V} \mathrm{d}E + \left(\frac{\partial S}{\partial n}\right)_{E,V} \mathrm{d}n + \left(\frac{\partial S}{\partial V}\right)_{E,n} \mathrm{d}V \qquad (10)$$

$$\mathrm{d}E = \left(\frac{\partial E}{\partial S}\right)_{n,V} \mathrm{d}S + \left(\frac{\partial E}{\partial n}\right)_{S,V} \mathrm{d}n + \left(\frac{\partial E}{\partial V}\right)_{S,n} \mathrm{d}V \qquad (11)$$

Thus, we define

$$T = \left(\frac{\partial E}{\partial S}\right)_{n,V} = \text{temperature} \; ; \qquad p = -\left(\frac{\partial E}{\partial V}\right)_{S,n} = \text{pressure} \; ;$$

$$\mu = \left(\frac{\partial E}{\partial n}\right)_{S,V} = \text{chemical potential} \; .$$

Moreover, upon solving equation (10) for dE and comparing the result with equation (11), we find

$$T = \frac{1}{(\partial S/\partial E)_{n,V}} \; ; \qquad \mu = -T(\partial S/\partial n)_{E,V} \; ; \qquad p = T(\partial S/\partial V)_{E,n} \; .$$

The concepts of temperature, chemical potential, and pressure are valid only for stable equilibrium states, and are very useful because, if two systems are in mutual stable equilibrium, then the temperature, chemical potential, and pressure of one system is equal to the temperature, chemical potential, and pressure of the other, respectively. In other words, we have temperature equality, chemical potential equality, and pressure equality.

It is noteworthy that either of the differentials

$$\mathrm{d}S = \mathrm{d}E/T - (\mu/T)\,\mathrm{d}n + (p/T)\,\mathrm{d}V \qquad (12)$$

$$\mathrm{d}E = T\,\mathrm{d}S + \mu\,\mathrm{d}n - p\,\mathrm{d}V \qquad (13)$$

is a mathematical relation between properties of neighboring stable equilibrium states, and not the result of any interactions. Temperature is measured on different scales. The same is true for chemical potential and pressure.

11. Lecture 11

11.1. ENERGY VERSUS ENTROPY GRAPHS

Because they are defined in terms of the values of the amount of the constituent, the volume, and a complete set of independent properties, states can in principle

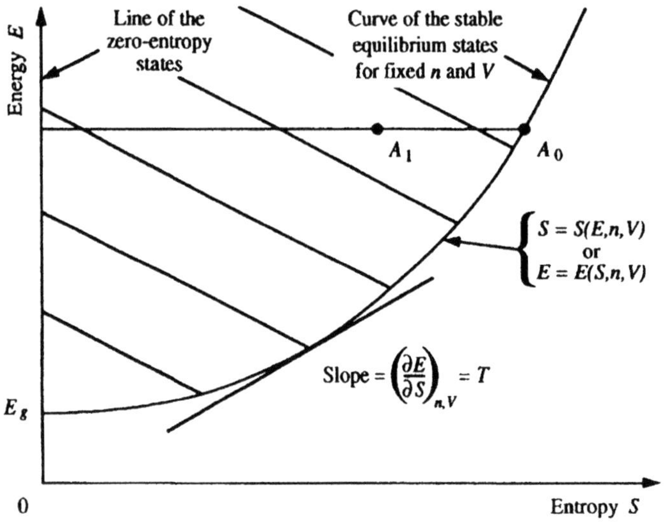

Figure 1. Typical energy versus entropy graph. All concepts of the introductory lectures, including work and heat interactions, can be illustrated on this graph.

be represented by points in a multidimensional geometrical space with one axis for the amount, one for the volume, and one for each independent property. Such a representation, however, would not be enlightening because the number of independent properties is infinite. Nevertheless, useful information can be summarized by first cutting the multidimensional space with a surface corresponding to given values of both the amount of the constituent and the volume, and then projecting the result onto a two-dimensional plane — a plane of two property axes. Here we discuss the energy versus entropy plane which illustrates many of the basic concepts of thermodynamics.

For given values of the amount of the constituent and the volume, we project the multi-dimensional state space of the system onto the E versus S plane. This projection includes both stable equilibrium states and other states that are not stable equilibrium. The laws of thermodynamics imply that the projection must have the shape of the cross-hatched area shown in Figure 1, that is, all the states that share the given characteristics have property values that project on the area between the vertical line denoted as the line of zero-entropy states, and the curve of the stable equilibrium states.

Each point either inside the cross-hatched area or on the vertical line $S_g = 0$ represents a large number of states. Each such state has the same values of amount of the constituent, volume, energy and entropy, but differing values of other properties, and is not a stable equilibrium state. It can be any state except a stable equilibrium state.

A point on the convex curve of stable equilibrium states represents one and only one state. For each of these states, the value of any property is uniquely determined by the values of the amount of the constituent, the volume, and either E or S of the point on the curve.

The zero-entropy states correspond to states contemplated in mechanics. All other states belong to the broad class of states implied and required by the second law or, more specifically, the existence of entropy as a property. So mechanics can be regarded as a special case of thermodynamics — zero-entropy physics.

The stable-equilibrium-state curve represents relations $S = S(E, n, V)$ and $E = E(S, n, V)$. Each stable equilibrium state has either the lowest energy for a given entropy, or the highest entropy for a given energy. So, stable equilibrium states can be regarded as either lowest energy physics for a given entropy, or highest entropy physics for a given energy.

Because each stable equilibrium state is unique, the temperature $T = (\partial E/\partial S)_{n,V}$ at each point on the curve. Temperature is not defined for states that are not stable equilibrium because then E, S, n, V are independent and, therefore, the partial derivative of E with respect to S is meaningless.

Starting from a stable equilibrium state A_0, the system cannot transfer energy to affect a solely mechanical effect on the environment without net changes in n and V because no state of lower energy exists that has an entropy equal to or greater than the entropy of A_0. This is a graphical illustration of the impossibility of PMM2. It is sometimes expressed as the nonexistence of a *Maxwellian demon*. The vertical distance of state A_1 from the stable equilibrium state curve is the adiabatic availability Ψ_1. The vertical distance of A_1 from a tangent to the curve of slope T_R is the available energy Ω_1^R of system A in state A_1 with respect to reservoir R at temperature T_R.

Generation of entropy occurs as an isolated system changes its state spontaneously from A_1 to A_0. All other aspects of thermodynamics can be illustrated on suitable E versus S, or E versus n, or E versus V graphs.

11.2. THE THIRD LAW OF THERMODYNAMICS

For the given values n, V, the energy E_g is the lowest energy — ground-state energy — for which the system can exist. It corresponds to zero entropy and zero temperature. If we were using quantum-theoretical concepts, we would be able to show that all ground-energy states have the same temperature T_g, and that the value of this temperature is equal to zero. Instead, we note that this important conclusion of the quantum-theoretical treatment cannot be drawn as a logical conclusion of the statements of the first and second laws, but here must be presented as an additional fundamental postulate. It is known as the *third law of thermodynamics* or the *Nernst principle* and can be stated as follows. For each given set of values of the amount of the constituent and the volume of a system, there exists one stable equilibrium state with zero temperature.

12. Lecture 12

12.1. INTERACTIONS

Interactions between a system and its environment can have a great variety of effects. Some result in exchange only of energy between the system and its environment at constant system volume. Other interactions result in exchanges of energy and entropy, but neither the amount of the constituent nor the value of the volume are affected. Still others may result in exchanges of energy, entropy, and the constituent, as well as in changes in values of the volume. Such exchanges result in changing the states of the system and its environment. In addition, if as a result of interactions a system is brought to a state that is not stable equilibrium, this state may evolve spontaneously toward a stable equilibrium state, thus causing further changes in the values of the properties of the system.

Knowing how much of the change of a property is due to exchanges with other systems and how much to spontaneous generation or destruction within the system is very important to understanding the performance of the system. For example, if the entropy of a system increases solely because of spontaneous generation — irreversibility — such an increase implies imperfections and is subject to improvement by redesign.

An ingenious method for identifying the amount of entropy generation by irreversibility is by specifying each interaction in terms of the net exchanges that it causes at the boundary between the system and its environment, and comparing these exchanges to the changes of properties of the system. For this reason we classify interactions into different categories, such as exchanges involving energy and the constituent but no entropy, and energy and entropy but not the constituent. Two types of interactions are work and heat. An interaction that results in a net exchange of energy between a system and its environment but no exchanges of entropy and the constituent is called a *work interaction*. The amount of energy exchanged is called *work*, and denoted by $W^{\rightarrow}$. Work $W^{\rightarrow}$ is positive if energy flows out of the system. An interaction that results in net exchanges of energy and entropy but no constituent between the system and an environment at fixed temperature T_Q, that is, a reservoir, and such that the ratio of the amount of energy exchanged and the amount of entropy is equal to T_Q, is called a *heat interaction*. The energy exchanged with the system is called *heat*, and denoted by $Q^{\leftarrow}$. Heat $Q^{\leftarrow}$ is positive if energy flows into the system.

12.2. PRACTICAL USE OF BALANCES

Interactions can be combined with the various balances to provide relations between changes of properties of a system and property flows at the boundary of the system. For example, if as a result of heat and work interactions system A changes its state from A_1 to A_2, the energy, entropy and mass balances are

$$(E_2 - E_1)_{\text{system } A} = E^{A\leftarrow} = Q^{A\leftarrow} - W^{A\rightarrow} \tag{14}$$

$$(S_2 - S_1)_{\text{system } A} = S^{A\leftarrow} + S_{\text{irr}} = \frac{Q^{A\leftarrow}}{T_Q} + S_{\text{irr}} \tag{15}$$

$$(m_2 - m_1)_{\text{system } A} = m^{A\leftarrow} = 0 \tag{16}$$

It is noteworthy that here the energy balance is just a very special consequence of the two laws of thermodynamics and, hence, cannot be called "the first law." Similarly, the entropy balance is just a very special consequence of the two laws of thermodynamics and, hence, cannot be called "the second law."

Work and heat are ingenious concepts. For given end states of a system, they allow the quantitative distinction between entropy generated by irreversibility and entropy exchanged via interactions with other systems. As such, these two concepts provide practical means for identifying opportunities to reduce the entropy generation by irreversibility and, hence, improve the performance of the system.

The identification of these opportunities would be missed if heat were defined as just any interaction that is not work, i.e., any nonwork interaction. The usefulness of balances 12 to 14 is increased if the initial and final states — end-states — of the system are assumed to be stable equilibrium states. Then, the differences $(E_2 - E_1)_{\text{system}}$ and $(S_2 - S_1)_{\text{system}}$ can be expressed in terms of temperatures and pressures, and many practical results can be computed.

13. Concluding Remarks

In our experience with undergraduate teaching in an engineering school, we manage to develop the preceding sequence in the first 12 hours of the course. After these lectures, we proceed with applications of the state principle and of energy, entropy, and mass balances to heat engines, simple systems, Euler and Maxwell relations, the phase rule, two-phase states, bulk-flow interactions, energy conversion devices, and other standard topics of thermodynamics. We find that exposing the undergraduate students to all the precise definitions and also to some of the key proofs, instead of just saying "it can be proved" on all occasions, has two advantages. First, the students gain a precise idea of the nature of the logic involved in thermodynamics and the explicit role of its laws. Second, the students build up much more confidence in all the results if they have seen the precise and rigorous logic that is behind at least for some of them. This is what is missing from traditional expositions.

References

1. Keenan, J.H. and Shapiro, A.H. (1947) History and exposition of the laws of thermodynamics, *Mech. Eng.*, **69**:915–921.
2. Bejan, A. (1988) *Advanced Engineering Thermodynamics*, John Wiley & Sons, New York, pp. 89–96.
3. Feynman, R. (1963) *Lectures on Physics*, Addison-Wesley, Vol. 1.
4. Tisza, L. (1966) *Generalized Thermodynamics*, MIT Press, p. 16.
5. Brush, S.G. (1976) *The Kind of Motion We Call Heat*, North Holland, Vols. 1 & 2.
6. Landau, L.D. and Lifshitz, E.M. (1980) *Statistical Physics*, Pergamon Press, p. 45.
7. Guggenheim, E.A. (1967) *Thermodynamics*, 7th Ed., North Holland, p. 10.
8. Keenan, J.H. (1941) *Thermodynamics*, John Wiley & Sons, p. 6.
9. Hatsopoulos, G.N. and Keenan, J.H. (1965) *Principles of General Thermodynamics*, John Wiley & Sons, p. xxiii.
10. Gyftopoulos, E.P. and Beretta, J.P. (1991) *Thermodynamics: Foundations and Applications*, Macmillan.
11. Timoshenko, S. and Young, D.H. (1948) *Advanced Dynamics*, McGraw-Hill, p. 1 and p. 106.
12. Hatsopoulos, G.N. and Keenan, J.H., *op. cit.*, p. 367.
13. Carnot, S. (1960) *Reflections on the Motive Power of Fire*, in E. Mendoza (ed.), Dover Publications.

PICTORIAL VISUALIZATION OF THE ENTROPY OF THERMODYNAMICS

E.P. GYFTOPOULOS
Massachusetts Institute of Technology
Department of Nuclear Engineering, Room 24-111
Cambridge, MA 02139, USA

1. Introductory Remarks

The term entropy was coined by Clausius [1]. He said: "I have felt it more suitable to take the names of important scientific quantities from the ancient languages in order that they may appear unchanged in all contemporary languages. Hence I propose that we call S the *entropy* of the body after the Greek word τροπή, meaning transformation. I have intentionally formed the word *entropy* to be as similar as possible to the word *energy* since the two quantities that are given these names are so closely related in their physical significance that a certain likeness in their names has seemed appropriate."

In the scientific literature, Clausius' perceptive and important observation that the energy and entropy of thermodynamics are closely related has not always been heeded. Since its inception, the concept of energy has been accorded a universal meaning that underlies our understanding of all physical phenomena. In contrast, the term entropy has been associated with many statistical concepts and mathematical expressions. Because these concepts and expressions differ from each other, the common name is a source of confusion [2,3] and has confounded understanding of many issues not only about the relations among various entropies but, more importantly, about both the physical meaning of the entropy of thermodynamics and the idea of transformation.

In addition to the issues raised by the association of the term entropy with different concepts, many scientists and engineers have expressed concerns about the completeness and clarity of the usual expositions of thermodynamics. For example, Obert [4] writes: "Most teachers will agree that the subject of engineering thermodynamics is confusing to the student despite the simplicity of the usual undergraduate presentation."

Intrigued and challenged by the issues and concerns just cited, over the past few decades we have made an attempt to clarify both the theoretical foundations of thermodynamics and the experimental evidence in support of these foundations, and have reached conclusions that differ from the interpretations presented in the scientific literature. We have discussed the new viewpoint first as a unified quantum theory of mechanics and thermodynamics in Refs. [5–10], and then as a nonquantal exposition of the foundations and applications of thermodynamics in Ref. [11].

A. Bejan and E. Mamut (eds.), Thermodynamic Optimization of Complex Energy Systems, 19–44.

In this paper, we recapitulate our findings about both the thermodynamic definition of entropy, and its quantum-theoretic expression. Then, we provide a pictorial visualization of entropy in terms of the quantum-theoretic probability density function in ordinary space, that is, in terms of the shape of the constituents of a system. Moreover, we show that the generation of entropy (irreversibility) is due to spontaneous changes or transformations of the shape of constituents as they try to conform to the external and internal forces of the system. The spontaneous changes can occur until the entropy assumes its largest value. At that condition, the spatial shape of the constituents yields zero values in any direction of both the translational momentum and the internal structure angular momentum of each atom or molecule. In other words, in thermodynamic equilibrium, all motions cease to exist — everything is at rest.

Many scientists and engineers interested in applications may feel that a general, rigorous, noncircular, and complete description of thermodynamics, and a quantum-theoretic explanation of the foundations of the subject are not important in solving practical problems. This view is only partially correct, and if insisted upon may deprive us of opportunities to make new discoveries and major practical contributions. Some of the reasons for the preceding comment are as follows: (*i*) The entropy of thermodynamics is a cornerstone of the profession of mechanical engineers as well as of all other engineering and scientific disciplines. For example, if it were not for the possible spontaneous increase of entropy, humanity would have no problems of exhaustion of natural resources of both energy and raw materials, no problems of pollution, and no need for so many professionals preoccupying themselves with such problems; (*ii*) The popular meaning of the word energy does not represent the concept of energy of thermodynamics but the concept of available energy or exergy, and available energy is decisively controlled by entropy rather than the energy defined in thermodynamics. So, understanding entropy is of paramount importance to all concerned about energy systems and processes; (*iii*) In our works we have expressed the laws of thermodynamics in terms of the concepts of space, time, and force or inertial mass only, and have proven many theorems of these laws without inconsistencies, incompleteness, circular arguments, and ambiguities. As a result, we have discovered that thermodynamics and quantum theory are facets of a new unified paradigm of physics, and that conventional quantum mechanics (the quantum mechanics taught to undergraduate students) is a special case of the unified theory (zero entropy physics); (*iv*) The rigorous exposition of thermodynamics helped us to discover that entropy is a nonstatistical property of the constituents of any system in any state. Moreover, the unified quantum theory of mechanics and thermodynamics helped us to determine the quantum- theoretic analytical expression for this important property; (*v*) As a result of the quantum-theoretic considerations, we have obtained both a pictorial visualization of the mathematical expression of entropy as a measure of the shape of the constituents of a system, and a correlation between spontaneous shape changes and entropy generation (irreversibility). Hopefully, the visualization and the correlation will stimulate interest in better understanding of the foundations of thermodynamics both without and with quantum theory; and (*vi*) Educators, practitioners, and students have a scientific and professional responsibility to pay more attention to the rigorous, noncircular, and nonconfusing understanding and practice of thermodynamics of all systems and all states. Such understanding and practice are lacking now. A wider and more frequent appreciation of both the beauty, generality, and power of thermodynamics, and the correct use of its fundamental principles is not a matter of presenting a new theory. It is an imperative that will help us address and solve many problems of great importance to humanity and its planet Earth.

2. The Entropy of Thermodynamics

Among the many novel results of the nonquantal exposition [11] is the recognition that the entropy of thermodynamics is a nonstatistical property of a system, in the same sense that inertial mass, or energy, or momentum is a nonstatistical property of a system, and that this entropy is well-defined for all systems, both macroscopic and microscopic, and all states, both thermodynamic equilibrium and not thermodynamic equilibrium.

Because the nonquantal exposition is not widely known yet, a brief summary of the laws of thermodynamics and some of their theorems is presented in Appendix A, including the definition of entropy by the relation

$$S_2 - S_1 = \frac{1}{c_R} \left[(E_2 - E_1) - \left(\Omega_2^R - \Omega_1^R\right)\right] \tag{1}$$

where S_i and E_i are the entropy and energy of state A_i of system A, respectively, Ω_i^R is the generalized available energy of state A_i with respect to reservoir R, and c_R a well-defined positive constant.

Based on the characteristic features of E_i and Ω_i^R listed in Appendix A, we find that the entropy of thermodynamics must conform to the following criteria: (*i*) be well-defined for every system (both macroscopic and microscopic), and every state (both stable equilibrium and not stable equilibrium); (*ii*) be independent of the reservoir R; (*iii*) be invariant in all reversible adiabatic processes, and increasing in all irreversible adiabatic processes; (*iv*) be additive; (*v*) be nonnegative, and vanish for all the states encountered in mechanics; (*vi*) for given values of energy, amounts of constituents, and (external) parameters, admit one and only one state that corresponds to the largest value of entropy; (*vii*) for given values of amounts of constituents, and parameters, the graph of entropy versus energy of stable equilibrium states be concave and smooth; (*viii*) for a composite C of two subsystems A and B, must be such that the constrained maximization procedure for C (criterion *vi*) yields identical thermodynamic potentials (for example, temperature, total potentials, and pressure) for all three systems A, B, and C; (*ix*) for stable equilibrium states, must reduce to relations that have been established experimentally and that express the entropy in terms of energy, amounts of constituents, and parameters, such as the relations for ideal gases.

In due course, we prove that the preceding reasonable and relatively simple criteria reject all but one of the analytical expressions for entropy that have appeared in the scientific literature. In fact, even the one and only acceptable expression requires a profound and radical reinterpretation of its independent variables.

3. Unified Quantum Theory of Mechanics and Thermodynamics

In many textbooks on quantum mechanics, probabilities associated with measurement results are derived from a normalized wave function $\psi(\mathbf{x})$ or, equivalently, from either a Dirac ket $|\psi\rangle$ or a projector $|\psi\rangle\langle\psi|$, where $\mathbf{x}$ are the spatial coordinates of the constituents of a system, and $\langle\psi|$ is the bra of ket $|\psi\rangle$ [12]. For example, for a system with one degree of spatial freedom x, the probability density function of measurement results of x is given by $|\psi(x)|^2 = \langle\psi|x\rangle\langle x|\psi\rangle = |\langle x|\psi\rangle|^2$.

In statistical quantum mechanics, probabilities associated with measurement results are derived from a density operator $\rho = \sum_i \alpha_i \rho_i$ which represents a mixture of quantal probabilities derived from projectors $\rho_i = |\psi_i\rangle\langle\psi_i|$ for $i = 1, 2, \ldots$, and statistical (informational) probabilities α_i which reflect the lack of information about some or all aspects of the state of a system.

In general, the foundations and theorems of the unified quantum theory of mechanics and thermodynamics differ from those of the ordinary expositions of quantum mechanics and statistical quantum mechanics. Among the many novel concepts and results, two deserve special emphasis at this point: (*i*) In contrast to statistical quantum mechanics, a novel concept of the unified quantum theory is that its postulates — quantum-theoretic and thermodynamic — do not apply to density operators that represent mixtures of quantal probabilities derived from projectors $\rho_i = |\psi_i\rangle\langle\psi_i|$, and statistical (informational) probabilities α_i that reflect the lack of information about some or all aspects of the state of a system. Instead, the unified theory avers that the laws of physics apply only to density operators each of which is construed as the seat of exclusively quantal probabilities, that is, only to a ρ that can be represented by a homogeneous ensemble of identical systems, identically prepared. *Homogeneous* is an ensemble in which the probabilities of results of measurements on any member are represented by the same density operator ρ as those on any other member. Accordingly, experimentally (in contrast to algebraically) the ensemble cannot be decomposed into statistical mixtures of projectors or other non-projector density operators. The concept of homogeneous ensemble was introduced by von Neumann [13]. But he assumed that it applies only to projectors ($\rho_i = \rho_i^2$), whereas here the concept is extended to all self-adjoint, nonnegative definite, linear, unit trace density operators. Each such operator satisfies the relation $\rho \geq \rho^2$; and (*ii*) As already indicated, in contrast to the plethora of expressions for entropy that have been proposed in the scientific literature over the past fourteen decades, we find that only one satisfies the criteria listed in the preceding section. Because the unified quantum theory of mechanics and thermodynamics is not widely known yet, a brief summary of both the quantum-theoretic postulates and some of their theorems is presented in Appendix B, including a discussion of the fact that, of the known expressions for entropy S, the only acceptable is

$$S = -\mathrm{k}\,\mathrm{Tr}\,[\rho \ln \rho] \tag{2}$$

where k is Boltzmann's constant, Tr stands for the trace of the operator that follows, and ρ is a density operator which can be represented solely by a homogeneous ensemble. It is noteworthy that, if ρ cannot be represented by a homogeneous ensemble, then $-\mathrm{k}\,\mathrm{Tr}\,[\rho \ln \rho]$ does not represent the entropy of thermodynamics.

For given values of energy, parameters, and amounts of constituents, if ρ is a projector (wave function) then $S = 0$, if ρ is not a projector but corresponds to a state which is not stable equilibrium (not thermodynamic equilibrium) then S has a positive value smaller than the largest possible for the given specifications, and if ρ corresponds to the unique stable equilibrium state, then S has the largest value of all the entropies of the system which share the given values of energy, parameters, and amounts of constituents.

4. Pictorial Visualization of Entropy

In many textbooks, the probability density function associated with measurement results of the spatial coordinates of the constituents of a system is interpreted as the spatial shape of the constituents of the system [14,15], and the shape is used to calculate atomic, molecular, and ionic radii [16], and to explain the formation of various compounds. The probability density function of the spatial coordinates enters also in the evaluation of the entropy S (equation (2)) and, therefore, we can think of entropy as a measure of the spatial shape of the constituents of a system, and thus achieve a pictorial visualization of a concept that has been puzzling scientists and engineers over more than a century.

A few graphs of surfaces of constant probability density are shown in Figure 1. They are derived from probability density functions which represent energy eigenfunctions (energy eigenprojectors) of the electron in a hydrogen atom. They have been calculated by Brandt and Dahmen [15]. In the notation $\rho_{n\ell m}$ of each shape, ρ denotes the probability density function as a function of the Cartesian coordinates x, y, z of the electron, subscript n denotes the energy quantum number and appears in the energy eigenvalue relation

$$\varepsilon_n = -13.61/n^2\ (eV)\ , \tag{3}$$

subscript ℓ denotes the angular momentum quantum number $(0 \leq \ell \leq n-1)$ and appears in the square of the total angular momentum eigenvalue relation

$$L^2 = \hbar^2 \ell(\ell+1)\ , \tag{4}$$

L^2 is the square of the total angular momentum, and $\hbar$ Planck's constant divided by 2π, and subscript m denotes the z-component of the angular momentum quantum number $(-\ell \leq m \leq \ell)$ and appears in the eigenvalue relation

$$L_z = m\hbar \tag{5}$$

where L_z is the z-component of the angular momentum. The probability density function equals the square of the absolute magnitude of the electron energy eigenfunction as a function of x, y, z.

If the normalized wave function $\psi(x,y,z)$ of the electron is not an energy eigenfunction, and has energy $E_{\text{electron}} = \langle\psi|H_{\text{electron}}|\psi\rangle$, then the probability distribution function in x, y, z-space is again given by $|\psi|^2$, and a graph of the electron shape can be made in a manner analogous to that described for energy eigenfunctions.

Regardless of whether a wave function is an eigenfunction of a Hamiltonian operator H, or another operator that represents an observable, or not an eigenfunction of any operator, the probability density function can be visualized as a shape in Cartesian space, and the value of the entropy S of any such shape equals zero.

For a nonzero value of S, we must consider a density operator which does not correspond to a projector $|\psi\rangle\langle\psi|$ or wave function ψ. The choices are infinite. For our purposes, we have selected ρ^0_{electron} which corresponds to the stable equilibrium state of the electron of a hydrogen atom, at electron energy E_{electron} [17]. Because the energy eigenstates of the electron in a hydrogen atom are degenerate, the

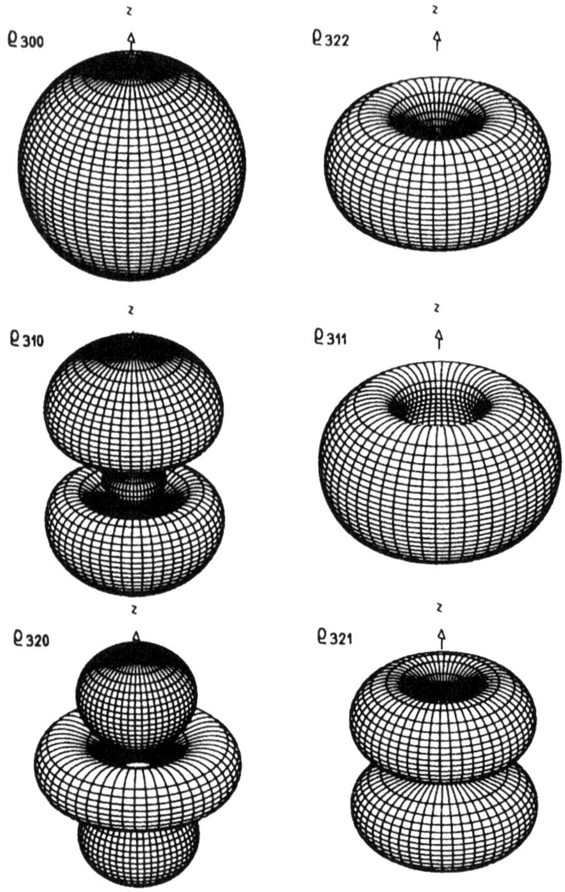

Figure 1. Surfaces of constant probability density $\rho_{3\ell m} = 0.0002$ in full x, y, z-space of the electron in a hydrogen atom [15].

probability density function ρ^0_{electron} of the electron in a stable equilibrium state of energy E_{electron} is given by the relation (see Appendix B)

$$\rho^0_{\text{electron}} = \sum_n \rho^0_{nn}\, \rho_n \tag{6}$$

where

$$\rho^0_{nn} = \frac{g_n \exp(-\beta\varepsilon_n)}{\sum_n g_n \exp(-\beta\varepsilon_n)} \tag{7}$$

$$\rho_n = \frac{\sum_\ell \sum_m \rho_{n\ell m}}{g_n} \tag{8}$$

$g_n = n^2$ and is the degeneracy of the energy eigenvalue ε_n, β is determined by the energy by means of the relation

$$E_{\text{electron}} = \sum_n \rho^0_{nn}\, \varepsilon_n \tag{9}$$

and $\rho_{n\ell m}$ is the probability density of the energy eigenstate with quantum numbers n, ℓ, m ($0 \leq \ell \leq n-1$ and $-\ell \leq m \leq \ell$). As shown in Appendix B, $\beta = 1/kT$, where T is the temperature of the electron. Because β is determined by E_{electron}, it follows that T is also determined solely by E_{electron}.

Using graphs such as in Figure 1, we find

$$g_3 = 9\ ; \qquad \rho_3 = \frac{1}{9}\left(\rho_{300} + \rho_{310} + \rho_{320}\right) + \frac{2}{9}\left(\rho_{311} + \rho_{321} + \rho_{322}\right) \tag{10}$$

$$g_4 = 16\ ; \qquad \rho_4 = \frac{1}{16}\left(\rho_{400} + \rho_{410} + \rho_{420} + \rho_{430}\right) + \frac{2}{16}\left(\rho_{411} + \rho_{421} + \rho_{422} + \rho_{431} + \rho_{432} + \rho_{433}\right) \tag{11}$$

For the purposes of this paper, Dahmen and Stroh [18] graphed ρ_3, ρ_4 and ρ_n for any value of n. As any quantum-thermodynamicist might have expected, ρ_n is spherically symmetric for $n = 1, 2, \ldots$ (Figure 2) and, therefore, the probability density function ρ^0_{electron} or the corresponding electron shape around the proton is spherically symmetric. The spherical shape has at least two important implications. First, the angular momentum of the electron in any direction equals zero. Zero angular momentum means that the electron is not moving around the proton nucleus of the hydrogen atom, as the term stable equilibrium suggests and electromagnetic theory requires for equilibrium to be achieved. It is noteworthy that there exists an infinite number of spherical shapes of the electron around the proton which correspond to the same value E_{electron} of the energy but an upper limit of n in equation (6) smaller than infinity. Each such shape is an unstable equilibrium probability distribution function, and upon a minute perturbation tends spontaneously to the shape ρ^0_{electron}.

The second important implication is that the value of the entropy that corresponds to the shape ρ^0_{electron} is larger than that determined by any other density operator $\rho \neq \rho^0_{\text{electron}}$ and such that $\text{Tr}\,\rho\, H_{\text{electron}} = \text{Tr}\,\rho^0_{\text{electron}}\, H_{\text{electron}}$ (see, however, Ref. 17).

If the change of shape from ρ to ρ^0_{electron} occurs, the change from a low value $S(\rho)$ to the largest value $S(\rho^0_{\text{electron}})$ is spontaneous, and the process is irreversible. There are, however, spontaneous shape changes such as either from one wave function to another, or from ρ_1 to $\rho_2 \neq \rho_1$ but at constant values of energy ($\text{Tr}\,\rho_1\, H_{\text{electron}} = \text{Tr}\,\rho_2\, H_{\text{electron}}$), amounts of constituents, and parameters, for which $S(\rho_2) = S(\rho_1)$ and, then, the process is reversible.

Completion of the discussion of a hydrogen atom in a finite-size box requires also the consideration of the shapes arising from the translational degrees of freedom of the proton nucleus. The energy eigenvalues, eigenfunctions, and shapes

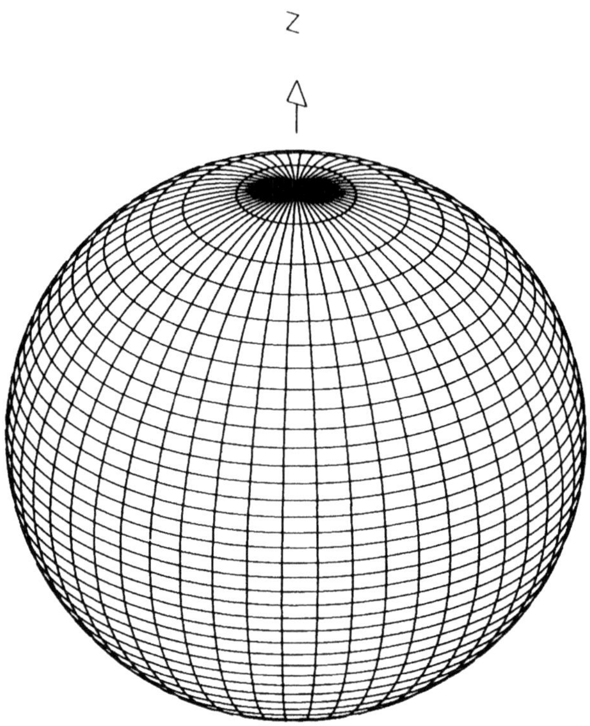

Figure 2. Surface of constant probability density ρ_n in full x, y, z-space of the electron in a hydrogen atom, arbitrary units for any value of n [18].

of a structureless proton in a box, and the procedure for combining the translational shapes with the electron shapes are discussed in many texts, including Hatsopoulos and Gyftopoulos [19]. For the sake of brevity, we will not repeat this discussion here except for only one comment. If the probability density function of the proton is derived from an energy eigenfunction, the value of the speed of the proton in any direction is zero (see Appendix B). So, in a stable equilibrium state, both the electron velocities and the proton velocities are all zero — nothing moves, that is, the concept of stable equilibrium in the unified theory is the same as the concept of stable equilibrium in classical mechanics. However, there exists an important difference between the two theories. Whereas in classical mechanics the stable equilibrium state of a system corresponds to the lowest energy of the system, in the unified theory there exists one stable equilibrium state for each set of values of energy, amounts of constituents, and parameters (see Appendix A).

5. Concluding Remarks

In closing, we wish to reiterate that the entropy of thermodynamics is: (*i*) a nonstatistical property of the constituents of a system, in the same sense that inertial mass is such a property; (*ii*) valid for any system, both macroscopic and microscopic, including a system with only one particle; (*iii*) valid for any state, thermodynamic equilibrium or not; and (*iv*) a measure of the geometric shape of the constituents of the system, even if the system consists of only one particle. Moreover, irreversibility is due to the spontaneous change of the shape of the constituents as they try to conform to the external and internal forces of the system.

Acknowledgements

I am very thankful to Professors Doctors S. Brandt and H.D. Dahmen, and to Mr. T. Stroh for Figures 1 and 2.

Appendix A. Thermodynamics

A.1. GENERAL REMARKS

Gyftopoulos and Beretta [11] have composed a novel exposition in which all basic concepts of thermodynamics are defined completely and without circular arguments in terms of the mechanical concepts of space, time, and force or inertial mass. Many of these definitions are new. The order of introduction of concepts and postulates is: system (types and amounts of constituents, forces between constituents, and external forces or parameters); properties; states; the first law; energy (without work and heat); energy balance; classification of states in terms of time evolutions; stable equilibrium states; second law (without temperature, heat, and entropy); generalized available energy; entropy of any state, stable equilibrium or not, in terms of energy and generalized available energy and not in terms of temperature and heat; entropy balance; fundamental relation for stable equilibrium states only; temperature, total potentials, and pressure in terms of energy, entropy, amounts of constituents and parameters for stable equilibrium states only; the third law; work in terms of energy; and heat in terms of energy, entropy, and temperature. All concepts and postulates are valid for all systems (both macroscopic and microscopic), all states (both stable equilibrium and not stable equilibrium), and require no statistical probabilities.

A.1.1. *Definition*

We define general thermodynamics or simply thermodynamics as the study of motions of physical constituents (particles and radiations) resulting from externally applied forces, and from internal forces (the actions and reactions between constituents). This definition is identical to that given by Timoshenko and Young about mechanical dynamics [20]. However, because of the second law, the definition encompasses a much broader spectrum of phenomena than mechanical dynamics.

A.1.2. *Kinematics: Conditions at an Instant in Time*

In kinematics we give verbal definitions of the terms system, property, and state so that each definition is valid without change in any physical theory, and involves no statistics attributable to lack of information. The definitions include innovations. To the best of our knowledge, they contradict neither any theoretical principle nor any experimental result.

A *system* is defined as a collection of *constituents* confined by a nest of *internal intermolecular forces* and *external forces* or *parameters*. Without modification, this definition applies to all paradigms of physics. The term paradigm is used in the sense of Kuhn [21].

Everything that is not included in the system is the *environment*.

For a system with r constituents, we denote their amounts by the vector $\boldsymbol{n} = \{n_1, n_2, \ldots, n_r\}$. For a system with external forces described by s parameters we denote the parameters by the vector $\boldsymbol{\beta} = \{\beta_1, \beta_2, \ldots, \beta_s\}$. One parameter may be volume, $\beta_1 = V$.

At any instant in time, the amount of each constituent, and the parameters of each external force have specific values. We denote these values by $\boldsymbol{n}$ and $\boldsymbol{\beta}$ with or without additional subscripts.

By themselves, the values of the amounts of constituents and of the parameters at an instant in time do not suffice to characterize completely the condition of the system at that time. We also need the values of a set of independent properties at the same instant in time. The value of an independent property can be varied without affecting the values of other properties. Each *property* is an attribute that can be evaluated at any given instant in time (not as an average over time) by means of a set of measurements and operations that are performed on the system and result in a numerical value — the *value of the property*. This value is independent of the measuring devices, other systems in the environment, and other instants in time.

For a given system, the values of the amounts of the constituents, the values of the parameters, and the values of a complete set of independent properties encompass all that can be said about both the system at an instant in time, and the results of any measurements that may be performed on the system at that same instant in time. We call this complete characterization of the system at an instant in time the *state* of the system. This definition of state is novel and, without change, applies to any branch and any paradigm of physics.

A.2. DYNAMICS: CHANGES OF STATE IN TIME

The state of a system may change in time either spontaneously due to the internal and external forces, or as a result of interactions with other systems, or both. The relation that describes the evolution of the state of an isolated system — *spontaneous changes of state* — as a function of time is the *equation of motion*. Certain time evolutions obey *Newton's equation* which relates the total force F on each system particle to its inertial mass m and acceleration a so that $F = ma$. Other evolutions obey the *time-dependent Schroedinger equation*, that is, the quantum-mechanical equivalent of Newton's equation. Other experimentally observed time evolutions, however, do not obey either of these two equations. So the equations of motion that we have are incomplete. The discovery of the complete equation of motion that describes all physical phenomena remains a subject of research at the frontier of science — one of the most intriguing and challenging problems in physics [2, 5, 8–10, 22–24]. Many features of the equation of motion have already been discovered. These features provide not only guidance for the discovery of the complete equation but also a powerful alternative procedure for analyses of many time-dependent, practical problems. Two of the most general and well-established features are captured by the consequences of the first and second laws of thermodynamics discussed later.

A.3. ENERGY AND ENERGY BALANCE

Energy is a concept that underlies our understanding of all physical phenomena, yet its meaning is subtle and difficult to grasp. It emerges from a fundamental principle known as the first law of thermodynamics. The ***first law*** asserts that any two states of a system may always be the initial and final states of a change (weight process) that involves no net effects external to the system except the change in elevation between z_1 and z_2 of a weight, that is, a mechanical effect. Moreover, for a given weight, the value of the expression $Mg(z_1 - z_2)$ is fixed only by the end states of the system, where M is the mass of the weight, and g the gravitational acceleration.

The main consequence of this law is that every system A in any state A_1 has a property called ***energy***, with a value denoted by the symbol E_1 (Ref. 11, Sec. 3.4, pp. 32–33). The energy E_1 can be evaluated by means of a weight process that connects A_1 and a reference state A_0 to which is assigned an arbitrary reference value E_0 so that

$$E_1 = E_0 - Mg(z_1 - z_0) \tag{A-1}$$

Energy is shown to be an additive property (Ref. 11, Sec. 3.6, pp. 34–35), that is, the energy of a composite system is the sum of the energies of its subsystems. Moreover, it is also shown that energy has the same value at the final time as at the initial time if the system experiences a zero-net-effect weight process, or remains invariant in time if the process is spontaneous (Ref. 11, Sec. 3.7, pp. 35–37). In either of the last two processes, $z_2 = z_1$ and $E(t_2) = E(t_1)$ for time t_2 greater than t_1, that is, energy is ***conserved***. Energy conservation is a time-dependent result. In Ref. 11, this result is obtained without use of the general equation of motion.

Energy can be exchanged between systems by means of interactions. Denoting by $E^{A\leftarrow}$ the amount of energy exchanged between the environment and system A in a process that changes the state of A from A_1 to A_2, we can derive the ***energy balance***. This derivation is based on the additivity of energy and energy conservation (Ref. 11, Sec. 3.8, pp. 37–38), and reads

$$(E_2 - E_1)_{\text{system } A} = E^{A\leftarrow} \tag{A-2}$$

In words, the energy change of a system must be accounted for by the energy transferred across the boundary of the system. The energy $E^{A\leftarrow}$ crossing the boundary of A is positive if energy flows from the environment to system A.

A.4. TYPES OF STATES

Because the number of independent properties of a system is infinite even for a system consisting of a single particle with a single translational degree of freedom — a single variable that fixes the configuration of the system in space — and because most properties can vary over a range of values, the number of possible states of a system is infinite. The discussion of these states is facilitated if they are classified into different categories according to time evolutions. This classification brings forth many important aspects of physics, and provides a readily understandable motivation for the introduction of the second law of thermodynamics.

The classification consists of: unsteady states; steady states; nonequilibrium states; and equilibrium states (Ref. 11, Sec. 4.1, pp. 53–58). An ***equilibrium state*** is one that does not change as a function of time while the system is isolated —

a state that does not change spontaneously. An *unstable equilibrium state* is an equilibrium state that may be caused to proceed spontaneously to a sequence of entirely different states by means of a minute and short-lived interaction that has only an infinitesimal effect on the state of the environment. A *stable equilibrium state* is an equilibrium state that can be altered to a different state only by interactions that leave net effects in the environment of the system. These definitions are identical to the corresponding definitions in mechanics but include a much broader spectrum of states than those encountered in mechanics. The broader spectrum is due to the second law discussed later.

Starting either from a nonequilibrium state or from an equilibrium state that is not stable, energy can be extracted from a system and affect a mechanical effect without leaving any other net changes in the state of the environment. In contrast, experience shows that, starting from a stable equilibrium state, no energy can be extracted from the system that would result in the mechanical effect just cited. This impossibility is one of the most striking consequences of the first and second laws of thermodynamics. It is consistent with innumerable experiences.

A.5. GENERALIZED AVAILABLE ENERGY

The existence of stable equilibrium states is not self-evident. It was first recognized by Hatsopoulos and Keenan [25] as the essence of all correct statements of the second law. Gyftopoulos and Beretta (Ref. 11, Ch. 4, pp. 53–66) concur with this recognition, and state the second law as follows (simplified version): Among all the states of a system with a given value of energy, and given values of the amounts of constituents and the parameters, there exists one and only one stable equilibrium state. The second law cannot be derived from the laws of mechanics. Within mechanics, the stability analysis yields that among all the allowed states of a system with fixed values of amounts of constituents and parameters, the only stable equilibrium state is that of lowest energy. In contrast the second law avers the existence of a stable equilibrium state for each value of the energy.

The existence of stable equilibrium states for various conditions of a system has many theoretical and practical consequences. One consequence is that, starting from a stable equilibrium state of any system, no energy is available to affect a mechanical effect while the values of both the amounts of constituents and the parameters of the system experience no net changes (Ref. 11, Sec. 4.5, pp. 64–65). This consequence is often referred to as the impossibility of a perpetual motion machine of the second kind (PMM2). In some expositions of thermodynamics, it is taken as the statement of the second law. In the new exposition, it is only one aspect of both the first and the second laws.

At fixed values of both the amounts of constituents and the parameters, another consequence is that not all states of a system can be changed to the state of minimum energy by means of solely a mechanical effect. This is a generalization of the impossibility of a PMM2, and is represented by a property called *adiabatic availability*, and denoted by Ψ (Ref. 11, Secs. 5.2 to 5.4, pp. 73–76). The concept of adiabatic availability can be generalized to the concept of *generalized adiabatic availability* (Ref. 11, Secs. 5.5 and 5.6, pp. 77–80). The latter represents the optimum amount of energy that can be exchanged between a system and a weight in a weight process in which the respective initial and final values of the amounts of constituents and/or the parameters differ. Generalized adiabatic availability differs from energy. Like energy, this property is well defined for any system in any state. Unlike energy, it is not additive (Ref. 11, Sec. 5.3.8, p. 75).

In striving to define an additive property that captures the important features of generalized adiabatic availability, Gyftopoulos and Beretta introduce a special reference system, called a *reservoir*, and discuss the possible weight processes that the composite of a system and the reservoir may experience. Thus, they disclose a third consequence of the first and second laws, that is, a limit on the optimum amount of energy that can be exchanged between a weight and a composite of a system and a reservoir R — the optimum mechanical effect. They call the optimum value *generalized available energy* (Ref. 11, Sec. 6.8, pp. 95–97), denote it by Ω^R, and show that it is additive (Ref. 11, Sec. 6.9.6, pp. 98–99). It is a generalization of the concept of motive power of fire first introduced by Carnot because he assumed that both systems of the composite acted as reservoirs with fixed values of their respective amounts of constituents and parameters, whereas Gyftopoulos and Beretta do not use this assumption. The definition of a reservoir is given in the article on the foundations of thermodynamics by E. P. Gyftopoulos in this volume.

If the net exchanges between a system and its environment involve only energy, the process experienced by the system is called *adiabatic*. For an adiabatic process of system A only, it is shown that the energy change $E_1 - E_2$ of A and the generalized available energy change $\Omega_1^R - \Omega_2^R$ of the composite of A and reservoir R satisfy the relations (Ref. 11, Sec. 6.9, pp. 97–99):

If the adiabatic process of A is reversible: $$E_1 - E_2 = \Omega_1^R - \Omega_2^R \tag{A-3}$$

If the adiabatic process of A is irreversible: $$E_1 - E_2 < \Omega_1^R - \Omega_2^R \tag{A-4}$$

A process is *reversible* if both the system and its environment can be restored to their respective initial states. A process is *irreversible* if the restoration just cited is impossible.

It is noteworthy that energy and generalized available energy are defined for any state of any system, regardless of whether the state is steady, unsteady, equilibrium, nonequilibrium, or stable equilibrium, and regardless of whether the system has many degrees of freedom or one degree of freedom, or whether the size of the system is large or small.

A.6. ENTROPY AND ENTROPY BALANCE

A system A in any state A_1 has many properties. Two of these properties are: energy E_1, and generalized available energy Ω_1^R with respect to a given auxiliary reservoir R. These two properties determine a third property called *entropy*, and denoted by the symbol S. It is a property in the same sense that inertial mass is a property, or energy is a property, or momentum is a property. For a state A_1, S_1 can be evaluated by means of an auxiliary reservoir R, a reference state A_0, with energy E_0 and generalized available energy Ω_0^R, to which is assigned a reference value S_0, and the expression

$$S_1 = S_0 + \frac{1}{c_R}\left[(E_1 - E_0) - \left(\Omega_1^R - \Omega_0^R\right)\right] \tag{A-5}$$

where c_R is a well-defined positive constant that depends on the auxiliary reservoir R only. Entropy S is shown to be independent of the reservoir (Ref. 11, Sec. 7.4, pp. 108–112), that is, S is a property of system A and the reservoir is auxiliary and is used only because it facilitates the definition of S. It is also shown that S can be assigned absolute values that are nonnegative, and that vanish for all the states encountered in mechanics (Ref. 11, Sec. 9.8, pp. 137–138).

Because energy and generalized available energy satisfy relations (A-3) and (A-4), the entropy defined by equation (A-5) remains invariant in any reversible adiabatic process of A, and increases in any irreversible adiabatic process of A. These conclusions are valid also for spontaneous processes and for zero-net-effect interactions. The latter features are known as *the principle of nondecrease of entropy*. Both a spontaneous process and a zero-net-effect interaction are special cases of an adiabatic process of system A.

The entropy created during an irreversible process as a state changes in time is called *entropy generated by irreversibility*. It is positive. The entropy nondecrease is a time-dependent result. In the exposition of thermodynamics in Ref. 11, this result is obtained without use of the general equation of motion. Because both energy and generalized available energy are additive, equation (A-5) implies that entropy is also additive (Ref. 11, Sec. 7.2.2, pp. 103–104).

Like energy, entropy can be exchanged between systems by means of interactions. Denoting by $S^{A\leftarrow}$ the amount of entropy exchanged between the environment and system A in the course of a process that changes the state of A from A_1 to A_2, we derive a very important analytical tool, the *entropy balance* (Ref. 11, Sec. 7.3, pp. 106–108), that is,

$$(S_2 - S_1)_{\text{system } A} = S^{A\leftarrow} + S_{\text{irr}} \tag{A-6}$$

where S_{irr} is positive or at least zero and represents the entropy generated spontaneously within system A during the time interval from t_1 to t_2 required to affect the change from state A_1 to state A_2. Spontaneous entropy generation within a system occurs if the system is in a state that is not stable and the system forces precipitate the natural tendency towards stable equilibrium. The entropy $S^{A\leftarrow}$ crossing the boundary of A is positive if entropy flows from the environment to system A. The dimensions of S depend on the dimensions of both energy and c_R. It turns out that the dimensions of c_R are independent of mechanical dimensions, and are the same as those of temperature. Temperature is defined later.

A.7. STABLE EQUILIBRIUM STATES

It is shown that among the many states of a system that have given values of the energy E, the amounts of constituents $\boldsymbol{n}$, and the parameters $\boldsymbol{\beta}$, the entropy of the unique stable equilibrium state that corresponds to these values is larger than that of any other state with the same values $E, \boldsymbol{n}, \boldsymbol{\beta}$ (Ref. 11, Sec. 8.2, pp. 119–120). Moreover, the entropy of a stable equilibrium state must be a function solely of $E, \boldsymbol{n}, \boldsymbol{\beta}$, that is,

$$S = S(E, \boldsymbol{n}, \boldsymbol{\beta}) \tag{A-7}$$

Equation (A-7) is called the *fundamental relation* (Ref. 11, Sec. 8.3, pp. 120–124). For states that are not stable equilibrium, S depends on more variables than $E, \boldsymbol{n}, \boldsymbol{\beta}$, that is, equation (A-7) is not valid.

The fundamental relation is concave with respect to energy (Ref. 11, Sec. 9.4, pp. 131–132), that is,

$$\left(\frac{\partial^2 S}{\partial E^2}\right)_{n,\beta} \leq 0 \tag{A-8}$$

and analytic in each of its variables $E, \boldsymbol{n}, \boldsymbol{\beta}$ (Ref. 11, Sec. 8.3, pp. 120–124).

Moreover, the fundamental relation is used to define other properties of stable equilibrium states, such as *temperature* T (Ref. 11, Ch. 9),

$$\frac{1}{T} = \left(\frac{\partial S}{\partial E}\right)_{n,\beta} \tag{A-9}$$

total potentials μ_i for i = 1, 2, ..., r (Ref. 11, Ch. 10, pp. 147–151)

$$\mu_i = -T\left(\frac{\partial S}{\partial n_i}\right)_{E,n,\beta} \qquad \text{for i} = 1, 2, \ldots, \text{r} \tag{A-10}$$

and *pressure* p (Ref. 11, Ch. 11, pp. 157–162)

$$p = T\left(\frac{\partial S}{\partial V}\right)_{E,n,\beta} \qquad \text{for } \beta_1 = V = \text{volume}\ . \tag{A-11}$$

Detailed discussion of properties of stable equilibrium states are given in Ref. 11, Chapters 8–11.

A.8. COMMENT

The concept of entropy introduced here differs from and is more general than that in all textbooks except Ref. 11. It does not involve the concepts of temperature and heat; it is not restricted to large systems; it applies to both macroscopic and microscopic systems, including a system with one spin, or a system with one particle with only one (translational) degree of freedom; it is not restricted to stable (thermodynamic) equilibrium states; it is defined for both stable equilibrium and not stable equilibrium states because energy and generalized available energy are defined for all states; and most certainly, it is not statistical — it is a property of matter and not a measure of the lack of information of an observer.

A.9. INTERACTIONS

Work, heat, bulk flow, and diffusion interactions, and their use in the balance equations are discussed in Refs. [11] and [26] and will not be repeated in this brief review.

Appendix B. Unified Quantum Theory of Mechanics and Thermodynamics

B.1. STATISTICAL DESCRIPTIONS

Ever since the enunciation of the first and second laws of classical thermodynamics by Clausius more than 130 years ago, the question of the relation between classical thermodynamics and mechanics has been the subject of intense investigations and controversy. Invariably, Maxwell's seminal ideas prevail [27], that is, "...the molecules in a vessel full of air at uniform temperature are moving with velocities by no means uniform ...", and "In dealing with masses of matter, while we do not perceive the individual molecules, we are compelled to adopt what I have described as the statistical method of calculation, ...". Though deeply rooted in classical mechanics, these ideas play a major role in conventional quantal explanations as well.

Specifically, in statistical quantum mechanics, the dominant view currently held about the physical significance of classical thermodynamics is based on the interpretation of a "thermodynamic equilibrium state" as a composite that best describes the knowledge of an observer possessing only partial information about the "actual state" of a macroscopic system. The "ith actual state" at any instant in time is defined as the set of quantal probabilities derivable from an energy eigenprojector $\rho_i = |\epsilon_i\rangle\langle\epsilon_i|$, or from an energy eigenfunction ϵ_i, where $|\epsilon_i\rangle$ is an energy eigenket of conventional quantum mechanics. Because observers are uncertain about the actual state, they consider all possible "actual states" and assign to each of them a statistical probability α_i. The assignment is achieved by using a hypothesis in addition to the laws of conventional quantum mechanics. The combination of the two types of probabilities is a mixture characterized by an overall density operator $\rho = \sum_i \alpha_i \rho_i$, where $\sum_i \alpha_i = 1$, and $\rho > \rho^2$. The theories that have evolved pursuant to the view just cited are called informational, though the general idea is the foundation of each statistical interpretation of thermodynamics proposed to date.

If at an instant in time, ρ_i for $i = 1, 2, \ldots$, and ρ are represented by ensembles of identical systems, von Neumann has shown that the ensemble for each ρ_i is and must be homogeneous [13], that is, each member of the ensemble is assigned the same ρ_i as any other member, whereas the ensemble for ρ is heterogeneous, that is, only a fraction α_i of the members of the ρ ensemble is assigned the projector ρ_i (Figure B-1). In addition to representing the fractions of the ρ_i's in ρ, the statistical probabilities α_i enter in the evaluation of an informational measure of uncertainty, a subjective entropy, such as $S = -k \sum_i \alpha_i \ln \alpha_i$, where k is Boltzmann's constant.

B.1.1. *Comments*

Statistical theories of thermodynamics yield many correct and practical results. For example, they yield the canonical, grand-canonical, Boltzmann, Bose-Einstein, and Fermi- Dirac distributions, and predict the equality of temperatures of systems in mutual stable equilibrium, the Maxwell relations, and the Gibbs equation [28,29]. Mutual stable equilibrium is defined in Ref. 11, Sec. 6.2, pp. 86–87. (See also article on the foundations of thermodynamics by E.P. Gyftopoulos in this volume.) Despite these successes, the premise that entropy is a subjective characteristic of the knowledge of a partially informed observer rather than a property of a system leaves much to be desired in the light of many accurate, reproducible and nonstatistical experiences, such as the mixing of hot and cold substances, the characteristics of an internally discharging electricity storage battery, the Peltier effect, and chemical reactions. In all these experiences, entropy plays a dominant and decisive role which is entirely independent of whether an observer is informed or misinformed. As pointed out by Schroedinger and others [30–32], the conceptual foundations of statistical interpretations of thermodynamics are not on solid ground. For example, they seem to require abandonment of the concept of state

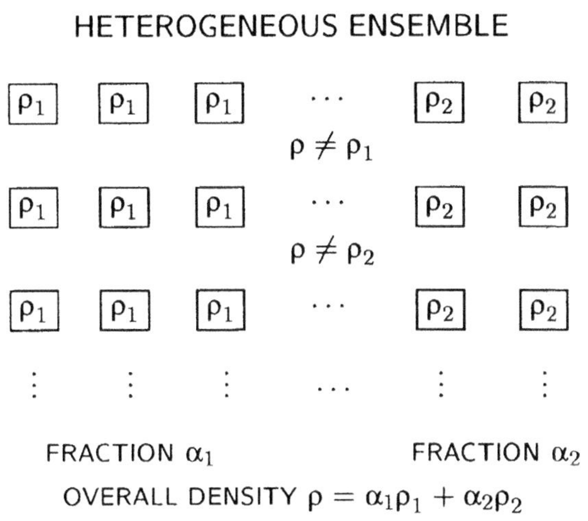

Figure B-1. Representation of a heterogeneous ensemble.

of a system, a cornerstone of traditional physical thought. Again, they foreclose opportunities for the development of a sound theory of nonequilibrium. The reason for the foreclosure is that each statistical theory considers either Newton's equation or Schroedinger's equation as the relation that specifies the evolution of the "actual state" in time, but faces insurmountable conceptual difficulties to propose deterministic equations for the evolutions of the α_i's in time. In fact, any argument that considers the α_i's as indices of ignorance but proposes a deterministic evolution of this ignorance in time is an oxymoron. Though the successes of mechanics, equilibrium thermodynamics, and the *mathematical formalism* of statistical mechanics leave no doubt about the validity of the numerical results, the need for a coherent physical theory capable of encompassing these same results within a sound unified conceptual framework continues to be an interesting challenge.

B.2. THE UNIFIED THEORY

Intrigued by the experiences, ideas, and concerns just cited, Hatsopoulos and Gyftopoulos [5–8] have proposed a resolution of the dilemmas and paradoxes that have preoccupied generations of physicists over more than a century in their attempts to rationalize the relation between mechanics and thermodynamics. The resolution differs from all statistical interpretations of thermodynamics, in general, and from Maxwell's explanation, in particular. This resolution eliminates the need for the statistical estimates α_i, and is the quantum-theoretic underpinning of the exposition of thermodynamics summarized in Appendix A, that is, the exposition which asserts that thermodynamics is a general, nonstatistical or noninformational theory of all physical phenomena. The basis of the resolution is a unified quantum theory of mechanics and thermodynamics which without modification encompasses all systems (both microscopic and macroscopic), and all states (both thermodynamic equilibrium and not thermodynamic equilibrium). The key

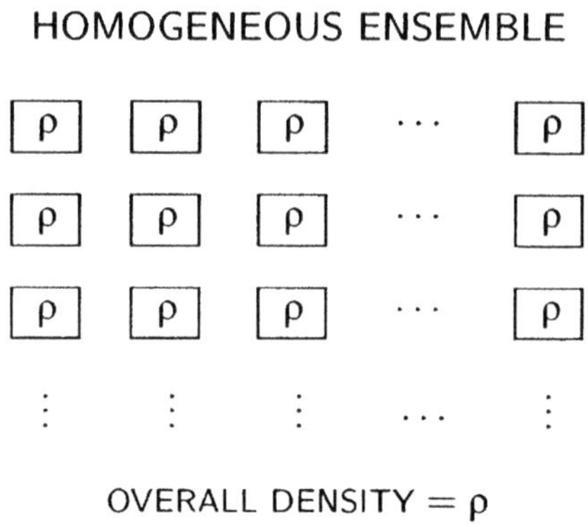

Figure B-2. Representation of a homogeneous ensemble.

for the elimination of the statistical probabilities α_i is the recognition that the only density operators $\rho > \rho^2$ that are subject to the laws of physics (quantum theoretic and thermodynamic) are those that can be represented by a homogeneous ensemble. In such an ensemble, every member is assigned the same ρ as any other member (Figure B-2) and experimentally (in contrast to algebraically) ρ cannot be decomposed into a statistical mixture of either projectors or density operators different from ρ. The impossibility of decomposition is analogous to von Neumann's conclusion that a projector cannot be decomposed into a statistical mixture of states of classical mechanics. Moreover, and perhaps more importantly, the extension of the concept of homogeneity to density operators $\rho > \rho^2$ is accomplished without radical modifications of the quantum-theoretic postulates and theorems about observables, measurement results, values of observables, and densities or probabilities of measurement results. Key concepts — definitions, postulates, and theorems — of the unified theory are discussed briefly below. They are included here because of the emphasis that must be given to the concept of homogeneous or unambiguous ensemble.

B.3. KINEMATICS

The term *kinematics* refers to the definitions of the terms system, property, and state, all at an instant in time.

B.3.1. *System*

The meaning of the term *system* is discussed in Appendix A. The quantum-theoretic representation of a system is as follows.

B.3.2. *System Postulate*

To every system there corresponds a complex, separable, complete, inner product space, a Hilbert space $\mathcal{H}$. The Hilbert space of a composite system of two distinguishable subsystems 1 and 2, with associated Hilbert spaces $\mathcal{H}^1$ and $\mathcal{H}^2$, respectively, is the direct product space $\mathcal{H}^1 \otimes \mathcal{H}^2$ [33].

B.3.3. *Homogeneous or Unambiguous Ensemble*

At an instant in time, an ensemble of identical systems is called ***homogeneous*** or *unambiguous* [8] only if upon subdivision into subensembles in any conceivable way short of measurements, each subensemble yields in every respect measurement results — spectra of values and frequency of occurrence of each value within a spectrum — identical to the corresponding results obtained from the ensemble. For example, the spectrum of energy measurement results and the frequency of occurrence of each energy measurement result obtained from any subensemble are identical to the spectrum of energy measurement results and the frequency of occurrence of each energy measurement result obtained from an independent ensemble that includes all the subensembles.

B.3.4. *Preparation*

A *preparation* is a reproducible scheme used to generate one or more homogeneous ensembles for study.

B.3.5. *Property*

The meaning of the term *property* is discussed in Appendix A. It refers to any attribute of a system that can be quantitatively evaluated at an instant in time by means of measurements and specified procedures. All measurement results and procedures are assumed to be precise.

Without any modifications the meanings of the concepts of homogeneous ensemble, preparation, and property are valid in all paradigms of physics.

B.3.6. *Observable*

From the definition just cited, it follows that each property can be observed, that is, evaluated. Traditionally, however, in quantum theory, a property is called an *observable* only if it conforms to the following mathematical representation.

B.3.7. *Correspondence postulate*

Some linear Hermitian operators A, B, ... on Hilbert space $\mathcal{H}$, which have complete orthonormal sets of eigenvectors, correspond to observables of a system [33].

As explained by Park and Margenau, the content of this postulate is slightly different from that of its analogues in typical axiomatics inspired by the work of von Neumann. In its original form, the correspondence postulate included both of the following statements: (*i*) every Hermitian operator corresponds to a physical observable; and (*ii*) every observable has a Hermitian operator representative. Superselection rules introduced by Wick et al. [34] exclude certain Hermitian operators from being observable. By replacing the word *every* in statement (*i*) by the word *some*, superselection rules are satisfied. Compatibility of simultaneous measurements introduced by Park and Margenau [33] excludes certain observables from corresponding to Hermitian operators. In addition, in a unified theory of mechanics and thermodynamics other properties are observable, such as temperature, but correspond to no Hermitian operators. By replacing the word *every* in statement (*ii*) by the word *some* the asymmetry between observables and operators is embraced. It is clear that the correspondence postulate as stated earlier accommodates both the asymmetry between operators and observables and the asymmetry between observables and operators.

B.3.8. *Measurement Act*
A *measurement act* is a reproducible scheme of measurements and operations on a member of an ensemble. The result of such an act is a precise — error free — number associated with an observable.

If a measurement act is applied to each and every member of a homogeneous ensemble, the results conform to the following mathematical representation.

B.3.9. *Mean-value Postulate*
If a measurement act of an observable represented by Hermitian operator A is applied to each and every member of a homogeneous ensemble, there exists a linear functional m(A) of A such that the value of m(A) equals the arithmetic mean of the ensemble of A measurements, that is,

$$m(A) = \langle A \rangle = \textstyle\sum_i a_i / N \qquad \text{for } N \to \infty \tag{B-1}$$

where a_i is the error-free measurement result of the measurement act applied to the ith member of the ensemble, and $\langle A \rangle$ another notation for m(A) [33].

B.3.10. *Mean-value Theorem*
For each of the mean-value functionals m(A) of a system at an instant in time, there exists the same Hermitian operator ρ such that [33]

$$m(A) = \langle A \rangle = \mathrm{Tr}\,[\rho A] \tag{B-2}$$

The operator ρ is known as the *density operator* or the *density of measurement results of observables*. The concept of the density operator was introduced by von Neumann [35] as a statistical average of projectors. In contrast, here ρ is restricted to homogeneous ensembles and, therefore, it is exclusively quantum-theoretic. The operator ρ is proven to be Hermitian, nonnegative-definite, unit trace and, in general, not a projector [8,22,36], that is,

$$\rho > 0 \; ; \qquad \mathrm{Tr}\,\rho = 1 \; ; \qquad \rho \geq \rho^2 \; . \tag{B-3}$$

B.3.11. *Probability Theorem*
If a measurement act of an observable represented by operator A is applied to each and every member of a homogeneous ensemble characterized by ρ, the probability or frequency $W(a_n)$ that the results will yield eigenvalue a_n is given by the relation

$$W(a_n) = \mathrm{Tr}\,[\rho A_n] \tag{B-4}$$

where A_n is the projection onto the subspace belonging to a_n,

$$A|\alpha_n\rangle = a_n|\alpha_n\rangle \qquad \text{for } n = 1, 2, \dots \; . \tag{B-5}$$

and $|\alpha_n\rangle$ the nth eigenket of operator A.

B.3.12. *Measurement Result Theorem*
The only possible result of a measurement act of the observable represented by A is one of the eigenvalues of A (equation B-5).

Though the statements of the mean-value postulate, and the probability and measurement result theorems are practically the same as those given by Park and Margenau [33], here the contents of the statements differ from those of Park and Margenau because of the restriction of ρ to a homogeneous ensemble. The importance and necessity of this restriction in the unified theory cannot be overemphasized.

B.4. COMMENT ON THE PICTORIAL REPRESENTATION OF A HOMOGENEOUS ENSEMBLE

Because no conceivable decomposition of a homogeneous ensemble short of measurements can yield component subensembles with different measurement results — spectra of values and frequency of occurrence of each value — the density operator ρ can be assigned to each member of the homogeneous ensemble as shown in Figure B-2. This assignment introduces no ambiguities, that is, a homogeneous ensemble can be safely thought of as consisting of identical systems each of which has the same operators $A, B, \ldots$ that represent observables, the same values $\langle A \rangle, \langle B \rangle, \ldots$ of observables, and the same density operator ρ. In fact, this unambiguous characterization of a member at an instant in time is all that can be said at that time.

B.4.1. *State*

The meaning of the term state is discussed in Appendix A. In essence, it is all that can be said about a system at an instant in time. In view of the discussion of the meaning of Figure B-2, the mathematical representation of state in the unified theory consists of a set of Hermitian operators $A, B, \ldots$ that correspond to a complete set of independent observables — the value of an independent observable can be varied without affecting the values of other observables — and the relations

$$\begin{aligned} \langle A \rangle &= \mathrm{Tr}\,[\rho A] = \textstyle\sum_i a_i/N \\ \langle B \rangle &= \mathrm{Tr}\,[\rho B] = \textstyle\sum_i b_i/N \\ &\;\;\vdots \qquad\quad \vdots \qquad\quad \vdots \end{aligned} \tag{B-6}$$

In these relations, either the density operator ρ is specified *a priori* and the values of the observables are calculated, or the values of all the independent observables $\sum_i a_i/N$, $\sum_i b_i/N, \ldots$ are measured, and a unique density operator is calculated. The density operator is unique because equations (B-6) are linear in ρ.

Because the idea is sometimes overlooked, it is very important to emphasize that the quantum-theoretic value of an observable is always determined by the corresponding equation in set (B-6) and not by any particular measurement result. A measurement act yields an eigenvalue. But no correct quantum postulate or theorem asserts that this eigenvalue was necessarily the value of the observable prior to measurement. An ensemble of measurement results of an observable is needed in order to ascertain its value.

B.4.2. *Dynamics*

Under the heading of dynamics there is only one concept, the equation of motion. As discussed in Refs. [5,11], the equation of motion of conventional quantum mechanics in the Schroedinger picture is incomplete because it prescribes only a unitary evolution in time of a projector $\rho_i = \rho_i^2$ for $i = 1, 2, \ldots$, and such an evolution corresponds to a reversible adiabatic process only. In a unified theory, the equation of motion must account for unitary changes in time of density operators that are not projectors ($\rho > \rho^2$), for reversible adiabatic processes that are not unitary and, of course, for irreversible processes. Until a complete equation of motion is universally accepted by the scientific community, three postulates provide a partial substitute for the purposes of the unified theory — equation (B-7) and the first and second laws of thermodynamics. The substitute is partial because it covers only some of the requirements of the unified theory.

B.4.3. *Limited Dynamical Postulate*

Hatsopoulos and Gyftopoulos [5] postulate that unitary transformations of ρ in time obey the relation

$$\frac{d\rho}{dt} = -\frac{i}{\hbar}[H\rho - \rho H] \tag{B-7}$$

where H is the Hamiltonian operator of the system. The unitary transformation of ρ satisfies the equation

$$\rho(t) = U(t, t_0)\, \rho(t_0)\, U^+(t, t_0) \tag{B-8}$$

where U^+ is the Hermitian conjugate of U and, if H is independent of t,

$$U(t, t_0) = \exp[-(i/\hbar)(t - t_0)\, H] \tag{B-9}$$

and, if H is explicitly dependent on t,

$$\frac{dU(t, t_0)}{dt} = -(i/\hbar)\, H(t)\, U(t, t_0) \tag{B-10}$$

Though equation (B-7) is well known in the literature as the von Neumann equation, here it must be postulated for the following reason. In statistical quantum mechanics [37], the equation is derived as a statistical average of Schroedinger equations, each of which describes the evolution in time of a projector ρ_i in the statistical mixture represented by ρ, and each of which is multiplied by a time independent statistical probability α_i. In the unified theory, ρ is not a mixture of projectors and, therefore, cannot be derived as a statistical average of projectors. It is noteworthy that the dynamical postulate is limited or incomplete because all unitary evolutions of ρ in time correspond to reversible adiabatic processes. But not all reversible adiabatic processes correspond to unitary evolutions of ρ in time [7], and not all processes are reversible.

B.4.4. *The First and Second Laws of Thermodynamics*

A partial relief to the limitations of the incomplete dynamical postulate just cited is provided by adding to equation (B-7) two more statements, the first law and the second law of thermodynamics. These statements are given in Appendix A. The quantum-theoretic postulates and theorems, and the two laws of thermodynamics provide the conceptual framework for the exposition of the unified quantum theory of mechanics and thermodynamics, a theory that applies to all systems and all states. Moreover, the quantum-theoretic concepts lurk behind every aspect of the exposition of thermodynamics summarized in Appendix A.

It is noteworthy that the third law of thermodynamics is not needed because it is inherent in the quantum theoretic foundations.

B.4.5. *Entropy*

On the basis of the new exposition of thermodynamics presented in Ref. [11] and summarized in Appendix A, and the unified theory presented in Refs. [5–8], and summarized in this appendix, in Ref. [38] we prove that of all the expressions for entropy S that have been proposed in the literature the only one that satisfies all the necessary criteria is given by the relation

$$S = -\mathrm{k}\,\mathrm{Tr}\,[\rho \ln \rho] \tag{B-11}$$

provided that ρ is exclusively represented by a homogeneous ensemble. If ρ is represented by a heterogeneous ensemble, then equation (B-11) does not represent the entropy of thermodynamics.

For given values of energy, amounts of constituents, and parameters, if ρ is a projector then $S = 0$, and if ρ corresponds to the unique stable equilibrium state required by the second law then S has the largest value of all the entropies of states that share the given values of energy, amounts of constituents, and parameters. If, as it is usually done for projectors, we interpret a density operator ρ as the shape of constituents of a system, then the entropy of the system is a special measure of the shape with values ranging from zero to a maximum for each set of values of energy, amounts of constituents, and parameters.

If we adopt the measure of shape interpretation for entropy, an interesting concomitant ensues. Let us assume that the paradigm of the unified quantum theory was conceived prior to that of classical mechanics, and that a physicist wished to approximate quantum theoretic results by classical concepts. We can safely predict that he would have done an excellent job because for macroscopic systems with highly degenerate eigenkets, densities of measurement results of practically all observables can be approximated by the Dirac delta function $\delta(\mathbf{q}-\mathbf{q}_0)\delta(\mathbf{p}-\mathbf{p}_0)$ of space coordinates $\mathbf{q}$ and momenta $\mathbf{p}$. Though highly accurate, such an approximation would be inadequate because it does not include the concept of shape of the constituents of the system and, therefore, provides neither the mathematical representation for the concept of entropy as a property of the constituents, nor the possibility of change of this mathematical representation over a range of values. This is another aspect of the inadequacy of classical mechanics to accommodate the concepts of thermodynamics.

B.4.6. *Density Operator of a Stable Equilibrium State*

We can find the density operator ρ^0 of a thermodynamic or stable equilibrium state A_0 of system A by maximizing the entropy S subject to the constraints

$$\mathrm{Tr}\,\rho = 1 \qquad \text{and} \qquad \langle \mathrm{H} \rangle = \mathrm{Tr}\,[\rho \mathrm{H}] = \text{given value } E\ . \tag{B-12}$$

For simplicity, we assume that the system has only volume as a parameter, and only one constituent with an amount n equal to an eigenvalue of the number operator of the constituent. Moreover, we use a different subscript i even for orthonormal projectors that correspond to the same eigenvalue.

The constrained maximization solution is proven to be [39]

$$\rho^0 = \frac{\exp(-\beta \mathrm{H})}{\mathrm{Tr}\,\exp(-\beta \mathrm{H})} = \textstyle\sum_{\mathrm{i}} \rho^0_{\mathrm{ii}}\, |\epsilon_{\mathrm{i}}\rangle\langle\epsilon_{\mathrm{i}}| \tag{B-13}$$

where

$$H|\epsilon_i\rangle = \varepsilon_i|\epsilon_i\rangle \; ; \qquad \rho^0_{ii} = \exp(-\beta\varepsilon_i)/\textstyle\sum_i \exp(-\beta\varepsilon_i)$$

and β is determined by the value of the energy E because

$$\langle H\rangle = E = \mathrm{Tr}[\rho^0 H] = \textstyle\sum_i \rho^0_{ii}\, \varepsilon_i$$

We can show that

$$\beta = \left[(\partial S/\partial E)_{\varepsilon,n}\right]_0 / k \tag{B-14}$$

where the subscript "ε" stands for fixed values of all the energy eigenvalues ε_1, $\varepsilon_2, \ldots$, and the subscript "0" for state A_0, that is, the partial derivative is taken along the stable equilibrium state locus for fixed values of parameters (fixed ε) and fixed amount of the constituent n at state A_0. But for stable equilibrium states, the partial derivative $\left[(\partial S/\partial E)_{\varepsilon,n}\right]_0$ is defined as the inverse temperature of A_0. Accordingly

$$\beta = 1/kT_0 \tag{B-15}$$

B.4.7. *Translational Velocity of a Molecule*

We consider a system A in a stable equilibrium state A_0 with energy E, number of molecules $n \geq 1$, and volume V. For such a state, the value $\langle p_k\rangle$ of the momentum of a single molecule in the spatial direction x_k is given by the relation

$$\begin{aligned} \langle p_k\rangle &= \mathrm{Tr}\,[\rho^0 p_k] = \textstyle\sum_m \sum_n \rho^0_{nn}\, \langle\epsilon_m|\epsilon_n\rangle\langle\epsilon_n|p_k|\epsilon_m\rangle \\ &= \textstyle\sum_m \sum_n \rho^0_{nn}\, \delta_{nm}\, \langle\epsilon_n|p_k|\epsilon_m\rangle \\ &= \textstyle\sum_m \rho^0_{mm}\, \langle\epsilon_m|p_k|\epsilon_m\rangle = 0 \end{aligned} \tag{B-16}$$

where the third of these equations results from the fact that the energy eigenkets are orthonormal and, therefore,

$$\langle\epsilon_m|\epsilon_n\rangle = \delta_{nm} = \text{Kronecker delta} \tag{B-17}$$

and the last equation from the relation

$$\langle\epsilon_m|p_k|\epsilon_m\rangle = 0 \qquad \text{for all k and m}\,. \tag{B-18}$$

The proof of equation (B-18) is straightforward. First, we observe that the Hamiltonian operator H of the system and the momentum operator p_k of a particular molecule satisfy the commutation relation

$$[x_k, H] = i\hbar p_k/M \tag{B-19}$$

where M is the mass of the molecule. Next, upon defining for all k and m

$$(\Delta x_k)^2_m = \langle\epsilon_m|x_k^2|\epsilon_m\rangle - \langle\epsilon_m|x_k|\epsilon_m\rangle^2 \tag{B-20}$$

$$(\Delta H)^2_m = \langle\epsilon_m|H^2|\epsilon_m\rangle - \langle\epsilon_m|H|\epsilon_m\rangle^2 \tag{B-21}$$

we can readily prove that [40]

$$(\Delta x_k)_m \, (\Delta H)_m \geq \hbar |\langle \epsilon_m | p_k | \epsilon_m \rangle| / 2M \qquad \text{(B-22)}$$

But for a system with a finite extension L_x along the coordinate axis of x_k, and an energy eigenket $|\epsilon_m\rangle$, we have

$$0 < (\Delta x_k)_m < L_k \qquad \text{and} \qquad (\Delta H)_m = 0 \qquad \text{(B-23)}$$

and so equality (B-18) is proved.

References

1. Clausius, R. (1976) In *The Second Law of Thermodynamics*, J. Kestin (ed.), Dowden, Hutchinson and Ross, Inc., pp. 186–187.
2. Wehrl, A. (1978) General properties of entropy, *Rev. Mod. Phys.*, **50**, 2, 221–260.
3. Kline, S.J. (1997) The semantics and meaning of the entropies, Report CB-1, Dept. of Mech. Eng., Stanford University.
4. Obert, E.F. (1960) *Concepts of Thermodynamics*, McGraw-Hill, New York.
5. Hatsopoulos, G.N. and Gyftopoulos, E.P. (1976) A unified quantum theory of mechanics and thermodynamics. Part I. Postulates, *Found. Phys.*, **6**, 15–31.
6. Hatsopoulos, G.N. and Gyftopoulos, E.P. (1976) A unified quantum theory of mechanics and thermodynamics. Part IIa. Available energy, *Found. Phys.*, **6**, 127–141.
7. Hatsopoulos, G.N. and Gyftopoulos, E.P. (1976) A unified quantum theory of mechanics and thermodynamics. Part IIb. Stable equilibrium states, *Found. Phys.*, **6**, 439–455.
8. Hatsopoulos, G.N. and Gyftopoulos, E.P. (1976) A unified quantum theory of mechanics and thermodynamics. Part III. Irreducible quantal dispersions, *Found. Phys.*, **6**, 561–570.
9. Beretta, G.P., Gyftopoulos, E.P., Park, J.L., and Hatsopoulos, G.N. (1984) Quantum thermodynamics. A new equation of motion for a single constituent of matter, *Nuovo Cimento*, **82B**, 2, 169–191.
10. Beretta, G.P., Gyftopoulos, E.P., and Park, J.L. (1985) Quantum thermodynamics. A new equation of motion for a general quantum system, *Nuovo Cimento*, **87B**, 1, 77–97.
11. Gyftopoulos, E.P. and Beretta, G.P. (1991) *Thermodynamics: Foundations and Applications*, Macmillan, New York.
12. Shankar, R. (1994) *Principles of Quantum Mechanics*, 2nd Ed., Plenum Press, New York.
13. von Neumann, J. (1955) *Mathematical Foundations of Quantum Mechanics*, Princeton University Press, New Jersey.
14. Leighton, R.B. (1959) *Principles of Modern Physics*, McGraw-Hill, New York.
15. Brandt, S. and Dahmen, H.D. (1995) *The Picture Book of Quantum Mechanics*, 2nd Ed., Springer-Verlag, New York.
16. Slater, J. (1963) *Quantum Theory of Molecules and Solids*, McGraw-Hill, New York.
17. There is an inconsistency between the calculation of the eigenstates of the proton and those of the electron of a hydrogen atom. The former are calculated for a proton confined in a finite size container, whereas the latter are calculated for a Coulomb potential extending over all space. As a result the operator ρ^0_{electron} is not well defined because Tr $\exp(-\beta H_{\text{electron}})$ is unbounded, and the entropy of the electron is infinite. For details see: Conway, J.B. (1985) *A Course in Functional Analysis*, Springer-Verlag, New York.
18. Dahmen, H.D. and Stroh, T. (1998) Private communication.
19. Hatsopoulos, G.N. and Gyftopoulos, E.P. (1979) *Thermionic Energy Conversion*, Vol. II, MIT Press, Cambridge, Massachusetts.

20. Timoshenko, S. and Young, D.H. (1948) *Advanced Dynamics*, McGraw-Hill, New York.
21. Kuhn, T. (1970) *The Structure of Scientific Revolutions*, 2nd Ed., Chicago University Press, Chicago.
22. Park, J.L. and Simmons, R.F. (1983) In *Old and New Questions of Physics, Cosmology, and Theoretical Biology: Essays in Honor of Wolfgang Yourgrau*, A. van der Merwe (ed.), Plenum Press, New York, pp. 289–308.
23. Korsch, H.J. and Steffen, H. (1987) Dissipative quantum dynamics, entropy production and irreversible evolution towards equilibrium, *J. of Phys.*, **A20**, 3787–3803.
24. Lindblad, G. (1976) On the generators of quantum dynamical semigroups, *Commun. Math. Phys.*, **48**, 119–130.
25. Hatsopoulos, G.N. and Keenan, J.H. (1965) *Principles of General Thermodynamics*, Wiley, New York.
26. Gyftopoulos, E.P. (1997) Fundamentals of analyses of processes, *Energy Conv. Mgmt*, **38**,15–17, 1525–1533.
27. Maxwell, J.C. (1871) *Theory of Heat*, Longmans, Green and Co., London.
28. McQuarrie, D.A. (1973) *Statistical Thermodynamics*, Harper and Row, New York.
29. Callen, H.B. (1985) *Thermodynamics and an Introduction to Thermostatics*, 2nd Ed., Wiley, New York.
30. Schroedinger, E. (1935) Discussion of probability relations between separated systems, *Proc. Cambridge Philos. Soc.*, **31**, 555–563, and (1936) Probability relations between separated systems, *Proc. Cambridge Philos. Soc.*, **32**, 446–452.
31. Park, J.L. (1968) Nature of quantum states, *Amer. J. Phys.*, **36**, 211–226, and (1988) Thermodynamic aspects of Schrödinger's probability relations, *Found. Phys.*, **18**, 2, 225–244.
32. Jancel, R. (1963) *Foundations of Classical and Quantum Statistical Mechanics*, Pergamon Press, Oxford, England, 1963.
33. Park, J.L. and Margenau, H. (1968) Simultaneous measureability in quantum theory, *Int. J. Theor. Phys.*, **1**, 3, 211–283.
34. Wick, G.C., Wightman, A.S., and Wigner, E.P. (1952) The intrinsic parity of elementary particles, *Phys. Rev.*, **88**, 101–105.
35. von Neumann, J., *op. cit.*, pp. 313–316.
36. Jauch, J.M. (1973) *Foundations of Quantum Mechanics*, Addison Wesley, Reading, MA.
37. Tolman, R.C. (1938) *The Principles of Statistical Mechanics*, Oxford University Press, London, Great Britain.
38. Gyftopoulos, E.P. and Çubukçu, E. (1997) Entropy: Thermodynamic definition and quantum expression, *Phys. Rev. E*, **55**, 4, 3851–3858.
39. Katz, A. (1967) *Principles of Statistical Mechanics*, W.H. Freeman, San Francisco, CA.
40. Jauch, J.M. *op. cit.*, pp. 160–163.

THERMODYNAMIC OPTIMIZATION OF INANIMATE AND ANIMATE FLOW SYSTEMS

A. BEJAN
Duke University
Department of Mechanical Engineering and Materials Science
Durham, NC 27708-0300, USA

1. An Old and Prevalent Natural Phenomenon

Thermodynamic optimization is, literally, the search for the best thermodynamic performance subject to present-day constraints. This philosophy is very old. It has been with us throughout the history of the heat engine. Its reach however, is much broader and more *permanent*. We can be sure that the performance of power plants will continue to improve in time, in the same way that, in time, a sudden rain fall will generate an efficient flow structure (dendritic channels), not Darcy flow alone.

In the broader sense, optimization of performance has been the driving force behind the development of all machines, starting with the wheel and the pulley. Physiologists and geophysicists too are coming to the conclusion that optimization of performance has been the generating mechanism for geometric form and size in all plants, animals and natural flow structures.

If thermodynamic optimization is such an old and prevalent *natural* phenomenon, then what is new? New is the streamlining of the study of thermodynamic optimization into a simple and coherent discipline that is now recognized as an integral part of contemporary thermodynamics. The constant streamlining, simplification and generalization of individual observations and applications has always been the flow we call science, or knowledge. Exergy analysis, thermodynamic optimization* and thermoeconomics are the main methods of this new discipline. Thermodynamicists have developed these methods for the past one hundred years or more. Significant growth has occurred in the wake of the energy crisis, as is shown by the chronological sequence of books written in this field beginning with 1982 [1 - 10].

In this chapter I review the most basic aspects of thermodynamic optimization, with particular emphasis on why this method is different than pure thermodynamics. I then rely on some of the simplest models in order to stress that the method also requires *modeling*, which is a construction that precedes *optimization*. I conclude with illustrations from my most recent applications of the method, in a sequence that continues in my accompanying chapter on Constructal Flow Geometry Optimization.

* Thermodynamic optimization is also practiced under the equivalent titles of entropy generation minimization (EGM), finite-time (finite-size, finite-resources), and endoreversible (exoirreversible) thermodynamics. There is absolutely no fundamental difference between these lines of work: see sections 3 and 8.

A. Bejan and E. Mamut (eds.), Thermodynamic Optimization of Complex Energy Systems, 45–60.

2. Thermodynamics Combined with Heat Transfer and Fluid Mechanics

It is instructive to begin with a brief look at why in thermodynamic optimization we must rely on heat transfer and fluid mechanics, not just thermodynamics. Consider the most general system-environment configuration, namely a system that operates in the unsteady state. Its instantaneous inventories of mass, energy, and entropy are M, E, and S. The system experiences the net work transfer rate $\dot{W}$, heat transfer rates ($\dot{Q}_0$, $\dot{Q}_1, \ldots \dot{Q}_n$) with n + 1 temperature reservoirs (T_0, $T_1, \ldots, T_n$), and mass flow rates ($\dot{m}_{in}$, $\dot{m}_{out}$) through any number of inlet and outlet ports. Noteworthy in this array of interactions is the heat transfer rate between the system and the environmental (atmospheric) temperature reservoir, $\dot{Q}_0$, on which we focus shortly.

The thermodynamics of the system consists of accounting for the first law and the second law [4],

$$\frac{dE}{dt} = \sum_{i=0}^{n} \dot{Q}_i - \dot{W} + \sum_{in} \dot{m}h - \sum_{out} \dot{m}h \tag{1}$$

$$\dot{S}_{gen} = \frac{dS}{dt} - \sum_{i=0}^{n} \frac{\dot{Q}_i}{T_i} - \sum_{in} \dot{m}s + \sum_{out} \dot{m}s \geq 0, \tag{2}$$

where h is shorthand for the sum of specific enthalpy, kinetic energy, and potential energy of a particular stream at the boundary. In Eq. (2) the total entropy generation rate $\dot{S}_{gen}$ is simply a definition (notation) for the entire quantity on the left-hand side of the inequality sign. We shall see that it is advantageous to decrease $\dot{S}_{gen}$, and this can be accomplished only by changing at least one of the quantities (properties, interactions) specified along the system boundary.

We select the environmental heat transfer $\dot{Q}_0$ as the interaction that is allowed to float as $\dot{S}_{gen}$ varies. Historically, this choice was inspired (and justified) by applications to power plants and refrigeration plants, because the rejection of heat to the atmosphere was of little consequence in the overall cost analysis of the design. Eliminating $\dot{Q}_0$ between Eqs. (1) and (2) we obtain

$$\dot{W} = -\frac{d}{dt}(E - T_0 S) + \sum_{i=1}^{n}\left(1 - \frac{T_0}{T_i}\right)\dot{Q}_i + \sum_{in} \dot{m}(h - T_0 s) - \sum_{out} \dot{m}(h - T_0 s) - T_0 \dot{S}_{gen} \tag{3}$$

The power output in the limit of reversible operation ($\dot{S}_{gen} = 0$) is

$$\dot{W}_{rev} = -\frac{d}{dt}(E - T_0 S) + \sum_{i=1}^{n}\left(1 - \frac{T_0}{T_i}\right)\dot{Q}_i + \sum_{in} \dot{m}(h - T_0 s) - \sum_{out} \dot{m}(h - T_0 s) \tag{4}$$

In engineering thermodynamics each of the terms on the right-hand side of Eq. (4) is recognized as an exergy of one type or another [2, 8], and the calculation of $\dot{W}_{rev}$ is known as exergy analysis. Subtracting Eq. (3) from Eq. (4) we arrive at the Gouy-Stodola theorem,

$$\dot{W}_{rev} - \dot{W} = T_0 \dot{S}_{gen} \tag{5}$$

In Eq. (5) $\dot{W}_{rev}$ is *fixed* because all the heat and mass flows (other than $\dot{Q}_0$) are fixed.

Pure thermodynamics (e.g., exergy analysis) ends, and EGM begins, with Eq. (5).

The lost power ($\dot{W}_{rev} - \dot{W}$) is always positive, regardless of whether the system is a power producer (e.g., power plant) or a power user (e.g., refrigeration plant). To minimize lost power when $\dot{W}_{rev}$ is fixed is the same as maximizing power output in a power plant, and minimizing power input in a refrigeration plant. This operation is also equivalent to minimizing the total rate of entropy generation. If $\dot{W}_{rev}$ is not held fixed while the system design is being changed, then Eq. (5) and the calculation of $\dot{S}_{gen}$ lose their usefulness.

The critically new aspect of the EGM method—the aspect that makes the use of thermodynamics insufficient, and distinguishes EGM from exergy analysis—is the *minimization* of the calculated entropy generation rate. To minimize the irreversibility of a proposed design, the analyst must use the relations between temperature differences and heat transfer rates, and between pressure differences and mass flow rates. He or she must express the thermodynamic nonideality of the design $\dot{S}_{gen}$ as a function of the physical characteristics of the system, namely to finite dimensions, shapes, materials, finite speeds, and finite-time intervals of operation. For this the analyst must rely on heat transfer and fluid mechanics principles, in addition to thermodynamics. Only by varying one or more of the physical characteristics of the system, can the analyst bring the design closer to the operation characterized by minimum entropy generation subject to size and time constraints.

3. The Method and Its History

To appreciate the origins and history of the method of thermodynamic optimization, it is important to note that the field of low temperature refrigeration was the first where irreversibility minimization became an established method of optimization and design [1]. It is easy to prove that the power required to keep a cold space cold is equal to the total rate of entropy generation times the ambient temperature, with the observation that the entropy generation rate includes the contribution made by the leakage of heat from T_0 into the cold space [1]. The structure of a cryogenic system is in fact dominated by components with heat transfer irreversibility, that leak heat, e.g., mechanical supports, radiation shields, electric cables, and counterflow heat exchangers. The minimization of entropy generation along a heat leak path consists of optimizing the path in harmony with the rest of the refrigerator of liquifier.

Figure 1 shows a mechanical support of length L that connects the cold end of the machine (T_L) to room temperature (T_H). The rate of entropy generation inside the support shown as a vertical column is [1, 9],

$$\dot{S}_{gen} = \int_{T_L}^{T_H} \frac{\dot{Q}}{T^2} dT \qquad (6)$$

where it is important to note that the heat leak $\dot{Q}$ is allowed to vary with the local temperature T. The local heat leak decrement $d\dot{Q}$ is removed by the rest (the internal part) of the installation, which is modeled as reversible. The heat leak is also related to the local temperature gradient and conduction cross-section A,

$$\dot{Q} = kA \frac{dT}{dx}, \qquad (7)$$

where the thermal conductivity k(T) decreases toward low temperatures. Rearranged and integrated, Eq. (7) places a size constraint on the unknown function $\dot{Q}(T)$,

$$\frac{L}{A} = \int_{T_L}^{T_H} \frac{k}{\dot{Q}} dT \qquad (8)$$

According to variational calculus, the heat leak function that minimizes the $\dot{S}_{gen}$ integral (6) subject to the finite-size constraint (8) is $\dot{Q}_{opt} = (\lambda k)^{1/2} T$. The Lagrange multiplier λ is determined by substituting $\dot{Q}_{opt}$ into the constraint (8):

$$\dot{Q}_{opt} = \left(\frac{A}{L} \int_{T_L}^{T_H} \frac{k^{1/2}}{T} dT \right) k^{1/2} T, \qquad (9)$$

$$\dot{S}_{gen,min} = \frac{A}{L} \left(\int_{T_L}^{T_H} \frac{k^{1/2}}{T} dT \right)^2 . \qquad (10)$$

Equation (6) was provided by thermodynamics and Eq. (7) by heat transfer: together they prescribe the optimal design (9, 10), which is characterized by a certain distribution of intermediate cooling effect $(d\dot{Q}/dT)_{opt}$. Any other design, $\dot{Q}(T)$, will generate more entropy and will require more power in order to maintain the cold end of the support at T_L. Quantitative and older examples from cryogenic engineering are given in [1, 3, 4].

Together, Eqs. (6) and (8) illustrate the backbone of the method of thermodynamic optimization subject to a physical constraint, as it was practiced in engineering before 1974 when the preceding example was published [11]. Key advances had been made as early as the 1950s in heat exchanger optimization [12], cryogenics [13, 14], solar thermal energy conversion [15], and power maximization in power plants [16 - 24]. These advances outlined the field before the energy crisis, long before the growth that occurred in the 1980s and 1990s. For example, Curzon and Ahlborn's first physics paper on power maximization [25] is a rediscovery of the idea published 18 years earlier by Chambadal [16] and Novikov [19]. Physics papers followed Curzon and Ahlborn in the 1980s: there is absolutely no difference [26, 27] between this line of work and the constrained thermodynamic optimization developed by the pioneers and illustrated in this chapter (see also section 8).

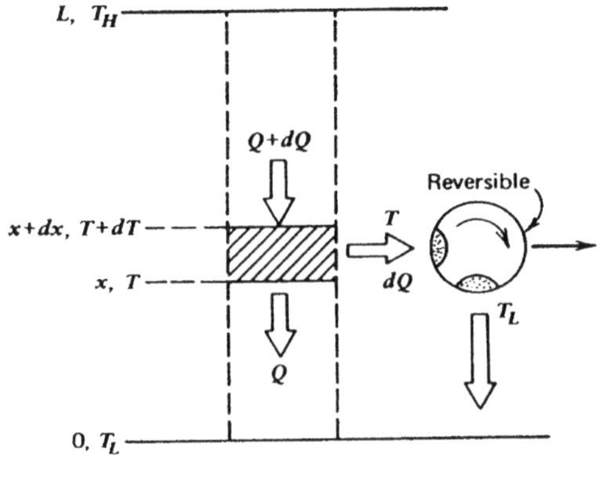

Figure 1. Mechanical support with variable heat leak and intermediate cooling [1].

4. Maximum Power from Flow between Two Reservoirs

Let us consider the peer-refereed and published criticism [28, 29] of Chambadal, Novikov, and Curzon and Ahlborn's power plant model, and their maximization of power output. We consider this model in a sequence of three analogs: mechanical [9], electrical [9], and thermodynamic, Fig. 2.

Imagine that we have access to two infinitely large reservoirs of a fluid at different pressures (P_H, P_L), and that we install a turbine between the reservoirs such that the stream $\dot{m}$ produces the power $\dot{W}$ [9]. The stream flows with pressure drops ($P_H - P_{HC}$ and $P_{LC} - P_L$) along the two ducts that connect the turbine to the reservoirs. The actual pressure difference across the turbine ($P_{HC} - P_{LC}$) is smaller than the overall pressure difference ($P_H - P_L$). The simplest flow resistance model is represented by the linear relations

$$P_H - P_{HC} = R_H \dot{m} \qquad\qquad P_{LC} - P_L = R_L \dot{m} \tag{11}$$

where the flow resistances (R_H, R_L) are related to the physical characteristics of the ducts and fluid. For simplicity, we assume that the fluid is incompressible (density ρ) and that the turbine is reversible. The power output is simply $\dot{W} = \dot{m}\,(P_{HC} - P_{LC})/\rho$: this quantity can be maximized by varying $\dot{m}$, that is by imagining—as needed—trickles or gushes of fluid, depending on what yields a higher $\dot{W}$. We can do this because we imagined that the reservoirs are infinite, i.e., the size of $\dot{m}$ has no absolute effect on the reservoirs and our *ability* to access them. It is easy to show that maximum power output occurs when the pressure difference across the turbine is exactly half of the overall pressure difference,

$$(P_{HC} - P_{LC})_{opt} = \frac{1}{2}(P_H - P_L) \tag{12}$$

The electrical analog of this example is a reversible electrical motor connected to two constant-voltage lines (V_H, V_L). The voltage drop across the motor is $V_{HC} - V_{LC}$.

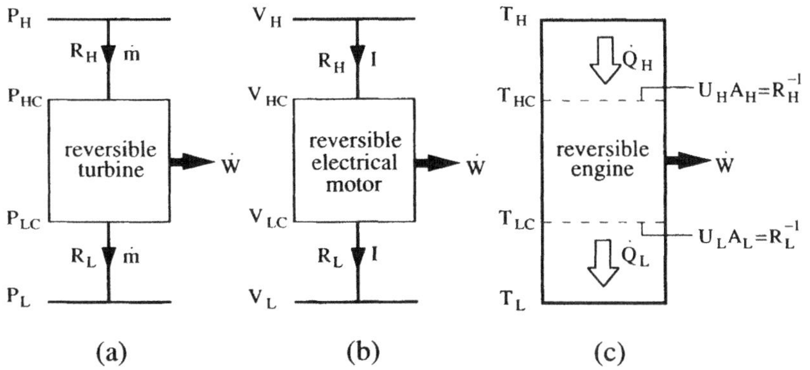

Figure 2. Mechanical, electrical and thermodynamic systems producing power while bridging the gap between two reservoirs of pressure, voltage and temperature.

If the resistances of the conductors between motor and reservoirs are R_H and R_L, then we may use the linear model $V_H - V_{HC} = R_H I$ and $V_{LC} - V_L = R_L I$, where I is the current. The power delivered by the motor, $I(V_{HC} - V_{LC})$, can be maximized by selecting the steady-state current that we need. We can do this because we have free access to V_H and V_L, and because V_H and V_L are insensitive to I. Maximum power occurs when the current is such that the equivalent of Eq. (12) holds, $(V_{HC} - V_{LC})_{opt} = (V_H - V_L)/2$.

The thermodynamic analog of this dream is the reversible heat engine sandwiched between two infinitely large entities at different temperatures (T_H, T_L). The heat input ($\dot{Q}_H$) and the heat rejection require finite temperature differences,

$$T_H - T_{HC} = R_H \dot{Q}_H \qquad T_{LC} - T_L = R_L \dot{Q}_L \tag{13}$$

where the thermal resistances (R_H, R_L) are determined by the dimensions and materials of the systems that effect the two thermal contacts, T_H-engine and engine-T_L. Internal engine reversibility means that $\dot{Q}_H / T_H = \dot{Q}_L / T_L$. The power output $\dot{W} = \dot{Q}_H (1 - T_{LC} / T_{HC})$ can be maximized by varying $\dot{Q}_H$ freely. It turns out that the optimum heat input $\dot{Q}_{H,opt}$ is such that maximum power is reached when

$$\left(\frac{T_{HC}}{T_{LC}}\right)_{opt} = \left(\frac{T_H}{T_L}\right)^{1/2} \tag{14}$$

The three-way analogy is complete. The questioning of the thermodynamic optimum, Eq. (14), concerns its *realism*. In the optimization of a power plant, is the heat input—the fuel burnt—a degree of freedom? Do we have access to an infinite reservoir of $\dot{Q}_H$ in the same way that we can attach toy turbines and toy electrical motors to the water and power lines in the house? The answer is no in power plant designs where $\dot{Q}_H$ is an expensive commodity (e.g., fossil fuel). We may contemplate situations in which the power plant can be placed in the vicinity of two "environments" at different temperatures (e.g., solar power, geothermal, hydroelectric applications), but even then the access to these temperature reservoirs is limited by economic and geophysical constraints. An example of these limitations is given in the next section.

The thermodynamic optimum of Eq. (15) has also been reported as an efficiency at maximum power $\eta = \dot{W}_{max} / \dot{Q}_{H,opt}$, which is the conclusion of Chambadal [16], Novikov [19] and Curzon and Ahlborn [25],

$$\eta = 1 - \left(\frac{T_L}{T_H}\right)^{1/2} \tag{15}$$

This result is questionable not only for the free-$\dot{Q}_H$ assumption on which it is based, but also conceptually: to use an efficiency (a figure of merit) ratio $\dot{W} / \dot{Q}_H$ makes no sense when the denominator varies freely. It has been shown that the free-$\dot{Q}_H$ modeling error is also responsible for the observation that in Curzon and Ahlborn's model the operating point of maximum power is not the same as the point of minimum entropy generation [29], when the entropy generation is summed only over the space between T_H and T_L in Fig. 2c. The reason is clear if we re-examine Eq. (5): when $\dot{Q}_H$ is free to vary, the term $\dot{W}_{rev} = \dot{Q}_H (1 - T_L / T_H)$ is also free to vary.

Gyftopoulos [28] pointed out that the approximate agreement between Eq. (15) and the efficiencies of old power stations is simply a coincidence. It turns out that agreement with the reported efficiencies can also be reached when we hold $\dot{Q}_H$ *fixed*, and use the same power plant model as in Fig. 2c [8, 9]. For better description, we use the thermal conductances $U_H A_H = R_H^{-1}$ and $U_L A_L = R_L^{-1}$, where U_H and U_L are the overall heat transfer coefficients based on the two heat transfer contact areas, A_H and A_L, respectively. If the overall size of the heat exchange equipment is constrained [4]

$$U_H A_H + U_L A_L = UA \qquad \text{(constant)} \qquad (16)$$

then the power output of the model can be maximized with respect to how UA is split into $U_H A_H$ and $U_L A_L$. The optimal way is $(U_H A_H)_{opt} = (U_L A_L)_{opt}$, and the maximum power output is represented by the maximum efficiency (i.e., maximum power per unit of $\dot{Q}_H$):

$$\eta_{max} = 1 - \left(\frac{T_L}{T_H}\right) \Big/ \left(1 - \frac{4\dot{Q}_H}{T_H\, UA}\right) \qquad (17)$$

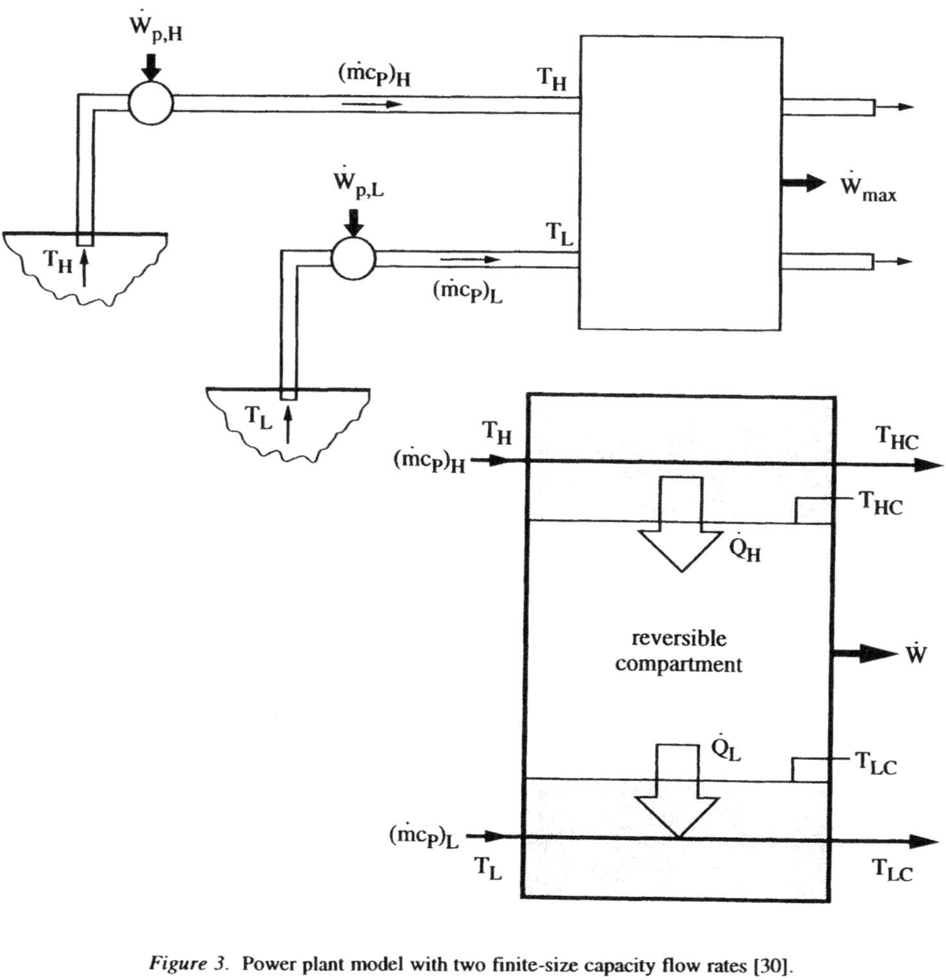

Figure 3. Power plant model with two finite-size capacity flow rates [30].

This efficiency is expectedly lower than the Carnot efficiency associated with the same temperature extremes. When the dimensionless group $\dot{Q}_H / (T_H UA)$ is a constant of order 0.1, the η_{max} values agree well with the reported efficiencies of ten power plants [10]. The constancy of the $\dot{Q}_H / (T_H UA)$ group makes sense because in the real world both $\dot{Q}_H$ and UA must scale with the overall size of the power plant.

Is there a class of power plants in which the heat input $\dot{Q}_H$ is as free to vary as the heat rejection rate $\dot{Q}_L$? The trouble with finding an answer to this question in Fig. 2c is that in that model the physical origin of $\dot{Q}_H$ is not shown: Fig. 2c is incomplete as a model of a power plant. A possible alternative is the power plant model shown in the lower-right detail of Fig. 3 [30]. Here the heat input $\dot{Q}_H$ is drawn from a hot single-phase stream of original temperature T_H, namely $\dot{Q}_H = C_H (T_H - T_{HC})$, where $C_H = (\dot{m}c_p)_H$ is the capacity rate of the stream. It is assumed that the heat exchanger between the hot stream and the reversible compartment is sufficiently large such that the outlet temperature of the hot stream is practically equal to the hot-end temperature of the reversible compartment, T_{HC}.

The cold end of the power plant is modeled similarly. The heat transfer rate $\dot{Q}_L$ is rejected to a cold stream of original temperature T_L and capacity rate $C_L = (\dot{m}c_p)_L$. When the cold-end heat exchanger is sufficiently large, the outlet temperature of the cold stream is nearly equal to the cold-end temperature of the reversible compartment, such that the heat rejection rate is $\dot{Q}_L = C_L (T_{LC} - T_L)$. Models similar to Fig. 3 have been recognized in the literature. For example, a model with one stream at the hot end was proposed in [4], p. 382.

It can be shown that the models of Fig. 2c and 3 are analogous analytically: the thermal conductances $U_H A_H$ and $U_L A_L$ are replaced by the finite capacity rates C_H and C_L. The optimization with respect to variable $\dot{Q}_H$ leads to Eq. (14) and

$$\dot{W}_m = \frac{\left(T_H^{1/2} - T_L^{1/2}\right)^2}{C_H^{-1} + C_H^{-1}} \tag{18}$$

which shows that the maximum power $\dot{W}_m$ can be increased further by increasing C_H and C_L. Equation (18) also shows that the maximum power output is limited by the smaller of the two capacity flow rates: to see this, imagine that $C_H << C_L$ such that Eq. (19) approaches the limit $\dot{W}_m \cong C_H (T_H^{1/2} - T_L^{1/2})^2$. The search for a larger power output sends us in the direction of larger streams (C_H,C_L), i.e., in the direction of *large fluid reservoirs* of temperatures T_H and T_L, and, especially, large flow rates. On the practical aspects of this limit we focus next.

Consider now the complete power plant model shown in the upper part of Fig. 3, where the two-stream model has been connected to the fluid reservoirs T_H and T_L. It is useful to think of Fig. 3 as the simplest possible model of a geothermal or ocean thermal energy conversion (OTEC) power plant. Even when the fluid reservoirs are infinite, the flow rates cannot be increased indefinitely: the pumping power $(\dot{W}_{p,H} + \dot{W}_{p,L})$ must be subtracted from the power output of the power plant $(\dot{W}_m)$. Let us assume that the two fluids are incompressible liquids, such that the pumping power requirement of each fluid can be expressed as $\dot{W}_p = \dot{m}\Delta P / (\rho\eta_{II})$. In this expression η_{II} is the second law efficiency of the pump (assumed constant; also known as pump isentropic efficiency, η_p), $\dot{m}$ is the mass flow rate, ρ is the density of the respective incompressible liquid, and ΔP is the pressure drop along the duct, $\Delta P = r\,\dot{m}^n$. In this expression r is the effective resistance to fluid flow, and n covers the range from laminar flow (n = 1) to turbulent flow in the fully rough regime ($n \cong 2$). We treat r and n as two constants that are known from the design of the duct. The net power output of the installation is $\dot{W}_{net} = \dot{W}_m - \dot{W}_{p,H} - \dot{W}_{p,L}$, where $\dot{W}_m$ is the maximized power output shown in Eq. (18). It can be shown that $\dot{W}_{net}$ becomes

$$\dot{W}_{net} = \frac{\Delta T\, C_H C_L}{C_H + C_L} - R_H C_H^{n+1} - R_L C_L^{n+1} \tag{19}$$

where $\Delta T = \left(T_H^{1/2} - T_L^{1/2}\right)^2$, and $R_{H,L} = \left(r / \rho \eta_{II}\, c_P^{n+1}\right)_{H,L}$. The net power output has maxima with respect to both C_H and C_L: the expressions for the optimal capacity rates [30] show that the flow path with the lower R value should be assigned to the stream with the larger C value. The net power output for $C_{H,opt}$ and $C_{L,opt}$ is

$$\dot{W}_{net,m} = n\left[\Delta T / (n+1)\right]^{(n+1)/n}\left[R_H^{1/(n+2)} + R_H^{1/(n+2)}\right]^{-(n+2)/n} \tag{20}$$

This power output was maximized three times: with respect to $\dot{Q}_H$ (or T_{HC}) in the internal design of the model of Fig. 3, and with respect to C_H and C_L. Equation (20) shows that $\dot{W}_{net,m}$ is larger when R_H and R_L are smaller, and that the maximum net power is limited by the larger of the two flow resistances.

5. Maximum Power from a Stream of Hot Exhaust

Another way to provide a freely varying heat input to a power plant is by placing the hot surface $T_s(x)$ of the power cycle in contact with a hot stream that is being dumped into the ambient, Fig. 4 [31]. The length traveled by the hot stream (L) is proportional to the heat transfer area, A = p L, where p is the heat transfer area per unit of flow path

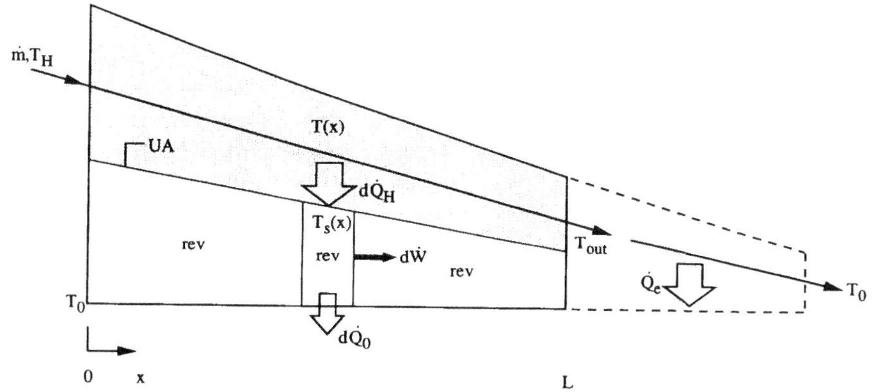

Figure 4. Power plant model with unmixed hot stream in contact with a nonisothermal heat transfer surface [31].

length. The power producing compartment is a succession of many infinitesimal reversible compartments of the kind shown in the center of the figure. The infinitesimal power output is $d\dot{W} = [1 - T_0 / T_s(x)]\, d\dot{Q}_H$, where the temperature is plotted on the vertical, and $d\dot{Q}_H = \dot{m}c_p dT$. The heat transfer through the heat exchanger area is $d\dot{Q}_H = [T(x) - T_s(x)]\, Updx$. Combining these equations and integrating from x = 0 to x = L = A/p while treating U as a constant we arrive at the finite-area constraint and the total power output:

$$\int_{T_{out}}^{T_H} \frac{dT}{T - T_s} = \frac{UA}{\dot{m}c_p} = N_{tu} \tag{21}$$

$$\dot{W} = \int_{T_{out}}^{T_H} \left(1 - \frac{T_0}{T_s}\right) \dot{m}c_p dT \tag{22}$$

An alternate route to calculating the power output $\dot{W}$ is to apply the Gouy-Stodola theorem to the larger system (extended with dashed line) in Fig. 4: $\dot{W} = \dot{W}_{rev} - T_0 \dot{S}_{gen}$. The reversible-limit power output $\dot{W}_{rev}$ corresponds to the reversible cooling of the stream from T_H all the way down to T_0. The entropy generation rate $\dot{S}_{gen}$ is the total amount associated with the larger system, and is due to two sources: the temperature difference T - T_s, and the finite temperature difference required by the external cooling rate $\dot{Q}_e = \dot{m}c_p (T_{out} - T_0)$. These two contributions are represented by the two terms in the expression

$$\dot{S}_{gen} = \int_{T_{out}}^{T_H} \left(\frac{1}{T_s} - \frac{1}{T}\right) \dot{m}c_p dT + \left(\dot{m}c_p \ln \frac{T_0}{T_{out}} + \frac{\dot{Q}_e}{T_0}\right) \tag{23}$$

To maximize $\dot{W}$ is equivalent to minimizing $\dot{S}_{gen}$, because $\dot{W}_{rev}$ is fixed. The entropy generation expression (23) is subject to the size constraint (21). There are two degrees of freedom in the minimization of $\dot{S}_{gen}$: the shape of the surface temperature function $T_s(x)$, and the place of this function on the temperature scale (i.e., closer to T_H or T_0). The second degree of freedom is alternately represented by the value of the exhaust temperature T_{out}.

It was shown in [31] that the optimal hot-stream temperature distribution T(x) is exponential in x, and so is the temperature $T_S(x)$ along the hot end of the system that converts the heating into mechanical power. At any x, the temperature difference across the heat exchanger is proportional to the local absolute temperature. The optimal solution can be implemented in practice by using a single-phase stream in place of the $T_s(x)$ surface: this stream runs in counterflow relative to the hot stream $\dot{m}$. The counterflow imbalance (the ratio between the capacity flow rates of the two streams) is the result of thermodynamic optimization.

6. Optimum Breathing and Heart Beating

An illustration of *natural* thermodynamic optimization subject to global constraints is provided by the characteristic, finely tuned frequencies of physiological processes such as breathing and heart beating [10, 32]. Consider a simple model of how the lung works. During the inhaling time t_1 the chest cavity experiences the volume increase V, while drawing in atmospheric air (T_0, P_0) through the nasal passages with the time averaged mean velocity U_1. This flow is made possible by the lower-than-atmospheric pressure $(P_0 - \Delta P_1)$ maintained inside the lungs during the inhaling process. The pressure difference varies monotonically with the mean inhaling velocity, $\Delta P_1 = r U_1^n$, where r is the fluid resistance of the air passages, and n varies from $n \cong 1$ in laminar flow to $n \cong 2$ in turbulent flow. Mass conservation during the t_1-long process requires that $\rho_0 U_1 A_f t_1 = \rho_1 V$, where A_f is the flow cross-sectional area of the nasal passages, ρ_0 is the air density at atmospheric conditions, and ρ_1 is the density of the inhaled air (evaluated at $P_0 - \Delta P_1$ and body temperature).

The exhaling process is characterized by the excess pressure ΔP_2 inside the chest cavity, $\Delta P_2 = r U_2^n$, where r is the same nasal flow resistance, and U_2 is the mean exhaling velocity. The duration of the exhaling process is t_2, therefore mass conservation requires that $\rho_1 U_2 A_f t_2 = \rho_1 V$. The total work done by the thorax muscles during the inhaling and exhaling cycle ($t_1 + t_2$) is $W = (\Delta P_1 + \Delta P_2)V$. The cycle-averaged power consumed by the thorax muscles is $\dot{W} = W / (t_1 + t_2)$. By combining these equations with the assumption that $\rho_0 \approx \rho_1$, the average power reduces to

$$\dot{W} = r \frac{V^{n+1}}{A_f^n} \cdot \frac{t_1^{-n} + t_2^{-n}}{t_1 + t_2} \tag{24}$$

Equation (24) shows that the power requirement decreases monotonically as either t_1 or t_2 increases. Effortless breathing corresponds to the longest inhaling and exhaling strokes possible. It turns out, however, that both t_1 and t_2 must be finite, as required by the chief function of the breathing process: to facilitate the transfer of oxygen between the freshly inhaled air and the vascularized tissue beneath the surface of the pulmonary passage. The transfer of O_2 is by mass diffusion on both sides of the surface. The scale analysis of mass diffusion begins with writing that during the inhaling interval t_1 the O_2 mass flux is of order $j \sim D\Delta C / \delta$, where $\delta \sim (Dt_1)^{1/2}$ is the mass penetration distance associated with t_1. We find that the oxygen mass transferred during the t_1 interval is $m = j\, A\, t_1$, or that $m \sim A t_1\, D\Delta C / (Dt_1)^{1/2}$, where A is the total contact (mass transfer) area of the air passages. The important quantity is the mass transfer rate averaged over one breathing cycle, $\dot{m} = m / (t_1 + t_2)$, or $\dot{m} \sim A D^{1/2} \Delta C\, t_1^{1/2} / (t_1 + t_2)$. This $\dot{m}$ result places a constraint on the inhaling and exhaling time intervals, forcing both to be finite,

$$\frac{t_1^{1/2}}{t_1 + t_2} \sim \frac{\dot{m}}{A D^{1/2} \Delta C} = K \qquad \text{(constant)} \tag{25}$$

Finally, consider the behavior of the average power requirement, Eq. (24), subject to the mass transfer rate constraint (25). Eliminating t_2 and solving $\partial \dot{W} / \partial t_1 = 0$ we obtain the optimal inhaling time $t_{1,opt}$ for minimum breathing power consumption:

$$\left(1 - K t_{1,opt}^{1/2}\right)\left[\left(\frac{1}{K t_{1,opt}^{1/2}} - 1\right)^n + 1\right] \sim \frac{n}{2n+1} \qquad (1 < n < 2) \tag{26}$$

The chief implication of this result is that the scale of $t_{1,opt}$ is K^{-2}. In view of Eq. (25), we also conclude that $t_{1,opt}$ and $t_{2,opt}$ are of the same order of magnitude, which is in agreement with the large volume of observations catalogued and correlated in the biology literature [10]. The breathing time intervals are proportional to

$$\left(t_1, t_2\right)_{opt} \sim K^{-2} = \left(\frac{A D^{1/2} \Delta C}{\dot{m}}\right)^2 \tag{27}$$

Equation (27) also explains why the breathing interval increases with the size of the animal. Specifically, Eq. (27) predicts that the breathing time intervals must vary as $(A / \dot{m})^2$. We need two relations: (i) the relation between contact area (A) and body size (M), and (ii) the relation between metabolic rate (or $\dot{m}$) and body size. Relation (ii) may be derived by modeling the animal as an open system in the steady state [10], which leads to $\dot{m}$ = (constant) $\times M^{2/3}$. The numerous empirical data are correlated even better by $\dot{m}$ = (constant) $\times M^{0.76}$.

For the relation (i) between mass transfer area and body size, we proceed theoretically by arguing that the thickness of the tissue penetrated by mass diffusion during the time scale $t_{1,opt}$ (or $t_{2,opt}$) is $\delta \sim (D t_{1,opt})^{1/2}$. The total volume of the tissue penetrated by mass diffusion during this time is $V \sim A\delta$. Finally, since the volume V must be proportional to the body size M, we conclude that $A\delta$ is proportional to M, i.e., that A = (constant) $\times$ $M / t_{1,opt}^{1/2}$. Substituting this relation and $\dot{m} \sim M^{0.76}$ into Eq. (27) we predict that the breathing time increases with the body size, and that this increase is relatively slow, $(t_1, t_2)_{opt}$ = (constant) $\times$ $M^{0.24}$. This relation agrees amazingly well with the empirical correlation (allometric law) known for t_1 as a function of M in physiology [10].

7. Optimal Solar Collector Temperature

There are many directions in which the method of thermodynamic optimization has been extended and perfected, notably in the fields of cryogenics, heat transfer, storage systems, power generation, refrigeration, and solar energy conversion. These and other directions are illustrated in this book. Solar energy conversion has been studied under

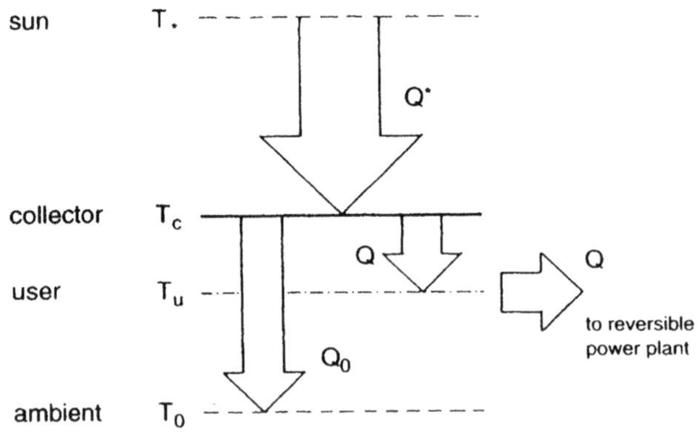

Figure 5. Solar power plant model with collector-ambient heat loss and collector-power cycle heat exchanger [33].

the banners of two fundamental problems. One is concerned with establishing the theoretical limits of converting thermal radiation into work, or calculating the exergy content of radiation. The other problem deals with the delivery of maximum power from a solar collector of fixed size [33]. This problem has also been solved in many subsequent applications [1, 9, 10, 34], which are united by an important design feature: the collector operating temperature can be optimized.

This optimization opportunity is illustrated with the help of Fig. 5. A power plant is driven by a solar collector with convective heat leak to the ambient. The heat leak is assumed to be proportional to the collector-ambient temperature difference, $\dot{Q}_0 = (UA)_c(T_c - T_0)$. The internal heat exchanger between the collector and the hot end of the power cycle (the user) is modeled similarly, $\dot{Q}_0 = (UA)_i(T_c - T_u)$. There is an optimal coupling between the collector and the power cycle such that the power output is maximum. This design is presented by the optimal collector temperature [1, 33]

$$\frac{T_{c,opt}}{T_0} = \frac{\theta_{max}^{1/2} + R\theta_{max}}{1+R} \tag{28}$$

where $R = (UA)_c / (UA)_i$ and $\theta_{max} = T_{c,max} / T_0$ is the maximum (stagnation) temperature of the collector. This optimum has its origin in the trade-off between the Carnot efficiency of the reversible part of the power plant $(1 - T_0/T_u)$ and the heat loss to the ambient, $\dot{Q}_0$. The power output is the product $\dot{Q}(1 - T_0 / T_u)$ When $T_c < T_{c,opt}$ the Carnot factor is too small. When $T_c > T_{c,opt}$ the heat input $\dot{Q}$ drawn from the collector is too small, because the heat loss to the ambient $\dot{Q}_0$ is large.

Corresponding optimal couplings have been identified in solar-driven power plants of many power-cycle designs, extraterrestrial power plants, and refrigeration systems [9].

8. Finite Time and Endoreversible: New Names for Old Ideas

The current state of thermodynamic optimization subject to size and time constraints is reviewed at length in [9] and [34 - 36]. Additional examples are given in this book in the chapter on Constructal Flow Geometry Optimization. The earlier work was reviewed in [1 - 4] and [26, 27].

The published record is very clear. It shows that by the end of the 1970s the method had been used by many independent schools in Europe and North America, over the course of three decades, and in several fields: cryogenics, heat transfer, power, refrigeration, storage systems, solar energy, and education. Steady-state and time-dependent systems had been optimized. Size and time constraints had been used. Optimization criteria had called for entropy generation minimization and power output maximization in power plants, as well as power input minimization in refrigeration plants. Endoreversible models had been used at the macroscopic level (e.g., [16, 19]) and at the infinitesimal level (e.g., Fig. 1). The method had become a self-standing graduate course with its own textbook [1].

On this rich background came in 1983 Andresen's surprising claim that "the field of finite-time thermodynamics was started in 1975 by Professor R. Stephen Berry, Peter Salamon and myself." This claim is false, not because Andresen et et al.'s first paper was actually in 1977, not in 1975, and not because it was merely a follow up to Curzon and Ahlborn 1975 paper [25], which it referenced. It is false because of Andresen's own definition of the supposedly new field: "Finite-time thermodynamics is the extension of traditional thermodynamics to deal with processes which have explicit time or rate dependencies. These constraints, of course, imply a certain amount of loss, or entropy production [37]." Andresen is clearly referring to an already established method: the combined thermodynamics & heat transfer optimization method illustrated here in section 3. My own way of recognizing the existence of this field (this in 1982, before Andresen's claim), was through the triangle diagram of Fig. 6.

Some of the work done by followers of Curzon and Ahlborn has been criticized for specific technical reasons in the peer-refereed literature, in both physics [29, 34, 39] and engineering [28, 38]. One aspect of this type formed the subject of section 4 in this chapter. Endoreversible—the assumption that the generation of entropy is located at the internal interfaces between the assumed building blocks of the system—is similar to the old concept of local thermodynamic equilibrium [40]: the foundation of all of our heat transfer and fluid mechanics. The local thermodynamic equilibrium model is very clear in Fig. 1 (the shaded element), which is why this model was and continues to be used routinely in the thermodynamic optimization of heat transfer systems [1], either macroscopic (e.g., Fig. 5) or distributed infinitesimally (e.g., Fig. 1). The place for future discussions of such aspects is the peer-refereed literature. These discussions, however, tend to obscure the much more important issue, which is the purported novelty of the method behind the work of Curzon and Ahlborn's followers.

It is also worth noting that none of this work represents "new physics." Each result such as Eq. (9) and Eq. (14) is case specific. It corresponds specifically not only to the type of system that is being modeled (e.g., insulation vs. power plant)

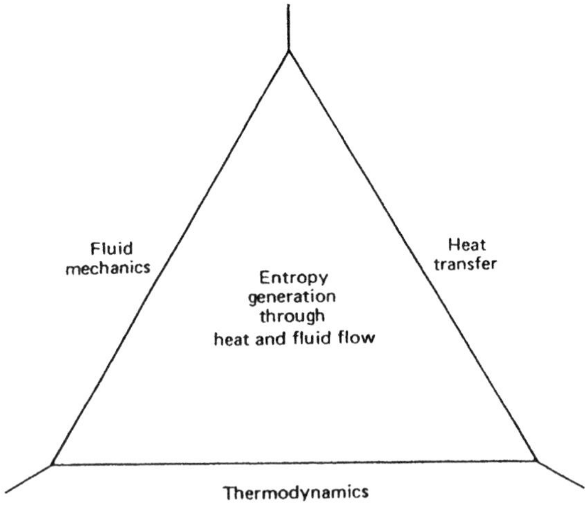

Figure 6. The interdisciplinary field covered by thermodynamic optimization in constrained systems with heat and fluid flow irreversibilities [1].

but also the personal talent and taste for simplicity and beauty exhibited by the modeler. All the physics that is needed continues to be covered by old thermodynamics, heat transfer and fluid mechanics, combined. The models and the associated optimization results occupy more and more the field of applications. This is the work that along with Andresen et al.'s more recent contribution continues to fill the triangular area depicted in Fig. 6.

New physics are the new principles that are waiting to be formulated in order to cover in a predictive sense the natural phenomena that so far have evaded determinism. One new principle is the constructal law [10] of geometric shape and structure in natural systems far from internal equilibrium. This direction is outlined in this book in the chapter on Constructal Flow Geometry Optimization. This new extension of thermodynamics covers the physics behind concepts such as optimization, purpose, necessity (the fittest), time direction and fractals, and unites physics with engineering and biology.

Acknowledgement. This work was supported by the National Science Foundation and the Air Force Office of Scientific Research.

References

1. Bejan, A. (1982) *Entropy Generation through Heat and Fluid Flow*, Wiley, New York.
2. Moran, M. J. (1982) *Availability Analysis: A Guide to Efficient Energy Use*, Prentice-Hall, Englewood Cliffs.
3. Feidt, M. (1987) *Thermodynamique et Optimisation Énergetique des Systèmes et Procedés*, Technique et Documentation, Lavoisier, Paris.
4. Bejan, A. (1988) *Advanced Engineering Thermodynamics*, Wiley, New York.
5. Sieniutycz, S. and Salamon, P., eds. (1990) *Finite-Time Thermodynamics and Thermoeconomics*, Taylor and Francis, New York.
6. De Vos, A. (1992) *Endoreversible Thermodynamics of Solar Energy Conversion*, Oxford University Press, Oxford.
7. Radcenco, V. (1994) *Generalized Thermodynamics*, Editura Tehnica, Bucharest.
8. Bejan, A., Tsatsaronis, G. and Moran, M. (1996) *Thermal Design and Optimization*, Wiley, New York.
9. Bejan, A. (1996) *Entropy Generation Minimization*, CRC Press, Boca Raton.
10. Bejan, A. (1997) *Advanced Engineering Thermodynamics*, second ed., Wiley, New York.
11. Bejan, A. and Smith, J. L., Jr. (1974) Thermodynamic optimization of mechanical supports for cryogenic apparatus, *Cryogenics* **14**, 158-163.
12. McClintock, F. A. (1951) The design of heat exchangers for minimum irreversibility, Paper no. 51-A-108, Winter Annual Meeting, ASME.
13. Grassmann, P. and Kopp. J. (1957) Zur gunstigsten Wahl der Temperaturdifferenz und der Wärmeübergangszahl in Wärmeaustauschern *Kältetechnik* **9**, 306-308.
14. Martynovskii, V. S., Cheilyakh, V. T. and Shnaid, T. N. (1971) Thermodynamic effectiveness of a cooled shield in vacuum low-temperature insulation, *Energ. Transp.* **2**; also in Ahern, J. E. (1980) *The Exergy Method of Energy Systems Analysis*, Wiley, New York.
15. Müser, H. (1957) Thermodynamische Behandlung von Electronenprozessen in Halbleiter-Randschichten, *Z. Phys.* **148**, 380-390.
16. Chambadal, P. (1957) *Les Centrales Nucleaires*, Armand Colin, Paris.

17. Chambadal, P. (1958) Le choix du cycle thermique dans une usine génératrice nucleaire, *Revue Générale de l'Électricité* **67**, 332 - 345.
18. Chambadal, P. (1963) *Evolution et Applications du Concept d'Entropie*, section 30, Dunod, Paris.
19. Novikov, I. I. (1957) The efficiency of atomic power stations, *Atomnaya Energiya* **3**, 409; English translation in (1958) *J. Nuclear Energy II* **7**, 125 - 128.
20. Vukalovich, M. P. and Novikov, I. I. (1972) *Thermodynamics*, Mashinostroenie, Moscow.
21. Novikov, I. I. and Voskresenskii, K. D. (1977) *Thermodynamics and Heat Transfer*, Atomizdat, Moscow.
22. Novikov, I. I. (1984) *Thermodynamics*, Mashinostroenie, Moscow.
23. El-Wakil, M. M. (1962) *Nuclear Power Engineering*, McGraw-Hill, New York.
24. El-Wakil, M. M. (1971) *Nuclear Energy Conversion*, International Textbook Company, Scranton, PA.
25. Curzon, F. L. and Ahlborn, B. (1975) Efficiency of a Carnot engine at maximum power output, *American Journal of Physics* **43**, 22-24.
26. Bejan, A. (1994) Engineering advances on finite time thermodynamics, *American Journal of Physics* **62**, January, 11-12.
27. Bejan, A. (1996) Notes on the history of the method of entropy generation minimization (finite time thermodynamics), *Journal of Non-Equilibrium Thermodynamics* **21**, 239-242.
28. Gyftopoulos, E. P. (1997) Fundamentals of analyses of processes, *Energy Conversion and Management* **38**, 1525-1533.
29. Bejan, A. (1996) Models of power plants that generate minimum entropy while operating at maximum power, *American Journal of Physics* **64**, 1054-1059.
30. Ikegami, Y. and Bejan, A. (1998) On the thermodynamic optimization of power plants with heat transfer and fluid flow irreversibilities, *Journal of Solar Energy Engineering* **120**, 139-144.
31. Bejan, A. and Errera, M. R. (1998) Maximum power from a hot stream, *International Journal of Heat and Mass Transfer* **41**, 2025-2036.
32. Bejan, A. (1997) Theory of organization in nature: pulsating physiological processes, International *Journal of Heat and Mass Transfer* **40**, 2097-2104.
33. Bejan, A., Kearney, D. W. and Kreith, F. (1981) Second law analysis and synthesis of solar collector systems, *Journal of Solar Energy Engineering* **103**, 23-30.
34. Bejan, A. (1996) Entropy generation minimization: the new thermodynamics of finite-size devices and finite-time processes, *Journal of Applied Physics* **79**, 1191-1218.
35. Sieniutycz, S. and Shiner, J. S. (1994) Thermodynamics of irreversible processes and its relation to chemical engineering: second law analyses and finite time thermodynamics. *Journal of Non-Equilibrium Thermodynamics* **19**, 303-348.
36. Bejan, A. (1996) Method of entropy generation minimization, or modeling and optimization based on combined heat transfer and thermodynamics, *Revue Générale de Thermique* **35**, 637-646.
37. Andresen, B. (1983) Finite time thermodynamics, Physics Laboratory II, University of Copenhagen, p. 3.
38. Moran, M. J. (1998) On second-law analysis and the failed promise o finite-time thermodynamics, *Energy* **23**, 517-519.
39. Sekulic, D. P. (1998) A fallacious argument in the finite time thermodynamics concept of endoreversibility, *Journal of Applied Phyiscs* **83**, 4561-4565.
40. Berg, C. A. (1992) Conceptual issues in energy efficiency, in Valero, A. and Tsatsaronis, G., eds., Proceedings of the International Symposium on Efficiency, Costs, Optimization and Simulation of Energy Systems (ECOS'92), Zaragoza, ASME, New York, pp. 7-16.

CONSTRUCTAL FLOW GEOMETRY OPTIMIZATION

A. BEJAN
Duke University
Department of Mechanical Engineering and Materials Science
Durham, NC 27708-0300, USA

1. Optimal Size

In the field of heat transfer, the method of thermodynamic optimization or entropy generation minimization (EGM) brings out the inherent competition between heat-transfer and fluid-flow irreversibilities in the optimization of devices subjected to overall constraints. Consider the most elementary heat exchanger passage [1] which is represented by a duct of arbitrary cross-section (A) and arbitrary wetted perimeter (p). The engineering function of the passage is specified in terms of the heat transfer rate per unit length (q') that is to be transmitted to the stream ($\dot{m}$); that is, both q' and $\dot{m}$ are fixed. In the steady state, the heat transfer q' crosses the temperature gap ΔT formed between the wall temperature ($T + \Delta T$) and the bulk temperature of the stream (T). The stream flows with friction in the x direction; hence, the pressure gradient $(-dP/dx) > 0$.

Taking as thermodynamic system a passage of length dx, the first law and the second law state that

$$\dot{m}\, dh = q'\, dx \qquad \text{and} \qquad \dot{S}'_{gen} = \dot{m}\frac{ds}{dx} - \frac{q'}{T + \Delta T} \geq 0, \tag{1}$$

where $\dot{S}'_{gen}$ is the entropy generation rate per unit length. Combining these statements with the assumption that the fluid is single phase ($dh = T\, ds + v\, dP$), we obtain

$$\dot{S}'_{gen} = \frac{q'\,\Delta T}{T^2(1 + \Delta T / T)} + \frac{\dot{m}}{\rho T}\left(-\frac{dP}{dx}\right) \cong \frac{q'\,\Delta T}{T^2} + \frac{\dot{m}}{\rho T}\left(-\frac{dP}{dx}\right) \geq 0 \tag{2}$$

The denominator of the first term on the right-hand side has been simplified by assuming that the local temperature difference ΔT is negligible when compared with the local absolute temperature T. The heat exchanger passage is a site for both flow with friction and heat transfer across a finite ΔT; the $\dot{S}'_{gen}$ expression thus has two terms, each accounting for one irreversibility mechanism, $\dot{S}'_{gen} = \dot{S}'_{gen,\Delta T} + \dot{S}'_{gen,\Delta P}$.

A remarkable feature of the $\dot{S}'_{gen}$ expression (2) and of many like it for other simple devices is that a proposed design change (for example, making the passage narrower) induces changes of opposite signs in the two terms of the expression. An optimal trade-

A. Bejan and E. Mamut (eds.), Thermodynamic Optimization of Complex Energy Systems, 61–72.

off then exists between the two irreversibility contributions, an optimal design for which the overall measure of exergy destruction ($\dot{S}'_{gen}$) is minimal while the system continues to serve its specified function (q', $\dot{m}$). This trade-off becomes visible analytically when heat-transfer and fluid-friction information [2] is added to Eq. (2), i.e., when ΔT and $(-dP/dx)$ are replaced by the Nusselt number and friction factor expressions that characterize the duct. For example, it was shown [1] that when the cross-section is round and the flow is fully developed and turbulent, Eq. (2) can be arranged as

$$N_S = \frac{\dot{S}'_{gen}}{\dot{S}'_{gen,\,min}} = 0.856\left(\frac{Re}{Re_{opt}}\right)^{-0.8} + 0.144\left(\frac{Re}{Re_{opt}}\right)^{4.8} \tag{3}$$

where the ratio on the left-hand side is known as the entropy generation number N_S [1], and $Re = 4\dot{m}/(\pi\mu D)$. Equation (3) shows that $\dot{S}'_{gen}$ has a sharp minimum with respect with the size of the duct, D.

The literature contains a growing number of internal flow configurations that have been subjected to the same type of thermodynamic optimization: reviews can be found in [3, 4]. External flow configurations are governed by the same competition between heat-transfer and fluid-flow irreversibilities. The overall entropy generation rate associated with external flow is

$$\dot{S}'_{gen} = \frac{\dot{Q}_B(T_B - T_\infty)}{T_\infty T_B} + \frac{F_D U_\infty}{T_\infty} \tag{4}$$

where $\dot{Q}_B, T_B, T_\infty, F_D,$ and U_∞ are, respectively, the instantaneous heat transfer rate between the body and the fluid reservoir, the body surface temperature, the fluid reservoir temperature, the drag force, and the relative speed between body and reservoir. This remarkably simple result proves again that poor thermal contact (the first term) and fluid friction (the second term) contribute hand in hand to degrading the thermodynamic performance of the external convection arrangement. One area in which Eq. (4) has found application is the problem of selecting the size and number (density) of fins for the design of extended surfaces [1].

Most of the literature on the thermodynamic optimization of heat transfer devices is devoted to more complex systems that incorporate the elemental features (internal, external) mentioned in this section. Examples include heat exchangers of several configurations (counterflow, parallel flow, cross-flow) with single-phase or two-phase flow [3, 4].

2. Optimal Internal Spacings

Even simpler is the sizing of a system in which a single transport mechanism causes the irreversibility, for example, heat transfer. When the heat current is imposed, minimizing the entropy generation means minimizing the resistance to heat flow.

For example, in the cooling of electronic packages, both the volume and the heat generation rate that is distributed uniformly over that volume are fixed. The heat current is removed by a single-phase stream, with natural or forced convection. The geometric arrangement of components can then be optimized such that the hot-spot temperature is minimal. The plate-to-plate spacing (or number of columns) is free to vary. If the spacing is too large, however, there is not enough heat-transfer area and the hot-spot temperature becomes high; when the spacing is too small, the coolant flow rate decreases and the hot-spot temperature is again high. Between the two lies an optimal spacing—an optimal package architecture—that minimizes the thermal resistance between the system and the environment. This method of geometric optimization of internal architecture is known as the intersection of the asymptotes method [5, 6].

In this section we illustrate this principle for natural convection cooling, Fig. 1. We assume that the flow is laminar, the board surfaces are sufficiently smooth to justify the use of heat transfer results for natural convection over vertical smooth walls, and the maximum (hot spot) temperature T_{max} is the scale that represents the temperature at every point on the board surface. The method consists of two steps. In the first, the two extremes are identified: the total heat transfer rate $(\dot{Q})$ expressions in the small-D limit and the large-D limit. In the second step the two asymptotes are intersected for the purpose of locating the D value that maximizes $\dot{Q}$.

In the D → 0 limit we can use with confidence $\dot{Q}_1 = \dot{m}_1 c_p (T_{max} - T_\infty)$ for the heat transfer rate extracted by the coolant from one of the channels of spacing D. Note that in each channel T_∞ is the inlet temperature, T_{max} is the outlet temperature, and $\dot{m}_1$ is the mass flow rate through a single channel. To calculate $\dot{m}_1$ we rely on the classical solution for Hagen-Poiseuille chimney flow between two vertical plates [5]

$$U = g\beta(T_{max} - T_\infty)D^2 / (12\nu) \tag{5}$$

$$\dot{m}_1 = \rho DWU = \rho W g\beta(T_{max} - T_\infty)D^3 / (12\nu) \tag{6}$$

where W is the third dimension of the volume, and U is the vertical velocity averaged across the D-wide channel. The fluid is modeled as Boussinesq incompressible with the

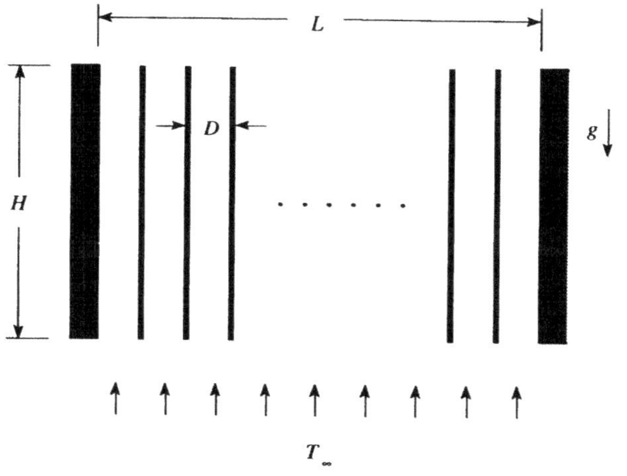

Figure 1. Fixed volume with a variable number of vertical plates cooled by natural convection [5, 6].

density ρ, kinematic viscosity ν, and coefficient of volumetric thermal expansion β. The remaining parameters are defined in Fig. 1. The total rate of heat transfer removed from the package is $\dot{Q} = n\dot{Q}_1$, where $n = L/D$, hence

$$\dot{Q} = \rho c_P WL g\beta(T_{max} - T_\infty)^2 D^2 / (12\nu) \tag{7}$$

Consider next the opposite limit in which D is large enough so that it exceeds the thickness of the thermal boundary layer that forms on each vertical surface [5], namely $D > H[g\beta H^3(T_{max} - T_\infty)/(\alpha\nu)]^{-1/4}$ where α is the fluid thermal diffusivity, and the Prandtl number is of order 1 or greater. In this limit the center region of the board-to-board spacing is occupied by fluid of temperature T_∞. The number of distinct boundary layers is $2n = 2L/D$. The heat transfer rate through one boundary layer is $\bar{h}HW(T_{max} - T_\infty)$ for which the average heat transfer coefficient $\bar{h}$ is furnished by the correlation for laminar flow [2, 5], namely $\bar{h}H/k = 0.517 Ra_H^{1/4}$, where $Ra_H = g\beta(T_{max} - T_\infty)H^3/(\alpha\nu)$. The total rate of heat transfer extracted from the entire package is 2n times larger than $\bar{h}HW(T_{max} - T_\infty)$,

$$\dot{Q} = 2(L/D)HW(T_{max} - T_\infty)(k/H)0.517 Ra_H^{1/4} \tag{8}$$

Figure 2 shows that the two asymptotes (7, 8) intersect above what would be the peak of the actual $\dot{Q}(D)$ curve. It is not necessary to know the actual $\dot{Q}(D)$ relation. The optimal spacing D_{opt} for maximum $\dot{Q}$ is approximately

$$D_{opt}/H \cong 2.3\left[g\beta(T_{max} - T_\infty)H^3/(\alpha\nu)\right]^{-1/4} \tag{9}$$

This estimate reproduced within 20 percent the optimal spacing deduced based on more exact and lengthier methods, such as the maximization of the $\dot{Q}(D)$ relation, or the

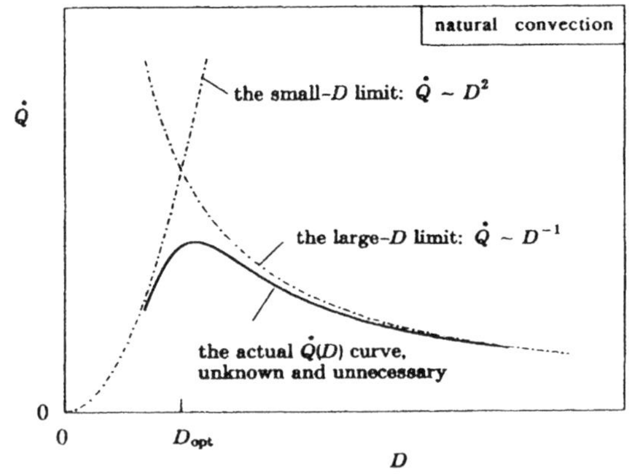

Figure 2. The maximization of the total heat transfer rate removed by natural convection from a fixed volume [5, 6].

finite-difference simulations of the complete flow and temperature fields in the package. An order of magnitude estimate for the maximum heat transfer rate can be obtained by substituting D_{opt} in Eqs. (7) and (8): $\dot{Q}_{max} < 0.45k(T_{max} - T_{\infty})(LW/H)Ra_H^{1/2}$. The unequal sign is a reminder that the peak of the actual $\dot{Q}(D)$ curve falls under the intersection of the two asymptotes (Fig. 2). The intersection of asymptotes method has been used to determine the optimal spacing of elemental features of several other shapes [5]. One example is the spacing between horizontal cylinders in a fixed volume (for example, tubes in a compact heat exchanger).

A completely analogous geometric optimization rules the internal spacings of volumes bathed by forced convection. This problem was solved more recently [7], and for the sake of conciseness we review only its main result. The analysis can also be found in [5, 6]. We assume that the coolant is forced to flow by the pressure difference Δp, and that the plate length in the direction of flow is L. The intersection of the small-D and large-D asymptotes yields the optimal spacing formula

$$D_{opt} / L \cong 2.7\left[\Delta p \cdot L^2 / (\mu\alpha)\right]^{-1/4} \tag{10}$$

This result agrees within 20 percent with the value obtained by locating the maximum of the actual $\dot{Q}(D)$ curve, and is adequate when the board surface can be modeled as isothermal. For uniform-flux surfaces, the 2.7 factor is replaced by 3.2 on the right side of formula (10).

Compare (10) with (9) and notice the symmetry between the nondimensional optimal spacings for forced convection and natural convection. Note further that the role that in the natural convection formula is played by Ra_H, in forced convection is played by the new dimensionless pressure drop number [5] defined as $\Pi_L = \Delta p \cdot L^2 / (\mu\alpha)$. This and a related dimensionless group were also identified in external forced convection, thermodynamic optimization, and electronic cooling, and named *Bejan number*, Be [8].

If the stack is exposed to (surrounded by) a flow approaching with the velocity U_{∞} parallel to L, we may approximate Δp with $(1/2)\rho U^2$ in (10) to estimate the optimal spacing $D_{opt} / L \cong 3.2\, Pr^{-1/4}\, Re_L^{-1/2}$, where $Re_L = U_{\infty}L/\nu$. This estimate is supported by numerical studies of stacks with several types of distributions of heat generation on the board surfaces (uniform, flush mounted sources, protruding sources, etc.) [9]. The same method was applied recently to determine the optimal spacing between cylinders in cross-flow in a fixed volume [5], and to predict the formation of regularly placed cracks in solids that shrink upon cooling or drying [10].

3. Volume-to-Point Flow, and Constructal Tree Networks

The research frontier in heat exchangers and the cooling of electronics is being pushed in the direction of smaller and smaller package dimensions. There comes a point where miniaturization makes convection cooling impractical, because the ducts through which the coolant must flow take up too much space. The only way to channel the generated heat out of the electronic package is by conduction. This conduction path will have to be very effective (of high thermal conductivity, k_p), so that the temperature difference between the hot spot (the heart of the package) and the heat sink (on the side of the package) will not exceed a certain value. Conduction paths also take up space. Designs with fewer and smaller paths are better suited for the miniaturization evolution. The fundamental problem that was proposed in [11] is: "Consider a finite-size volume in which heat is being generated at every point and which is cooled through a small patch (heat sink) located on its boundary. A finite amount of high conductivity (k_p) material is available. Determine the optimal distribution of k_p material through the given volume such that the highest temperature is minimized."

The discovery made in [11] is purely geometric: every portion of the given volume can have its shape optimized, such that its resistance to heat flow is minimal. This design principle applies at any volume scale, and to other forms of transport (fluid, electric charge, mass species). The volume-to-point path was determined in a sequence of steps consisting of shape optimization and subsequent construction (assembly, grouping). Very important is the time arrow of this construction, which points from small to large. It starts from the smallest building block (elemental system), and proceeds toward larger building blocks (assemblies, constructs). It was shown that determinism vanishes if the direction is reversed, from large to small. To emphasize the link between determinism and the direction from small to large, and as a reminder that theory runs against fractal thinking, the geometric optimization principle proposed in [11] and summarized in [12] was named "constructal theory."

Consider a volume (V) that generates heat volumetrically at the uniform rate q''', which is collected over the volume and led to a heat sink on its boundary. For simplicity, assume that the heat-flow geometry is two-dimensional over the area A, hence V = AW where W is the third dimension of the volume. Most of A is occupied by the material of low conductivity k_0, which generates q'''. A small fraction of A (namely A_p) is to be covered by high-conductivity material.

We regard A as a collection of subsystems of various sizes (A_0, A_1, A_2,...), and begin with the smallest such system (A_0), which is shown in Fig. 3. The defining feature of this "elemental system" is that its heat current ($q_0 = q'''A_0W$) is collected by a single k_p blade. The other important feature is that the size A_0 is known and fixed (dictated by manufacturing constraints) while the shape H_0/L_0 may vary. The objective of the following analysis is to determine the optimal shape such that the overall resistance between A_0 and the heat sink [the D_0 patch located at (0,0)] is minimal.

An analytical solution is possible in the limit $\tilde{k} >> 1$, where $\tilde{k} = k_p/k_0$ is a known constant of the conducting composite. Symmetry and the requirement to minimize resistance are why the D_0-thin blade is placed on the longer of the two axes of symmetry of A_0 (note $H_0 > L_0$ in Fig. 3). When $\tilde{k} >> 1$, heat flows vertically through the k_0 material and horizontally through the k_p material. This allows us to develop the expression for the hot-spot temperature $T(L_0, \pm H_0/2)$ as a sum of two terms: the temperature drop through k_0 from (L_0, $H_0/2$) to (L_0,0), plus the temperature drop through k_p from (L_0,0) to (0,0). The result is [11, 12]:

$$\frac{\Delta T_0 k_0}{q''' A_0} = \frac{1}{8} \times \frac{H_0}{L_0} + \frac{k_0 H_0}{2k_p D_0} \times \frac{L_0}{H_0} \qquad (11)$$

Next, we note that the ratio D_0/H_0 is another constant of the conductive composite, because it accounts for the proportion in which k_p material is allocated to A_0, namely

$A_{p0}/A_0 = D_0L_0/(H_0L_0)$. We call this constant $\phi_0 = D_0/H_0 << 1$. With this, Eq. (11) shows clearly that the overall volume-to-point resistance can be minimized with respect to the shape of the elemental system (H_0/L_0):

$$\left(H_0 / L_0\right)_{opt} = 2\left(\tilde{k}\,\phi_0\right)^{-1/2} < 1 \qquad \left(\frac{\Delta T_0\, k_0}{q'''\, A_0}\right)_{min} = \frac{1}{2\left(\tilde{k}\,\phi_0\right)^{1/2}} \tag{12, 13}$$

An important feature of (12, 13) is that the resistance through the k_0 material matches the resistance through the k_p material: the two terms have the same value on the right-hand side of Eq. (13). This principle of equipartition of resistance applies at every volume scale.

As we proceed toward larger scales, we can only assemble elemental volumes of size A_0, which have to be connected so that their heat currents can be collected into a single stream, Fig. 4. This first construct (A_1) contains n_1 elemental volumes, $A_1 = n_1A_0$. The number n_1 (or the shape H_1/L_1) is a degree of freedom. The elemental heat currents are collected by a new blade of k_p material and thickness D_1.

When n_1 is large, the first construct conducts heat in almost the same way as the elemental volume. The only difference is that the k_0 regions of Fig. 4 are now replaced by a composite material. The effective thermal conductivity of this composite in the vertical direction is $k_1 = k_pD_0/H_0$. The hot-spot temperature difference ΔT_1 occurs between the far corner (L_1, $H_1/2$) and the heat sink (0,0). Nothing has changed, except the notation in Eq. (11) which now reads:

$$\frac{\Delta T_1\, k_1}{q'''\, A_1} = \frac{1}{8} \times \frac{H_1}{L_1} + \frac{k_1 H_1}{2k_p D_1} \times \frac{L_1}{H_1} \tag{14}$$

The geometric optimization too is a repeat of what went on at the elemental level. First, it is convenient to rewrite Eq. (14) in a form that shows the effect of n_1,

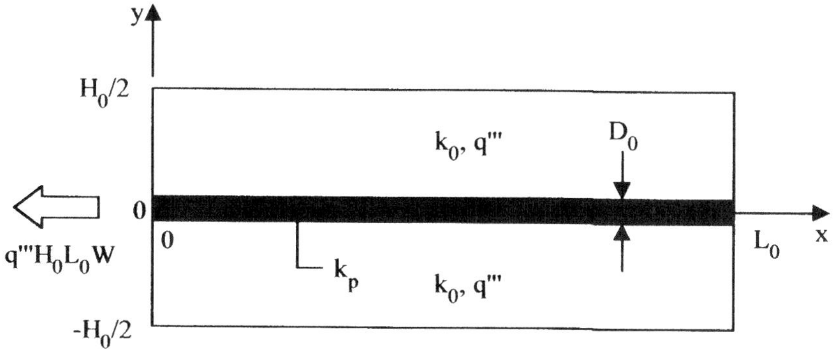

Figure 3. Elemental volume with volumetric heat generation and one high-conductivity path along its axis of symmetry [11, 12].

$$\frac{\Delta T_1 k_0}{q''' A_1} = \frac{1}{4n_1\left(\tilde{k}\phi_0\right)^{1/2}} + \frac{n_1}{2^{3/2}\tilde{k}\left(\tilde{k}\phi_0\right)^{1/4}\tilde{D}_1} \tag{15}$$

where $\tilde{D}_1 = D_1 / A_0^{1/2}$. This expression shows how the overall volume-to-point resistance of the first construct depends on n_1: Resistance can be minimized through optimal growth,

$$n_{1,opt} = \frac{\left(\tilde{k}\tilde{D}_1\right)^{1/2}}{2^{1/4}\left(\tilde{k}\phi_0\right)^{1/8}} >> 1 \qquad \left(\frac{\Delta T_1 k_0}{q''' A_1}\right)_{mm} = \frac{1}{2^{3/4}\left(\tilde{k}\tilde{D}_1\right)^{1/2}\left(\tilde{k}\phi_0\right)^{3/8}} \tag{16, 17}$$

Beginning with the first construct, the overall resistance can be minimized for a second time. When the total k_p volume is constrained, $A_{p1} = D_1 L_1 + n_1 A_{p0}$, the parameters $\tilde{D}_1$ and ϕ_0 cannot be increased independently. Writing $\phi_1 = A_{p1}/A_1$ (= constant), and noting that $A_{p0} = \phi_0 A_0$ and $H_1 = 2L_{1,opt}$, we find that the A_{p1} constraint provides a relation between $\tilde{D}_1$ and ϕ_0, namely $\phi_1 = \tilde{D}_1 / [2^{1/2}(\tilde{k}\phi_0)^{1/4}] + \phi_0$. Minimizing (17) subject to the ϕ_0 we find, in order, $\phi_{0,opt} = \phi_1 / 2$, $\tilde{D}_{1,opt} = 2^{-3/4}\tilde{k}^{1/4}\phi_1^{5/4}$, and

$$n_{1,opt} = \left(\tilde{k}\phi_1 / 2\right)^{1/2} >> 1 \qquad \left(\frac{\Delta T_1 k_0}{q''' A_1}\right)_{mm} = \frac{1}{\tilde{k}\phi_1} \tag{18, 19}$$

where the subscript $(\ )_{mm}$ indicates the result of double minimization. Combining $\tilde{D}_{1,opt}$ with $D_0 = \phi_0 H_{0,opt}$, and using the $H_{0,opt}$ result listed in the preceding section, we find the optimal ratio between the thicknesses of the k_p blades: $(D_1/D_0)_{opt} = (\tilde{k}\phi_1/2)^{1/2}$. The unexpected geometric feature that follows from this double optimization is that the external aspect ratio is exactly equal to the integer 2, namely

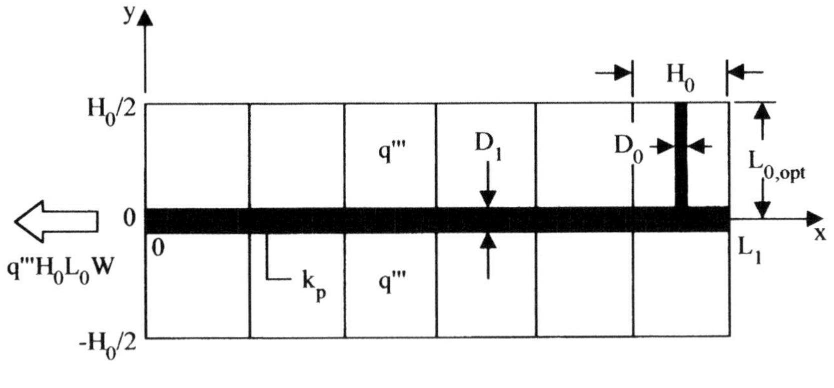

Figure 4. The first construct: a large number of elemental volumes (Fig. 3) connected to a central high-conductivity path [11, 12].

$(H_1 / L_1)_{opt} = 2$; in other words, it is independent of the design parameters of the conductive composite built into the first assembly ($\tilde{k}$, ϕ_1).

To fill an even larger volume, we connect a large number ($n_2 = A_2/A_1$) of first constructs and arrive at the second construct shown in Fig. 5. The question, again, is how many building blocks to assemble into a second construct, so that we continue on the path that we started—a path where resistance is minimized geometrically at every step. The analysis begins with the analog of Eq. (15), which in the case of the overall thermal resistance of the second construct is an expression that can be rearranged as follows,

$$\frac{\Delta T_2 \, k_0}{q''' \, A_2} = \frac{1}{n_2 \, \tilde{k} \, \phi_1} + \frac{n_2}{2^{3/2} \, \tilde{k} \, \tilde{D}_2} \tag{20}$$

where $\tilde{D}_2 = D_2 / A_1^{1/2}$. The temperature difference ΔT_2 is measured between the hot-spot (the upper right corner in Fig. 5) and the heat sink (the left end of the D_2 blade). The minimization of (20) with respect to n_2 yields

$$n_{2,opt} = 2^{3/4} \left(\tilde{D}_2 / \phi_2\right)^{1/2} \qquad \left(\frac{\Delta T_2 \, k_0}{q''' \, A_2}\right)_{min} = \frac{2^{1/4}}{\tilde{k}\left(\phi_1 \tilde{D}_2\right)^{1/2}} \tag{21, 22}$$

The second optimization consists of balancing ϕ_1 against $\tilde{D}_2$ while taking into account the k_p-material constraint of the second construct, $A_{p2} = D_2 L_2 + n_2 A_{p1}$, which, after we define $\phi_2 = A_{p2}/A_2$, becomes $\phi_2 = 2^{-1/2} \tilde{D}_2 + \phi_1$. The results of minimizing (22) for a second time are $\phi_{1,opt} = \phi_2 / 2$, $\tilde{D}_{2,opt} = 2^{-1/2} \phi_2$, and

$$n_{2,opt} = 2 \qquad \left(\frac{\Delta T_2 \, k_0}{q''' \, A_2}\right)_{mm} = \frac{2}{\tilde{k} \, \phi_2} \tag{23, 24}$$

as well as $(H_2 / L_2)_{opt} = 1$ and $(D_2 / D_1)_{opt} = 2$. The conclusion that $n_{2,opt} = 2$ is important because it is the first time that *dichotomy* (bifurcation, pairing) has been *predicted*. The best second construct is the one that contains the smallest number of

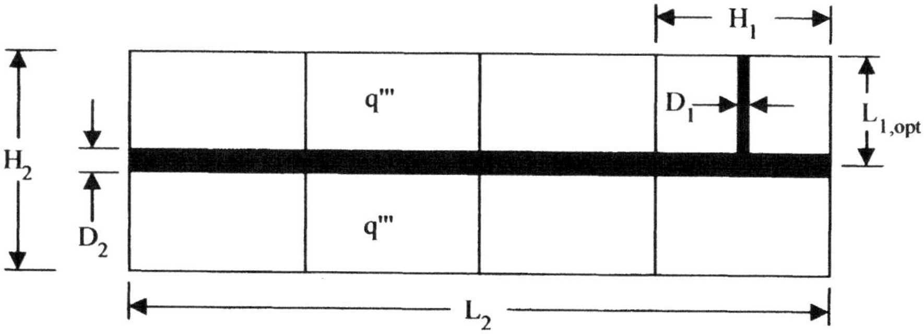

Figure 5. Second construct containing a large number of first constructs [11, 12].

first constructs, and that number is 2. Contrary to the present theory, dichotomy is assumed at the start of tree network analyses, for example, in physiology [13]. In constructal theory the integer 2 is a result, not an assumption.

This single principle of geometric optimization and subsequent construction (growth) is repeated toward stepwise larger volume scales. The end result is shape and structure—the optimized architecture—of the composite (k_0/k_p) that connects the sink point to the finite-size volume. The infinity of points of the given volume is "connected" to the sink because at the smallest volume scale the transport is volumetric, by thermal diffusion through the low-conductivity material. At larger scales the transport is via channels (streams) of high conductivity. Diffusion comes first and streams later.

Astonishing is not that the high-conductivity channels form a tree (a network without loops) but that each feature of the tree is deterministic, the result of a single principle of optimization. This conclusion runs against the currently accepted doctrine that natural structures are nondeterministic, the result of chance and necessity. The discovery then is not the tree but the principle that generates this structure.

4. Shape and Structure in Natural Flow Systems

The hand-in-glove geometry formed by slow (disorganized) and fast (organized) flow regimes unites all the optimized volume-to-point flows. The slow regime is the "glove," and the fast regime is the "hand." In the flow of oxygen through a mammal, the slow, shapeless flow is volumetric mass diffusion through the tissues, while the faster regime is stream flow through blood vessels and bronchial passages. The length scale of the smallest volume element from which constructs begin is known. It is the distance to which oxygen diffuses into the tissue during one breathing time interval or heart beat.

The same hand-in-glove geometry characterizes all turbulent flows. Viscous diffusion in the smallest volume elements is accompanied by the macroscopic structure of faster streams known empirically as "eddies." The length scale of the smallest volume element (or the smallest eddy) is known from the distance of viscous diffusion during the first roll [5]. Pure diffusion is replaced by the first eddy, in other words, by diffusion combined with the first stream, when the stream motor can advance the mixing process farther and faster than diffusion alone. The evolution of constructs of stepwise larger sizes can be observed *in time* by looking at any shear layer, jet or plume. The oldest structure (shape) in the field of vision is the smallest eddy, which is swept intermittently from the laminar tip section of the flow. The youngest structure is the largest eddy, which is a construct of smaller constructs of all the preceding sizes.

Most natural tree networks accommodate the flow of fluid or electricity. In fluid trees the small-scale volumetric flow is by slow diffusion (for example, Darcy flow in the wet banks of the smallest rivulet), while the larger-scale flow is organized into faster conduits [12]. In the electric tree visible during lightning, new features are three-dimensionality and the unsteady (one shot) nature of the flow. Additional electric tree networks are found in living systems, for example, the neural dendrites of the nervous system. Street patterns and urban growth have been predicted based on a kinematics analog of the resistance minimization shown in this chapter. That analog is the minimization of the travel time between one point and a finite-size area (an infinity of points) [12].

The resistance minimization principle is raised to the rank of law by the structure of all the tree networks anticipated by constructal theory. This law can be summarized as follows [11, 12]: "For a finite-size system to persist in time (to live), it must evolve in such a way that it provides easier access to the imposed currents that flow through it." This statement has two parts. First, it recognizes the natural tendency of imposed currents to construct shapes, i.e. paths of optimal access through constrained open systems. The second part accounts for the improvements of these paths, which occurs in an identifiable direction that can be aligned with time itself.

This formulation of the principle refers to a system with imposed steady flow, as in the heat flow examples discussed in this chapter. If the system discharges itself to one point in unsteady fashion, then the constructal minimization of volume-to-point resistance is equivalent to the minimization of the time of discharge, or the maximization of the speed of approach to internal equilibrium (uniformity, zero flow). If the volume is unbounded, the constructs compound themselves and continue to spread indefinitely. Complexity continues to increase in time because "growth" is the path to resistance minimization. Examples are the jet in a fluid reservoir, and the dendritic crystal that grows into a subcooled liquid [12].

All the structured phenomena mentioned at various stages in this chapter, including the round cross-section of the blood vessel and the proportionality between river width and depth, can be predicted based on constructal theory [12]. The theoretical heat-tree networks [11] were subjected to numerical tests and refinements in [14]. Constructal trees for fluid flow through a heterogeneous porous medium were developed in [15, 16]. Analogous trees of streets that provide minimum-time travel between an area and one point are described in [17-19]. Constructal theory also accounts for the onset, structure and heat transfer rate of turbulent flow [12, 20] and Bénard convection [21], and for the structure in cracked solids that are cooled volumetrically [10]. Three-dimensional constructal networks are presented in [16, 18, 22]. The time-dependent discharge between a volume and one point—the lightning tree problem—was added to constructal theory in [23]. The deterministic development in time of a river drainage basin is presented in [24]. Reviews of the applications of constructal theory are provided in [12, 17, 25].

Acknowledgment. This work was supported by the National Science Foundation.

References

1. Bejan, A. (1982) *Entropy Generation through Heat and Fluid Flow*, Wiley, New York.
2. Bejan, A. (1993) *Heat Transfer*, Wiley, New York.
3. Bejan, A. (1996) *Entropy Generation Minimization*, CRC Press, Boca Raton, FL.
4. Bejan, A. (1996) Entropy generation minimization: the new thermodynamics of finite-size devices and finite-time processes, *Journal of Applied Physics* **79**, 1191-1218.
5. Bejan, A. (1995) *Convection Heat Transfer*, second edition, Wiley, New York.
6. Bejan, A. , Tsatsaronis, G. and Moran, M. (1996) *Thermal Design and Optimization*, Wiley, New York.

7. Bejan, A. and Sciubba, E. (1992) The optimal spacing of parallel plates cooled by forced convection, *International Journal of Heat and Mass Transfer* **35**, 3259-3264.
8. Petrescu, S. (1994) Comments o the optimal spacing of parallel plates cooled by forced convection, *International Journal of Heat and Mass Transfer* **37**, 1283.
9. Bejan, A. (1996) Geometric optimization of cooling techniques, chapter 1 in Kim, S. J. and Lee, S. W., eds., *Air Cooling Technology for Electronic Equipment*, CRC Press, Boca Raton, FL.
10. Bejan, A. Ikegami, Y. and Ledezma, G. A. (1997) Constructal theory of crack pattern formation for fastest cooling, *International Journal of Heat and Mass Transfer* **41**, 1945-1954.
11. Bejan, A. (1997) Constructal-theory network of conducting paths for cooling a heat generating volume, *International Journal of Heat and Mass Transfer* **40**, 799-816.
12. Bejan, A. (1997) *Advanced Engineering Thermodynamics*, second edition, Wiley, New York.
13. Thompson, D'A. W. (1942) *On Growth and Form*, Cambridge University Press, Cambridge, UK.
14. Ledezma, G. A., Bejan, A. and Errera, M. R. (1997) Constructal tree networks for heat transfer, *Journal of Applied Physics* **82**, 89-100.
15. Bejan, A. and Errera, M. R. (1997) Deterministic tree networks for fluid flow: geometry for minimum flow resistance between a volume and one point, *Fractals* **5**, 685-695.
16. Bejan, A. (1997) Constructal tree network for fluid flow between a finite-size volume and one source or sink, *Revue Générale de Thermique* **36**, 592-604.
17. Bejan, A. (1996) Street network theory of organization in Nature, *Journal of Advanced Transportation* **30**, 85-107.
18. Bejan, A. and Ledezma, G. A. (1998) Streets tree networks and urban growth: optimal geometry for quickest access between a finite-size volume and one point, *Physica A* **255**, 211-217.
19. Bejan, A. and Tondeur, D. (1998) Equipartition, optimal allocation, and the constructal approach to predicting organization in nature, *Revue Générale de Thermique* **37**, 165-180.
20. Bejan, A. (1998) Questions in fluid mechanics: natural tree-shaped flows, *Journal of Fluids Engineering* **120**, 429-430.
21 Nelson, R. A., Jr. and Bejan, A. (1998) Constructal optimization of internal flow geometry in convection, *Journal of Heat Transfer* **120**, 357-364.
22. Ledezma, G. A. and Bejan, A. (1998) Constructal three-dimensional trees for conduction between a volume and one point, *Journal of Heat Transfer* **120**, 977-984.
23. Dan, N. and Bejan, A. (1998) Constructal tree networks for the time-dependent discharge of a finite-size volume to one point, *Journal of Applied Physics*, **84**, 3042-3050.
24. Errera, M. R. and Bejan, A. (1998) Deterministic tree networks for river drainage basins, *Fractals* **6**, 245-261.
25. Bejan, A. (1998) Constructal optimization of paths for heat transfer, in Tupholme, G. E. and Wood, A. S., eds., *Mathematics of Heat Transfer*, Clarendon Press, Oxford, UK.

FUNDAMENTALS OF EXERGY ANALYSIS AND EXERGY-AIDED THERMAL SYSTEMS DESIGN

M.J. MORAN
The Ohio State University
Department of Mechanical Engineering
Columbus, OH 43210, USA

1. Introduction

A close reading of the early literature of the second law of thermodynamics originating from the time of Carnot shows that second law reasoning in today's engineering sense - - optimal conversion of inputs to end use - - was present from the beginning. Interest in this theme occasionally flickered in the interim, but it was kept alive and now forms the basis for contemporary applications of engineering thermodynamics. The bonds that have been forged in recent years between thermodynamics and economics have made the theme of optimal use more explicit and have allowed second law methods to be even more powerful and effective than ever before for the design and analysis of thermal systems.

The development of second law analysis owes considerably to the efforts of a great many towering figures in the history of science whose contributions have advanced our present understanding immeasurably. Among these are Carnot, Clausius, Gibbs, Kelvin, and Planck. Significant contributions to engineering thermodynamics have been provided by the likes of Stodola, Bosnjakovic, Keenan, and Obert. Still, the bulk of the literature of second law analysis in engineering has appeared in the last quarter-century, and most of those who have contributed remain active somewhere over the globe, including Szargut, Gyftopoulos, Gaggioli, El-Sayed, Valero, Tsatsaronis, Bejan, and many more.

Today, second law analysis has several offshoots, including *exergy analysis, thermoeconomics, entropy-generation minimization,* and *finite-time thermodynamics* (or endoreversible thermodynamics). The first three are well-embedded in contemporary thermodynamics. Many experts attest to their pedagogical value. These methods are increasingly being used in engineering practice and even finding their way into handbooks [1]. As they are now familiar, the mere use of such methods is no longer sufficiently novel to stir interest; rather, it is the results elicited from them that have become the focus. In short, these methods have achieved a level of maturity, and the spotlight is now on what can be achieved with them [2]. Some additional evolution surely will occur, but as we begin a new millennium we can regard these methods as established, and expect them to be used increasingly for the design and analysis of thermal systems. By comparison, finite-time (endoreversible) thermodynamics faces an uncertain future, stemming from fundamental flaws, over-reliance on highly simplified models and lack of engagement with real-world considerations [3].

The present discussion provides an introduction to thermal system design from a second law perspective featuring exergy. Since exergy measures the true thermodynamic values of the work, heat, and other interactions between a system and its surroundings as well as the effect of irreversibilities within the system, exergy is also a rational basis for assigning costs. This is called *exergy costing.* Although the importance of considering

A. Bejan and E. Mamut (eds.), Thermodynamic Optimization of Complex Energy Systems, 73–92.

costs is emphasized throughout the current presentation, an indepth discussion of exergy costing is not provided, and readers are referred to [2].

Essential background material concerning engineering thermodynamics is reviewed in Sec. 2. Various aspects of exergy analysis are then surveyed in Sec. 3. Guidelines for improving thermodynamics effectiveness are listed in Sec. 4. The role of exergy analysis in thermal system design is considered in Sec. 5. Closing comments are provided in Sec. 6. References are listed in Sec. 7.

2. Engineering Thermodynamics Background

Engineering applications involving exergy are generally analyzed on a control volume basis. Accordingly, the control volume formulations of the mass, energy, and entropy balances presented in this section play important roles. These are provided here in the form of overall balances assuming one-dimensional flow. Equations of change for mass, energy, and entropy in the form of differential equations are also available in the literature [4].

For applications in which inward and outward flows occur, each through one or more ports, the extensive property balance expressing the conservation of mass principle takes the form

$$\frac{dm}{dt} = \sum_i \dot{m}_i - \sum_e \dot{m}_e \tag{1a}$$

where dm/dt represents the time rate of change of mass contained within the control volume, $\dot{m}_i$ denotes the mass flow rate at an inlet port, and $\dot{m}_e$ denotes the mass flow rate at an exit port.

Energy is a fundamental concept of thermodynamics and one of the most significant aspects of engineering analysis. Energy can be stored within systems in various macroscopic forms: kinetic energy, gravitational potential energy, and internal energy. Energy can also be transformed from one form to another and transferred between systems. Energy can be transferred by work, by heat transfer, and by flowing matter. The total amount of energy is conserved in all transformations and transfers. The extensive property balance expressing the conservation of energy principle (first law) takes the form

$$\frac{d(U + \mathrm{KE} + \mathrm{PE})}{dt} = \dot{Q} - \dot{W}_{cv} + \sum_i \dot{m}_i \left(h_i + \frac{\mathrm{V}_i^2}{2} + gz_i \right) - \sum_e \dot{m}_e \left(h_e + \frac{\mathrm{V}_e^2}{2} + gz_e \right) \tag{2a}$$

where U, KE and PE denote, respectively, the internal energy, kinetic energy, and gravitational potential energy of the overall control volume, respectively. The right side of Eq. (2a) accounts for transfers of energy across the boundary of the control volume.

Energy can enter and exit control volumes by work. Because work is done on or by a control volume when matter flows across the boundary, it is convenient to separate the work rate (or power) into two contributions. One contribution is the work rate associated with the force of the fluid pressure as mass is introduced at the inlet and removed at the exit. Commonly referred to as *flow work*, this contribution is accounted for by $\dot{m}_i(p_i v_i)$ and $\dot{m}_e(p_e v_e)$, respectively, where p denotes pressure and v denotes specific volume. The other contribution, denoted as $\dot{W}_{cv}$, includes all other work effects, such as those associated with rotating shafts, displacement of the boundary, and electrical effects. $\dot{W}_{cv}$ is considered *positive* for energy transfer *from* the control volume.

Energy can also enter and exit control volumes with flowing streams of matter. On a one-dimensional flow basis, the rate at which energy enters with matter at inlet i is $\dot{m}_i(u_i + V_i^2/2 + gz_i)$, where the three terms in parentheses account, respectively, for the specific internal energy, specific kinetic energy, and specific gravitational potential energy of the substance flowing through port i. In writing Eq. (2a), the sum of the specific internal energy and specific flow work at each inlet and exit is expressed in terms of the specific enthalpy $h(= u + pv)$. Finally, $\dot{Q}$ accounts for the rate of energy transfer by heat, and is considered *positive* for energy transfer *to* the control volume.

The second law of thermodynamics can be expressed in terms of entropy or, alternatively, in terms of exergy (Sec. 3). Like mass and energy, entropy can be stored within systems and transferred across system boundaries. However, unlike mass and energy, entropy is not conserved, but generated (or produced) by irreversibilities within systems. The extensive property balance for entropy takes the form

$$\underbrace{\frac{dS}{dt}}_{\text{rate of entropy change}} = \underbrace{\sum_j \frac{\dot{Q}_j}{T_j} + \sum_i \dot{m}_i s_i - \sum_e \dot{m}_e s_e}_{\text{rates of entropy transfer}} + \underbrace{\dot{S}_{gen}}_{\text{rate of entropy generation}} \tag{3a}$$

where dS/dt represents the time rate of change of entropy within the control volume. The terms $\dot{m}_i s_i$ and $\dot{m}_e s_e$ account, respectively, for rates of entropy transfer into and out of the control volume accompanying mass flow. $\dot{Q}_j$ represents the time rate of heat transfer at the location on the boundary where the instantaneous temperature is T_j, and $\dot{Q}_j/T_j$ accounts for the accompanying rate of entropy transfer. $\dot{S}_{gen}$ denotes the time rate of entropy generation due to irreversibilities within the control volume.

When applying the entropy balance in any of its forms, the objective is often to evaluate the entropy generation term. However, the value of the entropy generation for a given process of a system usually does not have much significance by itself. The significance is normally determined through comparison: The entropy generation within a given component might be compared with the entropy generation values of the other components included in an overall system formed by these components.

To evaluate the entropy transfer term of the entropy balance requires information about both the heat transfer and the temperature on the boundary where the heat transfer occurs. The entropy transfer term is not always subject to direct evaluation, however, because the required information is either unknown or not defined. In such applications, it may be convenient, therefore, to enlarge the system to include enough of the immediate surroundings that the temperature on the boundary of the enlarged system corresponds to the ambient temperature T_{amb}. The entropy transfer rate is then simply $\dot{Q}/T_{amb}$. However, as the irreversibilities present would not be just those for the system of interest but those for the enlarged system, the entropy generation term would account for the effects of internal irreversibilities within the system and external irreversibilities within that portion of the surroundings included within the enlarged system.

For control volumes at steady state, Eqs. (1a) - (3a) reduce, respectively, to read

$$\sum_i \dot{m}_i = \sum_e \dot{m}_e \tag{1b}$$

$$0 = \dot{Q} - \dot{W}_{cv} + \sum_i \dot{m}_i \left(h_i + \frac{V_i^2}{2} + gz_i \right) - \sum_e \dot{m}_e \left(h_e + \frac{V_e^2}{2} + gz_e \right) \tag{2b}$$

$$0 = \sum_j \frac{\dot{Q}_j}{T_j} + \sum_i \dot{m}_i s_i - \sum_e \dot{m}_e s_e + \dot{S}_{gen} \tag{3b}$$

The mass, energy and entropy balances generally must be supplemented by appropriate thermodynamic property data [1, 5]. For some applications a momentum equation expressing Newton's second law of notion for control volumes is also required. Data for transport properties, heat transfer coefficients, friction factors, and so on also may be needed for a comprehensive engineering analysis.

3. Exergy Analysis

In this section essential aspects of exergy analysis are surveyed, For further details, see [2, 6-9].

3.1. DEFINING EXERGY

An opportunity for doing work exists whenever two systems at different states are placed in communication because, in principle, work can be developed as the two are allowed to come into equilibrium. When one of the two systems is a suitably idealized system called an *exergy reference environment* or simply, an *environment,* and the other is some system of interest, *exergy* is the maximum theoretical useful work (shaft work or electrical work) obtainable as the systems interact to equilibrium, heat transfer occurring with the environment only. (Alternatively, exergy is the minimum theoretical useful work required to form a quantity of matter from substances present in the environment and to bring the matter to a specified state.) Exergy is a measure of the *departure* of the state of the system from that of the environment, and is therefore an attribute of the system and environment together. Once the environment is specified, however, a value can be assigned to exergy in terms of property values for the system only, so exergy can be regarded as an extensive property of the system.

Exergy can be destroyed and generally is not conserved. A limiting case is when exergy would be completely destroyed, as would occur if a system were to come into equilibrium with the environment *spontaneously* with no provision to obtain work. The capability to develop work that existed initially would be completely wasted in the spontaneous process. Moreover, since no work needs to be done to effect such a spontaneous change, the value of exergy can never be negative.

Although the term environment is selected to suggest kinship with the *natural environment*, the exergy reference environment used in any particular application is a thermodynamic model. Models with various levels of specificity are used to evaluate exergy. The exergy reference environment is typically regarded as composed of common substances existing in abundance within the Earth's atmosphere, oceans, and crust. The substances are in their stable forms as they exist naturally, and there is no possibility of developing work from interactions - physical or chemical - between parts of the environment. Although the intensive properties of the environment are assumed to be unchanging, the extensive properties can change as a result of interactions with other systems. Kinetic and potential energies are evaluated relative to coordinates in the environment, all parts of which are considered to be at rest with respect to one another.

For computational ease, the temperature T_0 and pressure p_0 of the environment are often taken as standard-state values, such as 1 atm and 25°C(77°F). However, these properties may be specified differently depending on the application. T_0 and p_0 might be taken as the average ambient temperature and pressure, respectively, for the location at which the system under consideration operates. Or, if the system uses atmospheric air, T_0 might be specified as the average air temperature. If both air and water from the natural surroundings are used, T_0 would be specified as the lower of the average temperatures for air and water. Further discussion of the exergy reference environment is provided in Sec. 3.3.

When a system is in equilibrium with the environment, the state of the system is called the *dead state.* At the dead state, the conditions of mechanical, thermal, and chemical equilibrium between the system and the environment are satisfied: the pressure, temperature, and chemical potentials of the system equal those of the environment, respectively. In addition, the system has no motion or elevation relative to coordinates in the environment. Under these conditions, there is no possibility of a spontaneous change within the system or the environment, nor can there be an interaction between them. At the dead state, the value of exergy is zero. Another type of equilibrium between the system and environment can be identified. This is a restricted form of equilibrium where only the conditions of mechanical and thermal equilibrium must be satisfied. This state of the system is called the *restricted dead state.* At the restricted dead state, the fixed quantity of matter under consideration is imagined to be sealed in an envelope impervious to mass flow, at zero velocity and elevation relative to coordinates in the environment, and at the temperature T_0 and pressure p_0.

3.2. CONTROL VOLUME EXERGY RATE BALANCE

Exergy balances can be written in various forms, depending on whether a closed system or control volume is under consideration and whether steady-state or transient operation is of interest. Owing to its importance for a wide range of applications, an exergy rate balance for control volumes at steady state is presented here.

Exergy can be transferred across the boundary of a control volume by three means: exergy transfer associated with work, exergy transfer associated with heat transfer, and exergy transfer associated with the matter entering and exiting the control volume. All such exergy transfers are evaluated relative to the environment used to define exergy. Exergy is also destroyed by irreversibilities within the control volume. At steady state

$$0 = \underbrace{\sum_j \dot{E}_{qj} - \dot{W}_{cv} + \sum_i \dot{E}_i - \sum_e \dot{E}_e}_{\text{rates of exergy transfer}} - \underbrace{\dot{E}_D}_{\text{rate of exergy destruction}} \tag{4a}$$

or

$$0 = \sum_j \left(1 - \frac{T_0}{T_j}\right)\dot{Q}_j - \dot{W}_{cv} + \sum_i \dot{m}_i e_i - \sum_e \dot{m}_e e_e - \dot{E}_D \tag{4b}$$

where $\dot{W}_{cv}$ is the work rate excluding the *flow work*. $\dot{Q}_j$ is the time rate of heat transfer at the location on the boundary of the control volume where the instantaneous temperature is T_j. The associated rate of exergy transfer is

$$\dot{E}_{qj} = \left(1 - \frac{T_0}{T_j}\right)\dot{Q}_j \tag{5}$$

In Eqs. (4), the subscripts i and e denote inlets and outlets, respectively. The exergy transfer rates at control volume inlets and outlets are denoted, respectively, as $\dot{E}_i = \dot{m}_i e_i$ and $\dot{E}_e = \dot{m}_e e_e$. Finally, $\dot{E}_D$ accounts for the time rate of exergy destruction due to irreversibilities within the control volume. The exergy destruction rate is related to the entropy generation rate by

$$\dot{E}_D = T_0 \dot{S}_{gen} \tag{6}$$

The specific exergy transfer terms e_i and e_e are expressible in terms of four components: physical exergy, e^{PH}, kinetic exergy e^{KN}, potential exergy e^{PT}, and chemical exergy e^{CH}:

$$e = e^{PH} + e^{KN} + e^{PT} + e^{CH} \tag{7}$$

As discussed in [2], the first three components are evaluated as follows:

$$e^{PH} = (h - h_0) - T_0(s - s_0) \tag{8a}$$

$$e^{KN} = \frac{V^2}{2} \tag{8b}$$

$$e^{PT} = gz \tag{8c}$$

where h and s denote, respectively, the specific enthalpy and specific entropy, and V and z denote velocity and elevation relative to coordinates in the environment, respectively. The subscript o denotes the restricted dead state.

3.3. CHEMICAL EXERGY

To evaluate the chemical exergy, the exergy component associated with the departure of the chemical composition of a system from that of the environment, the substances comprising the system are referred to the properties of a suitably selected set of environmental substances. For discussion of alternative sets of substances tailored to particular applications, see [6, 8]. Exergy analysis is facilitated, however, by employing a *standard environment* and a corresponding table of *standard chemical exergies.* Standard chemical exergies are based on standard values of the environmental temperature T_0 and pressure p_0 - for example, 298.15 K(25°C) and 1 atm, respectively. A standard environment also consists of a set of reference substances with standard concentrations closely reflecting the chemical makeup of the natural environment. The reference substances generally fall into three groups: gaseous components of the atmosphere, solid substances from the lithosphere, and ionic and nonionic substances from the oceans. The chemical exergy data of Table 1 correspond to two alternative standard exergy reference environments, called here model I and model II, that have gained acceptance for engineering evaluations.

Although the use of standard chemical exergies greatly facilitates the application of exergy principles, the term *standard* is somewhat misleading since there is no one specification of the environment that suffices for all applications. Still, chemical exergies calculated relative to alternative specifications of the environment are generally in good agreement. For a broad range of engineering applications the simplicity and ease of use of standard chemical exergies generally outweigh any slight lack of accuracy that might result. In particular, the effect of slight variations in the values of T_0 and p_0 about the values used to determine the standard chemical exergies reported in Table 1 can be neglected.

TABLE 1 Standard chemical exergy, e^{CH} (kJ / kmol) of various substances at 298 K and p_0.

Substance	Formula	Model I[a]	Model II[b]
Nitrogen	$N_2(g)$	640	720
Oxygen	$O_2(g)$	3,950	3,970
Carbon dioxide	$CO_2(g)$	14,175	19,870
Water	$H_2O(g)$	8,635	9,500
	$H_2O(l)$	45	900
Carbon (graphite)	C(s)	404,590	410,260
Hydrogen	$H_2(g)$	235,250	236,100
Methane	$CH_4(g)$	824,350	831,650
Ethane	$C_2H_6(g)$	1,482,035	1,495,840
Methanol	$CH_3OH(g)$	715,070	722,300
	$CH_3OH(l)$	710,745	718,000
Ethyl alcohol	$C_2H_5OH(g)$	1,348,330	1,363,900
	$C_2H_5OH(l)$	1,342,085	1,357,700

a. See [1, 10] p_0 = 1.019 atm.

b. See [1, 9] p_0 = 1.0 atm.

A common feature of standard exergy reference environments is a gas phase, intended to represent air, that includes N_2, O_2, CO_2, $H_2O(g)$, and other gases. Each gas i present in the gas phase is at temperature T_0 and partial pressure $p_i^e = x_i^e p_0$, where the superscript e denotes the environment and x_i^e is the mole fraction of gas i in the gas phase. Referring to the device at steady state shown in Fig. 1, the standard chemical exergy for a gas included in the environmental gas phase can be evaluated as follows: Gas i enters at temperature T_0 and pressure p_0, expands isothermally with heat transfer only with the environment, and exits to the environment at temperature T_0 and partial pressure $x_i^e p_0$. The maximum theoretical work per mole of gas i would be developed when the expansion occurs without internal irreversibilities. Accordingly, with energy and entropy balances together with the ideal gas equation of state, the chemical exergy per mole of i is

$$\bar{e}_i^{CH} = -\bar{R}T_0 \ln \frac{x_i^e p_0}{p_0}$$

$$= -\bar{R}T_0 \ln x_i^e \tag{9}$$

The details of the derivation are left as are exercise. As an application of Eq. (9) consider CO_2 with $x_{CO_2}^e = 3.3 \times 10^{-4}$ and $T_0 = 298.15\,K$, as in [9]. Inserting values gives $\bar{e}^{CH}_{CO_2} = 19{,}870\ \mathrm{kJ/kmol}$, which is the value listed for CO_2 in Table 1 (Model II).

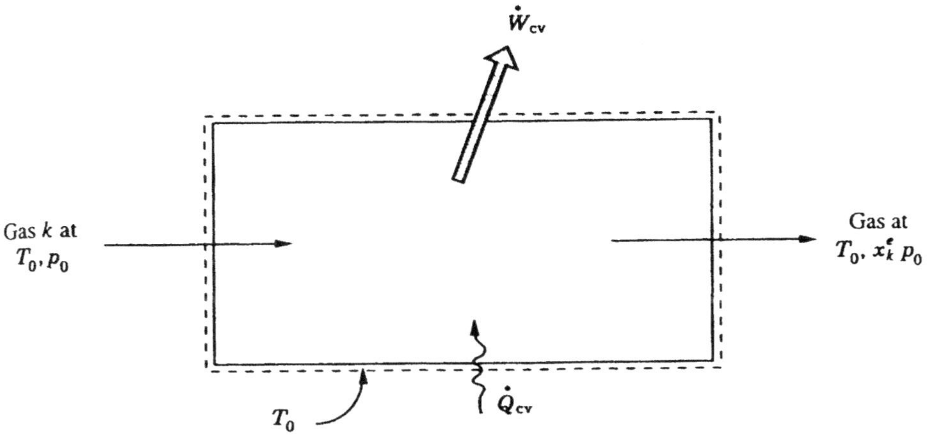

Figure 1. Device for evaluating the chemical exergy of a gas.

The chemical exergy of a mixture of n gases each of which is present in the environmental gas phase can be obtained similarly. We may think of a set of n devices such as shown in Fig. 1, one for each gas in the mixture. Gas i, whose mole fraction in the gas mixture at T_0, p_0 is x_i, enters at T_0 and the partial pressure $x_i\, p_0$. As before, the gas exits to the environment at T_0 and the partial pressure $x_i^e\, p_0$. Paralleling the previous development, the work per mole i is $-\bar{R}T_0 \ln\left(x_i^e / x_i\right)$. Summing over all components, the chemical exergy per mole of mixture is

$$\bar{e}^{CH}_{mixture} = -\bar{R}T_0 \sum x_i \ln \frac{x_i^e}{x_i}$$

This expression can be written alternatively with Eq. (9) as

$$\bar{e}^{CH}_{mixture} = \sum x_i \bar{e}_i^{CH} + \bar{R}T_0 \sum x_i \ln x_i \tag{10}$$

Equation (10) remains valid for mixtures containing gases other than those present in the reference environment and can be extended to mixtures (and solutions) that do not adhere to the ideal gas model [8]. In such applications, the terms $\bar{e}_i^{CH}$ are selected from a table of standard chemical exergies.

As a further illustration of the evaluation of chemical exergy, consider the case of hydrocarbon fuels. In principle, the standard chemical exergy of a substance can be evaluated by considering an idealized reaction of the substance with other substances for which the chemical exergies are known. For the case of a pure hydrocarbon fuel C_aH_b at T_0, p_0, refer to the system shown in Fig. 2, where all substances are assumed to enter and exit at T_0, p_0. For operation at steady state, heat transfer only with the environment, and no internal irreversibilities, a systematic application of extensive property balances results in the following alternative expressions [2]:

$$\bar{e}_F^{CH} = \overline{HHV}(T_0, p_0) - T_0\left[\bar{s}_F + \left(a + \frac{b}{4}\right)\bar{s}_{O_2} - a\bar{s}_{CO_2} - \frac{b}{2}\bar{s}_{H_2O}\right](T_0, p_0)$$
$$+ a\bar{e}_{CO_2}^{CH} + \left(\frac{b}{2}\right)\bar{e}_{H_2O}^{CH} - \left(a + \frac{b}{4}\right)\bar{e}_{O_2}^{CH} \qquad (11a)$$

$$\bar{e}_F^{CH} = \left[\bar{g}_F + \left(a + \frac{b}{4}\right)\bar{g}_{O_2} - a\bar{g}_{CO_2} - \frac{b}{2}\bar{g}_{H_2O}\right](T_0, p_0)$$
$$+ a\bar{e}_{CO_2}^{CH} + \left(\frac{b}{2}\right)\bar{e}_{H_2O}^{CH} - \left(a + \frac{b}{4}\right)\bar{e}_{O_2}^{CH} \qquad (11b)$$

Where the subscript F denotes the fuel, $\overline{HHV}$ denotes the higher heating value (H_2O is a liquid), $\bar{s}$ denotes absolute entropy, and $\bar{g}$ denotes the Gibbs function.

With Eqs. (11), the chemical exergy of C_aH_b can be calculated using the standard chemical exergies of O_2, CO_2, and $H_2O(l)$, together with selected property data: the fuel higher heating value and absolute entropies, or Gibbs functions of formation. As an application, consider the case of carbon. With $a = 1$, $b = 0$, Gibbs function data from [7] and standard chemical exergies for CO_2 and O_2 from Table 1 (Model II), Eq. (11b) gives $\bar{e}_C^{CH} = 410,280 \text{ kJ/kmol}$, which agrees closely with the value listed in Table I (Model II). Similarly, for the case of hydrogen ($a = 0$, $b = 2$) we have $\bar{e}_{H_2}^{CH} = 236,095 \text{ kJ/kmol}$. When tabulated chemical exergy data are lacking, as for example in the cases of coal, char, and fuel oil, the approach of Eq. (11a) can be invoked using a *measured* heating value and an *estimated* value of the fuel absolute entropy determined with procedures discussed in the literature. For an illustration, see [2].

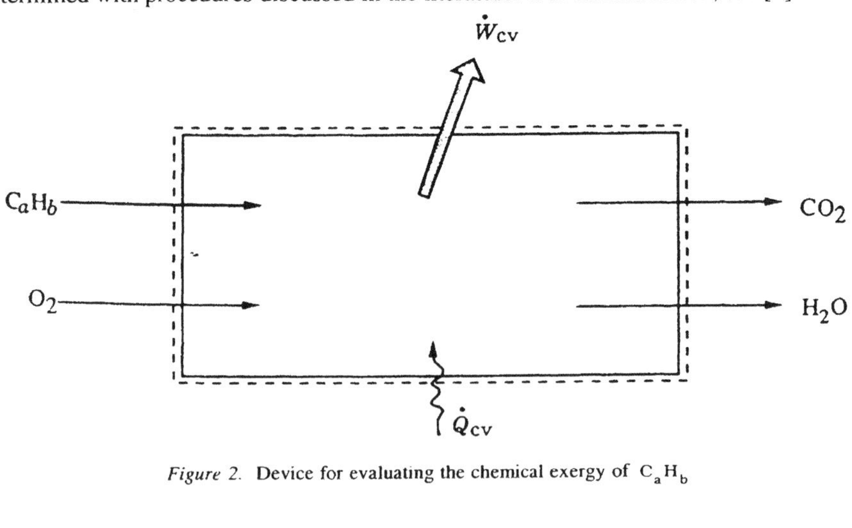

Figure 2. Device for evaluating the chemical exergy of C_aH_b

3.4. EXERGETIC EFFICIENCY

The exergetic efficiency (second law efficiency, effectiveness, or rational efficiency) provides a true measure of the performance of a system from the thermodynamic viewpoint. To define the exergetic efficiency both a *product* and a *fuel for* the system being analyzed are identified. The product represents the desired result of the system (power, steam, some combination of power and steam, etc.). Accordingly, the definition of the product must be consistent with the purpose of purchasing and using the system. The fuel represents the resources expended to generate the product and is not necessarily restricted to being an actual fuel such as a natural gas, oil, or coal. Both the product and the fuel are expressed in terms of exergy.

For a control volume at steady state whose exergy rate balance reads

$$\dot{E}_F = \dot{E}_P + \dot{E}_D + \dot{E}_L \tag{12}$$

the exergetic efficiency is

$$\varepsilon = \frac{\dot{E}_P}{\dot{E}_F} = 1 - \frac{\dot{E}_D + \dot{E}_L}{\dot{E}_F} \tag{13}$$

where the rates at which the fuel is supplied and the product is generated are $\dot{E}_F$ and $\dot{E}_P$ respectively. $\dot{E}_D$, and $\dot{E}_L$ denote the rates of exergy destruction and exergy loss, respectively. Exergy is destroyed by irreversibilities within the control volume, and exergy is lost from the control volume via stray heat transfer, material streams vented to the surroundings, and so on.

The exergetic efficiency shows the percentage of the fuel exergy provided to a control volume that is found in the product exergy. Moreover, the difference between 100% and the value of the exergetic efficiency, expressed as a percent, is the percentage of the fuel exergy wasted in this control volume as exergy destruction and exergy loss. To apply Eq. (13), decisions are required concerning what are considered as the fuel and the product. Table 2 provides illustrations for four common components. Similar considerations are used to write exergetic efficiencies for systems consisting of several components, as for example a power plant.

Exergetic efficiencies can be used to assess the thermodynamic performance of a component, plant, or industry relative to the performance of similar components, plants, or industries. By this means the performance of a gas turbine, for instance, can be gauged relative to the typical present-day performance level of gas turbines. A comparison of exergetic efficiencies for dissimilar devices - gas turbines and heat exchangers, for example - is generally not significant, however. The exergetic efficiency is generally more meaningful, objective, and useful than other efficiencies based on the first or second law of thermodynamics, including the thermal efficiency of a power plant, the isentropic efficiency of a compressor or turbine, and the effectiveness of a heat exchanger. The thermal efficiency of a cogeneration system is misleading because it treats both work and heat transfer as having equal thermodynamic value. The isentropic turbine efficiency does not consider that the working fluid at the outlet of the turbine has a higher temperature (and consequently a higher exergy that may be used in the next component) in the actual process than in the isentropic process. The heat exchanger effectiveness fails to identify the exergy destruction associated with the pressure drops of the heat exchanger working fluids.

TABLE 2. The Exergetic Efficiency for Selected Components at Steady State[a]

	Turbine or Expander	Extraction Turbine	Compressor, Pump, or Fan	Heat Exchanger[b]
Component	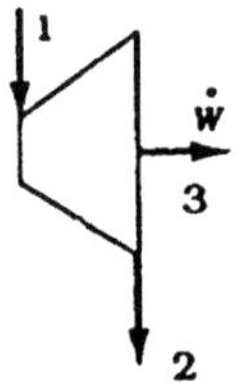	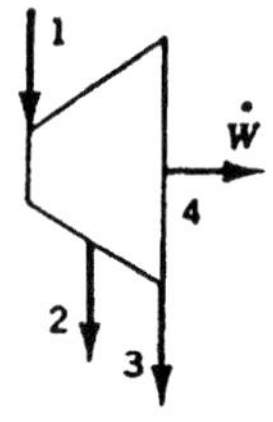	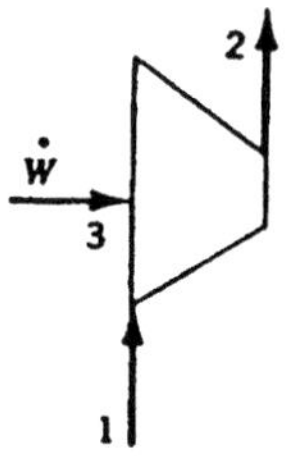	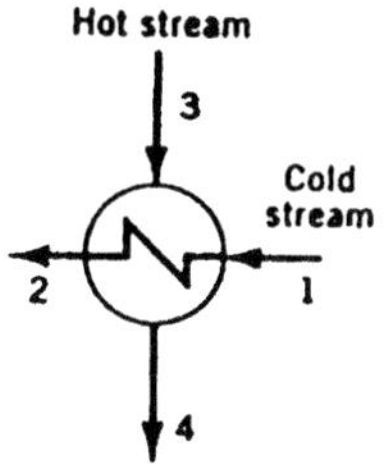
$\dot{E}_P$	$\dot{W}$	$\dot{W}$	$\dot{E}_2 - \dot{E}_1$	$\dot{E}_2 - \dot{E}_1$
$\dot{E}_F$	$\dot{E}_1 - \dot{E}_2$	$\dot{E}_1 - \dot{E}_2 - \dot{E}_3$	$\dot{W}$	$\dot{E}_3 - \dot{E}_4$
ε	$\dfrac{\dot{W}}{\dot{E}_1 - \dot{E}_2}$	$\dfrac{\dot{W}}{\dot{E}_1 - \dot{E}_2 - \dot{E}_3}$	$\dfrac{\dot{E}_2 - \dot{E}_1}{\dot{W}}$	$\dfrac{\dot{E}_2 - \dot{E}_1}{\dot{E}_3 - \dot{E}_4}$

a) For discussion, see [2].
b) This definition assumes that the purpose of the heat exchanger is to heat the cold stream $(T_1 \geq T_0)$.

3.5. CASE STUDY PROBLEM

As an illustration of exergy analysis, consider the gas turbine cogeneration system shown in Fig. 3, which produces process steam and a net power output. Table 3 provides exergy and exergy destruction data for the cogeneration system determined using concepts introduced in Secs. 3.2, 3.3; see [2] for details.

Using the approach of Sec. 3.4, the overall exergetic efficiency is calculated as the percentage of the fuel exergy supplied to the system that is recovered in the product of the system. Identifying the product of the cogeneration system as the sum of the net power developed and the net increase of the exergy of the feedwater.

$$\varepsilon = \frac{\dot{W}_{net} + \left(\dot{E}_9 - \dot{E}_8\right)}{\dot{E}_{10}}$$

$$= \frac{30\,\text{MW} + 12.75\,\text{MW}}{85\,\text{MW}} = 0.503(50.3\%) \tag{14}$$

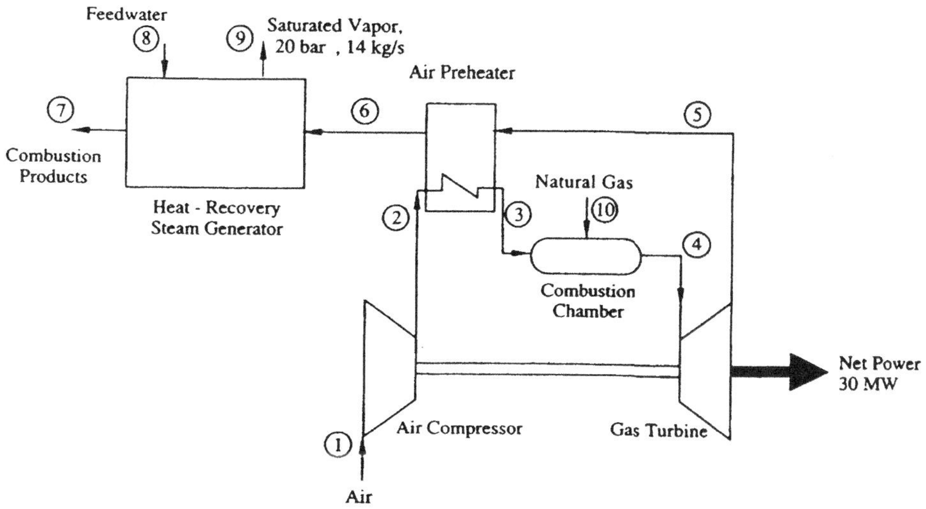

Figure 3. Gas turbine cogeneration system

where the exergy values are obtained from Table 3. The exergy carried out of the system at state 7, amounting to 3.27 percent of the fuel exergy, is regarded as a loss.

The exergy destruction and loss data summarized in Table 3 clearly identify the combustion chamber as the major site of thermodynamic inefficiency. The next most prominent site is the recovery steam generator (HRSG). Roughly equal contributions to inefficiency are made by the gas turbine, air compressor, and the loss associated with stream 7. The gas compressor is an only slightly small contributor.

The exergy destructions in these components stem from one or more of three principal internal irreversibilities associated, respectively, with chemical reaction, heat transfer, and friction. All three irreversibilities are present in the combustion chamber, where chemical reaction and heat transfer are both significant and friction is of secondary importance. For the HRSG and preheater the most significant irreversibility is related to the stream-to-stream heat transfer, with friction playing a secondary role. The exergy destruction in the adiabatic gas turbine and air compressor is caused mainly by friction.

Although combustion is intrinsically a significant source of irreversibility, a dramatic reduction in its effect on exergy destruction cannot be expected using ordinary means. Still, the inefficiency of combustion can be ameliorated by preheating the combustion air and reducing the air-fuel ratio. For the HRSG and preheater, the effect of the irreversibility associated with heat transfer tends to lessen as the minimum temperature difference between the streams, $(\Delta T)_{min}$, is reduced. The exergy destruction within the gas turbine and the air compressor decreases as the isentropic turbine and compressor efficiencies increase, respectively. The significance of the exergy loss associated with stream 7 reduces as the gas temperature reduces.

Although these considerations provide a basis for implementing practical engineering measures to improve the thermodynamic performance of the cogeneration system, such measures have to be applied judiciously. Measures that improve the thermodynamic performance of one component might adversely affect another, leading to

TABLE 3. Exergy and exergy destruction data for the gas turbine cogeneration system of Fig. 3.

Exergy Data:[1]

State	Substance	Mass Flow Rate (kg/s)	T (K)	p (bar)	$\dot{E}$ (MW)
1	Air	91.28	298.15	1.013	0.00
7	Combustion products	92.92	426.90	1.013	2.77
8	Water	14.00	298.15	20.000	0.06
9	Water	14.00	485.57	20.000	12.81
10	Methane	1.64	298.15	12.000	85.00

Exergy Destruction Data:

Component	MW	(%)[2]
Combustion chamber [3]	25.48	29.98
Heat recovery steam generator	6.23	7.33
Gas turbine	3.01	3.54
Air preheater	2.63	3.09
Gas compressor	2.12	2.49
Overall plant	39.47	46.43[4]

1. Determined relative to Table 1 (Model I).
2. Exergy destruction within the component as a percentage of the exergy entering the plant with the fuel.
3. Includes the exergy loss accompanying heat transfer from the combustion chamber, which is estimated as two percent of the lower heating value of the fuel.
4. An additional 3.27 percent of the fuel exergy is carried out of the plant as state 7 and is charged as a loss.

no net overall improvement. Moreover, measures to improve thermodynamic performance invariably have economic consequences. The objective in thermal system design normally is to identify the cost-optimal configuration, requiring consideration of both thermodynamic and economic imperatives, as discussed in Secs. 4.1 and 5.2.

4. Guidelines for Improving Thermodynamic Effectiveness

Various methods can be used to improve thermodynamic effectiveness. All such methods should achieve their objectives cost-effectively, of course. To improve thermodynamic effectiveness it is necessary to deal directly with inefficiencies related to exergy destruction and exergy loss. The following guidelines, called here *Sama's principles* [11,12], provide valuable advice for pursuing such an objective:

- There is nothing wrong with expending exergy if something useful is obtained in return.

- Some exergy destructions and exergy losses can be avoided, others cannot. Efforts should be centered on those that can be avoided.

4.1. OTHER GUIDELINES

The primary contributors to exergy destruction are chemical reaction, heat transfer, mixing, and friction, including unrestrained expansions of gases and liquids. To deal with them effectively, the principal sources of inefficiency not only should be understood qualitatively, but also determined quantitatively, at least approximately. Design changes to improve effectiveness must be done judiciously, however, for the cost associated with different sources of inefficiency can be different. For example, the unit cost of the electrical or mechanical power required to provide for the exergy destroyed owing to a pressure drop is generally higher than the unit cost of the fuel required for the exergy destruction caused by combustion or heat transfer.

Since chemical reaction is a significant source of thermodynamic inefficiency, it is generally good practice to minimize the use of combustion. In many applications the use of combustion equipment such as boilers is unavoidable, however. In these cases a significant reduction in the combustion irreversibility by conventional means simply cannot be expected, for the major part of the exergy destruction introduced by combustion is an inevitable consequence of incorporating such equipment. Still, the exergy destruction in practical combustion systems can be reduced by minimizing the use of excess air and by preheating the reactants as in Fig. 3. In most cases only a small part of the exergy destruction in a combustion chamber can be avoided by these means. Consequently, after considering such options for reducing the exergy destruction related to combustion, efforts to improve thermodynamic performance should focus on components of the overall system that are more amenable to betterment by cost-effective conventional measures. In other words, apply Sama's second principle.

Nonidealities associated with heat transfer also typically contribute heavily to inefficiency. Accordingly, unnecessary or cost-ineffective heat transfer must be avoided. Additional guidelines follow:

- The higher the temperature T at which a heat transfer occurs in cases where $T > T_0$, the more valuable the heat transfer and, consequently, the greater the need to avoid heat transfer to the ambient, to cooling water, or to a refrigerated stream.
- The lower the temperature T at which a heat transfer occurs in cases where $T < T_0$, the more valuable the heat transfer and, consequently, the greater the need to avoid direct heat transfer with the ambient or a heated stream.
- Exergy destruction associated with heat transfer between streams varies inversely with the temperature level. Accordingly, the lower the temperature level, the greater the need to minimize the stream-to-stream temperature difference.

Although irreversibilities related to friction, unrestrained expansion, and mixing are often secondary in importance to those of combustion and heat transfer, they should not be overlooked, and the following guidelines apply:

- Minimize the use of throttling; check whether power recovery expanders are a cost-effective alternative for pressure reduction.
- Avoid processes using excessively large thermodynamic driving forces (differences in temperature, pressure, and chemical composition). In particular, minimize the mixing of streams differing significantly in temperature, pressure, or chemical composition.
- The greater the mass rate of flow, the greater the need to use the exergy of the stream effectively.
- The lower the temperature level, the greater the need to minimize friction.

4.2. RELATION TO ENVIRONMENTAL IMPACT

For industries where energy resources are a major contributor to operating costs, an opportunity exists for cost savings by improving thermodynamic effectiveness using means such as discussed in Sec. 4.1. This is a well-known and largely accepted approach today. A related but less publicized aspect concerns effluent streams. The waste from a plant is often not an unavoidable result of plant operation but a measure of its inefficiency: The less efficient a plant, the more unusable by-products it produces, and conversely. Effluents not produced owing to greater efficiency require no costly cleanup and do not impose a burden on the environment.

Cleanup efforts have customarily featured an *end-of-the-pipe* approach that addresses the pollutants emitted from stacks, ash from incinerators, thermal pollution, and so on. Increasing attention is being given today to what goes *into the pipe,* however. This is embodied in the concept of *design for the environment* (DFE), in which the environmentally preferred aspects of a system are treated as design objectives rather than as constraints. The aim in DFE is to anticipate negative environmental impacts and engineer them out. In particular, efforts are directed to reducing the creation of waste and to managing materials better, using methods such as changing the process technology and/or plant operation, replacing input materials known to be sources of toxic waste with more benign materials, and doing more in-plant recycling. Exergy analysis, with its rational approach to costing and well-defined measures of true efficiency, is especially suited for use in DFE. Still, this is an aspect of exergy analysis that has lagged somewhat and deserves more development by the thermodynamics community.

A related area of application involving exergy that merits further development is in assessing environmental impact. An underlying idea is that the exergy of an effluent stream may serve as index of the stream's influence on nature. That is, the exergy of such a stream might be an indicator of its potential for driving damaging processes in the natural environment, Alternatively, the exergy of an effluent stream might be correlated to observed types of environmental damage. Though occasionally mentioned in the literature over the past two decades, exergy-aided assessment of environmental impact largely remains in its infancy. Recent efforts have suggested intriguing possibilities [13-16], but much remains to be done.

5. Thermal Systems Design

In this section, the life-cycle design process is considered together with the role of exergy analysis in design.

5.1. LIFE-CYCLE DESIGN PROCESS

Engineering design requires highly-structured critical thinking and active communication among the group of individuals, the design team, whose responsibility is the development of the design of a thermal system. The typical design project begins with an idea that something might be worth doing: a *primitive problem* proposed in recognition of a need or an economic opportunity. The *life-cycle* design process then involves several stages:

1. Understanding the problem: "What" not "How."

2. Concept development: "How" not "What."
3. Detailed design: Sizing, costing, optimization.
4. Project engineering: Equipment purchase, fabrication.
5. Service: Startup, operation, eventual retirement.

The influence of exergy analysis and exergy costing is felt the greatest at stages 2 and 3.

The primitive problem that engineers receive may be only a general statement of a need or opportunity. For example

> *Primitive problem:* To provide for a plant expansion, additional power and steam are required. Determine how the power and steam are best supplied.

A fundamental aspect of the design process makes it mandatory that knowledge about a project be generated as early as possible: At the outset, knowledge about the project and its potential solutions is at a minimum, but designers have the greatest freedom of choice because few decisions have been made. As the design process evolves, knowledge increases but design freedom decreases because earlier decisions tend to constrain what is allowable later. The early design effort should focus, therefore, on establishing an understanding of, and the need for, the project.

Accordingly, the first stage of the design process is to develop understanding. At this stage the object is to determine what qualities a system should possess and not how to achieve them. The question of how enters later. Specifically, the primitive problem must be defined, and the requirements to which the system must adhere must be determined. Not all requirements are equally important. The list of requirements normally can be divided into hard and soft constraints: musts and wants. The musts are essential and have to be met by the final design. The wants vary in importance and may be placed in rank order by the design team. Although the wants are not mandatory, some of the top-ranked ones may be highly desirable and useful in screening alternatives later. Each requirement should be expressed in measurable terms, for the lack of a measure often means that the requirement is not well understood. Moreover, only with numerical values can the design team evaluate the design as it evolves.

For the primitive problem above, let us assume for illustration purposes that the first stage of the design process results in the following requirements, though many others might apply: the power developed must be at least thirty megawatts. The steam must be saturated or slightly superheated at 20 bar and have a mass flow rate of at least fourteen kilograms per second. The system must adhere to all applicable environmental regulations. Subject to these, the total cost is to be minimized.

In the concept development stage, the design team shifts from the *what-to-do* phase to the *how-to-do-it* phase, and considers how the primitive problems can be solved. Experience plays an important role at this juncture, for successful solutions to similar problems can provide the basis for thinking about the current case. Artificial intelligence-based procedures such as recently reported [17] also may prove to be helpful.

The concept development stage can lead to a number, perhaps a great number, of plausible alternative solutions differing from one another in basic concept and detail. Because the methods of engineering analysis and optimization can only polish specific solutions, alternatives must be generated having sufficient quality to merit polishing.

Major design errors more often occur owing to a flawed alternative being selected than to subsequent shortcomings in analysis and optimization.

Alternative solution concepts for the current problem would include, but not necessarily be limited to the following: Produce all of the steam required, and

- generate the full electricity requirement, or
- generate a portion of the electricity requirement and purchase the balance of the electricity needed from a utility, or
- generate more than the electricity requirement, and sell the excess electricity to the utility.

Each cogeneration alternative might be configured using a steam turbine system, a gas turbine system, or a combined steam and gas turbine system, and the type of fuel (coal, natural gas, oil) introduced still more flexibility. Alternative concepts such as these must be screened to eliminate all but the most worthwhile.

5.2. EXERGY-AIDED DESIGN

After preliminary screening of the above alternative solution concepts, let us suppose for illustration purposes that the following have been retained for further screening and evaluation:

- Employ a coal-fired steam turbine cogeneration system.
- Employ a gas turbine cogeneration system using a natural gas as the fuel.
- Employ a combined steam and gas turbine cogeneration system using natural gas as the fuel.

These three alternatives, which are shown schematically in Fig. 1.6 of [2], would be subject to further screening and evaluation until a preferred design emerges. Exergy methods can be especially effective at this point of the design process. Exergy methods may enter the design process on a qualitative basis through design guidelines, as provided in Sec. 4, and on a quantitative basis through application of *thermoeconomics*.

Thermoeconomics combines thermodynamic and economic principles for engineering decision making. For instance, thermoeconomics can be applied to optimize an entire system or specific variables in a single component. Costing in thermoeconomics is based on exergy, which in most cases is the only rational basis for assigning costs to the exergy streams of a thermal system. Through thermoeconomic analysis, a system's real cost sources can be identified. Such information is useful in developing design changes that improve overall cost effectiveness.

When thermoeconomic principles are applied, the gas turbine cogeneration system emerges as the preferred alternative. In keeping with the design guidelines of Sec. 4, a thermodynamically improved version of this cogeneration configuration can be achieved by including an air preheater. The result is the case-study gas turbine cogeneration system of Fig. 3 and Table 3. This configuration provides the *starting point* for a thermoeconomic optimization.

A *thermodynamic* optimization aims at maximizing the exergetic efficiency: minimizing the exergy destruction and exergy loss. The objective of a *thermoeconomic* optimization, however, is to minimize costs, including costs related to exergy destruction and loss. A thermodynamically-optimal design is generally much different than a cost-optimal design. In particular, the total cost associated with the thermodynamically-optimal design is always higher than for the cost-optimal design. It is also noteworthy that the optimization of a thermal system seldom leads to a unique solution corresponding to a global mathematical optimum. Rather, acceptable alternative solutions may be feasible. Different solutions developed by different design teams may be equally acceptable and nearly equally cost effective. As a rule, the more complex the system being optimized, the larger the number of acceptable solutions.

Although the details are beyond the scope of the current presentation, the discussion of the cogeneration system case study initiated with the above primitive problem is brought to closure in [2] with the presentation of a cost-optimal design (see Sec. 9.5 of [2]). Referring again to life-cycle design, such a result provides a solid basis for completing the detailed design stage. Though important from an engineering perspective, the remaining aspects of the life-cycle design process are not detailed here for brevity, however.

6. Closure

Over the past two centuries, and especially in the final quarter of this century, the use of the second law for the analysis and design of thermal systems has evolved to a state of relative maturity. Today, engineers have powerful second-law methods at their disposal. Although such methods can be significant aids, it is worth observing that the successful completion of a thermal design project is a complex undertaking, requiring principles from many disciplines: engineering thermodynamics, fluid mechanics, heat and mass transfer, mechanical design, automatic controls, mathematics, and economics, to name just a few. Experience and intuition also play roles, and often a bit of luck is helpful.

7. References

1. Moran, M.J. (1998) Engineering Thermodynamics (Sec. 2) in F. Kreith (ed.), *Mechanical Engineering Handbook*, CRC Press, Boca Raton.
2. Bejan, A., Tsatsaronis, G. and Moran, M. (1996), *Thermal Design & Optimization*, John Wiley & Sons, New York.
3. Moran, M.J. (1998) On Second-Law Analysis and the Failed Promise of Finite-time Thermodynamics, *Energy Int. J.*, **23**, 517-519.
4. Bird, R.B., Stewart, W.E. and Lightfoot, E.N. (1960) *Transport Phenomena*, John Wiley & Sons, New York.
5. Moran, M.J. and Shapiro, H.N. (1996) *IT: Interactive Thermodynamics*, Software to accompany *Fundamentals of Engineering Thermodynamics*, 3rd ed., John Wiley & Sons, New York.
6. Moran, M.J. (1989) *Availability Analysis - A Guide to Efficient Energy Use*, ASME Press, New York.
7. Moran, M.J. and Shapiro, H.N. (1996) *Fundamentals of Engineering Thermodynamics*, 3rd ed., John Wiley & Sons, New York.
8. Kotas, T.J. (1995) *The Exergy Method of Thermal Plant Analysis* Krieger, Melbourne, FL.
9. Szargut, J., Morris, D.R., and Steward, F.R. (1988) *Exergy Analysis of Thermal, Chemical and Metallurgical Processes*, Hemisphere, New York.
10. Ahrendts, J. (1980) Reference states, *Energy Int. J.*, **5**, 667-677.

11. Sama, D.A. (1998) Department of Chemical Engineering, University of Massachusetts Lowell, Lowell, MA 01854 USA. Personal Communication.
12. Sama, D.A. (1995) The Use of Second Law of Thermodynamics in Process Design, *ASME Journal of Energy Resources Technology*, **117**, 179-185.
13. Rosen, M.A., and Dincer, I. (1997) On Exergy and Environmental Impact, *Int. J. Energy Research*, **21**, 643-654.
14. Creyts, J.C. and Carey, V.P. (1997) Use of Extended Exergy Analysis as a Tool for Assessment of the Environmental Impact of Industrial Processes, in *Proceedings of the ASME Advanced Energy Systems Division*, **AES-Vol.37**. ASME, New York, 129-137.
15. Ayres., R.U., Ayers, L.W., and Martinas, K. (1998) Exergy, Waste Accounting, and Life-Cycle Analysis, *Energy Int. J.*, **23**, 355-363.
16. Markarytchev, S.V. (1998) Environmental Impact Analysis of ACFB-Based Gas and Power Cogeneration, *Energy Int. J.* (to appear)
17. Sciubba, E. (1998) Toward Automatic Process Simulators, *ASME Journal of Engineering for Gas Turbines and Power*, **120**, 1-16.

STRENGTHS AND LIMITATIONS OF EXERGY ANALYSIS

G. TSATSARONIS
Institute for Energy Engineering
Technical University of Berlin
Marchstr. 18, 10587 Berlin, Germany

1. Introduction

An exergy analysis identifies the location, the magnitude and the sources of thermodynamic inefficiencies in a thermal system. This information, which cannot be provided by other means (e.g., an energy analysis), is very useful for improving the overall efficiency and cost effectiveness of a system or for comparing the performance of two systems. This paper deals with the use of exergetic variables, i.e., those variables calculated in an exergy analysis to characterize each component and some streams of the system. Their critical review will identify the strengths and the limitations of an exergy analysis. The following discussion is limited to systems at steady-state operation. The results can easily be extended to other systems.

The following variables are calculated in an exergy analysis for the *k*th component of a system:

- Exergy destruction rate

$$\dot{E}_{D,k} = T_0 \dot{S}_{gen,k} \quad (1)$$

- Exergy loss rate $\dot{E}_{L,k}$
- Exergy destruction ratio

$$y_{D,k} = \frac{\dot{E}_{D,k}}{\dot{E}_{F,tot}} \quad (2)$$

- Exergy loss ratio

$$y_{L,k} = \frac{\dot{E}_{L,k}}{\dot{E}_{F,tot}} \quad (3)$$

- Percentage of exergy destruction

$$y^*_{D,k} = \frac{\dot{E}_{D,k}}{\sum_l \dot{E}_{D,l}} \quad (4)$$

- exergetic efficiency

$$\varepsilon_k = \frac{\dot{E}_{P,k}}{\dot{E}_{F,k}} = 1 - \frac{\dot{E}_{D,k} + \dot{E}_{L,k}}{\dot{E}_{F,k}} \quad (5)$$

A. Bejan and E. Mamut (eds.), Thermodynamic Optimization of Complex Energy Systems, 93–100.

Many exergy analyses reported in the literature calculate the values of some of the above variables and reach some usually reasonable conclusions regarding system improvement but without checking whether the proposed changes would lead indeed to an improvement. In addition, little attention has been given in many exergy analyses to the structure of the system and the mutual interdependencies among its components. The reader of such articles could ask the well justified question "And now that I know how much exergy is destroyed in each component of a system what do I do?" The question remains often unanswered, and some critics, not always in good faith, use this situation to discredit the concept of exergy analysis. The purpose of this article is to shed some light on some aspects associated with this analysis. In the following, the strengths and limitations of exergy analysis applied to evaluate, optimize or develop a thermal system are briefly discussed.

2. Evaluation

The following three questions need to be answered when reviewing the exergetic variables used to evaluate the thermodynamic performance of system components:

1. Which variable best characterizes the performance of a component from the thermodynamic viewpoint?
2. Which variable should be used to compare the performance of similar components in the same system or in different systems?
3. Which variable should be used to compare the performance of dissimilar components?

To facilitate the following discussion, let us consider the system shown in Fig. 1, which consists of the components A and B. The fuel of component A ($\dot{E}_{F,A}$) is equal to the fuel of the total system ($\dot{E}_{F,tot}$). The product of A ($\dot{E}_{P,A}$) is the fuel of component B ($\dot{E}_{F,B}$), whereas the product of B ($\dot{E}_{P,B}$) is also the product of the overall system ($\dot{E}_{P,tot}$) and is kept constant. To further simplify the presentation, we also assume that there are no exergy losses:

$$\dot{E}_{L,A} = \dot{E}_{L,B} = 0 \tag{6}$$

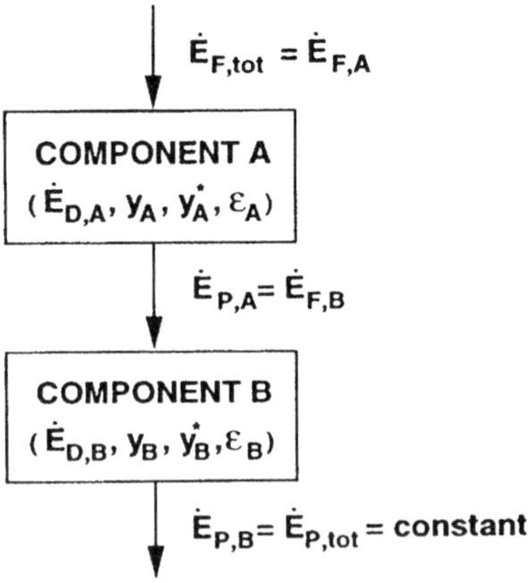

Figure 1. System in which the product of one component is the fuel of the next component.

Thus, all thermodynamic inefficiencies are caused by the exergy destruction within components A and B. Then, by applying the exergy balance

$$\dot{E}_{F,k} = \dot{E}_{P,k} + \dot{E}_{D,k} + \dot{E}_{L,k} \tag{7}$$

and Eqs (2), (4) and (5) to components A and B we obtain the following relations

$$\dot{E}_{D,A} = \frac{\dot{E}_{P,tot}}{\varepsilon_B}\left(\frac{1}{\varepsilon_A} - 1\right), \quad \dot{E}_{D,B} = \dot{E}_{P,tot}\left(\frac{1}{\varepsilon_B} - 1\right) \tag{8, 9}$$

$$y_{D,A} = 1 - \varepsilon_A, \quad y_{D,B} = \varepsilon_A\,(1 - \varepsilon_B) \tag{10, 11}$$

$$y^{*}_{D,A} = \frac{1 - \varepsilon_A}{1 - \varepsilon_A \varepsilon_B}, \quad y^{*}_{D,B} = \frac{(1 - \varepsilon_B)\varepsilon_A}{1 - \varepsilon_A \varepsilon_B} \tag{12, 13}$$

Equation (8) demonstrates that the rate of exergy destruction in component A depends not only on the efficiency of the same component (ε_A), but also on the exergetic efficiency of component B (ε_B). Thus, the rates of exergy destruction should be used very cautiously to characterize the performance of system components because, in general, a part of the exergy destruction occuring in a component is caused by the inefficiencies of the remaining system components (exogenous exergy destruction). The total exergy destruction within a component is the sum of the *exogenous* exergy destruction and the *endogenous* exergy destruction, i.e. the exergy destruction due exclusively to the component being considered assuming that all remaining components operate with exergetic efficiencies of 100%. In complex thermal systems it is very difficult and costly to separate these two parts of the exergy destruction within a system component. Only in the component where $\dot{E}_{P,tot}$ is generated is the exogenous exergy destruction zero (see Eq. (9) for component B of the system shown in Fig.1).

Similarly Eq. (11) shows that the exergy destruction ratio of component B depends on the exergetic efficiencies of both components A and B. Here only that component where $\dot{E}_{F,tot}$ is supplied to the entire system has a y_D value which is independent of the performance of the remaining components (see Eq. (10) for the component A). The cautiousness to be associated with the use of y_D is not reduced if the exergy destruction in the *k*th component is related to the product (instead of the fuel) of the overall system as the following equations demonstrate for the exergy destruction ratios for the components of the system shown in Fig. 1

$$y'_{D,k} = \dot{E}_{D,k} / \dot{E}_{P,tot} \tag{14}$$

$$y'_{D,A} = \frac{1}{\varepsilon_B}\left(\frac{1}{\varepsilon_A} - 1\right), \quad y'_{D,B} = \frac{1}{\varepsilon_B} - 1 \tag{15, 16}$$

The variable $y'_{D,k}$ provides a clear characterization only of the performance of the component in which $\dot{E}_{P,tot}$ is generated (see Eq. (16) for the component B). Neither does variable $y^{*}_{D,k}$ (Eq. (4)) provide an unambiguous characterization of the performance of the *k*th component as Eqs. (12) and (13) indicate for the system of Fig. 1.

The problems associated with the use of the variables $\dot{E}_{D,k}$, $y_{D,k}$, $y^{/}_{D,k}$ and $y^{*}_{D,k}$ during a thermodynamic evaluation of the kth component stem from the fact that these variables are affected by the structure of the system. In an attempt to consider these structural effects, Beyer [2-3] suggested the use of structural coefficients. The calculation of these coefficients requires the use of simulation software, and it is time consuming. These coefficients have not yet found appropriate attention in practical applications. Some structural effects are considered in the application of exergoeconomic techniques to the diagnosis of the operation of thermal systems [1, 12, 13].

To answer the questions formulated in the beginning of this section we can conclude that the only variable that unambiguously characterizes the performance of a component from the thermodynamic viewpoint is an appropriately defined exergetic efficiency. Details about the appropriate definition of exergetic efficiencies can be found in [4-6]. The exergetic efficiency should also be used to compare the performance of similar components operating under similar conditions in the same system or in different systems. For the comparison of dissimilar components the only variable that may be used (with the previously mentioned caveat in mind) is the exergy destruction ratio $y_{D,k}$ (or $y^{/}_{D,k}$ if $\dot{E}_{F,tot}$ remains constant).

In evaluating the thermodynamic performance of thermal systems it is always useful to know what part of the exergy destruction in each component could be avoided in a real process. This distinction between avoidable ($\dot{E}^{AV}_{D,k}$) and unavoidable ($\dot{E}^{UN}_{D,k}$) exergy destruction [7], although associated with many subjective decisions, has the potential to further facilitate an exergy-based evaluation and optimization procedure. By using a modified exergetic efficiency ε^{*}_{k} defined through

$$\varepsilon^{*}_{k} = \frac{\dot{E}_{P,k}}{\dot{E}_{F,k} - \dot{E}^{UN}_{D,k}} = 1 - \frac{\dot{E}^{AV}_{D,k} + \dot{E}_{L,k}}{\dot{E}_{F,k} - \dot{E}^{UN}_{D,k}} \tag{17}$$

with

$$\dot{E}_{D,k} = \dot{E}^{UN}_{D,k} + \dot{E}^{AV}_{D,k} \tag{18}$$

we are able to make a more realistic assessment of the potential for improving the kth component from the thermodynamic viewpoint. In addition, this modified exergetic efficiency enables the comparison of dissimilar components with respect to this potential.

However, the major contribution of an exergy analysis to the evaluation of a system comes through an exergoeconomic evaluation [1, 8] that considers not only the inefficiencies but also the costs associated with these inefficiencies and the investment expenditures required to reduce the inefficiencies.

3. Optimization

Before we discuss the use of exergetic variables for optimization purposes we should clearly state that a procedure with the final objective to maximize the thermodynamic efficiency in the *design of a new system* has no practical value and should be considered only in conjunction with other objectives such as, for example, the minimization of costs and pollutant emissions.

There is no structured procedure available for the iterative improvement ("optimization") of the efficiency of a thermal system. Only several guidelines and heuristic rules (e.g., [1, 9]) may be cautiously applied. However, they do not always lead to an improvement. In addition, a minimum experience and intuition are usually required to avoid misinterpretations or wrong applications of the rules and guidelines.

Usually in studies reporting the results of exergy analyses it is suggested to start the iterative process for improving the overall efficiency by improving the component with the highest destruction rate. It is apparent that this suggestion does not always lead to a thermodynamic improvement. For example, in the cogeneration system considered throughout Reference [1], the combustion chamber is the component with the highest exergy destruction rate with a y_D value of more than 30%. In the first workable design (the "base design") the exergy destruction rate is 25.75 MW and the exergetic efficiency is 79.8%, whereas in the thermodynamically optimal design these values are 23.81 MW and 78.8%, respectively (see Table 9.2 in [1]). Thus, if we started with the base design and just improved the efficiency of the combustion chamber, we would probably not improve the overall efficiency. An iterative approach in which the components are ranked according to their *avoidable* exergy destruction might provide an improvement of this iterative optimization procedure.

Here again the full potential of conducting an exergy analysis for optimization purposes is used when the exergy analysis is applied as the first step in an iterative exergoeconomic optimization procedure [1, 8] whose objective is to minimize the costs of generating the products of a system.

At this point a few words need to be said about two other related concepts that are also applied for optimization: the pinch analysis and the finite-time thermodynamics. A pinch analysis by providing an easy to use formal approach can be effectively applied for the evaluation and thermodynamic optimization of a *heat-exchanger network* and its integration with other components. For these applications a pinch analysis could nicely *complement* an exergy analysis. The elements of exergy analysis that have been incorporated into an extended so-called "pinch analysis" applied to *any* thermal system simply provide an additional proof of the superiority of exergy analysis in comparison with a "pinch analysis" when dealing with systems other than heat exchanger networks and/or with cost minimization. As far as the finite-time thermodynamics concepts are concerned, it is sufficient to say that their value for the optimization of real-world systems is practically zero.

4. Process Development

This area of exergy analysis applications has not yet received the attention it deserves. The decision-making process during the preliminary design phase may be significantly simplified if the exergetic efficiencies of components, processes, and/or subsystems are known. In this way the overall exergetic efficiency of various concepts can be roughly estimated, thus enabling a very quick indentification of the concepts that are expected to have efficiencies higher than a given minimum. This feature is particularly useful in the development stage of complex thermal systems.

In the following I would like to illustrate how exergy-based estimates assisted or perhaps guided me in developing a novel concept. At the end of 1981, I was evaluating various concepts for reducing the exergy destruction in combustion processes by avoiding the direct contact of oxygen with the fuel [10]. In this case, an oxygen carrier (e.g., CuO) reacts with the fuel, thus, providing the oxygen required for the combustion (e.g., CuO is reduced to Cu). The intermediate chemical component obtained in the previous reaction (e.g., Cu) is subsequently oxidized with air to provide the oxygen carrier (e.g., CuO).

One of the processes I was studying at that time was the following one in which for simplicity carbon (instead of coal) is used as fuel:

$$C + 2H_2SO_4 \rightarrow CO_2 + 2SO_2 + 2H_2O \qquad \text{(R1)}$$

$$2SO_2 + O_2 + 3.76N_2 \rightarrow 2SO_3 + 3.76N_2 \qquad \text{(R2)}$$

$$2SO_3 + 2H_2O \rightarrow 2H_2SO_4 \qquad \text{(R3)}$$

The net result obtained through the combination of reaction steps R1-R3 is the stoichiometric combustion of carbon with air:

$$C + O_2 + 3.76N_2 \rightarrow CO_2 + 3.76N_2 \qquad \text{(R4)}$$

The exergetic evaluation of the concept defined by reaction steps R1 through R3 showed that the exergy destruction in the process using reaction R1 is much smaller than the exergy destruction in the processes associated with reactions R2 and R3.

One day I attended a seminar presentation by a doctoral student who was studying the following hydrogen generation process, the so-called "Westinghouse" process:

$$2H_2O + SO_2 \rightarrow H_2 + H_2SO_4 \qquad \text{(R5)}$$

$$H_2SO_4 \rightarrow H_2O + SO_2 + \frac{1}{2}O_2 \qquad \text{(R6)}$$

Reaction R5 is conducted in an electrolytic cell. The net result of reactions R5 and R6 is the splitting of water through the supply of high-temperature (1150 K) thermal energy and small amounts of electric power:

$$H_2O \rightarrow H_2 + \frac{1}{2}O_2 \qquad \text{(R7)}$$

In this seminar presentation it was mentioned that the exergy destruction in the electrolytic step R5 is considerably lower than in step R6. Immediately after hearing that, I came up with the idea of combining the reaction steps from each of the above processes with the lowest exergy destruction, that is steps R1 and R5, into a new process:

$$C + 2H_2SO_4 \rightarrow CO_2 + 2SO_2 + 2H_2O \qquad \text{(R1)}$$

$$4H_2O + 2SO_2 \rightarrow 2H_2 + 2H_2SO_4 \qquad \text{(R5)}$$

net result:

$$C + 2H_2O \rightarrow CO_2 + 2H_2 \qquad \text{(R8)}$$

Before the presentation was over, by combining elements from the two original flow diagrams, I had already developed a rough flow diagram of what later became the HECAP process (Hydrogen and Electric Power generation using CArbonaceous materials and high-temperature Process heat), Fig. 2, [11]. The only thing I could be sure of at that moment was that the new concept, if it could be realized, should have a high overall thermodynamic efficiency. Subsequent evaluations confirmed the high exergetic efficiency (over 70%) and some significant environmental advantages of the new concept: no NO_X formation, flue gas consisting almost exclusively of CO_2, and no sulfur emissions to the environment regardless of the amount of sulfur contained in the carbonaceous fuel (e.g., coal). Later also the main difficulties associated with the realization of this concept became apparent: the lack of (a) gas turbines in which the gas products of reaction R1 could be expanded, (b) gas cleaning devices for the same gas mixture before expansion, and (c) an appropriate source for the high-temperature thermal energy that according to the plans at that time should be provided by a high-temperature nuclear reactor. The above case is a good example of how even qualitative information from an exergy analysis can assist an engineer in inventing a new process or concept. An energy analysis cannot provide a similar enhancement of creativity. Without the information about the relative exergy destruction there would be no incentive for combining steps R1 and R5 to develop a more complex process which uses two fuels (e.g., coal and high-temperature thermal energy) and generates two products (electricity and hydrogen).

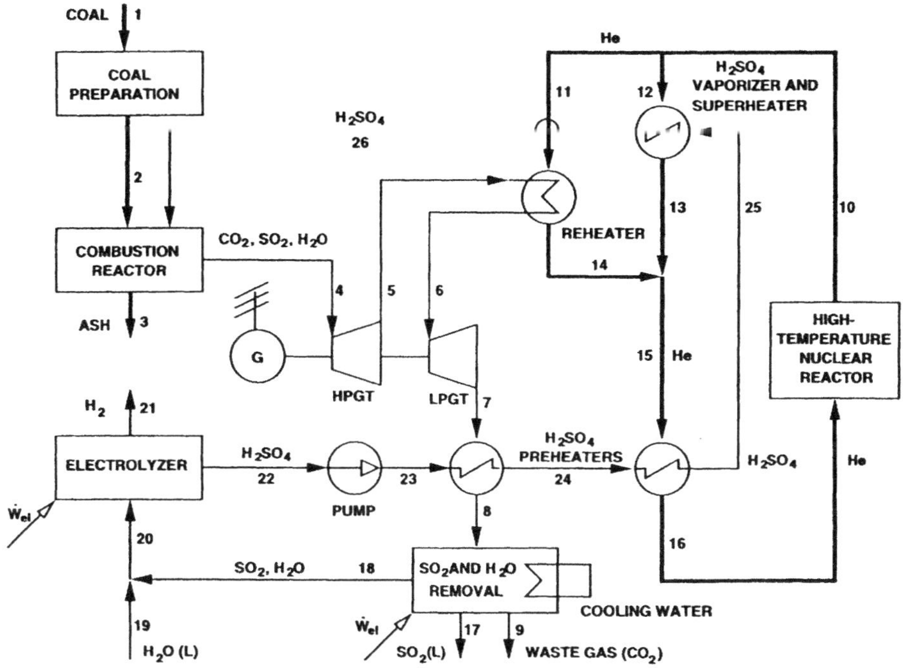

Figure 2. Flow diagram of the HECAP process.

5. Conclusions

Exergy analysis is without a doubt a powerful tool for developing, evaluating and improving thermal systems, particularly when this analysis is part of an exergoeconomic evaluation. However, the lack of a formal procedure in using the results from an exergy analysis is one of the major reasons for it not being very popular among practitioners. This represents simultaneously a weakness and a strength of the method. A strength because an exergy analysis can be conducted for *every* thermal system and because the engineers control the evaluation and optimization process and, thus, they could still use their intuition and creativity as opposed to a formal optimization procedure in which they do not know how and why specific optimization results were obtained. A specific method (e.g., pinch analysis) developed in a formal way for a specific system (e.g., heat-exchanger networks) always has the potential of gaining a wider acceptance among practitioners than the more general method of exergy analysis.

Future studies using exergetic and exergoeconomic variables need to focus more on the mutual interdependencies among the components of a thermal system, and on the calculation of the unavoidable exergy destruction for each component of a system. Today it is not sufficient any more to just present the results of an exergy analysis without using them to improve the system being considered.

Exergy analysis also has a high pedagogical value because it shows to the students what the real thermodynamic inefficiencies are and what causes them. The results from the analysis assist the students in developing ideas with respect to what can be done to reduce these inefficiencies. Teaching the concept of exergy analysis should focus on that strength and on the *appropriate* calculation, interpretation and use of the exergetic variables.

References

1. Bejan, A., Tsatsaronis, G., and Moran, M. (1996) *Thermal Design and Optimization*, J. Wiley, New York.
2. Beyer, J. (1970) Strukturuntersuchungen - notwendiger Bestandteil der Effektivitätsanalyse von Wärmeverbrauchersystemen *Energieanwendung* ***19****, 358-361.*
3. Beyer, J. (1974) Struktur wärmetechnischer Systeme und ökonomische Optimierung der Systemparameter *Energieanwendung* **23**, 274-279.
4. Tsatsaronis, G. (1995) On the efficiency of energy systems in Y.A. Göğüş, A. Öztürk and G. Tsatsaronis (eds.) *Proceedings of the International Conference on Efficiency, Costs, Optimization, Simulation and Environmental Impact of Energy Systems, ECOS '95*, Istanbul, Turkey, July 11-14, 53-60.
5. Lazzaretto, A. and Tsatsaronis, G. (1996) A general process-based methology for exergy costing in A.B. Duncan, J. Fiszdon, D. O'Neal and K. Den Braven (eds.) *Proceedings of the ASME Advanced Energy Systems Devision*, **36**, ASME, New York, 413-428.
6. Lazzaretto, A. and Tsatsaronis, G. (1997) On the quest for objective equations in exergy costing in M.L. Ramalingam, J.G. Lage, V.C. Mei, and J.N. Chapman (eds.), *Proceedings of the ASME Advanced Energy Systems Devision*, **37**, ASME, New York, 197-210.
7. Feng, X., Zhu, X.X. and Zheng, J.P. (1996) A practical exergy method for systems analysis, *Proceedings of the 1996 Intersociety Energy Converson Engineering Conference*, IEEE, New York, 2068-2071.
8. Tsatsaronis, G. (1998) Design optimization using exergoeconomics, Presentation at the NATO Advanced Study Institute *Thermodynamics and the Optimization of Complex Energy Systems*, July 13-24, 1998, Neptun, Constanza, Romania.
9. Szargut, J. and Sama, D.A. (1995) Practical rules of the reduction of energy losses caused by the thermodynamic imperfection of thermal processes, *Proceedings of ITEC '95, the Second International Thermal Energy Congress*, Agadir, Marocco, June 1995, 782-785.
10. Knoche, K.F. and Richter, H. (1968) Verbesserung der Reversibilität von Verbrennungsprozessen, *Brennstoff-Wärme-Kraft* **20**, 205-210.
11. Tsatsaronis, G. and Spindler, K. (1984) Thermodynamic analysis of a new process for hydrogen and electric power production by using carbonaceous fuels and high-temperature process heat, *Int. J. Hydrogen Energy* **9**, 595-601.
12. Lozano, M.A. and Valero, A. (1993) Theory of the exergetic cost, *Energy - The International Journal* **18**, 939-960.
13. Stoppato, A. and Lazzaretto, A. (1996) The exergetic analysis for energy systems diagnosis, *Proceedings of the Engineering Systems Design and Analysis (ESDA) Conference* PD-Vol. 73, **1**, ASME, New York, 191-198.

DESIGN OPTIMIZATION USING EXERGOECONOMICS

G. TSATSARONIS
Institute for Energy Engineering
Technical University of Berlin
Marchstr. 18, 10587 Berlin, Germany

1. Introduction

Exergoeconomics, this common branch of mechanical and chemical engineering, represents a unique combination of an exergy analysis and a cost analysis, to provide the designer or the operator of an energy-conversion plant with information not available through conventional energy, exergy or cost analyses but crucial to the design and operation of a cost-effective plant. Exergoeconomics may be defined as an exergy-aided cost-reduction method. In addition, exergoeconomics is a very powerful tool for understanding the interconnections between thermodynamics and economics, and, thus, the behavior of an energy conversion plant from the cost viewpoint [1].

Design optimization of a thermal system means the modification of the structure and the design parameters of a system to minimize the total levelized cost of the system products under boundary conditions associated with available materials, financial resources, protection of the environment, and government regulation, together with the safety, reliability, operability, availability and maintainability of the system. A truly optimized system is one for which the magnitude of every significant thermodynamic inefficiency (exergy destruction and exergy loss) is justified by considerations related to costs or is imposed by at least one of the above boundary conditions. A thermodynamic optimization, which aims at minimizing the thermodynamic inefficiencies, may be considered as a subcase of design optimization.

Appropriate formulation of the optimization problem is usually the most important and sometimes the most difficult step of a successful optimization study. In optimization problems we distinguish among independent variables, whose values are amenable to change; these are the decision variables and the parameters whose values are practically fixed by the particular application. In optimization studies, only the decision variables may be varied. The parameters are independent variables that are each given one specific and unchanging value in any particular model statement. The variables whose values are calculated from the independent variables using a mathematical model are the dependent variables.

We will initially consider an almost ideal case: A complete thermodynamic model and a complete economic model are available for the optimization of an energy system. In addition, the structure of the system might be considered as optimal, that is, either there are no alternative structures or each alternative structure is considered to be inferior to the present structure. Finally, we assume that an analytical or a numerical optimization technique (e.g., [2-4]) may be used to directly optimize the decision variables. Only in this ideal case, the calculation of exergy-based variables is unnecessary since the optimization can be conducted directly using the available models and techniques. In practical applications, however, thermal systems cannot usually be optimized as in this ideal case. The reasons include the following:

- Some of the input data and functions required for the thermodynamic and, particularly, the economic model might not be available or might not be in the required form. For

A. Bejan and E. Mamut (eds.), Thermodynamic Optimization of Complex Energy Systems, 101–115.

example, it is not always possible to express the purchased-equipment costs as a function of the appropriate thermodynamic decision variables.
- The analytical and numerical optimization techniques are applied to a specified structure of the thermal system. However, a significant decrease in the product costs may be achieved through changes in the structure of the system. It is not always practical to develop a mathematical optimization model for every promising design configuration of a system.
- Even if all the required information is available, the complexity of the system might not allow a satisfactory mathematical model to be formulated and solved in a reasonable time.

In these cases, that is, when any of the assumptions made for the ideal case is not fulfilled, the application of exergoeconomic techniques may provide significant benefits for the optimization process. The more complex the thermal system, the larger the expected benefits, particularly when chemical reactions are involved.

Exergoeconomics interacts with several other areas during the optimization procedure. Fig. 1 shows schematically some of these interactions: Exergoeconomics uses results from the synthesis, cost analysis and simulation of thermal systems and provides useful information for the evaluation and optimization of these systems as well as for the application of expert systems to improve the design and operation of such systems.

2. A General Exergoeconomic Methodology

This section presents the main features of a general methodology that can be used to evaluate and iteratively optimize the design of a thermal system. Sections 2.1 through 2.3 discuss the procedure for calculating the costs associated with all material and energy streams in the thermal system being considered. This information is used to calculate the characteristic variables of each system component that are discussed in section 2.4.

2.1. EXERGY COSTING

For a system operating at steady state there may be a number of entering and exiting streams as well as both heat and work interactions with the surroundings. Associated with these transfers of matter and energy are exergy transfers into and out of the system and exergy destructions caused by the irreversibilities within the system. Since exergy measures the true thermodynamic value of such effects, and costs should only be assigned to commodities of value, it is meaningful to use exergy as a basis for assigning costs in thermal systems. Indeed, exergoeconomics rests on the notion that exergy is the only rational basis for assigning costs to the interactions a thermal system experiences with its surroundings and to their sources of inefficiencies within it. We refer to this approach as *exergy costing*. In exergy costing a cost rate is associated with each exergy transfer. Thus, for entering and exiting streams of matter with associated exergy transfers $\dot{E}_i$ and $\dot{E}_e$, power $\dot{W}$, and the exergy tranfer associated with heat transfer $\dot{E}_q$ we write, respectively

$$\dot{C}_i = c_i \dot{E}_i, \quad \dot{C}_e = c_e \dot{E}_e \tag{1, 2}$$

$$\dot{C}_w = c_w \dot{W}, \quad \dot{C}_q = c_q \dot{E}_q \tag{3, 4}$$

Here c_i, c_e, c_w, and c_q denote average costs per unit of exergy, for example in \$ per gigajoule (\$/GJ).

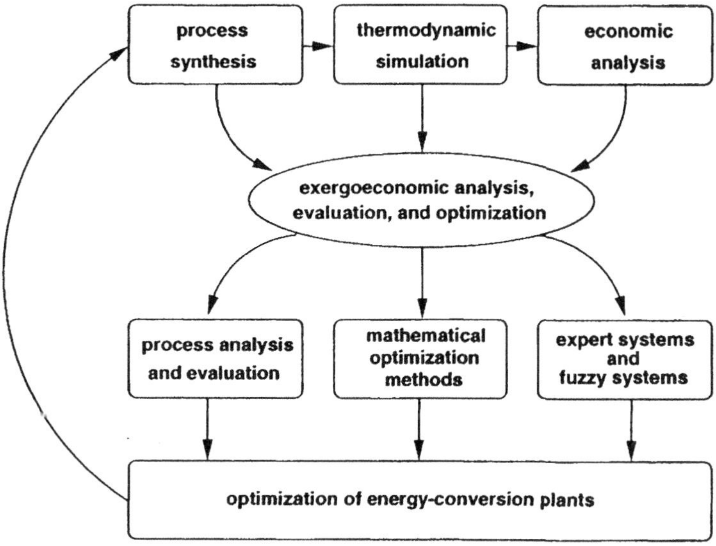

Figure 1. Interactions of exergoeconomics with other areas of engineering and optimization procedure.

Exergy costing does not necessarily imply that costs associated with streams of matter are related *only* to the exergy rate of each respective stream. Non-exergy related costs can also affect the total cost rate associated with material streams. Examples include the cost rates associated with a treated water stream at the outlet of a water treatment unit, an oxygen or nitrogen stream at the outlet of an air separation unit, a limestone stream supplied to a gasifier or fluidized-bed reactor, iron feedstock supplied to a metallurgical process, and an inorganic chemical fed into a chemical reactor.

Accordingly, when significant non-exergy-related costs occur, the total cost rate associated with the material stream *j*, denoted by $\dot{C}_j^{\mathrm{TOT}}$, is given by

$$\dot{C}_j^{\mathrm{TOT}} = \dot{C}_j + \dot{C}_j^{\mathrm{NE}} \tag{5}$$

where $\dot{C}_j$ is the cost rate directly related to the exergy of stream j (e.g., Eqs. (1) and (2)) whereas $\dot{C}_j^{\mathrm{NE}}$ is the cost rate due to nonexergetic effects. The term $\dot{C}_j^{\mathrm{NE}}$ represents a convenient way for charging non-exergy-related costs from one component to other components that should bear such costs. More details about $\dot{C}^{\mathrm{NE}}$ are given in [1] and [5].

2.2. COST BALANCE

Exergy costing involves cost balances usually formulated for each component separately. A cost balance applied to the *k*th system component shows that the sum of cost rates associated with all exiting exergy transfers equals the sum of cost rates of all entering exergy transfers plus the appropriate charges due to capital investment ($\dot{Z}_k^{\mathrm{CI}}$) and operating and maintenance expenses ($\dot{Z}_k^{\mathrm{OM}}$). The sum of the last two terms is denoted by $\dot{Z}_k$. Accordingly, for a component receiving a heat transfer and generating power, we write

$$\sum_e \dot{C}_{e,k} + \dot{C}_{w,k} = \dot{C}_{q,k} + \sum_i \dot{C}_{i,k} + \dot{Z}_k \tag{6}$$

This equation simply states that the total cost of the exiting exergy transfers equals the total expenditure to obtain them: the cost of the entering exergy streams plus the capital and other costs. Note that when a component receives power (as in a compressor or a pump) the term $\dot{C}_{w,k}$ would move with its positive sign to the right side of this expression. The term $\dot{C}_{q,k}$ would appear with its positive sign on the left side if there is a heat transfer *from* the component. Cost balances are generally written so that all terms are positive.

Introducing the cost rate expressions of Eqs. (1)-(4), Eq. (6) becomes

$$\sum_e \left(c_e \dot{E}_e\right)_k + c_{w,k} \dot{W}_k = c_{q,k} \dot{E}_{q,k} + \sum_i \left(c_i \dot{E}_i\right)_k + \dot{Z}_k \tag{7}$$

The exergy rates $\left(\dot{E}_e,\ \dot{W},\ \dot{E}_q, \text{ and } \dot{E}_i\right)$ exiting and entering the kth component are calculated in an exergy analysis [1] conducted at a previous stage. The term $\dot{Z}_k$ is obtained by first calculating the capital investment and O&M costs associated with the kth component and then computing the levelized values of these costs per unit of time (year, hour, or second) of system operation [1, 5-8].

The variables in Eq. (7) are the levelized costs [1] per unit of exergy for the exergy transfers associated with the kth component $\left(c_{e,k},\ c_{w,k},\ c_{q,k}, \text{ and } c_{i,k}\right)$. In analyzing a component, we may assume that the costs per exergy unit are known for all entering streams. These costs are known from the components they exit or, if a stream enters the overall system consisting of all components under consideration, from the purchase cost of this stream. Consequently, the unknown variables that need to be calculated with the aid of the cost balance for the kth component are the costs per exergy unit of the exiting streams $\left(c_{e,k}\right)$ and, if power or useful heat are generated in that component, the cost per unit of exergy associated with the transfer of power $\left(c_{w,k}\right)$ or heat $\left(c_{q,k}\right)$. Usually, some auxiliary equations are necessary to calculate these costs.

2.3. AUXILIARY COSTING EQUATIONS

Different approaches for formulating the auxiliary equations are suggested in the literature, each one having a certain degree of subjectivity (see, for example the references given in [9]). The method recommended for obtaining the least subjective auxiliary equations is detailed in [9] and [10] and summarized in the following:

The fuel and product used in the definition of the exergetic efficiencies of the components are formulated through the following procedure:

(1) The exergy carriers, consisting of all material and energy streams are identified.

(2) Depending on the purpose of the study, the total exergy or any of its components (e.g., thermal, mechanical and chemical exergies) are considered.

(3) The exergy streams are considered to be separated into two categories:
- continuous exergy streams, and
- interrupted exergy streams.

The purpose of the component being considered is used as a criterium to decide whether an exergy stream is continuous or interrupted.

(4) At this point the fuel and the product for each component are defined using the following criteria:
- For continuous exergy streams, every decrease in an exergy stream between the inlet and the outlet of the component is included in the fuel whereas every exergy increase between the inlet and the outlet of the component is included in the product.
- For interrupted exergy streams, the exergy at the inlet of the component belongs to the fuel, and the exergy at the outlet of the component is part of the product.

(5) The auxiliary costing equations are formulated as follows:

On the fuel side
- for continuous exergy streams the F rule is applied, whereas
- for interrupted exergy streams no cost equation is necessary.

On the product side
- for continuous exergy streams the P rule is used, and
- for interrupted exergy streams the P postulate is applied.

The F and P rules and the P postulate are presented in the following.

F rule

The F rule is based on the simple concept that the cost removal from a stream by a component must be equal to the cost at which the removed exergy was previously supplied to that stream.

For continuous exergy streams, the F rule is expressed as

$$c_{F,k}^{X} \sum_{y=1}^{NS} \left(\dot{E}_{in,k,y}^{X} - \dot{E}_{out,k,y}^{X} \right) = \sum_{y=1}^{NS} c_{in,k,y}^{X} \left(\dot{E}_{in,k,y}^{X} - \dot{E}_{out,k,y}^{X} \right) \tag{8}$$

where X represents the exergy form, k the component, and y the exergy stream. NS is the total number of exergy streams of the Xth exergy form entering the kth component. The average cost $c_{F,k}^{X}$ is defined by

$$c_{F,k}^{X} = \frac{\sum_{y=1}^{NS} (\dot{C}_{in,k,y}^{X} - \dot{C}_{out,k,y}^{X})}{\sum_{y=1}^{NS} (\dot{E}_{in,k,y}^{X} - \dot{E}_{out,k,y}^{X})} \tag{9}$$

$\dot{E}_{out,k,y}^{X}$ and $\dot{C}_{out,k,y}^{X}$ are the exiting exergy and cost rates, respectively, associated with the Xth exergy form of the yth material or energy stream, which enters the kth component with the respective values $\dot{E}_{in,k,y}^{X}$ and $\dot{C}_{in,k,y}^{X}$.

P rule

The P rule states that each exergy unit is supplied to all the streams associated with the product of the kth component at the same average cost. Thus, for continuous exergy streams y and x we write

$$c_{P,k,y}^{M} = c_{P,k,y}^{T} = c_{P,k,y}^{CH} = c_{P,k} \tag{10}$$

and

$$c_{P,k,x}^{M} = c_{P,k,y}^{M} = c_{P,k} \; ; \quad c_{P,k,x}^{T} = c_{P,k,y}^{T} = c_{P,k} \; ; \quad c_{P,k,x}^{CH} = c_{P,k,y}^{CH} = c_{P,k} \tag{11}$$

Here $c_{P,k}$ is the cost per exergy unit of the product for the kth component.

P postulate

The P postulate derives directly from the P rule and is applied when at least one interrupted exergy stream participates in the definition of the product: If both the Xth and Zth exergy forms of the *same* material stream (y) at the outlet are interrupted, the auxiliary equation to be used for this stream is

$$c_{out,k,y}^{X} = c_{out,k,y}^{Z} \tag{12}$$

For the same exergy form (X) of two *different interrupted* outlet streams (y and x) we write

$$c_{out,k,y}^{X} = c_{out,k,x}^{X} \tag{13}$$

If the Xth exergy form of the yth material stream is continuous and is part of the product and the Zth exergy form (of the xth material stream) being also part of the product is interrupted, then we use

$$c_{P,k,y}^{X} = c_{out,k,x}^{Z} \tag{14}$$

The approach just outlined for obtaining the auxiliary costing equations is general, objective, and illustrates the interdependency between the definition of exergetic efficiency (or the purpose) of a component and the formulation of the auxiliary equations.

By combining the cost balances for all system components with the auxiliary costing equations and by solving the obtained linear system of equations, we can calculate the cost associated with each material and energy stream in the system.

2.4. EXERGOECONOMIC EVALUATION

The exergoeconomic evaluation is conducted at the system component level using the variables presented in the following for the kth component. From the exergy analysis we know the

- rate of exergy destruction, $\dot{E}_{D,k}$

$$\dot{E}_{D,k} = \dot{E}_{F,k} - \dot{E}_{P,k} - \dot{E}_{L,k} \tag{15}$$

- exergetic efficiency ε_k

$$\varepsilon_k = \dot{E}_{P,k} / \dot{E}_{F,k} = 1 - \left(\dot{E}_{D,k} + \dot{E}_{L,k}\right) / \dot{E}_{F,k} \tag{16}$$

- and exergy destruction ratio y_k

$$y_{D,k} = \dot{E}_{D,k} / \dot{E}_{F,tot} \tag{17}$$

In addition, we calculate the following variables from the exergoeconomic analysis:

- cost per unit of fuel exergy

$$c_{F,k} = \dot{C}_{F,k} / \dot{E}_{F,k} \tag{18}$$

- cost per unit of product exergy

$$c_{P,k} = \dot{C}_{P,k} / \dot{E}_{P,k} \tag{19}$$

- cost rate associated with exergy destruction $\dot{C}_{D,k}$

$$\dot{C}_{D,k} = c_{F,k}\dot{E}_{D,k} \tag{20}$$

- cost rate associated with exergy losses $\dot{C}_{L,k}$

$$\dot{C}_{L,k} = c_{F,k}\dot{E}_{L,k} \tag{21}$$

- cost rate associated with capital investment $\dot{Z}_k^{CI}$
- cost rate associated with operating and maintenance expenses $\dot{Z}_k^{OM}$
- the sum $\dot{Z}_k$ of the last two cost rates

$$\dot{Z}_k = \dot{Z}_k^{CI} + \dot{Z}_k^{OM} \tag{22}$$

- relative cost difference r_k

$$r_k = \frac{c_{P,k} - c_{F,k}}{c_{F,k}} = \frac{1 - \varepsilon_k}{\varepsilon_k} + \frac{\dot{Z}_k}{c_{F,k}\dot{E}_{P,k}} \tag{23}$$

- exergoeconomic factor f

$$f_k = \frac{\dot{Z}_k}{\dot{Z}_k + c_{F,k}(\dot{E}_{D,k} + \dot{E}_{L,k})} \tag{24}$$

The values of $c_{F,k}$ depend on the relative position of the kth component in the system and on the interconnections between the kth component and the remaining components. As a general rule, the closer the kth component to the product (fuel) stream of the overall system, the larger (smaller) usually the value of $c_{F,k}$.

When appropriate cost functions can be developed [1, 8, 11], the cost optimal exergetic efficiency ε_k^{OPT} and the cost optimal relative cost difference r_k^{OPT} for the kth component can be approximated by

$$\varepsilon_k^{OPT} = 1 / (1 + F_k) \tag{25}$$

where

$$F_k = \left(\frac{(\beta + \gamma_k)B_k n_k}{\tau c_{F,k}\dot{E}_{P,k}^{1-m_k}} \right)^{\frac{1}{(n_k + 1)}} \tag{26}$$

and

$$r_k^{OPT} = (n_k + 1)F_k / n_k \tag{27}$$

Here β is the capital recovery factor, γ_k is a coefficient that accounts for the part of the fixed operating and maintenance costs depending on the total capital investment associated with the kth component (TCI_k) while B_k, n_k and m_k are constants used to express TCI_k as a function of ε_k and $\dot{E}_{P,k}$; τ is the average annual time of plant operation at the nominal capacity. Approaches using Eqs. (25) - (27) are still in the development stage.

An exergoeconomic analysis and evaluation identifies and compares the real cost sources in a system [given by Eqs. (20) - (22)], illustrates the cost formation process within a system, calculates separately the cost at which each product stream is generated, and, most importantly, assists in the effective cost optimization of a thermal system, particularly its design. Examples of the design optimization are given in [1, 6-8, and 11-16]. This point is further discussed in the next section.

3. An Iterative Exergoeconomic Procedure for Optimizing the Design of a Thermal System

The conventional approach in optimizing complex systems is to iteratively optimize sybsystems and/or ignore the influence of some structural changes and decision variables. Only energy-based principles, or the results from an exergy analysis at most, are used in the optimization procedure. An alternative to this approach is an iterative exergoeconomic optimization technique that consits of the following seven steps:

1. In the fist step a workable design is developed. The guidelines presented in [1,8, 11 and 16-18] may assist in developing a workable design that is relatively close to the optimal design. The use of these guidelines can reduce, therefore, the total number of iterations required.
2. A detailed exergoeconomic analysis and evaluation, and a pinch analysis are conducted for the design configuration developed in the previous step. The results are used to determine design changes that are expected to improve the design being considered. In this step, and in steps 3-7, we consider, however, only changes in the decision variables that affect both the exergetic efficiency and the investment costs. The remaining decision variables are optimized in step 8.
3. If the system has one or two components for which the sum of the cost rates ($\dot{Z}_k + \dot{C}_{D,k}$) is significantly higher than the same sum for the remaining components, the designs of these components are modified to approach the corresponding cost-optimal exergetic efficiency, Eqs. (25) and (26). To apply these equations, we assume that the costs per exergy unit remain constant for all inlet streams. This step is meaningful only for components in which each of the terms $\dot{Z}_k$ and $\dot{C}_{D,k}$ has a significant contribution to the costs associated with the respective component. If not, step 3 should either be omitted or, preferably, replaced by an efficiency maximization procedure when $\dot{C}_{D,k}$ is the dominating cost rate, or an investment cost minimization procedure when $\dot{Z}_k$ is the dominating cost rate.
   An alternative to using Eqs. (25) - (27) is to optimize one or two major decision variables for each of the most important components by graphically minimizing the relative cost difference (Eq. (23)) calculated for the components at various values of the corresponding decision variables. In this case it is unnecessary to assume that the costs per exergy unit remain constant for each inlet stream. This approach was used in [12] and [13] to optimize the gasification temperature in an integrated gasification-combined-cycle power plant.

4a. For the remaining components, particularly the ones having a relatively high value of the sum ($\dot{Z}_k + \dot{C}_{D,k}$), the relative deviations of the actual values from the cost-optimal values for the exergetic efficiency and relative cost difference are calculated:

$$\Delta\varepsilon_k = \frac{\varepsilon_k - \varepsilon_k^{OPT}}{\varepsilon_k^{OPT}} 100 \tag{28}$$

$$\Delta r_k = \frac{r_k - r_k^{OPT}}{r_k^{OPT}} 100 \tag{29}$$

Here ε_k and r_k are the actual values and ε_k^{OPT} and r_k^{OPT} are the cost-optimal values for the exergetic efficiency and the relative cost difference, respectively. The cost-optimal values are calculated using Eqs. (25) to (27) for the design being currently analyzed. The design of these remaining thermoeconomically important components is modified to reduce the values of $\Delta\varepsilon_k$ and Δr_k. If conflicting design changes are suggested from the evaluation of different components, the design changes for the components with the higher ($\dot{Z}_k + \dot{C}_{D,k}$) values prevail.

4b. If Eqs. (25) - (27) are not available, the exergoeconomic factor f_k should be used to identify the major cost source (capital investment or cost of exergy destruction).
   a. If the f_k value is high, investigate whether it is cost effective to reduce the capital investment of the *k*th component at the expense of the component efficiency.
   b. If the f_k value is low, try to improve the component efficiency by increasing the capital investment.
5. Eliminate any subprocesses that increase the exergy destruction or exergy loss without contributing to the reduction of capital investment or of fuel costs for other components.
6. Consider improving the exergetic efficiency of a component if it has a relatively low exergetic efficiency or relatively large values for the rate of exergy destruction, the exergy destruction ratio, or the exergy loss ratio.
7. Based on the results from steps 2-6, a new design is developed and the value of the objective function for this design is calculated. If in comparison with the previous design this value has been improved, we may decide to proceed with another iteration that involves steps 2 through 7. If, however, the value of the objective function is not better in the new design than in the previous one, we may either revise some design changes and repeat steps 2 through 7 or proceed with step 8.
8. In this step, we use an appropriate mathematical optimization technique to optimize the decision variables that affect the costs but not the exergetic efficiency. At the end of this step the cost-optimal design is obtained.
9. Finally, a parametric study may be conducted to investigate the effect on the optimization results of some parameters used and/or assumptions made in the optimization procedure.

When applying this methodology, it is important to recognize that the values of all thermoeconomic variables depend on the component types: heat exchanger, compressor, turbine, pump, chemical reactor, and so forth. Accordingly, whether a particular value is judged to be high or low can be determined only with reference to a particular class of components. It is also important to consider the effects of contemplated design changes in one component on the performance of the remaining components. These effects may be determined either by inspection of the system flowsheets or by using a simulation program.

The application of steps 1, 2, 4a, and 7 to a cogeneration system is presented in [11], which is one of four papers [11, 19-21] dealing with exergoeconomic optimization techniques applied to the same cogeneration system. This system is also used in the case study discussed in the following section to demonstrate the application of steps 1, 2, 4b, and 7. Examples of exergoeconomic evaluations and improvements of complex thermal systems are discussed in [12-15].

4. Case Study

Fig. 2 presents the base-case design of a cogeneration system that develops a net power output of 30 MW and provides 14 kg/s of saturated water vapor at 20 bar. The first five columns of Table 1 show relevant thermodynamic and economic data. The second column of Table 2 shows the purchased-equipment costs (PEC) for each component in the base-case design. These costs are obtained from the cost equations given in Appendix B of [1]. The remaining direct costs, as well as the indirect costs, are estimated using average factors. The total capital investment of the cogeneration system in the base case is estimated at approximately 46 million mid-1994 dollars. Table 7.9 in [1] summarizes the parameters and assumptions used in the economic analysis, which is based on the revenue-requirement method [22].

The year-by-year economic analysis results in the levelized annual costs for fuel (\$10.4 x 10^6), operating and maintenance (\$5.9 x 10^6), and carrying charges (\$10.5 x 10^6) for a levelization time period of 10 years. The values given in the previous sentence are the corresponding levelized current-dollar costs obtained for the base case. The levelized costs are used as input data for the thermoeconomic analysis and optimization. The cost flow rates in the system are obtained by dividing the levelized annual costs by the number of hours of system operation per year.

The methodology introduced above will now be applied to the case study cogeneration system. The objectives are to identify the effects of the design variables on the costs and suggest values of the design variables that would make the system more cost effective. The key design variables - the decision variables - for the cogeneration system are the compressor pressure ratio p_2/p_1, the isentropic compressor efficiency η_{sc}, the isentropic turbine efficiency η_{st}, the temperature of the air entering the combustion chamber T_3, and the temperature of the combustion products entering the gas turbine T_4.

4.1. FIRST ITERATION

The following nominal values of the decision variables correspond to the first workable design (base-case design) developed for the cogeneration system of Figure 2 and Tables 1, 2:

$$p_2/p_1 = 10, \ \eta_{sc} = \eta_{st} = 0.86, \ T_3 = 850K, \ T_4 = 1520K.$$

The last two columns of Table 1 and the last 10 columns of Table 2 summarize the values of the thermoeconomic variables calculated for each component of the cogeneration system for the base-case design. In accord with the presented methodology, the components are listed in order of descending value of the sum $\dot{C}_D + \dot{Z}$.

The combustion chamber, the gas turbine, and the air compressor have the highest values of the sum $\dot{Z} + \dot{C}_D$ and are, therefore, the most important components from the thermoeconomic viewpoint. The low value of the variable f for the combustion chamber shows that the costs associated with the combustion chamber are almost exclusively due to exergy destruction. A part of the exergy destruction in a combustion chamber can be avoided by preheating the reactants and by reducing the heat loss and the excess air, but this usually leads only to a small reduction in the exergy destruction. For simplicity, we assume here that the heat loss cannot be further reduced. The excess air is determined by the desired temperature T_4 at the inlet to the gas turbine. The temperature T_4 is a key design variable for it affects both the performance of the entire system (exergy destruction in the combustion chamber, gas turbine, air preheater, and HRSG, and exergy loss associated with stream 7) and the investment costs of the components.

An increase in the heat transfer rate in the air preheater achieved through an increase in temperature T_3 also results in a decrease of the exergy destruction in the combustion chamber. Thus, the temperature T_3 is also a key design variable because, in addition to the combustion chamber, it affects the exergy loss associated with stream 7 as well as the performance and

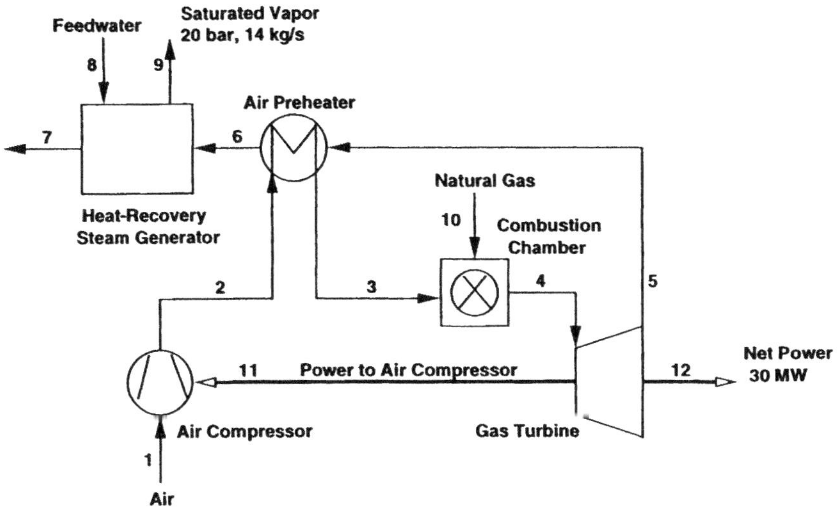

Figure 2. Base-case design of the cogeneration system.

investment costs of the air preheater and the heat recovery steam generator. Holding all other decision variables constant, the higher the temperature T_3 the smaller the average temperature difference in the air preheater and the heat-recovery steam generator. A decrease in the average temperature difference in these heat exchangers results in an increase in both the exergetic efficiency and the capital investment for each heat exchanger.

To summarize: By considering measures for reducing the high cost rate associated with the exergy destruction in the combustion chamber of the cogeneration system, two key design variables have been identified, temperatures T_3 and T_4. An increase in these temperatures reduces the $\dot{C}_D$ value for the combustion chamber and other components but increases their capital investment costs.

Turning next to the gas turbine, which has the second highest value of the sum $\dot{Z} + \dot{C}_D$, the relatively large value of factor f suggests that the capital investment and O&M costs dominate. The capital investment costs of the gas turbine depend on temperature T_4, pressure ratio p_2/p_1, and isentropic efficiency η_{st}. To reduce the high $\dot{Z}$ value associated with the gas turbine, we should consider a reduction in the value of at least one of these variables.

The air compressor has the highest f value and the second highest relative cost difference r among all components. Thus, we would expect the cost effectiveness of the entire system to improve if the $\dot{Z}$ value for the air compressor is reduced. This may be achieved by reducing the pressure ratio p_2/p_1 and/or the isentropic compressor efficiency η_{sc}.

The heat recovery steam generator has the lowest exergetic efficency and the highest r value among all the components. As the f value indicates, almost 45% of the relative cost difference is caused by the $\dot{Z}$ value in this component, with the remaining 55% caused by exergy destruction. Thus, we might conclude that a decrease of the exergy destruction in the HRSG could be cost effective for the entire system even if this would increase the investment costs associated with this component. The exergy destruction in the HRSG can be reduced by decreasing the values of T_6 and T_7. A decrease in the value of T_7 also results in a decrease in the exergy loss from the total system. In terms of the decision variables, temperatures T_6 and T_7 may be reduced by increasing T_5 and/or decreasing T_4 at fixed values of the remaining decision variables.

TABLE 1.Mass flow rate, temperature, pressure, exergy rate, and cost data for the streams of the cogeneration system

State	Stream	Mass Flow Rate $\dot{m}$ (kg/s)	Temperature T (K)	Pressure p (bar)	Exergy Flow Rate $\dot{E}$ (MW)	Cost Flow Rate $\dot{C}$ (S/h)	Cost per Exergy Unit c ($/GJ)
1	Air	91.28	298.1	1.01	0.000	0	0
2	Air	91.28	603.7	10.13	27.538	2756	27.80
3	Air	91.28	850.0	9.62	41.938	3835	25.40
4	Combustion products	92.92	1520.0	9.14	101.454	5301	14.51
5	Combustion products	92.92	1006.2	1.10	38.782	2026	14.51
6	Combustion products	92.92	779.8	1.07	21.752	1137	14.51
7	Combustion products	92.92	426.9	1.01	2.773	145	14.51
8	Water	14.00	298.1	20.00	0.062	0	0
9	Water	14.00	485.6	20.00	12.810	1256	27.23
10	Methane	1.64	298.1	12.00	84.994	1398	4.57
11	Power to air compressor	-	-	-	29.662	2003	18.76
12	Net power	-	-	-	30.000	2026	18.76

The relatively high value of f in the air preheater suggests a reduction in the investment costs of this component. This can be achieved by decreasing T_3.

. It should be noted, however, that changes suggested by the evaluation of this component should only be considered if they do not contradict changes suggested by components with a larger value of $\dot{C}_D + \dot{Z}$.

Summarizing the foregoing conclusions, the following changes in the design variables are expected to improve the cost effectiveness of the system:

- Increase the value of T_3 as suggested by the evaluation of the combustion chamber and HRSG.
- Decrease the pressure ratio p_2/p_1 (and thus p_4/p_5) and the isentropic efficiencies η_{sc}, and η_{st}, as suggested by the evaluation of the air compressor and gas turbine.
- Maintain T_4 fixed, since we get contradictory indications from the evaluation of the combustion chamber on one side and the gas turbine and HRSG on the other side.

4.2. SECOND ITERATION

Contemplating the effects of changes made in accord with the above list in the values of the design variables used in the first iteration, the following new values are selected for the second iteration:

TABLE 2. Values of the purchased-equipment costs (PEC) and the thermoeconomic variables for the base design case (T_3 = 850K; T_4 = 1520K; p_2/p_1 = 10; η_{sc} = η_{st} = 0.86)*

Component	PEC (10^6 $)	ε [%]	$\dot{E}$ [MW]	y_D [%]	c_F [$/GJ]	c_P [$/GJ]	$\dot{C}_D$ [$/h]	$\dot{Z}$ [$/h]	$\dot{C}_D + \dot{Z}$ [$/h]	r [%]	f [%]
Combustion Chamber	0.34	80.37	25.48	29.98	11.45	14.51	1050	68	1118	26.7	6.1
Gas Turbine	3.74	95.20	3.01	3.54	14.51	18.76	157	753	910	29.2	82.7
Air Compressor	3.73	92.84	2.12	2.50	18.76	27.80	143	753	896	48.2	84.0
HRSG	1.31	67.17	6.23	7.33	14.51	27.36	326	264	590	88.5	44.8
Air Preheater	0.94	84.55	2.63	3.09	14.51	20.81	137	189	326	43.4	57.9

* For the overall plant we have $\dot{C}_{P,tot}$ = $3617/h and $\dot{C}_{L,tot}$ = $\dot{C}_7$ = $145/h.

TABLE 3. Component data for the second iteration case: (T_3 = 870 K; T_4 = 1520 K; p_2/p_1 = 9; η_{sc} = 85%; η_{st} = 85%)*. The values given in parentheses refer to the third iteration case considered here (T_3 = 910K; T_4 = 1480K; p_2/p_1 = 7; η_{sc} = 0.83)**

Component	ε [%]	$\dot{E}_D$ [MW]	y_D [%]	c_F [$/GJ]	c_P [$/GJ]	$\dot{C}_D$ [$/h]	$\dot{Z}$ [$/h]	$\dot{C}_D + \dot{Z}$ [$/h]	r [%]	f [%]
Combustion Chamber	80.3 (81.3)	25.93 (27.47)	29.77 (29.92)	10.50 (9.42)	13.26 (11.71)	980 (931)	72 (55)	1052 (986)	26.3 (24.4)	6.8 (5.5)
Gas Turbine	94.9 (94.3)	3.18 (3.69)	3.66 (4.01)	13.26 (11.71)	16.97 (13.75)	152 (155)	647 (296)	799 (451)	28.0 (17.5)	81.0 (65.6)
Air Compressor	92.1 (90.5)	2.34 (2.99)	2.69 (3.25)	16.97 (13.75)	23.96 (18.38)	143 (148)	546 (324)	689 (472)	41.2 (33.6)	79.2 (68.7)
HRSG	66.6 (67.6)	6.40 (6.10)	7.35 (6.65)	13.26 (11.71)	25.60 (23.51)	305 (257)	261 (284)	566 (541)	93.1 (100.7)	46.1 (52.5)
Air Preheater	84.7 (85.6)	3.15 (4.97)	3.62 (4.90)	13.26 (11.71)	18.94 (16.53)	150 (190)	206 (275)	356 (464)	42.9 (41.2)	57.8 (59.2)
Overall Plant	49.1 (46.6)	41.01 (44.79)	47.09 (48.79)	4.57 (4.57)	21.80 (19.06)	675 (736)	1922 (1424)	2597 (2160)	377.0 (317.2)	74.0 (65.9)

* For the overall plant in the new design case we have $\dot{C}_{P,tot}$ = $3355/h and $\dot{C}_{L,tot}$ = $\dot{C}_7$ = $157/h.

** For the overall plant in this design case we have $\dot{C}_{P,tot}$ = $2934/h and $\dot{C}_{L,tot}$ = $\dot{C}_7$ = $167/h.

$$T_3 = 870K,\ T_4 = 1520K\ \text{(unchanged)},\ p_2/p_1 = 9,\ \eta_{sc} = 85\%,\ \eta_{st} = 85\%.$$

The new values of the thermoeconomic variables for each component are summarized in Table 3. In the new design, the combustion chamber, the gas turbine and the air compressor also have the highest values of the sum $\dot{C}_D + \dot{Z}$ and are, therefore, still the most important components from the thermoeconomic viewpoint. The high cost rate associated with the combustion chamber can be reduced by increasing the values of T_3 and T_4. In the evaluation of the cogeneration system we should, however, consider that the value of the combustion chamber will always be the highest among the $\dot{C}_D + \dot{Z}$ values for the components of the cogeneration system.

The gas turbine now has the highest f value. The reduction in this value from 82.7% in the base design to 81.0% in the new design is relatively small compared with a target value of below 75%. This observation suggests (1) a significant decrease in the values of η_{st} and/or p_2/p_1 - that is, a decrease that is greater than the decrease in these variables in the previous step: from 86% to 85% and from 10 to 9, respectively, and (2) a reduction in the values of T_4. Note that the decrease in T_4 value contradicts the corresponding suggestion from the combustion chamber.

The high values of the exergoeconomic factor f and the relative cost difference r for the air compressor suggest a decrease in the values of the decision variables p_2/p_1 and η_{sc}.

The anticipated increase in the exergetic efficiency of the HRSG (see first iteration) was not realized because of the interdependence of the components: The reduction in the values of p_2/p_1, η_{sc}, and η_{st} for the compressor and the turbine leads to an increase in the temperature differences (and therefore a decrease in the exergetic efficiency) of the HRSG. Thus, the HRSG thermoeconomic evaluation suggests that the T_3 value increases and the T_4 value decreases.

The relatively high value of f in the air preheater suggests a reduction in the T_3 value. As noted in the first iteration, however, changes suggested by the evaluation of this component should only be considered if they do not contradict changes suggested by components with a higher value of the sum $\dot{C}_D + \dot{Z}$.

Summarizing the foregoing suggestions from the thermoeconomic evaluation of each component, the following changes in the decision variables are expected to improve the cost effectiveness of the cogeneration system:

- Increase the value of T_3 as suggested by the evaluation of the combustion chamber and HRSG.
- Decrease the pressure ratio p_2/p_1 and the isentropic efficiencies η_{sc} and η_{st} as suggested by the evaluations of the air compressor and gas turbine.
- Decrease the temperature T_4 as suggested from the evaluations of the gas turbine and the HRSG.

4.3. THIRD ITERATION

To illustrate the effect of the suggested changes in the decision variables on the overall costs, we use the following new set of values for the design variables:

$$T_3 = 910K,\ T_4 = 1480K,\ p_2/p_1 = 7,\ \eta_{sc} = 0.83 \text{ and } \eta_{st} = 0.83.$$

The results from the thermoeconomic analysis for the last set of values are summarized in Table 3 by the values given in parentheses. A comparison of the corresponding values shown in Table 3 demonstrates the improvement in the cost effectiveness of the last design case. As a result of these changes, the value of the objective function $\dot{C}_{P,tot}$ is reduced from \$3355/h to \$2934/h. The $\dot{C}_7$ value has increased from \$157/h in the new case to \$167/h in the last case. This increase is, however, outweighed by the decreases in the values of $\dot{C}_{D,k} + \dot{Z}_k$.

4.4. ADDITIONAL ITERATIONS

Some additional iterations conducted in a similar way are necessary to further decrease the value of the objective function and/or establish a nearly-optimal design case. The cost-optimal values of the decision variables [1] are

$$T_3 = 910K,\ T_4 = 1463.0K,\ p_2/p_1 = 5.77,\ \eta_{sc} = 0.811 \text{ and } \eta_{st} = 0.845.$$

With these values we obtain for the objective function $\dot{C}_{P,tot}$ = $2870/h. For the cost-optimal case, the exergetic efficiency of the overall system is 45.0%, the cost rate associated with the exergy loss is $205/h, and the pinch temperature difference in the heat-recovery steam generator is 49.7 K.

References

1. Bejan, A., Tsatsaronis, G., Moran, M. (1996) *Thermal Design and Optimization*, J. Wiley, New York.
2. Reklaitis, G.V., Ravindran, A., Ragsdell, K.M. (1983) *Engineering Optimization*, J. Wiley, New York.
3. Edgar, T.F., Himmelblau, D.M. (1988) *Optimization of Chemical Processes*, McGraw-Hil, New York.
4. Papalambros, P.Y., Wilde, D.J. (1988) *Principles of Optimal Design-Modeling and Computation*, Cambridge University Press, Cambridge.
5. Tsatsaronis, G., Winhold, M., Stojanoff, C.G. (1986) *Thermoeconomic analysis of a gasification-combined-cycle power plant*, EPRI AP-4734, RP 2029-8, Final Report, Electric Power Research Institute, Palo Alto, CA.
6. Tsatsaronis, G., Winhold, M. (1984) *Thermoeconomic analysis of power plants*, EPRI AP-3651, RP 2029-8, Final Report, Electric Power Research Institute, Palo Alto, CA.
7. Tsatsaronis, G., Winhold, M. (1985) Exergoeconomic analysis and evaluation of energy conversion plants, *Energy-Int. J.* ***10***, 69-94.
8. Tsatsaronis, G. (1993) Thermoeconomic analysis and optimization of energy systems, *Prog. Energy Combust. Sci.* **19**, 227-257.
9. Lazzaretto, A. and Tsatsaronis, G. (1996) A general process-based methology for exergy costing, in A.B. Duncan , J. Fiszdon, D. O'Neal, and K. Den Braven (eds.), *Proceedings of the ASME Advanced Energy Systems Devision*, AES-Vol. 36, ASME, New York, 413-428.
10. Lazzaretto, A. and Tsatsaronis, G. (1997) On the quest for objective equations in exergy costing, in M.L. Ramalingam, J.G. Lage, V.C. Mei, and J.N. Chapman (eds.), *Proceedings of the ASME Advanced Energy Systems Devision*, AES-Vol. 37, ASME, New York, 413-428.
11. Tsatsaronis, G. and Pisa, J. (1994) Exergoeconomic evaluation and optimization of energy systems, *Energy-Int. J.* **19**, 287-321.
12. Tsatsaronis, G., Lin, L., Pisa, J., Tawfik, T. (1991) *Thermoeconomic design optimization of a KRW-based IGCC power plant*, Final Report, prepared for the US Department of Energy, Morgantown Energy Technology Center, DE-FC21-98MC26019
13. Tsatsaronis, G., Lin, L., Pisa, J., Tawfik, T. (1992) Optimization of an IGCC power plant - Parts I and II, in R.F. Boehm et al. (eds.), *Thermodynamics and the Design, Analysis and Improvement of Energy Systems*, AES-Vol. 27, ASME, New York, 37-67.
14. Tsatsaronis, G., Lin, L., and Tawfik, T.J. (1994) Economic evalutation of a KRW-based IGCC power plant, *Eng. Gas Turbines Power 116*, pp. 300-306.
15. Tsatsaronis, G., Krause, A., Lin,and L., Tawfik, T. (1992) *Thermoeconomic evaluation of the design of a pressurized fuidized-bed hydroretorting plant* , Final Report prepared for the Institute of Gas Technology and the Department of Energy, DE-AC21-87MC11089.
16. Lin, L., Tsatsaronis, G. (1993) Cost optimization of an advanced IGCC power plant concept design, in H.J. Richter (ed.) *Thermodynamics and the Design, Analysis and Improvement of Energy Systems*, AES-Vol. 10, ASME, New York, 156-166.
17. Linnhoff, B. et al. (1982) *A User Guide on Process Integration for the Efficient Use of Energy*, The Institution of Chemical Engineers, Rugby, Warks.
18. Sama, D.A. (1993) The use of the second law of thermodynamics, in J. Szargut, Z. Kolenda, G. Tsatsaronis, and A. Ziebik (eds.), *Proceedings of the International Conference ENSEC '93*, ASME, New York, 53-76.
19. Frangopoulos, C.A. (1994) Application of the thermoeconomic functional approach to the CGAM problem, *Energy-Int. J.* **19**, 323-342.
20. Valero, A., Lozano, M.A., Serra, L., and Torres, C. (1994) Application of the exergetic cost theory to the CGAM problem, *Energy-Int. J.* **19**, 365-381.
21. von Spakovksy, M.R. (1994) Application of engineering functional analysis to the analysis and optimization of the CGAM problem, *Energy-Int. J.* **19**, 343-364.
22. EPRI *Technical Assessment Guide* (TAG™) (1991) Electric Power Research Institute, TR-100281, 3, Revision 6.

ON-LINE THERMOECONOMIC DIAGNOSIS OF THERMAL POWER PLANTS

A. VALERO, L. CORREAS and L. SERRA
CIRCE - Center of Research for Power Plant Efficiency
María de Luna 3, 50015 Zaragoza, SPAIN

1. Introduction

Diagnosis is the identification of what is wrong with something that is not working properly. The main question that any diagnosis methodology must answer to, when applied to the design or operation of an energy system can be formulated as follows:

Where, how and which part of the consumed resources can be saved by keeping the quantity and specifications of the final product constant.

Thermoeconomic diagnosis is the only Second Law based technique devoted to operation analysis. The exergy balance of an installation allows us to allocate and calculate irreversibilities in the production process and to identify the equipment which affect overall efficiency and the reasons thereof. This information, although useful, has proved not to be enough. In practice, when attempting to achieve energy savings in an installation, we must consider that not all irreversibility can be avoided. Thus, the technical possibilities for exergy savings are always lower than the theoretical limit of thermodynamic exergy losses. Moreover, the local exergy savings which can be achieved in the different units or processes of an installation are not equivalent. The same decrease in the local irreversibility of two different plant components leads, in general, to different variations of the total plant energy consumption.

At this point it is necessary to introduce a concept that will help us to relate causes of irreversibility with their effects at a local and a global level. This is attained by means of a representation of the plant, not like its physical structure, composed of equipment that exchange material and energetic flows, but as a function of the objective that each component accomplishes in the utility. This representation, commonly called productive structure, contains the information about component interaction - or groups of components - by means of exchanging resources in order to obtain a certain product.

A. Bejan and E. Mamut (eds.), Thermodynamic Optimization of Complex Energy Systems, 117–136.

From this point we will be ready to know how any malfunction in the plant affects the additional consumption of resources, which is the final objective of the thermoeconomic diagnosis.

This paper first introduces the theory behind thermoeconomics and then explains an actual diagnosis system that has been developed according to the principles outlined.

2. Theoretical Background

2.1. EXERGETIC COST ANALYSIS

For a given *physical system* of components and flows, thermodynamic properties such as mass flow rate, pressure, temperature, composition, enthalpy, entropy can be determined. Using these properties and choosing a reference or dead state, the energy and exergy flows for the system can be found. If these flows are considered to be fuels, products or losses of the system's components (where the difference between fuels and products represent losses), the efficiency of each system component can be defined as the ratio of its products to its fuels.

These are the basic concepts used to analyze an energy system with energy and exergy balances. An energy balance indicates how the fuels' energy is distributed throughout the system as well as the location of losses which make the energy of the component products lower than that of their fuels. These energy losses, however, only represent losses from a component or the system and give no indication of losses due to the degradation of energy (irreversibilities) within each system component. In order to account for these losses, exergy balances are needed. These balances account for the quality of energy present and allow one to determine where and how much energy could be theoretically saved within the system itself.

A further enhancement of the more conventional analysis described above is the introduction of exergetic costs. These costs take into account the fact that the technical possibilities for exergy (and energy) savings are always lower than the theoretical thermodynamic limits and that the same decrease in local irreversibilities for two different system components leads, in general, to different variations in total system fuel consumption. Since exergy is a departure from equilibrium with the environment or reference state [1], it represents the minimum quantity of work needed to produce a product from the reference state. Thus, the exergy of a physical flow is a potential which does not depend on the process used to obtain this flow. Because real processes in energy systems are only possible with irreversibilities, the exergy of the resources of a component is always greater than that of its products. In other words, the exergy needed to produce a physical flow, i.e. the exergetic cost [1, 2], is always greater than the exergy which the flow contains. Minimizing the exergetic cost per unit of exergy, i.e. the unit exergetic cost of a flow, is a tool which can be used in the thermodynamic analysis and/or optimization of a system's operation. This unit cost for the product flow of a subsystem (or component) is set by both the subsystem's (or component's) internal irreversibilities and by the irreversibilities of the subsystems upstream. Thus, exergetic cost theory permits one to analyze how each subsystem is exergetically interconnected with other subsystems and explicitly determines, i.e. quantifies, which irreversibilities have the greater impact on the system's exergy efficiency, something not explicitly apparent from an exergy analysis alone.

A set of rules developed by Valero et al. [1, 2] allows one to rationally and uniquely calculate all the product exergetic unit costs as well as exergetic costs for all the subsystems or components which make up a system. They are as follows:

1. the exergetic cost is a conservative property;
2. all costs generated by the productive process must be included in the cost of the final products so that in the absence of external assessment, system losses to the environment have a zero exergetic cost;
3. in the absence of an external assessment, the exergetic cost of each flow entering the system is equal to its exergy;
4. if the output flow of a unit is part of the fuel of this unit, then it is understood that its unit exergetic cost is the same as that of the input;
5. if a component has a product composed of several flows, then the same unit exergetic cost is assigned to all of them;
6. if a component produces a residue, then its unit exergetic cost is negative and equal to the amount of external resources needed to dispose of it per unit of its exergy content.

Rule 1 can be justified by considering a subsystem with i fuels, j products and k losses to the environment. Applying this rule to the subsystem gives the following exergy and exergetic cost balances:

$$\sum_i F_i = \sum_j P_j + \sum_k I_k \tag{1}$$

$$\sum_i k^*_{F_j} \cdot F_i = \sum_j k^*_{P_j} \cdot P_j \tag{2}$$

where the k^* are unit exergetic costs values; F, P, I the exergy values of fuel, product, and irreversibilities respectively. The product amounts are set by downstream subsystems and losses to the environment while the exergetic unit costs of the product flows exiting the subsystem are set by both the subsystem's internal transformation process of fuel flow exergies and the fuels' exergetic unit costs. The latter are set by the irreversibilities of upstream subsystems. The transformation process depends on the internal irreversibilities of the subsystem in question. Since both input as well as transformation costs are directly transferred to the product flows, there is no internal cost generation. Exergetic cost is, thus, a conservative property.

To perform an exergetic cost analysis, it is useful to define a *productive or causal structure*, the counterpart to the physical structure used to calculate the energy and the exergy flows of a system. A productive structure is a schematic representation of the plant based on the fuel-product concept. Flows that connect components to one another are component fuels and products but not necessarily physical flows. When defining a productive structure, component fuels and products must be defined in terms of exergy.

Productive and dissipative units, junctions and bifurcations (branches)[1] are the elements which make up a productive structure. Productive units have the capacity for energy transformations which consume resources and produce one or more energy products required by a system's environment or by another unit. The goal of dissipative units is preparing a product, disposing of wastes or interacting with productive units in order, for example, to improve system efficiency, as is the case for the condenser[2].

The thermoeconomic model (mathematical representation of the productive structure) consists of a group of characteristic equations which express for all the components appearing in the productive structure each inlet flow as a function of the outlet flows and a set of internal parameters modeling the component behavior, i.e.:

$$Unit\ j:\ F_i = \kappa_{ij} \cdot P_j \tag{3}$$

$$Junction\ j:\ F_i = r_{ij} \cdot P_j \tag{4}$$

$$Branching\ point\ j:\ F_j = \sum_i P_i \tag{5}$$

where κ has the meaning of exergy consumption of the unit, and r is the equivalent parameter in the case of junctions, called the exergy ratio.

Considering the representation of the Exergetic Cost Theory, explained in [2, 3, 5], the characteristic equations can be expressed for unit *j* as

---

[1] Junctions and bifurcations are fictive components unless they happen to correspond to real components such as flow mixers and splitters. However, mixers should be considered as units rather than junctions as there is a degradation of exergy in the mixing process, resulting in an increase of entropy between mixers' input and output flows.

[2] The condenser permits the system (steam turbine cycle) to effectively comply with the Second Law of thermodynamics by transferring heat (with entropy as the carrier) to the environment in an efficient manner (i.e. at reduced pressure). This function of transferring heat to the environment is equivalent to the selling of negentropy by the condenser to the other components in the steam cycle turbine [4].

$$F_j = f_j(k_j, P_j) = k_j P_j \tag{6}$$

$$P_j = \omega_j + \sum_{i=1}^{n} q_{ji} \cdot F_i \tag{7}$$

where k_j is the unit exergetic consumption of component j, ω_j is a product that exits the plant, and q_{ji}, called production coefficients, are combinations of the r_{ji} coefficients. Eq. (6) informs about

- the productive function of each component, i.e. what is its production,
- what does the component need (fuel) to develop its productive purpose, and
- the thermodynamic efficiency of the process taking place in the component.

Equation (7) is the structural equation, that contains how the resources consumed by the plant are distributed through the plant components, i.e. it informs about how components are connected since a productive point of view.

In vectorial form, Eq. (6) and (7) transform into:

$$\mathbf{F} = \mathbf{K}_D \cdot \mathbf{P} \tag{8}$$

$$\mathbf{P} = \mathbf{P}_s + \langle \mathbf{PF} \rangle\, \mathbf{F} \tag{9}$$

The diagonal matrix $\mathbf{K}_D$ (n x n) contains the unit exergetic consumption of the productive and dissipative units, **<PF>** is a matrix (n x n) containing the coefficients q_{ji} from Eq. (7), and $\mathbf{P}_s$ are the products leaving the plant, expressed as vector.

Given that each unit and each junction has a primary product flow, the total number of these flows equals the total number of internal products whose unit exergetic cost k_j^* must be calculated.

Knowing the costs of the external resources k_e^* and applying the chain rule of derivatives to the set of characteristic equations the following system of equations is obtained:

$$\textit{Unit } j\text{: } k_{P,j}^* = \sum_i k_{F,i}^* \cdot k_i \tag{10}$$

$$\textit{Junction } j\text{: } k_{P,j}^* = \sum_i k_{F,i}^* \cdot r_{ij} \tag{11}$$

$$\textit{Branching point } j\text{: } k_{P,j}^* = k_{F,j}^* \tag{12}$$

from which the cost of all flows in the productive structure can then easily be calculated [7, 8].

2.2. MALFUNCTION AND IMPACT ON FUEL ANALYSIS

The cost calculations outlined in the previous section can be used to analyze the given state of a plant. However, when comparing two different states, such as those for design and off-design conditions, a better comparison can be made by performing a malfunction and impact on fuel analysis. This type of analysis, also known as the theory of perturbations [7, 9], determines how the costs of the resources vary when the internal parameters of an installation, the unit costs of resources consumed or the external demand vary.

For example, let us consider a productive structure with E external resources consumed, N productive and dissipative units and J junctions. Each primary flow is assigned a branch, from which the flow is distributed. Thus, the number of branches is equal to $M = E + N + J$.

Furthermore, the exergetic cost of the fuel used by a plant can be expressed as

$$C_0 = \sum_{j=1}^{E} k_{ej}^{*} \cdot F_{ej} \tag{13}$$

Applying Eq. (8) and Eq. (9) to Eq. (13), and deriving by the internal parameters of the units, x, we get:

$$\frac{\partial C_0}{\partial x} = {}^{t}k_F^{*} \frac{\partial K_D}{\partial x} P + {}^{t}k_P^{*} \frac{\partial < PF >}{\partial x} F \tag{14}$$

Also a equivalent formula in terms of product can be written:

$$\frac{\partial C_0}{\partial x} = \left({}^{t}c_e \frac{\partial K_D}{\partial x} + {}^{t}k_P^{*} \frac{\partial (< PF > K_D)}{\partial x} \right) \cdot P \tag{15}$$

And in terms of the unit exergy consumption:

$$\frac{\partial C_0}{\partial \kappa_{ij}} = k_{P,i}^{*} P_j \tag{16}$$

It means that the increase of resources cost due to a variation of the efficiency or the technical production coefficients is proportional to the cost of fuel used to obtain the product, and to the production of component.

So, we can express the additional cost incurred due to an infinitesimal perturbation in the state of the plant as:

$$dC_0 = \sum_{i=1}^{e} P_i \left(c_{e,i} dk_i + k_i dc_{e,i} \right) + \sum_{i,j=1}^{n} k_{P,i}^{*} P_j d\kappa_{ij} + \sum_{i=1}^{s} k_{P,n-s+i}^{*} d\omega_i \tag{17}$$

Here we like to focus the problems related with thermoeconomic diagnosis of the operation of thermal systems. That is, we wish to compare the real cost of the energy resources consumed with that which would have applied if the plant has the same external demand in design conditions. Therefore let $d\omega_i = 0$. We will also assume that $dc_{e,i} = 0$. So previous expression is simplified to:

$$dC_0 = \sum_{j=1}^{n}\left(c_{e,i}d\kappa_{0,j} + + \sum_{i,j=1}^{n} k^*_{P,i}d\kappa_{ij}\right)P_j \tag{18}$$

This equation allows to separate the additional fuel consumption dC_0, into the contributions of each unit and even each production coefficient. The performance test to be carried out on the installation should provide the actual values of the production coefficients, so we can calculate their deviation with regard to the design conditions. In a unit, the positive deviation of the productive coefficients $dk_{ij}P_j$, called malfunction, implies an additional fuel consumption to produce a given quantity of product.

The preceding Eq. (18) quantifies the system perturbations which affect fuel consumption. Together with the exergetic cost balances, Eq. (8), they form the basis for a diagnostic thermoeconomic analysis of an existing system operating under off-design conditions.

2.3. THERMOECONOMIC OPERATION DIAGNOSIS

Cost accounting is the technique that records, measures and reports about how much things cost. It has already been a long time since cost accounting became a recognized profession and today many techniques are applied to calculate costs of products and services, as well as to estimate the deviations from the pursued targets. Decision making at every level is always supported by an evaluation of the benefits for the organization.

In a parallel manner to monetary cost accounting, the energy cost accounting has been developed for systems in which energy plays an important role. Thermoeconomics, the branch of Thermodynamics that has adapted concepts of Economics, consists of a set of techniques devoted in general to energy saving, among which we remark operation diagnosis.

The thermoeconomic diagnosis of the operation of an energy system, explained and applied to actual utilities in [7] and [9], consists of identifying the degradation in the performance of the system and the assessment of its effects in terms of additional fuel consumption.

The application of the thermoeconomic diagnosis to complex plants becomes a very useful tool for the management, plant engineering, operation and maintenance departments. Moreover, if this tool has been integrated with the plant information management system, continuously processing plant data and spreading the generated information to the different users, the enhancement potential and derived benefits amplifies in several orders of magnitude.

Generally speaking, the possible applications of the thermoeconomic operation diagnosis system are outlined as follows:

- On-line operation costs, and economical assessment of the departure from the best attainable operation.
- Optimization of operation strategies as a function of the market's demand of electricity.
- Economical evaluation of corrective actions (opening turbines for maintenance, compressor cleaning, soot blowing)
- Objective assessment of the cost (not the price) of every product and subproducts of the plant.
- Assessment of operators efficacy.
- Continuous instrument monitoring.

Depending on the end user, and hence, on the decision level to which the diagnose is referred to, the system has to be customized. That is, an operators manager needs certain information and with a different frequency than a plant manager or a maintenance technician. It is important to remark that the system remains the same, only the way to present the results changes.

3. Description of the System

Valero and coworkers have already been installing and developing two real-time diagnosis systems, whose aim is to demonstrate the practical application of the Thermoeconomic Theories.

The first one was installed in Endesa's Teruel conventional pulverized-coal power plant, and focused its man-machine interface to operators, as explained in [10]. The scope of the second system covers not only operation but also will adapt to other potential users such as plant management or maintenance department. The application is currently being tested at the 330 MW-rated Combined Cycle located at Puertollano and will be expanded to cover the whole IGCC facility in the next phase. Puertollano IGCC is owned and operated by Elcogas, a consortium of the principal european producers. The Project has been supported by the EU Thermie Program for Clean Coal Technologies.

The design of the thermoeconomic diagnosis system has been accomplished in a modular fashion. The system requires in a general case four modules, which are depicted in Fig. 1:

1. Data Processing Utility: in connection with the plant data acquisition system, in charge of real-time data retrieval, checking against validation rules and historical data storage.

2. Performance Tests: a procedure, normally based on performance test codes, that determines the actual state of the plant with the higher attainable accuracy, with regard to the available instrumentation.
3. State of Reference: a validated model of the plant which represents the state of reference for any operation mode, ambient conditions or feedstock compositions.
4. Thermoeconomic Diagnosis Model: that allocates and assesses the increased consumption of resources compared to the one predicted by the state of reference and explains the underlying causes.

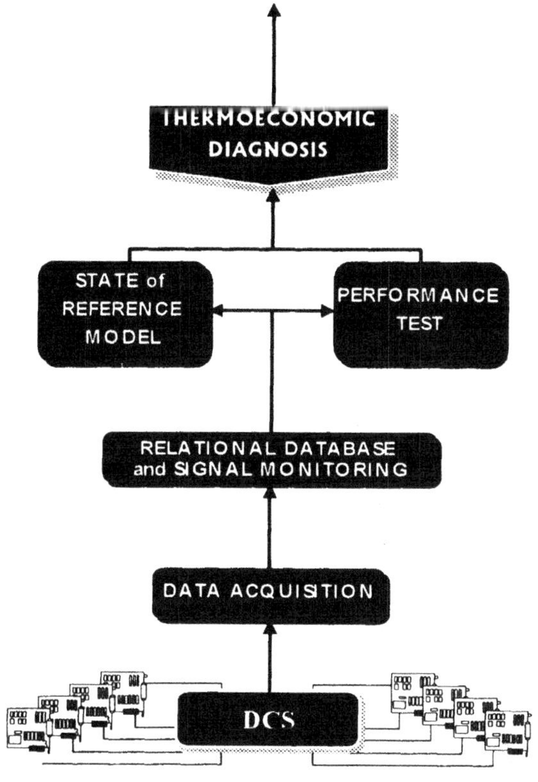

Figure 1. Modular design of the thermoeconomic diagnosis system.

Given a steady-state period, characterized by a certain operation mode, a set of raw instrument data are checked for consistency, validated and processed. These inform about the actual state of the plant, and allow to quantitative analyses through the following successive steps:

- Comparison with a reference case (target) obtained with the same operating conditions.

- Identification of inefficiencies or malfunctions (performance degradation) of sub-systems or components of the plant.
- Evaluation of the causes for the component malfunctions.
- Assessment of the extra-operating cost due to malfunctions, and the cost impact of appropriate maintenance actions.

3.1. DATA PROCESSING UTILITY

The first module performs the tasks concerning connection to the data sources, and retrieval and filtering of data sets that will be processed by the Performance Test module. Basically, it can be decomposed in:

1. Data source access, continuous performance monitoring and data sets retrieval.
2. Preprocessing and validation.
3. Data storage.

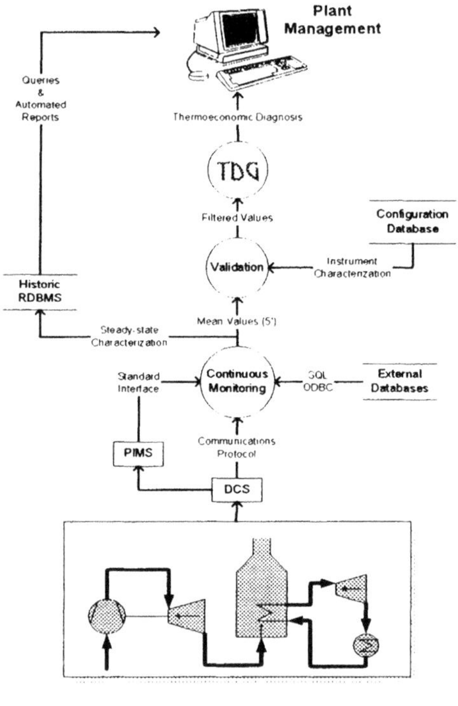

Figure 2. General architecture of data processing utility.

Main data sources usually are plant acquisition systems, or even the distributed control systems themselves, that perform a propietary communications protocol. Interfacing directly with them implies developing a tailor-made layer to interact. However, there are some commercial applications, database engines actually, that have already implemented various specific propietary communications protocols and allow standard queries through SQL or ODBC support, thus avoiding the development of a particular solution. Other existing databases (feedstock analysis, meteorology, emissions) should also be accessed.

The data retrieval tool must continuously monitor the plant performance, because it is necessary to discriminate steady states, where it makes sense to apply the thermoeconomic diagnosis. Surveillance of instantaneous values of the main variables determine stability, as a function of their high-frequency variability, as shown in Fig. 3. For a Combined Cycle, main variables are power production of both steam and gas turbine.

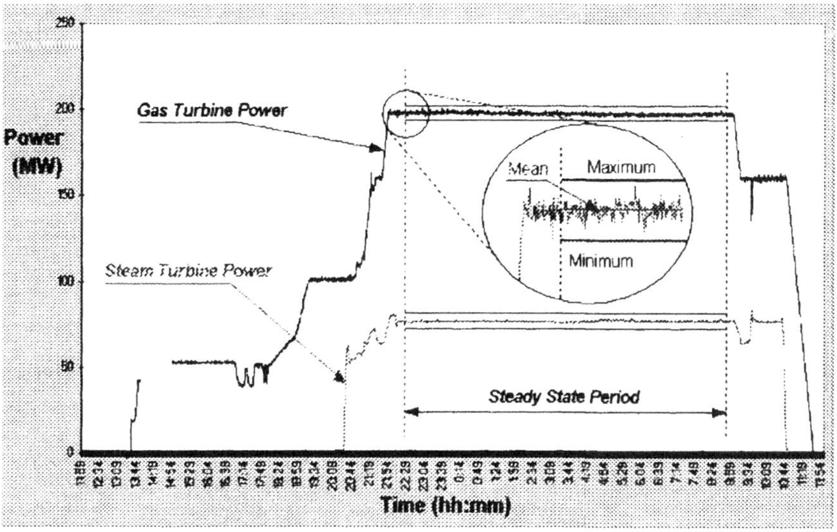

Figure 3. Steady-state period detection.

It is important to remark that raw data from plant instrumentation are measured values, and hence, not actual values. A certain accuracy and precision must be taken into account, depending on the instrument and its calibration, which may degrade as time goes by.

Mean values for raw data are generated over a period (every 5 to 10 minutes typically), and then undergo a series of tests that determine departures on actual instrument readings from the expected ones. First, every measurement is compared against the historically most probable value, as a function of operating mode.

Moreover, relationships between measurements allow to build validation rules. These involve some comparison between instrument readings, like e.g. a difference in pressure measurements upstream and downstream a certain equipment that usually is 1,3 bar, or the power consumption of a certain pump which is proportional to plant load. Validation rules must be particularly formulated for every utility, depending on available instrumentation. The most important feature is their sensibility to detect biases in single instruments.

As a result of these tests, the measurements that have failed any of the validation rules are marked as suspicious and recorded. Depending on the severity of the departure from the expected value, like a null value, the data could be replaced by their most probable value for the actual operating mode, so as not to introduce a clearly inconsistent input to the calculations thereafter. At this point, the preprocessed data set is ready to be sent to the Performance Test module.

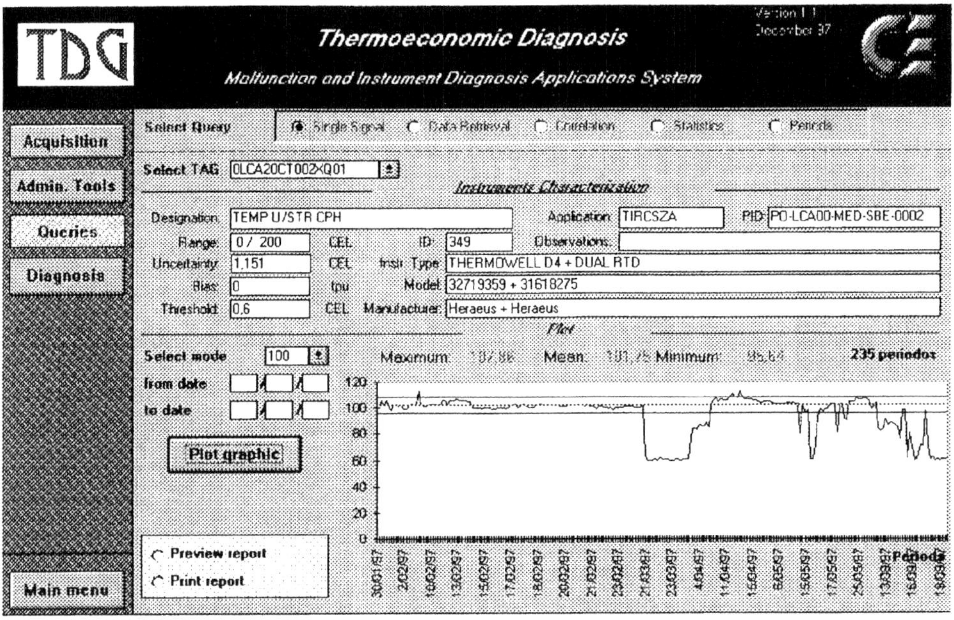

Figure 4. Instrumentation monitoring.

The Diagnosis system features a historical database complementary to the real-time database from which is eventually fed, that is not reliable for long term storage. One must remark that typically control systems handle rates of data of over some thousands of measurements refreshed at a frequency of some seconds. Actually, not all this information is relevant in the long term, and thus can be somewhat volatile. But it is necessary to record all the processed information, in order to conduct analysis of tendencies, very useful to maintenance or operation, or for energy cost accounting with an adequate periodicity for a financial department. In this sense, a complete record for steady-state as well as for transient periods, containing all the necessary information characterizing it, is stored in a relational database.

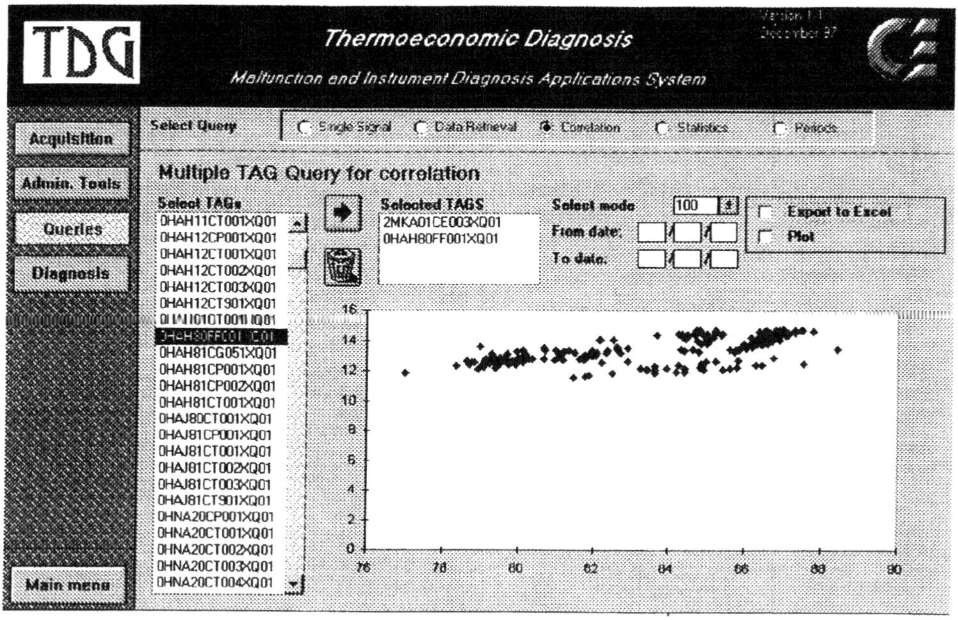

Figure 5. Relationships between measurements.

3.2. PERFORMANCE TESTS

The Performance Test module assures the calculation of a complete set of equipment parameters that reflect the actual state of operation. There are mainly two possible strategies:

- Conventional: based on the election of a minimum set of reliable instruments, generally normalized by recognized International Codes, such as ASME, ISO or DIN. It is not a matter of this paper to deepen in the widely known Performance Test Codes for gas or steam turbines, steam generators, or HRSGs, [11-15].
- Data reconciliation: applies error optimization techniques [16, 17], in order to take advantage of all the available plant information. There are usually some instrumentation redundancies, normally very little, but anyway useful so that combining all measurements, weighted with an estimation of their accuracy, the global uncertainty decreases. This method requires an exact knowledge and monitoring of all available instrumentation: instrument type, measuring conditions, signal processing.... Combining these factors yields an estimation

of the individual uncertainties. A further fine-tuning of the estimations, based on the operating experience, is however necessary.

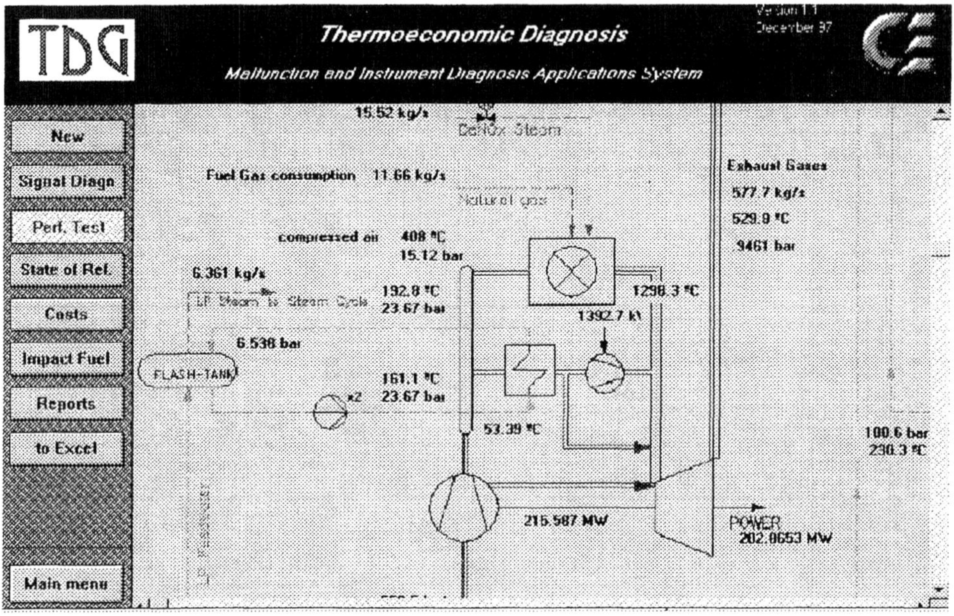

Figure 6. Performance test results.

Global plant performance monitoring calculated by means of a direct method does not require a big amount of input measurements, that on the other hand are usually reliable enough (e.g. net power and natural gas mass flow). Thus, a data reconciliation procedure is not justified in this case.

However, taking into account that the thermoeconomic diagnosis tries to find out a causality, equipment by equipment, from a doubtful instrumentation, is very convenient to minimize error propagation through a more robust calculus procedure, that does not rely on just the exact amount of information to solve the energy and mass balances. In the case of a steam generator, gas mass flow is not measured and thus should be obtained from an indirect method. The more information, the more accurate result would be obtained.

Another advantage of data reconciliation techniques is the open equation procedure implicitly implemented to solve the constraints and energy and mass balances, that permits the developer to focus in the set of constraints itself and not in the converging criteria.

3.3. STATE OF REFERENCE MODEL

For a given operating conditions, the Diagnosis system requires two set of values that characterize the thermodynamic behavior of the plant components: a so called state vector, from the validated measurements, and another one with the state of reference of the plant. The first one is obtained from the Performance Test module, and the second from a model that provides the State of Reference, that is, the better possible operation of the plant for a given conditions.

The State of Reference model consists of a model of the plant, that emulates the physical behavior of the components for every usual operating mode and react against changes in boundary conditions, such as ambient conditions or feedstock composition.

The model has been elaborated by means of an open-equation procedure with non-linear constraints, which is solved as a whole with a quasi-Newton algorithm. This method assures a perfect coherence of energy and mass balances and composition, and a fast convergence, because the initial guesses, taken from the results of the Performance Test module, are sufficiently close to the solution.

As a first approach, the model is based on design data, but has to be validated against actual data of the considered optimum plant performance. Generally, Acceptance Tests set a very adequate reference to compare with.

3.4. EXERGETIC COST AND IMPACT ON FUEL ANALYSIS

As it was already mentioned, the analysis of the interactions between the various components of a utility develops into the so called productive structure. This contains the production objective for each component (e.g. heating or evaporating water, increasing gas pressure,...) related to all resources consumed (e.g. temperature decrease of another fluid, electric power, pressure losses to overcome resistance), and to other secondary effects, the ones that are somehow beneficial (subproducts, like e.g. temperature rise when compressing a gas), or those that need even more resources to be disposed or eliminated (residuals, like e.g. SO_2 from combustion or entropy increase in a steam cycle).

The productive structure of a combined cycle is showed in Fig. 7, as an example of the great complexity that arises in any real utility.

Exergetic Cost Theory allows us to calculate exergetic costs not only for every product going out of the plant, but for every stream or every process that takes place, as it can be seen in Fig. 8.

Monetary costs, even taking into account pay-off and O&M costs, are calculated easily, as a direct application of the Thermoeconomic Theory.

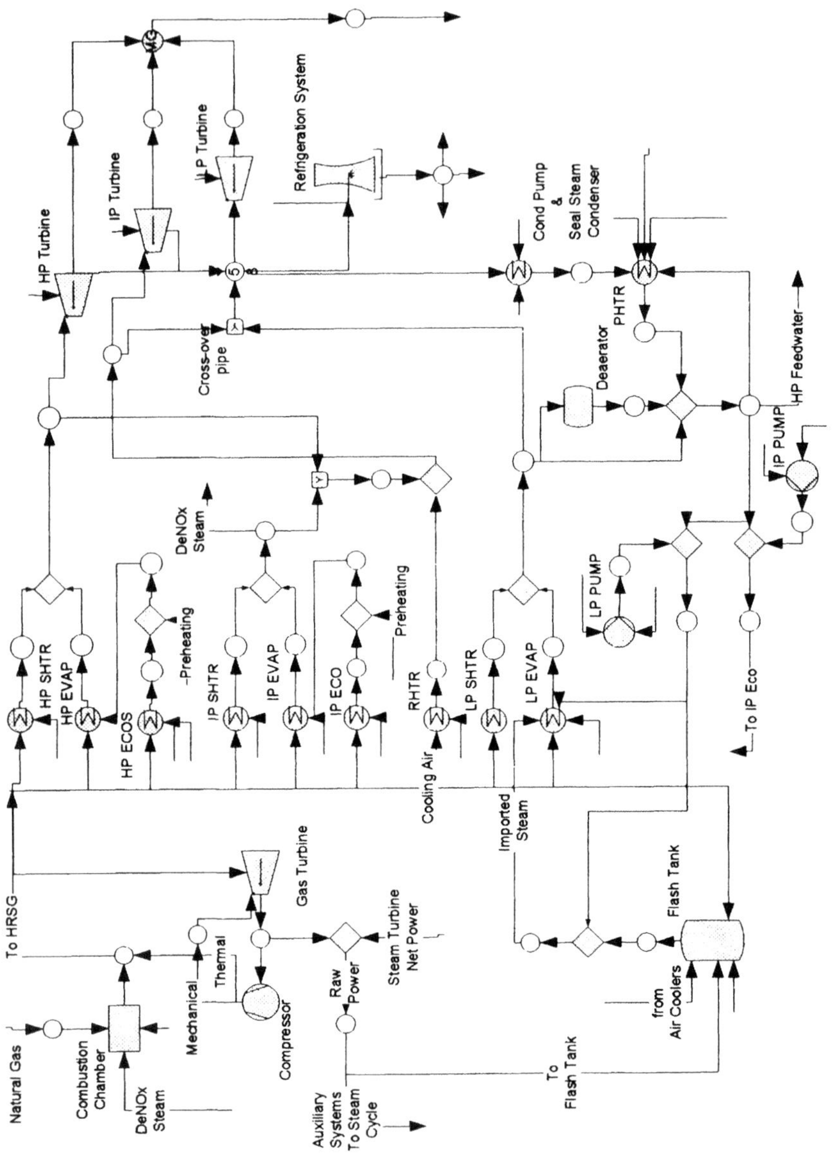

Figure 7. Productive structure of the combined cycle.

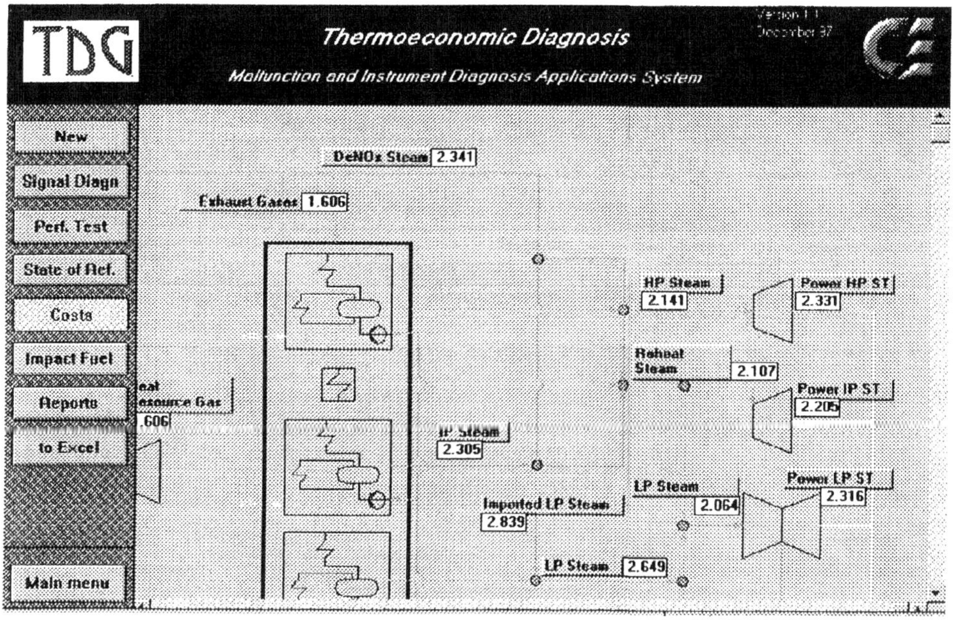

Figure 8. Cost calculation results.

The assessment of the higher specific consumption compared to the reference case is obtained calculating impact on fuel at the component level, from which interpretation the adequate corrective actions should be adopted. Fig. 9 shows impact on fuel disaggregated by components for two actual cases, where a deep evaluation of two operation strategies is conducted.

This working case compares two slight different operation strategies, based on a higher ratio of fuel gas to steam for DeNOx abatement, and the set point at a heater inlet (changed from 100° C to 60° C). Moreover, ambient conditions varied from one state to the other, what affected a different condenser pressure. The disaggregated results revealed the following effects:

- Set point change, that evidently enhanced heat generator efficiency, implies saving 0,4 MW.
- Energetic cost, and hence economic, of requiring a higher steam consumption for DeNOx abatement has been assessed as 2,2 MW. This allows an economical optimization that meets adequate emission levels and maximal efficiency.
- The biasing effect of comparing two states in two slight different ambient conditions is isolated and quantified. A worse condenser vacuum affects the low pressure steam turbine, assessed as 1,9 MW.

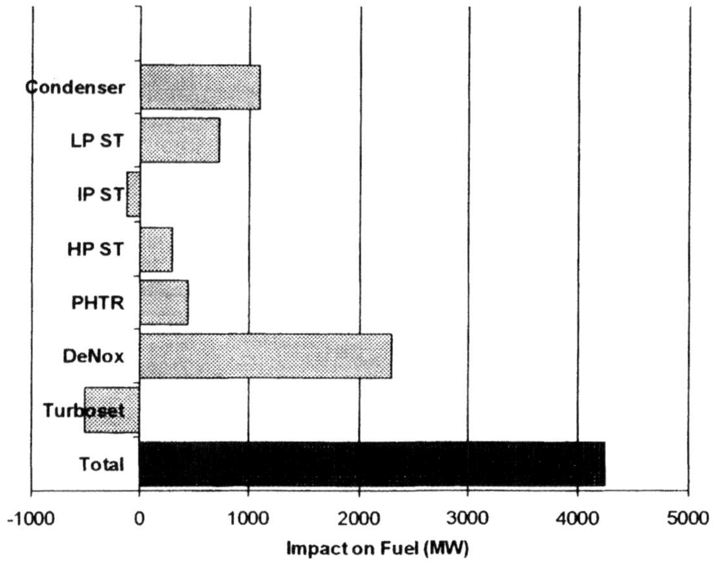

Figure 9. Disaggregated impacts for a comparison of two operation strategies.

Generally speaking, the system offers a wide variety of possible applications, as mentioned above. Is the final user who determines the scope of the tool, that basically must satisfy his needs of organizing the adequate information at the corresponding level. Evidently, already cost increase allocation and quantification at a component level is a step ahead the systems considered state-of-the-art.

4. Results and Conclusions

This paper has presented the motivation of applying a on-line thermoeconomic diagnosis system to enhance energetic resources utilization and thermal power plant management, as well as the main features of the system. Two relevant conclusions must be outlined.

First, quality of input data must be absolutely assured, by means of a measurements monitoring system and historical data analysis supported by an adequate data base. Moreover, it has been demonstrated that data reconciliation techniques are the most suitable for calculations based on process signals, that have not necessarily a good quality. This characteristic is common to all tools that are fed with measurement data, and hence the application needs certain techniques, which have been developed in recent years.

Second, the methodology that supports thermoeconomic diagnosis is unique, because of having a different approximation than other applications.

Thermoeconomic approximation is based on *a priori* analysis, proactive, that is, it previews the phenomena to be diagnosed. The fundamental feature consists of the detailed analysis of the equipment interactions. The systematic cause-effect relationships are materialized into the productive structure. Once given confident input data, malfunction causes and quantification are obtained.

The practical implementation at the Combined Cycle has demonstrated that thermoeconomic diagnosis is becoming a promising industrial tool that integrates the technician's point of view with the interest on enhancing utility's benefits.

For the research team at CIRCE, the thermoeconomic diagnosis system implies successfully reaching the summit of a research field that began more than ten years ago, with the development of the Exergetic Cost Theory and the Structural Theory, theories that could become industrial practice.

Acknowledgments. The authors would like to acknowledge the management of Elcogas and of Endesa's Teruel power plant for their faithful collaboration at every moment, specially to Dr. Manuel Treviño, chief executive officer of Elcogas, and Dr. Javier Pisa, head of the Technology Group at Elcogas, for his constant and active support to this idea. Last but not least, we acknowledge all the coworkers and researchers within the development of the thermoeconomic diagnosis, Ing. Angel Martínez, Ing. Pedro Casero, Ing. Luis M. Blasco, Dr. Scarpellini and the collaboration of Dr. Miguel A. Lozano, Dr. Javier Collado and Ing. Juan L. Bartolomé.

5. References

1. Lozano, M. A. (1987). Metodología para el análisis exergético de calderas de vapor en centrales térmicas, Doctoral Thesis. Universidad de Zaragoza, Zaragoza, Spain.
2. Lozano, M. A. and Valero, A. (1993). Theory of the exergetic cost, *Energy* Vol. 18, No. 9, pp. 939-960.
3. Valero, A., Muñoz, M. and Lozano, M. A. (1986). A general theory of exergy saving, ASME. AES, Vol. 2-3 *Computer-Aided Engineering and Energy Systems* Vol. 3: Second Law Analysis and Modeling, pp. 1-21 Editors: R.A. Gaggioli Book No. H0341C
4. Lozano, M. A. and Valero, A. (1993). Thermoeconomic analysis of gas turbine cogeneration systems, ASME. *Winter Annual Meeting Symposium on thermodynamics and the design, analysis and improvement of energy systems*. November 28-December 3, Session II General Thermodynamics & Energy Systems
5. Valero, A., Lozano, M. A., Serra, L. and Torres C. (1994). Application of the exergetic cost theory to the CGAM problem, *Energy* Vol. 19, No. 3, pp 365-381.
6. Torres, C. (1991). Exergoeconomía simbólica. Metodología para el análisis termoeconómico de los sistemas energéticos, Doctoral Thesis, Universidad de Zaragoza, Zaragoza, Spain.
7. Lozano, M. A., Bartolomé, J.L., Valero and A., Reini, M. (1994). Thermoeconomic diagnosis of energy systems, Flowers 94, *Florence World Energy Research Symposium*. July 6-8, Florence, Italy.

8. Valero, A., Serra, L. and Lozano, M. A. (1993). Structural Theory of Thermoeconomics, ASME. *Winter Annual Meeting Symposium on thermodynamics and the design, analysis and improvement of energy systems*. November 28-December 3, Session II General Thermodynamics & Energy Systems.
9. Schwarcz, P., Lozano, M. A., Von Spakovsky, M.R. and Valero, A. (1997). Diagnostic Analysis Of A PFBC Power Plant Using A Thermoeconomic Methodology. TAIES'97: *Thermodynamic Analysis and Improvement of Energy Systems* Proceedings of Intl. Conference June 10-13, Beijing, China Eds. Cai, Ruixian, M.J. Moran, Zhang, S., Xiao, Y. World Pubs. Corp. Beijing, China, ISBN 7-5062-3264-Z/TK.6 pp. 240-249.
10. Valero, A., Lozano, M. A. and Bartolomé, J.L. (1995). On Line Monitoring Of Power Plants Performance, Using Exergetic Cost Techniques. *Exergoeconomical Analysis and Optimization in Chemical Engineering*, EUROTHERM Seminar No. 41, July 17-18, Aachen, Germany.
11. ASME (1964). *Performance Test Codes: Steam Generating Units, PTC 4.1*. The American Society of Mechanical Engineers.
12. ASME (1970). *Performance Test Codes Report: Simplified Procedures for Routine Performance Tests of Steam Turbines*. The American Society of Mechanical Engineers.
13. ASME (1976). *Performance Test Codes: Steam Turbines, PTC 6*. The American Society of Mechanical Engineers.
14. ASME (1981). *Performance Test Codes: Gas Turbine Heat Recovery Steam Generators. PTC 4.4.* American Society of Mechanical Engineers.
15. British Standard (1989). *British Standard Specification for Gas Turbine acceptance tests. ISO 2314.*
16. Stephenson, G.R. and Shewchuk, C.F. Reconciliation of Process Data with Process Simulation (1986). *AIChE Journal*, Vol. 32, No. 2, pp. 247-254.
17. Heyen, G., Marechal, E. and Kalitventzeff, B. (1996). Sensitivity Calculations and Variance Analysis in Plant Measurement Reconciliation. *Computers Chemical Engineering* Vol. 20 Sup. S539-S544.

EXERGY IN THE THERMAL SYSTEMS ANALYSIS

J. SZARGUT
Technical University of Silesia
Institute of Thermal Technology
Konarskiego 22, 44 100 Gliwice, Poland

1. Introduction

Exergy analysis has been developed as a result of application of the second law of thermodynamics to the investigation of the thermodynamic imperfection of industrial thermal processes. Precursors of exergy analysis were Gouy [1] and Stodola [7]. They formulated independently the law determining the loss of ability to perform work due to the thermodynamic irrevesibility. The term *exergy* has been introduced by Rant [4] whose works initiated a very intensive development of this field of technical thermodynamics. In sixties the exergy analysis has been developed mainly in Europe, in eighties the interest in this analysis increased distinctly in America. The bibliography of exergy analysis contains thousands of papers, among them many monographic books.

2. Definition of Exergy and Exergy Losses

Various kinds of energy differ in their quality characterized by their ability to perform work. *Exergy is a measure of quality of various kinds of energy.* It has been defined as a maximum ability to perform work in the existing natural environment. Very clear definition of exergy has been proposed by Riekert [5], with small supplementation of the author of present paper:

Exergy expresses the minimal consumption of work necessary to produce the investigated material with required parameters from commonly appearing components of the surroundings nature, in reversible processes, using the environment as a single source of heat.

Similarly as in the case of energy, it should be distinguished between the exergy of matter crossing the control boundary of the system and exergy of matter contained within the boundary (difference similar to that between internal energy and enthalpy) [8].

Exergy does not fulfill the law of conservation. Every irreversible process causes an irrecoverable loss of exergy δB, which can be determined by means of the Gouy-Stodola law:

$$\delta B = T_0 \Sigma \Delta S \tag{1}$$

A. Bejan and E. Mamut (eds.), Thermodynamic Optimization of Complex Energy Systems, 137–150.

where T_0 is the environmental temperature and $\Sigma \Delta S$ is the sum of entropy increases of all kinds of matter taking part in the process.

As stated Sama [6] an exergy loss has to be accepted in order to ensure a limited value of the investment expenditures. E.g. a limited area of heat transfer in a heat exchanger can by attained, if there exists some temperature difference between the heating and heated fluid in every point of the heat transfer surface. But every temperature difference during the heat transfer evokes some exergy loss. Hence *exergy loss should always have some economical justification.* If such a justification does not exist, it indicates that the exergy loss results only from an error in engineering art.

On the other hand, every exergy loss causes some decrease of the useful effect of the process or an increase of the consumption of driving means (if the required useful effect is given). Hence, the exergy loss indicates always the possibility of thermodynamic improvement of the process, but the profitability of such an improvement should be checked by means of an economical analysis (because is usually connected with some increase of investment expenditures).

Main causes of exergy losses are: friction, heat transfer with finite temperature difference, diffusion, combustion. The lower is the temperature in the process, the greater is the deleterious influence of irreversibility. E.g. in processes realized below the environmental temperature exergy loss due to the friction is greater than the heat of friction. The analysis of exergy losses is not easy, because the change of irreversibility in some link of the system evokes changes of exergy losses also in other links of the system.

The exergy losses can be divided into internal and external ones. *Internal exergy losses* are connected with the internal,parts of the system, appear inside the control boundary of the system and can be easily calculated by means of Eq. (1). *External exergy losses* result from the discharge of waste products of the process to the environment. Exergy of the waste products is destroyed in the environment. Hence the external exergy loss equals to the exergy of the considered waste product. The calculation of external exergy loss by means of Eq. (1) is difficult, it is easier to determine it by mean of the exergy of waste product. The appearance of external exergy losses indicates that the primary causes of exergy losses act often far from the place of appearance of exergy loss. Often it is difficult to decide, which is the cause of exergy loss. E.g. the external exergy loss with combustion gases flowing out from the boiler can be understood as a result irreversible phenomena inside the boiler or as a result of the lack of utilization after the boiler.

3. Exergy Balance and Exergy Efficiency

Exergy is exempt from the law of conservation, so the exergy balance must be closed by the internal exergy loss:

$$B_d = \Delta B_s + \Sigma \Delta B_q + B_{au} + B_{aw} + L + \delta B_i \qquad (2)$$

where B_d is the exergy delivered to the system with the streams of matter, B_{au}, B_{aw}

denote, respectively, the exergy carried out from the system with useful products and waste products, ΔB_s is the increase of exergy of the system, ΔB_q denotes the increase of exergy of the heat source being in contact with the system, L is the work performed by the system and δB_i denotes the internal exergy loss, due to the irreversibilities inside the system.

Exergy carried out with waste products expresses the external exergy losses. Increase of exergy of the external heat source can be determined by means of the Carnot-factor:

$$\Delta B_q = -Q\frac{T-T_0}{T} \tag{3}$$

where Q is the heat taken from the external heat source and T is the temperature of heat source measured on the system boundary.

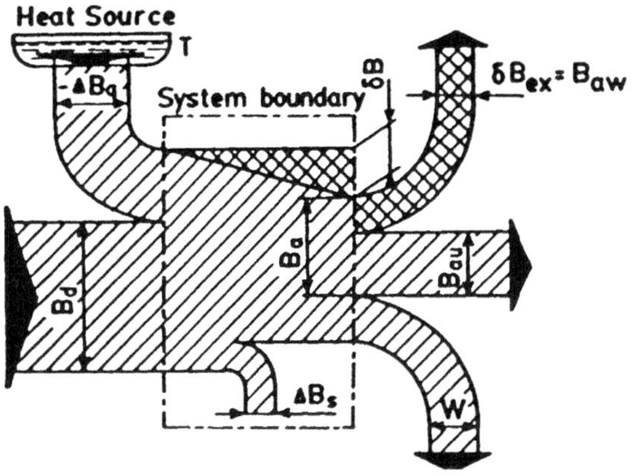

Figure 1. Band diagram of the exergy balance.

For a steady-state process in which flow velocities, chemical compositions and thermal parameters are constant at all points of the system, exergy of the system remains constant and exergy balance can be related to the time unit:

$$\dot{B}_d = \dot{B}_{au} + \dot{B}_{aw} + \Sigma\Delta\dot{B}_q + \dot{L} + \delta\dot{B}_i \tag{4}$$

where the dotted terms denote quantities per time unit.

Exergy balance can be presented in the form of band diagram, Fig.1 [2]. The width of the bands is proportional to the exergy values. The band of the internal exergy loss has a width increasing inside the system boundary from zero to the value resulting from Eq.(1).

Ratio of the useful exergetic effect to the consumption of driving exergy expresses the *exergetic efficiency* called also rational efficiency or effectiveness:

$$\eta_B = \frac{B_{au} - B_{dr} + \Delta B_{su} + \Sigma \Delta B_{qu} + L_u}{B_{df} - \Sigma \Delta B_{qf} - L_f} \tag{5}$$

where B_{df}, B_{dr} denote, respectively, the exergy of delivered driving materials (e.g. fuels) and of nondriving raw materials, ΔB_{su} is the useful exergy increase of the system, $-\Delta B_{qf}, \Delta B_{qu}$ denote, respectively, the exergy decrease of the source of driving heat and increase of the serviced source of heat, $-L_f, L_u$ are, respectively, the driving work and useful performed work.

Exergetic efficiency is always smaller than one. In some processes its value is negative. In such a case the *degree of thermodynamic perfection* η_p can be calculated in order to characterize the irreversibility of the process. It can be defined as a ratio of the carried out useful exergy to the delivered exergy. When calculating η_p the *transit exergy* should be omitted. It comprises the exergy components not taking part in the thermodynamic changes (e.g. the chemical exergy of the solvent when analyzing the chemical processes in solutions).

The main task of exergy analysis is to detect and quantify the causes of thermodynamic imperfection. The mentioned causes are characterized by the connected exergy losses.

Szargut and Sama [16] elaborated 20 practical rules facilitating the reduction of exergy losses. The mentioned rules take also into account the mutual influence of the irreversibility of processes appearing in the large network.

Influence of exergy losses appearing in particular parts of the process on its useful effect can be examined by means of the *coefficients of sensibility*:

$$\sigma_k = -\frac{\partial B_u}{\partial(\delta B_k)} = -\frac{\partial \eta_B}{\partial \varepsilon_k} \tag{6}$$

where B_u, η_B are, respectively, the useful exergy effect and exergy efficiency of the process, $\delta B_k, \varepsilon_k$ are, respectively, the total and relative exergy loss in the k-th part of the process.

4. Economic Applications of Exergy

Exergy losses appearing in particular links of the analyzed system are not of the same importance. In course of thermodynamic processes the specific economic value of exergy increases. E.g. in a steam power plant the specific economic value of fuel exergy is smaller than that of live steam, and the last one is smaller than the specific economic value of electricity produced. Hence the exergy losses are the more harmful, the farther is located the analyzed link in the chain of partial processes.

The cited statement stimulated many authors to propose economic applications of exergy. The new field of exergy analysis, called *thermoeconomics* has been developed

[20],[21],[22],[23]. However exergy is a thermodynamic concept, not an economic one [9]. It should not be assumed, that the economic value is proportional to the exergy. E.g. specific exergy of pure gold is smaller than specific exergy of pure iron.

The economic applications of exergy usually base upon the determination of specific cost of exergy in particular points of the considered system. However such a determination is not possible without additional assumptions, if the system contains cycles. Consider e.g. a simple power plant containing a steam boiler, a turbine, a condenser and a pump. Four independent economic balances can be formulated for this system, e.g. three balances for particular links and a balance for the entire system. The balance of entire system enables us to calculate the specific cost of electricity produced (the investment costs and the cost of fuel are given).There remain three independent balances with four unknown quantities (specific cost of exergy of steam before and after the turbine, specific cost of water exergy before and after the pump). Hence if the specific costs of exergy in all the characteristic points of the system are calculated, some conscious or unconscious assumption has been made.

The determination of the specific cost of exergy is usually used for the partition of production costs between the products of cogeneration. Such a partition can be made by means of purely economic tools i.e. using the *principle of avoided costs*. In every cogeneration process a major product can be distinguished. Its demand determines the location of the plant and the production rate. The additionally fabricated product can be called a by-product if it substitutes a major product of another specialized process. The production cost of the by-product can be determined according to the production costs avoided in the substituted specialized process.

The principle of the avoided expenditures can be also used for determination of the consumption of driving exergy burdening the fabrication of major product of a 4cogeneration (combined) process:

$$b_{fm} = \frac{1}{G_m}\left(B_f - G_b \frac{b_b}{\eta_{Bs}}\right) \tag{7}$$

where b_{fm} is the specific consumption of driving exergy burdening the major product, G_m, G_b are, respectively, the amount of the major product and by-product, B_f is the overall consumption of driving exergy, b_b denotes the specific exergy of the by-product and η_{Bs} is the exergetic efficiency of the avoided process of separate fabrication of by-product.

Despite the mentioned objections, exergy can be used as an auxiliary tool for solving some economical problems. E.g. the heat lost from the pipeline of live steam in a steam power plant has a better quality, than the mean quality of heat delivered inside the boiler. Also the economic value of the lost heat is greater than the mean value. It can be approximately evaluated by means of exergy:

$$k = k_m \frac{T - T_0}{T} \frac{T_m}{T_m - T_0} \tag{8}$$

where k_m, k are, respectively, the mean specific cost of the heat delivered inside the boiler and local specific cost of the heat lost from the pipeline and T_m, T are, respectively, the mean temperature during heat absorption inside the boiler and local temperature in the pipeline.

5. Cumulative Consumption of Exergy and Partial Exergy Losses

Consumption of exergy connected with the fabrication of the considered product appear not only in the plant fabricating this product, but also in plants delivering semi-finished products and raw materials for the final production process. Total consumption of exergy of natural resources connected with the fabrication of the considered product and appearing in all the links of the network of production processes, has been called *cumulative consumption of exergy CExC* [12]. Analysis of CExC can be made in a global scope or in scope of a particular country. In the last case the influence of the import of semi-finished product on the domestic consumption of exergy should be investigated. The calculations can comprise all the natural resources of exergy or only the unrestorable resources. The last case is usually more interesting. The CExC-values can be calculated separately for particular kinds of primary exergy or can comprise the sum of primary exergy. Szargut [17] proved, that *it is not necessary to take into account the human labor when calculating the CExC-value*, because it would denote a double introduction of the same quantity. The mentioned statement results e.g. from the analysis of a simple system presented in Fig.2. The system contains coal mines, the industry and the society. The society consumes immediately a part C_k of the coal production and the entire production of industry. The part C_i of coal production consumes the industry. The consumption of human labor amounts to L_i in the industry and L_c in coal mines.

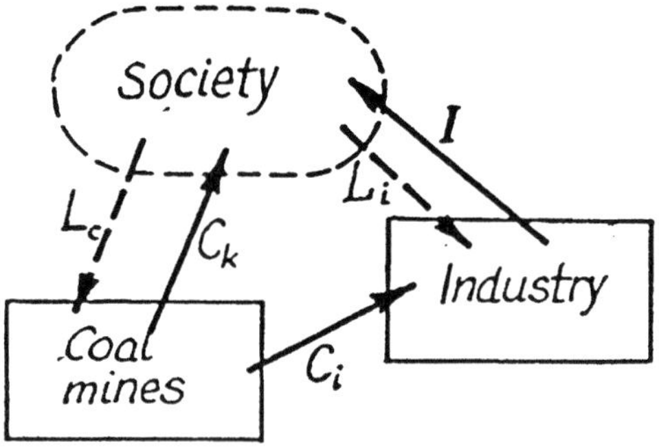

Figure 2. Scheme of a simplified system.

If we take into account the CExC-value burdening the human labor, the CExC-values characterizing the unit of coal and of industrial product are:

$$b_c^* = b_c + l_c r,$$
$$b_i^* = \frac{C_i}{I} b_c^* + l_i r \tag{9}$$

where b_c^*, b_i^* are, respectively, the CExC-values burdening a unit of coal and of industrial product, b_c is the specific exergy of coal, l_c, l_i denote, respectivelyu, the specific consumption of human labor in coal mines and in industry, r is the CExC-value burdening an unit of human labor and C_i, I denote, respectively, the consumption of coal in the industry and amount of industrial products.
From eqs (9) it results:

$$b_i^* = \frac{C_i}{I}(b_c + l_c r) + l_i r = \frac{C_i}{I} b_c + r(l_c \frac{C_i}{I} + l_i) \tag{10}$$

Total consumption of the primary exergy is:

$$b_i^* I + b_c^* C_k = C b_c + (L_c + L_i) r \tag{11}$$

where C is the total production of coal.
On the other hand it is evident, that the total consumption of primary exergy is $C b_c$.
So the balance of CExC is closed, if the quantity r does not appear in eqs.(8). Hence the mentioned statement is correct.

CExC can be calculated by means of the *process analysis* (sequence method) or by means of the *set of balance equations*. The process analysis (sequence method) begins in the final process of the fabrication of the considered product and goes through the processes of production of semi-finished products, fabrication of machines and equipment until the extraction of raw materials from natural resources. The sequence method enables us to analyze only one product, but can be applied if the connections between the considered links of the technological network are week (the consumption of the considered product in preceding links is small). The method of balance equations is more general, but requires more work.

The balance equation of CExC bases upon the statement, that the CExC-value of the product of considered link results from the CExC-values burdening the semi-finished products and by-product and from the exergy of natural resources extracted from nature in the considered link. If the CExC-value are calculated in a scope of a particular country, the set of balance equations has a form:

$$b_j^* + \sum_i f_{ij} b_i^* = \sum_i a_{ij} b_i^* + \sum_r a_{rj} b_r^* + \beta_j \tag{12}$$

where b_j^*, b_i^*, b_r^* are, respectively, the specific CExC-values burdening the major product of j-th and i-th process and the r-th imported semi-finished product, a_{ij}, a_{rj} are, respectively, the coefficients of consumption of i-th domestic and r-th imported semi-finished product per unit of j-th product, f_{ij} is the coefficient of the by-production expressed in units of the substituted i-th major product, per unit of the j-th product, and β_j denotes the exergy of the natural resources extracted from nature in the j-th process, per unit of the j-th product.

An exact method of calculation of b_r^* has been formulated by Szargut [13] and based upon the statement, that the financial means for import are gained by export. Hence the CExC-value burdening the monetary unit of the import equals to that of export. An exact method is rather complicated. An approximate method can be based upon the assumption that the structure of export is similar to that of domestic production. From this assumption it results:

$$b_r^* = D_r \frac{B}{NGP} \tag{13}$$

where D_r is the specific monetary cost of the imported material, B is the total consumption of exergy of unrestorable natural resources in the considered country, per year and NGP is the national gross product in monetary units per year.

The difference between the CExC-value and the exergy of the considered product can be partitioned into *partial exergy losses PEL* appearing in particular links of the network of production processes [14]. Such a partition is possible if the calculation of CExC comprises *all the kinds of exergy* extracted from nature, i.e. taken from 2 unrestorable or restorable, domestic or foreign sources:

$$b_j^* = b_j + \sum_k \overline{\delta b_{kj}} \tag{14}$$

where b_j^* is the specific cumulative consumption of exergy burdening the j-th product, and $\overline{\delta b_{kj}}$ is the partial exergy loss connected with the fabrication of the j-th product and appearing in the k-th link of the network of production processes.

Analysis of the partial exergy losses indicates the possibilities to decrease the CExC-value. It can be attained by improving the links with great PEL-values, by changing the technology of production of some semi-finished products or by changing the kind of some semi-finished products. The calculation of partial exergy losses can be made only in a global scope or for products fabricated without distinct consumption of the imported materials and energy carriers.

The coefficients of consumption a_{ij}, β_j appearing in Eq.(12) relate to the unit of major j-th product together with by-products. E.g. in a heat-and-power plant the coefficient of coal consumption relates to the unit of useful heat together with the cogenerated amount of electricity. In order to calculate the partial exergy losses the *net coefficients of consumption* of semi-finished products an natural resources should be determined:

$$A_{ij} = a_{ij} - \sum_p f_{pj} A_{ip} \tag{15}$$

$$\beta_{Nj} = \beta_j - \sum_p f_{pj} \beta_{Np} \tag{16}$$

where A_{ij},A_{ip} are, respectively, the coefficients of a net consumption of i-th product per unit of the j-th and p-th major product and β_{Nj}, β_{Np} are, respectively, the coefficients of the immediate net consumption of natural resources per unit of the j-th and p-th major product.

Equations (15) and (16) can be presented in a matrix form:

$$\boldsymbol{A} = \boldsymbol{a}(\boldsymbol{E} + \boldsymbol{f})^{-1} \ , \ \boldsymbol{B}_N^T = \boldsymbol{B}^T(\boldsymbol{E} + \boldsymbol{f})^{-1} \tag{17}$$

where $\boldsymbol{E}$ is the unitary diagonal matrix.

The coefficients of net consumption can be used for calculation of the CExC-values similarly as the coefficients of gross consumption:

$$\boldsymbol{b}^{*T} = \boldsymbol{B}_N^T(\boldsymbol{E} - \boldsymbol{A})^{-1} \tag{18}$$

We can check the correctness of eqs (17) introducing them into Eq.(18):

$$\boldsymbol{b}^{*T}[\boldsymbol{E} - \boldsymbol{a}(\boldsymbol{E} + \boldsymbol{f})^{-1}] = \boldsymbol{B}^T(\boldsymbol{E} + \boldsymbol{f})^{-1} \tag{19}$$

Multiplying Eq.(19) by the matrix $(\boldsymbol{E} + \boldsymbol{f})$ we obtain a formula (12) with omission of the imported semi-finished products.

The local net exergy loss δb_{Nj} related only to the unit of major product of the j-th process, results from the exergy balance of the j-th process. It contains the internal and external exergy losses:

$$\sum_i A_{ij} b_i + \beta_{Nj} = b_j + \delta b_{Nj} \tag{20}$$

or in a matrix form:

$$\boldsymbol{b}^T(\boldsymbol{E} - \boldsymbol{A}) = \boldsymbol{B}_N^T - \boldsymbol{\delta B}_N^T \tag{21}$$

Partial net exergy losses result from the cumulative consumption of particular products per unit of the considered product:

$$\overline{\delta b_{Nkj}} = S_{Nkj} \delta b_{Nk} \tag{22}$$

where S_{Nkj} is the cumulative net consumption of k-th product per unit of j-th product. The value of S_{Nkj} results from the matrix relation:

$$S_N = (E - A)^{-1} \tag{23}$$

The sum of partial exergy losses fulfills Eq.(14), but only then, when the balance equation (12) does not contain the CExC-values of imported products.

Example 1. In order to calculate the partial exergy losses connected with the production of coke and liquid pig iron, a simplified system comprising a coal mine, a iron-ore mine, a source of natural gas, a power plant, a coke plant, a ore-sinter plant and blast furnace plant has been investigated [15]. A coke plant is equipped by a dry quenching installation and a steam turbine driving an electric generator. The mentioned plants have only week connections with other industrial plants. Four by-products appear in the considered system: coke-oven gas, tar and electricity produced in the coke plant and the blast-furnace gas. It has been assumed that the coke-oven gas and tar substitute the natural gas in the proportion 0.514 kmol n.g./kmol c-o.g. and 0.0499 kmol n.g./kg tar. The basic part of the blast-furnace gas substitutes a natural gas in the proportion 0.114 kmol n.g./kmol b-f.g. The peak part of the produced blast-furnace gas is burnt in the boiler house and substitutes a hard coal in the proportion 3.62 kg coal/kmol b-f g. The consumption of steel in coal mines has been substituted by means of the consumption of pig iron.

Following values of the specific exergy and of the coefficients of consumption and by-production have been assumed (the mentioned values of the substitution ratio have been taken into account):

1) *Coal mine* $b_1 = 24.7$ MJ/kg,

$a_{11} = 0.0058$ kg/kg, $a_{31} = 0.000041$ kmol/kg, $a_{41} = 0.175$ MJ/kg, $a_{71} = 0.004$ kg/kg.

2) *Iron-ore mine* $b_2 = 0.35$ MJ/kg,

$a_{32} = 0.00023$ kmol/kg, $a_{42} = 0.0735$ MJ/kg

3) *Extraction of natural gas* $b_3 = 802$ MJ/kmol,

$a_{43} = 4.17$ MJ/kmol.

4) *Power plant* $b_4 = 1$ MJ/MJ, $a_{14} = 0.156$ kg/MJ.

5) *Coke plant* $b_5 = 28.95$ MJ/kg,

$a_{15} = 1.60$ kg/kg, $a_{35} = 0.00458$ kmol/kg, $f_{35} = 0.0112$ kmol/kg,

$a_{45} = 0.0958$ MJ/kg, $f_{45} = 0.391$ MJ/kg.

6) *Ore sinter plant* $b_6 = 0.73$ MJ/kg,

$a_{26} = 0.827$ kg/kg, $a_{36} = 0.0002247$ kmol/kg, $a_{46} = 0.1073$ MJ/kg,

$a_{56} = 0.0612$ kg/kg.

7) *Blast furnace plant* $b_7 = 8.75$ MJ/kg liquid pig iron,

$a_{17} = 0.0366$ kg/kg, $f_{17} = 0.0214$ kg/kg, $a_{27} = 0.206$ kg/kg,
$a_{37} = 0.00441$ kmol/kg, $f_{37} = 0.00549$ kmol/kg, $a_{47} = 0.0568$ MJ/kg,
$a_{57} = 0.571$ kg/kg, $a_{67} = 1.594$ kg/kg.

Calculation results of the partial exergy losses connected with the production of coke and pig iron, are presented in Table 1.

TABLE 1. Partial net exergy losses
connected with the production of coke and liquid pig iron (in % of CExC).

Material	Process	Coefficient of consumption			Partial net exergy loss
		gross $a_{ij} - f_{ij}$	net A_{ij}	unit	%
Coke	coal mine	1.600	1.539	kg/kg	1.803
	iron ore mine	-	-	-	0.007
	natural gas extraction	-0.00602	0.0011	kmol/kg	0.002
	coke plant				
	power plant	-	-	-	11.128
	sinter plant	-0.2952	0.0677	MJ/kg	2.918
	blast furnace plant	-	-	-	0.048
		-	-	-	0.160
	cumulative degree op perfection	-	-	-	83.934
Pig iron	coal mine	0.0152	0.0366	kg/kg	1.802
	iron ore mine	0.206	0.206	kg/kg	1.577
	natural gas extraction	-0.00108	-0.0038	kmol/kg	0.077
	coke plant				
	power plant	0.571	0.571	kg/kg	10.251
	sinter plant	0.0568	0.0327	MJ/kg	6.684
	blast furnace plant	1.594	1.594	kg/kg	10.351
		-	-	-	34.315
	cumulative degree of perfection	-	-	-	34.943

6. Ecological Cost

The inevitable depletion of unrestorable natural resources is very dangerous for the future existence of mankind. Exergy can be accepted as the common quality measure of all the natural resources. Therefore the cumulative exergy consumption of unrestorable natural resources, termed *ecological cost,* has been proposed as a measure of their depletion [10],[11],[17],[18]. Set of Equations (12) can be used for the calculation of ecological cost but the quantity β_j should comprise only the unrestorable resources and additionally the deleterious ecological impact of waste products should be taken into account.

The waste products of every process have usually a deleterious impact on the environment. Three kinds of this impact can be named: the corrosion of buildings, machines and equipment, the reduction of agricultural and forest production and the damages in human health. The corrosion is most important. All the mentioned negative effects should be compensated by an additional consumption of useful products substituting the lost useful products or preventing the possible damages [17]. The determination of the coefficients expressing the additional consumption of useful products is very difficult. Also the assumption, that the ecological losses are proportional to the exergy of waste products is not justified. E.g. CO has e relatively great exergy, but does not cause any corrosion of useful products. Therefore Szargut [18] proposed to evaluate the deleterious impact of waste products by means of their monetary index of harmfulness. The additional exergy consumption of unrestorable natural resources due to the emission of waste products (e.g. in kJ/kg) can be expressed as:

$$\xi_k = \frac{B\sigma_k}{NGP + \sum_k P_k \sigma_k} \tag{24}$$

where B - see Eq.(13), σ_k is the monetary index of harmfulness of the k-th waste product, and P_k denotes the annual emission of the k-th waste product in the country.
Eq.(24) takes into account the reduction of national product due to the emission of waste products (the last term in the denominator).

After the introduction of the coefficients ξ_k the set of balance equations determining the value of specific ecological cost takes the form:

$$\rho_j + \sum_i (f_{ij} - a_{ij})\rho_i = \sum_s b_{sj} + \sum_k p_{kj}\xi_k + \sum_r a_{rj}\rho_r \tag{25}$$

where ρ_j, ρ_i denotye, respectively, the specific ecological cost of the major product of j-th and i-th process, b_{sj} is the exergy consumption of the s-th unrestorable natural resource, per unit of the j-th product, p_{kj} denotes the emission of the k-th waste product per unit of the j-th product and ρ_r is the specific ecological cost of the imported r-th semi-finished product, Eq.(13).
Set of Eqs.(25) can be separately formulated for a group of production processes having week connections with the remaining part of the network of production processes.

The ecological cost of the imported semi-finished products (e.g. fuels) is smaller than their exergy, if in the exported goods prevail the highly processed products or products obtained with a great participation of restorable exergy resources (agricultural or forest products).

Example 2. The annual exergy consumption of domestic unrestorable natural resources in Poland contains the exergy B_{hc} of hard coal, B_{bc} of brown coal, B_{ng} of own natural gas, B_{cr} of own crude oil, B_S of sulfur and B_{Cu} of copper ore. In 1995 the mentioned values were:

$$B = B_{hc} + B_{bc} + B_{ng} + B_{cr} + B_S + B_{Cu} =$$

$$= 3315 \cdot 10^9 + 579 \cdot 10^9 + 145 \cdot 10^9 + 22 \cdot 10^9 + 47 \cdot 10^9 + 15 \cdot 10^9 = 4123 \cdot 10^9 \text{MJ/a}$$

The national gross product was $119.8 \cdot 10^9$ \$/a. The ecological cost per unit of national gross product was 34.4 MJ/\$. The specific cost of the imported natural gas was 2.24 \$/kmol. So the ecological cost of the imported gas was 77.1 MJ/kmol (about 10 times smaller than the specific exergy). Hence the import of natural gas distinctly decreases the depletion of the domestic unrestorable natural resources [19].

Example 3. The annual emission of the main waste products (SO_2, NO_x and dust) was in 1995 in Poland:

$$P_{SO_2} = 2337 \cdot 10^3, P_{NO_x} = 1120 \cdot 10^3, P_d = 1308 \cdot 10^3 \text{t/a}.$$

Following monetary indices of harmfulness have been accepted:

$$\sigma_{SO_2} = 1500, \sigma_{NO_x} = 1500, \sigma_p = 300 \text{ \$/t}.$$

The indices of ecological cost of waste products, resulting from Eq.(24) are:

$$\xi_{SO_2} = 49305, \xi_{NO_x} = 49305, \xi_p = 10190 \text{ MJ/t}.$$

7. Final Remarks

The ecological cost can be used for the optimization of the design- and operational parameters of thermal plants and installations from the view point of the minimization of depletion of unrestorable natural resources. However some problems should be additionally solved, e.g.:

1. The methodology of optimization should be developed. The problem of the necessity to introduce some kind of discount calculation has to be solved [3].
2. The minimization of ecological cost would prefer simple processes with a great consumption of human work, but a small consumption of natural fuels. The human labor does not appear in the calculation of ecological cost. Presumably the consumption of human work should appear as a limitation in the optimization of the ecological cost.
3. It has to be concluded, how to take into account the consumption of natural resources recoverable with some delay (e.g. the consumption of timber).

A most effective method of the economy of unrestorable natural resources results from the utilization of restorable resources. It should be however not forgotten, that in the processes of utilization of restorable resources appears also some consumption of unrestorable resources.

8. References

1. Gouy G. (1889) Sur l'énergie utilisable. *Journal de physique* II, **8**, 501—518.
2. Grassmann P. (1959) Die Exergie und das Flußbild der technisch nutzbaren Leistung, *Allg. Wärmetechn.* **9**, No 4/5, 79—86.
3. Iantovski E. (1997) What is exergonomics? (on a branch of engineering thermodynamics), *Proc. FLOWERS'97*, Florence, 1163—1174.
4. Rant Z. (1956) Exergie, ein neues Wort für "technische Arbeitsfähigkeit", *Forsch. Ing.- Wes.* **22**, No 1, 36—37.
5. Riekert L. (1974) The efficiency of energy utilization in chemical processes, *Chem. Eng. Sci.* No 29, 48—51.
6. Sama D. (1992) A common sense 2nd law approach to heat exchanger network design, *Proc. ECOS'92*, Zaragoza, 329—338.
7. Stodola A. (1898) Die Kreisprozesse der Gasmaschine, *Z.d.VDI* **32**, No 38, 1086—1091.
8. Szargut J. (1956) Potential balance of physical processes resulting from the 2nd law of thermodynamics (in Polish), *Arch. Budowy Maszyn* **3**, No 3, 231—276.
9. Szargut J. (1967) Grenzen für die Anwendungsmöglichkeiten des Exergiebegriffs, *Brennstoff-Wärme-Kraft* **19**, No 6, 309—313.
10. Szargut J. (1978) Minimization of the consumption of natural resources, *Bull. Pol. Acad., Sci. Techn.*, No 6, 611—615.
11. Szargut J. (1986) Application of exergy for the calculation of ecological cost, *Bull. Pol. Acad., Sci. Techn.*, No 7-8, 475—480.
12. Szargut J. and Morris D.R. (1987) Cumulative exergy consumption and cumulative degree of perfection of chemical processes, *Energy Research*, No 11, 245—261.
13. Szargut J. (1987) Influence of the imported goods on the cumulative energy indices. *Bull. Pol. Acad., Sci. Techn.* **35**, No 9-10, 591—595.
14. Szargut J. (1988) Exergy losses in the chains of technological processes. *Bull. Polish Acad., Sci. Techn.* **36**, No 7-8, 513—521.
15. Szargut J. and Majza E. (1989) Thermodynamic imperfection and exergy losses of the production of pig iron and steel, *Arch. Hutnictwa*, No 2, 197—216.
16. Szargut J. and Sama D.A. (1995) Practical rules of the reduction of energy losses caused by the thermodynamic imperfection of thermal processes, *Proc. The Second Intern. Thermal Energy Congress*, Agadir, 782—785.
17. Szargut J. (1995) Exergy and ecology, *Proc. of the Workshop" Second Law Analysis of Energy Systems"*, Roma, 11—19
18. Szargut J. (1997) Depletion of the unrestorable natural exergy resources, *Bull. Pol. Acad., Techn. Sci.* **45**, No 2, 241—250.
19. Szargut J. and Stanek W. (1998) Influence of the import of natural gas on the depletion of unrestorable domestic natural resources (in Polish), *Gospodarka Paliwami i Energią*, No 2, 2—4.
20. Tribus M. and Evans R. (1962) Thermoeconomics, *UCLA Report*, No 52-63.
21. Tsatsaronis G. and Pisa J. (1992) Exergoecomomic evaluation and optimization of energy systems, application to the CGAM problem, *Proc. ECOS'92*, Zaragoza.
22. Valero A., Torres C. and Serra L. (1992) A general theory of thermoeconomics, part I, structural analysis, *Proc. ECOS'92*, Zaragoza 1992, 137—145.
23. Valero A. and Lozano M.A. (1992) A general theory of thermoeconomics, part II, the relative free energy function, *Proc. ECOS'92*, Zaragoza, 147—154.

ALLOCATION OF FINITE ENERGETIC RESOURCES VIA AN EXERGETIC COSTING METHOD

E. SCIUBBA
Dipartimento di Meccanica e Aeronautica,
Università di Roma 1,
via Eudossiana 18 - 00184 Roma, Italy

1. Introduction

This paper describes a method for the assessment of an "exergetically optimal" resource allocation and distribution. *Allocating* a certain resource R means to make it available in a given amount at a finite or infinite number of "consumption centres", and to find the optimal siting of the production and/or of the consumption points, so that the resource is distributed as needed, under the specified constraints, and its distribution pattern is such that it minimises the overall *exergy* consumption caused by the technological chain of processes the resource undergoes from its extraction to its final use (including disposal of residues). Two general formulations are possible: in the discrete problem, the number of consumption sites as well as that of the production centres is finite, and usually rather small. In the continuum problem, while the number of production sites remains small, the possible consumption points are distributed uniformly over the entire region of interest, so that in effect the resource must be made available everywhere in the domain. In both cases, the practical implementation of the method consists in the minimisation of an objective function that contains the extraction, preparation, transportation, distribution and environmental costs of the resource: *the important novel feature here is the fact that these costs are expressed in exergetic units*, i.e., they are represented by the amount of *exergy* that is expended for that particular operation. The constraints are topological (the orography and the geography of the region), technical and environmental, and are imposed in the form of *exergetic penalty functions.* For the continuum formulation, a similarity with a diffusion problem can be invoked to extend the mathematical analogy to the solution of a general dual dispersion equation: the reader is referred to a companion paper [10].

2. Problem Position

Consider a generic geographical area *D* in which N Production Centres P_i (i=1,2, ...N) for a single resource R are located (Fig. 1)[1]. In the same area, but of course in different sites, there will also be M Consumption Points C_j (j=1,2,...M) for that resource. In the

[1] The map of Romania has been downloaded via Internet from MapQuest™. The location of Production and Consumption centers is arbitrary and hypothetical, and serves only as an example of application of the method proposed in this lecture.

A. Bejan and E. Mamut (eds.), Thermodynamic Optimization of Complex Energy Systems, 151–162.

entire course of this Lecture, we shall make the following assumptions:

- each one of the N Production Centres is characterised by a specific production cost $g_i\,(x,y,m_i)$ whose structure is known;
- the transportation cost of the resource from Production Centre to Consumption Site is a function of the equivalent distance d_{ij} and of the quantity of resource made available at Consumption Site, m_{ij};
- the equivalent distance d_{ij} is a computable function of the actual distance r_{ij}, of the oro-geographic conditions and of the actual transportation line and means;
- all "costs" are to be intended as exergetic, and not monetary, costs. The specific exergetic equivalent cost we shall refer to is the total "exergetic content" of the unit mass of resource, whereby by *exergetic content we mean the sum of the thermodynamic exergy of the untapped resource and of all exergetic expenditures incurred in during the processes of extraction, refinement, transportation, packaging and final distribution (including disposal of residues)*;
- transportation mass losses are negligible;
- the problem is at steady-state.

The following three formulations shall be considered:

1. given the $C_j\,(x,y,m_j)$ find the optimal siting for the P_i's for a specified $m_{tot} = \sum_{j=1}^{M} m_j$;
2. given the $P_i(x,y,m_i)$, find the optimal siting of the C_j's for the prescribed $m_{tot} = \sum_{i=1}^{N} m_i$;
3. given $C_j(x,y,m_j)$ and $P_i(x,y,m_i)$, find the optimal resource allocation (the distribution of the m_j's among the consumption sites) such that the overall exergetic cost is minimised and the mass flow rate constraints are abided by.

A fundamental distinction must be made depending on the demand structure for the resource R: if the demand is *concentrated*, i.e., if there are few and relatively wide apart Consumption Sites, a discrete formulation of the problem is in order. If, on the other hand, the demand for R is *distributed*, i.e., almost everywhere in D there are potential Consumption Sites for R, then a continuum approach is needed. The two cases are discussed in Sections 3 and [10] respectively.

3. The Discrete Problem: Concentrated Consumption Sites

3.1. OPTIMAL SITING OF THE PRODUCTION CENTRES FOR A KNOWN DISTRIBUTION OF CONSUMPTION SITES

3.1.1. One Production Centre, one Consumption Site

The *specific exergetic* cost k of resource R at location C is given by the sum of the production, transportation and "external" costs:

$$k = g(x_p, y_p, m) + f(d_{pc}, m) + e(d, m) \tag{1}$$

Since by hypothesis f, g and e are known or computable (see Appendix 2), it will be always possible to draw iso-cost curves on the map of D. An *inhibitory* constraint may be active (there is a certain area A_c around the Consumption Site in which no Production Centre may be located), and the solution is obvious: *P must be localised on a point lying on the lowest-k curve AND not contained in the forbidden area* A_c. A simple (hypothetical) example is shown in Fig. 2.

3.1.2. Multiple Production Points, one Consumption Site

The *total exergetic* cost *K* of resource R at location C is given by:

$$K = \sum_{i=1}^{N} m_i \left[g(x_i, y_i, m_i) + f(d_{ic}, m_i) + e(d_{ic}, m_i) \right] \qquad (2)$$

where m_i is the mass flow rate reaching C from P_i. Since *f*, *g* and *e* are known or computable, the problem can be restated:

$$K = \sum_{i=1}^{N} A_i m_i \qquad (3)$$

where A_i is a function of (x_i, y_i) and (x_c, y_c) only, through d_{ic}, g_i and e_{ic}. A mass conservation constraint applies:

$$m_c = \sum_{i=1}^{N} m_i \qquad (4)$$

This formulation corresponds to a linear programming problem in its canonical form (the minimum of *K* must be found). Again, taking into consideration possible inhibitory constraints, the solution in this case is: *the* P_i *'s must be localised on any set of N points which lie on the lowest K-curve and are not contained in any of the forbidden areas* A_g. A simple example is shown in Fig. 3. Notice that in this case additional constraints on the locations of the P_j's relative to each other might be imposed ("*proximity*" constraints, see Appendix 1).

3.1.3. Multiple Production Points, multiple Consumption Sites

The total exergetic cost *K* of the resource at each location C_j is given by :

$$K_j = \sum_{i=1}^{N} m_{ij} \left[g(x_i, y_i, m_j) + f(d_{ij}, m_j) + e(d_{ij}, m_j) \right] \qquad (5)$$

where $m_j = \sum_{i=1}^{N} m_{ij}$. The total exergetic cost is found by summing over the index j:

$$K_{tot} = \sum_{i=1}^{N} \sum_{j=1}^{M} m_{ij} \left[g(x_i, y_i, m_j) + f(d_{ij}, m_j) + e(d_{ij}, m_j) \right] \qquad (6)$$

and the problem is again one of linear programming (the minimum of K_{tot} must be found). There are two types of mass-conservation constraints in this case:

$$m_i = \sum_{j=1}^{M} m_{ij} \qquad \text{(N additional constraints)} \qquad (7)$$

$$m_j = \sum_{i=1}^{N} m_{ij} \qquad \text{(M additional constraints)} \qquad (8)$$

and N congruency constraints, that require that for each Production Site $\sum_{j=1}^{M} m_{ij} \le m_{i,max}$. Again, possible inhibitory and proximity constraints must also be taken into account. The solution prescribes that *the P_i's must be localised on any N of the points which lie on the lowest-K_{tot} curve AND are not contained on any of the forbidden areas A_g*.

3.2. Optimal siting of the Consumption Sites for a specified distribution of the Production Centres

The general discrete allocation problem formulated here is clearly symmetric with respect to "sources" and "sinks": if in the expressions hitherto computed for the exergetic cost the indices *p* and *c* (or *i* and *j*) are inverted, the corresponding formulae give the solution of the symmetric problem, i.e., the one in which the location and strength (productivity) of the Production Centres are both known, and the solution consists in finding the optimal siting for the Consumption Points. Therefore, no specific discussion of this case is offered here, and the reader is referred to the previous Section (and to [3] for a more detailed treatment).

3.3.- Optimal allocation of a known amount of resource generated at N Production Sites and distributed to M Consumption Centres

The problem here is to find the optimal combination of the m_{ij}, i.e. to determine how much resource ought to be generated from each production centre P_i and transported to each consumption site C_j. The total cost is again given by equation 6), and the same constraints given by equations 7) and 8) apply. The N congruency constraints are the same as well, so that the Linear Programming problem is formally identical to that illustrated in Section 2.1.3.

4. Conclusions

A new formulation of the resource allocation problem has been outlined in this paper. Several simplifying assumptions have been made, most of them aimed at linearising the problem. The following conclusions can be drawn:

1. For resources that admit of *localised Consumption Sites*, discrete optimisation methods based on linear programming techniques can be efficiently employed to achieve at least a quasi-optimal allocation; such resources are mineral ores, fossil fuels, and in general all those materials that must undergo substantial transformations before being suitable for industrial or commercial use;
2. For resources that admit of *uniformly distributed Consumption Sites*, the method can still be applied, and admits of a very powerful and elegant mathematical formulation based on the dual form of the dispersion equation. The details are developed in [10]. The resources amenable to be treated under a dispersion analogy are typically those that can be consumed as-such: food, commercial goods and the like, water, electricity or district utilities;
3. It must be stressed that the possibility of practical application for this entire approach depends on the availability of a coherent and reliable cost basis. The cost functions employed must allow for the consistent incorporation of different expenses occurred in distinct locations and in diverse phases of the extraction-transportation-distribution chain; moreover, they must allow for the quantification

of physically different "costs" (value of the Gibbs' function for a fuel, energy used for transportation, extraction costs of a resource, etc.) in a homogeneous way. This cannot be achieved using monetary costs. The exergetic cost adopted here (first proposed by [12,13,14]) not only leads to the formulation of a proper, complete and homogeneous objective function, but *it evaluates the flux of resource for its value, and not for its price*: the difference is substantial, and carries far-reaching implications if correctly applied to the analysis of an industrial system.

Since a hierarchy of cases can be easily constructed (a resource R from N wells is allotted among M pre-treatment stations; then, each one of these stations becomes the "producer" of an R-derivative which is distributed among M_1 chemical industries, each one of which becomes a "producer" of several types of further derivatives, and so on until final consumer production/distribution is reached), the method described in this Lecture can be repeatedly applied to the entire chain of resource extraction, (production), transformation and distribution, to investigate the possibility of existence of "optimal paths" in the resource use.

5. References

1. Ayres R. (1995): *Life cycle analysis: a critique*, Res. conservation and recycling, n.14.
2. Azzarone F.,Langer P. (1996): *Optimal Resource Allocation Model*, TR-CIRCUS 010996, Univ. of Roma.
3. Azzarone F.,Sciubba E. (1995): *Analysis of the energetic and exergetic sustainability of complex systems*, in ASME AES-35.
4. Evans R.L. (1980): *Thermo-economic isolation and Essergy Analysis*, Energy, vol.5 n.8-9.
5. Gaggioli R.A.,El-Sayed Y. (1987: *A critical review of second-law costing methods*, in ASME vol.I00236.
6. Gong M.,Wall G. (1997): *On Exergetics, Economics and optimization of technical processes to meet environmental conditions*, Proc. TAIES'97, Beijing (China).
7. Marchuk G.I. (1986): *Mathematical models in environmental problems*, North Holland.
8. Oni A.A.,Cambel A.B. (1985): *Generalised thermodynamic approach to scarce energy resources allocation through the disjointed incrementalism algorithm*, in ASME-AES-1.
9. Paolocci S.,Sciubba E. (1994): *A new criterion for power plant siting based on a dual formulation of the dispersion of pollutants in the atmosphere*, ASME-AES vol.33.
10. Sciubba E. (1998): *A novel exergetic costing method for determining the optimal allocation of scarce resources*, Proc. Contemporary Problems of Thermal Engineering, A.Ziebik ed., Gliwice (Poland).
11. Szargut J. (1997): *Depletion of Unrestorable Natural Exergy Resources*, Bull.Polish Ac.of Sci.
12. Szargut J.,Morris D.R.,Stewart F.R. (1988): *Exergy analysis of thermal, chemical and metallurgical processes*, Hemisphere,Washington, DC.
13. Valero A.,Lozano M.A.,Munoz M. (1986): *A general theory of exergy saving, Part 1: on the exergetic cost*, in ASME-AES-2/3.
14. Wall G. (1986): *Exergy: a useful concept within resource accounting*, Ph.D.Th., Chalmers T.U., Göteborg, Sweden.
15. Ziebik J.,Gwozdz J.,Presz K. (1994): *Matrix method for calculating the unit costs of energy carriers as a co-ordination procedure in the optimisation of industrial energy systems*, in ASME PD-64/3.

Appendix 1. The Solution Method

To begin with, let us observe that the general formula given in equation 6 can be applied to the simple cases of only one Production Centre or only one Consumption Site by proper contraction of the corresponding index. Therefore, we will take equation 6 to represent the general solution to the optimal resource allocation problem studied here. This equation is non-linear (the cumulative yearly resource production appears in the specific cost functions), but can be quasi-linearised by devising a proper quasi-continuous formulation for the cost functions (see Appendix 2). The constraints considered here are of three different types:

1) *mass conservation* (strong constraint):
   1a - the total mass flow of resource from each Production Centre P_i must be equal to the sum of the (at most) M mass flow rates supplied by P_i to the various C_j it is connected to;
   1b - conversely, the total mass flow of resource into each Consumption Site C_j must be equal to the sum of the (at most) N mass flow rates supplied to C_j by the various P_i it is connected to;
2) *congruence* (weak constraint): the maximum extractive capacity of each Production Centre $m_{i,max}$ shall not be exceeded;
3) *minimum distance of any Production Centre from any Consumption Site* (weak constraint). These are the so-called *inhibitory constraints*, and set a lower bound to the physical distance r_{ij} between any P_i and any C_j;
4) *minimum distance between different Production Points* (weak constraint). These are the so-called *proximity constraints*, and set a lower bound to the physical distance between each pair of P_i's.

The solution procedure can then be outlined as follows:

1. the region under examination is discretised according to two distinct criteria: first, areas of similar oro-geographical conformation are detected, in which the *penalty factor* γ (a coefficient ≥ 1 by which the physical distance r_{ij} must be multiplied to find the equivalent distance d_{ij}) can be assumed to take a constant value; second, areas of similar *environmental fragility* (see Appendix 2) are identified, in which the marginal environmental exergetic cost takes constant values. The mesh adopted in the final calculations is then the finer of the two;
2. the cost functionals are computed, with the formulae reported in Appendix 2;
3. the constraints are imposed in discrete form. Weak constraints are always imposed in their equivalent strong form (e.g.: both "less than or equal to" and "greater than or equal to" are imposed as "equal to");
4. a standard linear programming algorithm is employed to find the minimum of the exergetic cost expressed by equation 6.

The errors deriving from the discretisation procedure are unknown and difficult to estimate *a priori*: therefore, the solution algorithm yields in practice only a quasi-optimal allocation. These computational errors have two different sources: first, they are caused by the functional approximations introduced by the quasi-linearisation of the cost functions; second, they are originated by the discretisation approximations (like for instance that of applying a constant penalty factor to a relatively large and non-homogeneous geographical area).

Appendix 2. The Cost Functions

A2.1. THE EXTRACTION/REFINING/PRODUCTION COST

This cost is associated with each one of the N production centres, and is assumed to be of the form:

$$g_i = \frac{c_{1,i}}{m} + c_{2,i} + c_{3,i} + c_{4,i} \qquad \text{(A.1)}$$

both g, c_2, c_3 and c_4 are expressed in kJ/kg, and c_1 in kW.

A2.1.1. Fixed costs, c_1

The fixed costs c_1 include land-, factory buildings-, infrastructures and facilities costs. In real life, they show a marked dependence on the total production rate of resource (m_R) and to the global yearly output μ. In the discretisation employed here, they have though taken to be constant (Fig. 4).

A2.1.2. Cost of the Resource, c_2

An intrinsic specific cost (per unit mass) c_2 can be attributed to the resource R itself, to represent its exergetic value in the conditions in which R is extracted. A distinction is necessary between renewable and non-renewable resources: while the marginal exergetic value of a renewable resource can be taken to be independent of the mass consumed, a non-renewable resource has a marginal exergetic cost that increases with the mined amount (Fig. 5).

A2.1.3. Process costs, c_3

In real resource tapping operations, process costs c_3 are a function of both total yearly yield (μ) and maximum productivity (m_R). Here, a stepwise linearisation of c_3 with m_R and μ has been adopted (Fig. 6), which amounts to "discretising" scale effects.

A2.1.4. Operational costs, c_4

These costs represent the amount of exergy expended in the chain of mechanical, thermal and chemical processes that bring the resource R from its "natural" to its "final" state. For a given technology, scale effects can be assumed to be negligible: a constant function of m_R is proposed here.

A2.2. THE TRANSPORTATION/DISTRIBUTION COST

This cost pertains to the distribution network, and has been assumed to have the structure:

$$f_i = t_{1,i} + t_{2,i} + t_{3,i} d_{ij} \qquad \text{(A.2)}$$

i.e., the functional dependence has been split into a fixed (capital) cost t_1, a "friction" effect t_2 and a distance/difficulty cost t_3: f, t_1 and t_2 are expressed in kJ/kg, t_3 is in kJ/(kg*m)=m/s^2. Notice that $d_{ij}=\gamma r_{ij}$, the *penalty factor* γ depending on the oro-geographical conditions of the region. The total transportation cost for N Production Centres serving M Consumption Sites is:

$$F = \sum_{i=1}^{N} \sum_{j=1}^{M} m_{ij} f_{ij} \qquad \text{(A3)}$$

A2.2.1. Fixed cost, t_1

This cost represents the (specific) exergy spent to set up the transportation/distribution network. It has been assumed here to depend only on the transportation means, but in reality it varies (at times discontinuously) with both the total annual yield and with the mass flow rate of the resource on that specific route (Fig. 7).

A2.2.2. "Friction" cost, t_2

This cost coefficient depends on the means of transportation employed and on the technological level of the distribution chain. As an example, Fig. 8 represents the functional dependence of t_2 on m_{ij} for a hypothetical transport of the same resource via barge and by truck. Notice that there is an implicit correlation between the penalty factor γ and t_2, as shown by the fact that the transport by barge in the upstream direction carries a higher exergetic cost than that in the downstream direction, similar in principle to the difference between the exergetic costs of transportation by truck on flat (cheaper) or hilly (more expensive) terrain shown in Fig. 9.

A2.2.3. Distance cost, t_3

This cost coefficient depends mainly on the physical distance between the Production and the Consumption Sites. A functional dependence on the type and state of terrain, climate, etc. must be also considered, by introducing the corrected (d_{ij}) instead of the physical (r_{ij}) distance. This marginal cost coefficient is also dependent (for the same resource) on the type of transportation means employed.

A2.3. THE ENVIRONMENTAL COST, e

Production and transportation costs are not the only expenditures that are incurred in when allocating a resource. There are other costs, unaccounted for as yet, namely the environmental pollution generated by the exploitation of natural resources, and the costs associated with it (which may include social- and health effects as well). An advantage of the allocation method just described is that it allows a quantitative estimate of environmental effects to be included in the objective function. To begin with, it has been assumed here that the *harmful effects of the release of waste products into the environment can be expressed quantitatively in terms of the exergetic value of the actions necessary to nullify their effects*. This is not a new assumption: it has been proposed -with slightly different meanings- by several Authors [1,2,6,10,11,14]. A discussion on the methods of quantification of the exergetic value of a waste stream is outside of the limits of this Lecture: we shall assume that such a quantification can be carried out and that it produces univocal results, and adopt a heuristic criterion, by dividing the task into three sub-tasks:

a) determine the chemical and physical nature of the released waste stream;
b) estimate the portion of the biosphere that is directly affected by this release;
c) calculate the already existing concentration of waste products of any kind in the geographical area directly affected by the "new" release.

The proposed cost assessment procedure is then the following steps. First, the environmental costs are considered to be directly related to the clean-up costs, i.e., to the costs of the re-establishment of the pristine environmental situation: this "pristine" environmental situation is taken to be the one existing in the region D prior to the introduction of the Production Centres P_i and of the corresponding transportation lines. Second, the environmental cost is split in two terms: one is a direct measure of the exergetic costs of the clean-up, and the other is a quantification of the present state of (environmental) affairs in the region. This last cost component quantifies the concept that the cumulative effects of a "new" pollution source are more costly if the region is already polluted. It is clear that a higher mass flow rate of "new" releases would produce higher restoration costs, and it is also clear that release of wastes in areas which have already been contaminated by previous releases carries an extra cost penalty: this can be quantified by introducing a coefficient of *environmental fragility*, and assuming that the exergetic cost of environmental damage be a function of site fragility and of the mass flow rate of the new release (Fig. 10). The environmental cost can therefore be computed with the following formula:

$$e_i = \frac{\Delta A_D a_{o,j}}{m_{ij}} + b_j \qquad \text{(A.4)}$$

where ΔA_D is the area of the portion of D on which the cost is being calculated, a_{oj} is the fragility coefficient (kW/m^2) and b_j (kJ/kg) is the marginal cost of the new release. The total cost for N Production Centres serving M Consumption Sites is:

$$E = \sum_{i=1}^{N}\sum_{j=1}^{M} m_{ij} e_{ij} \qquad \text{(A.5)}$$

and can be derived as the sum of several "route" integrals (Fig. 10).

In the method presented here, *the environmental cost has been considered as a constraint for the allocation procedure*: all other parameters being the same, the siting of the Production Centres (considered as concentrated pollution sources) is optimal when they are located in the least fragile areas. If several areas of equal fragility are available, then the cost function is the discriminating parameter (see [2] for an example of application).

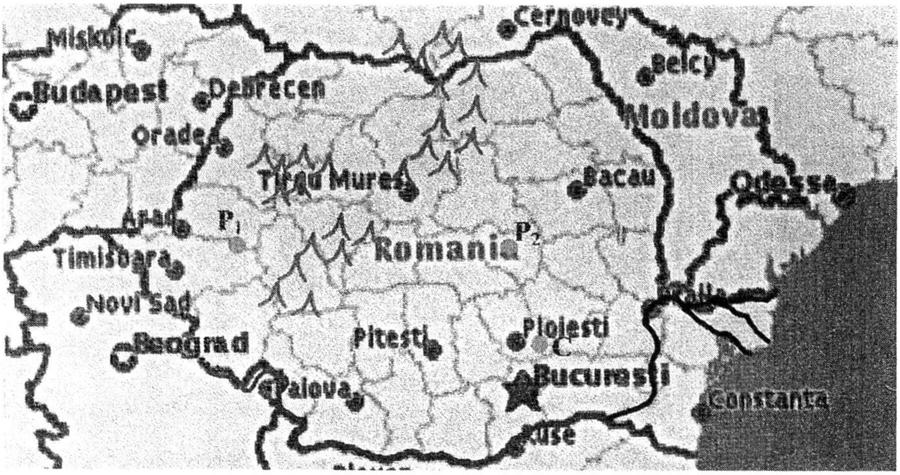

Figure 1. Problem position (two possible production centers shown).

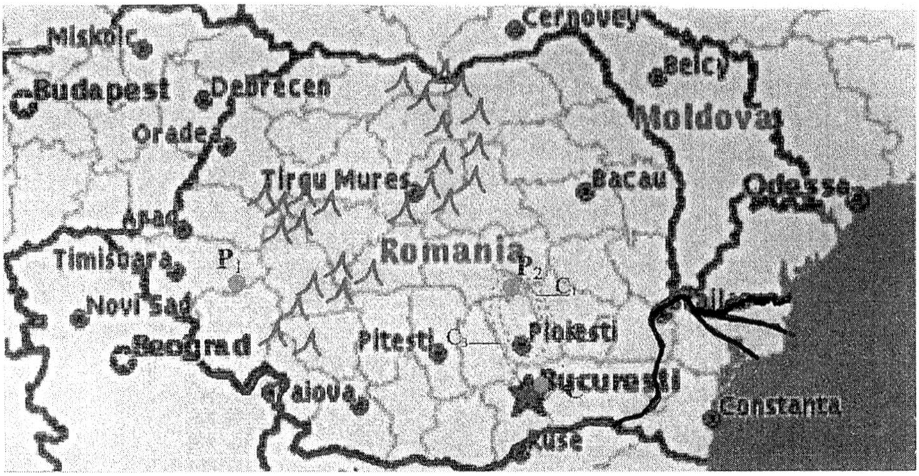

Figure 2. One production center, one consumption site: iso-cost curves ($c_{i+1} > c_i$).

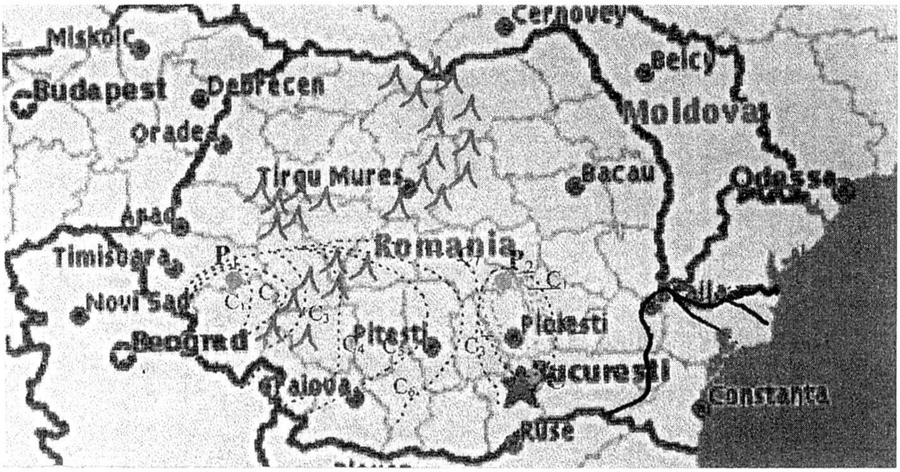

Figure 3. Two production centers, one consumption site: iso-cost curves.

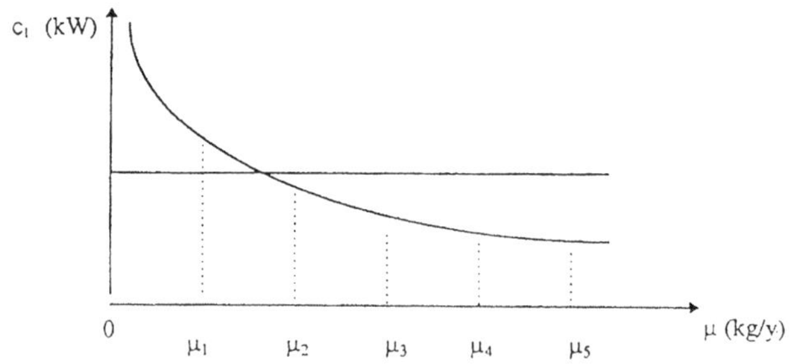

Figure 4. Fixed cost coefficient c_i as a function of the yearly mass output of resource.

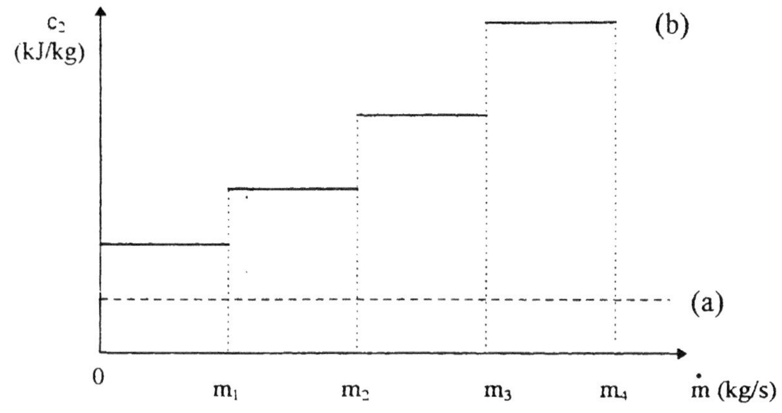

Figure 5. Resource cost coefficient c_2 for a renewable (a) and non-renewable (b) resource.

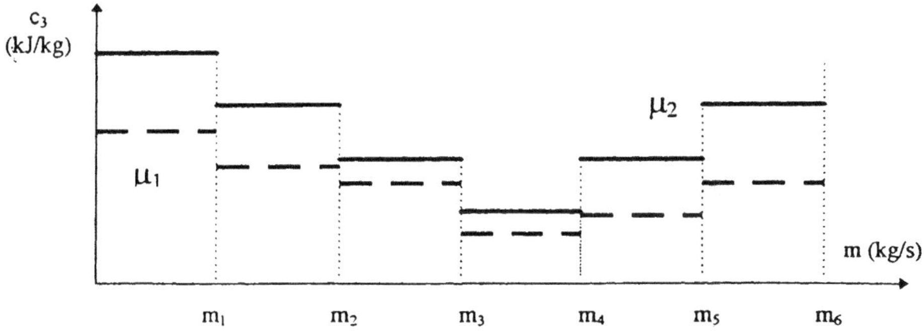

Figure 6. Process cost coefficient c_3 as a function of the production mass flow rate and of the yearly output ($\mu_1 > \mu_2$).

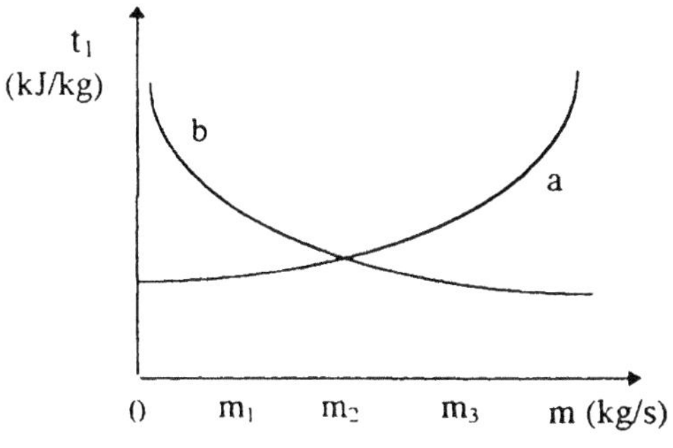

Figure 7. Fixed cost coefficient t_1 for transportation by pipeline (a) and by truck (b)

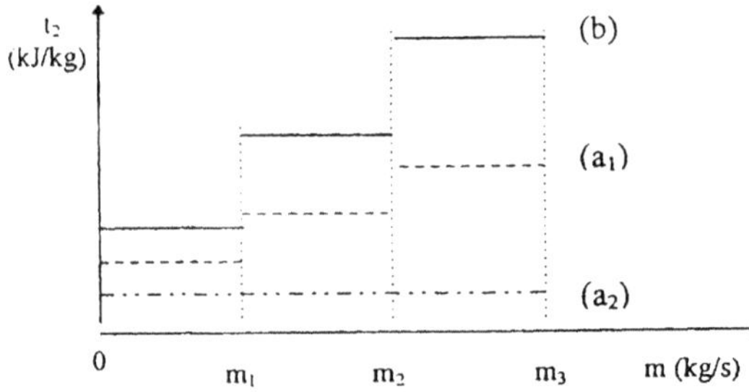

Figure 8. "Friction" cost t_2 for transportation by barge up- (a_1) and downstream (a_2), and by truck (b)

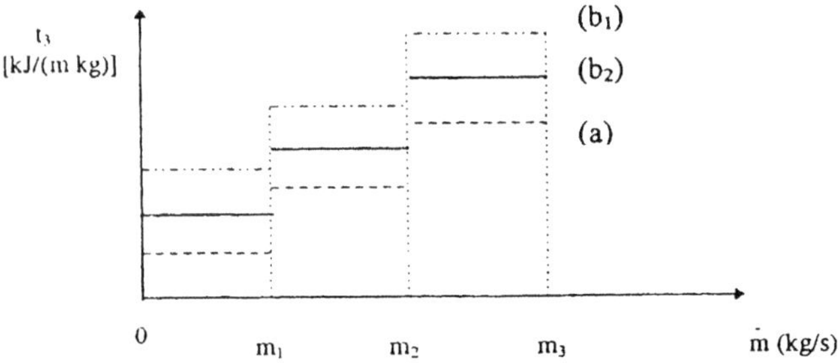

Figure 9. Distance cost coefficient t_3 for transportation by barge (a) and by truck on hilly (b_1) and flat (b_2) terrain

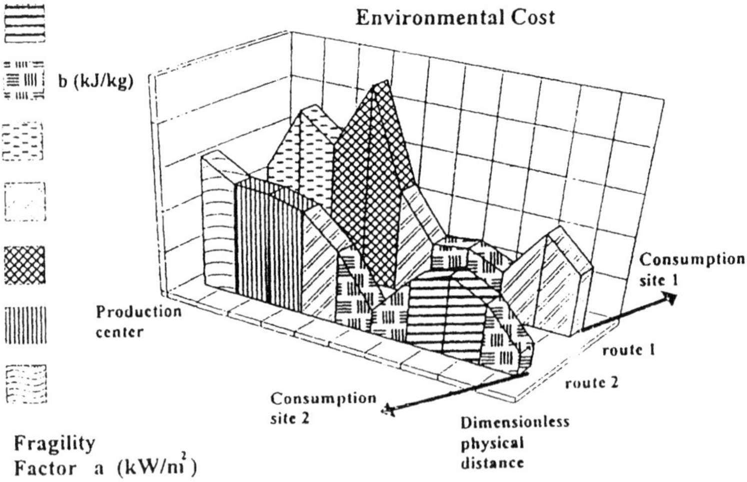

Figure 10. Example of calculation of the environmental cost for two different routes.

OPTIMISATION OF TURBOMACHINERY COMPONENTS BY CONSTRAINED MINIMISATION OF THE LOCAL ENTROPY PRODUCTION RATE

E. SCIUBBA
Dipartimento di Meccanica ed Aeronautica,
Università di Roma 1 - "La Sapienza"
via Eudossiana 18 - 00184 Roma, Italy

1. Introduction

This Lecture presents some reflections on the practical measure of thermodynamic irreversibilities affecting the interactions of a process fluid with a part of a turbomachine: the main contention here being that *the entropy generation rate is the only thermodynamic quantity that can be used to correctly estimate viscous and thermal losses, and to compare both the absolute and the relative performance of different configurations.* An entropic analysis (i.e., the calculation of the entropy generation rates associated with the various types of "losses") can be performed at a bulk, at a microscopic and at a macroscopic level. The *bulk* level is the one usually employed in lumped formulations, where either the entire system or portions thereof are considered as a control volume, and the entropy generation rate is computed via the usual thermodynamic expressions derived by direct manipulation of the constitutive relations. The *microscopic* or *local* level is the method advocated here: it consists of separating the two different forms of irreversible entropy production in fluid flow, due to thermal (s_t) and viscous (s_v) effects respectively, and of expressing them as functions of the local velocity- and temperature gradients and of the local temperature. The *macroscopic* or *global* level consists in integrating the results of the microscopic analysis over the domain of interest: it produces a value of the total entropy generation rate quantitatively very close to that obtained by the bulk method, but different in principle from it, because calculations at the global level take explicitly into account the non-uniformity of the medium.

This lecture describes in detail a general procedure by which cases of practical engineering relevance can be numerically evaluated in order to obtain useful *quantitative* design information about the relative performance of different configurations. An optimisation procedure is presented, whose objective function is the global entropy generation rate and which contains appropriate integral design

A. Bejan and E. Mamut (eds.), Thermodynamic Optimization of Complex Energy Systems, 163–186.

constraints, like the maximum material temperature (weak constraint, applies to both fixed and rotating passages) and the overall useful enthalpy drop of the working fluid in the passage (strong constraint, applies to rotating passages and to heat exchanging elements). The discontinuous optimisation applications implemented to-date are limited to a trial-and-error technique, but the thermodynamic basis upon which they rest can be rigorously formalised, and this is one of the objectives of the Lecture. The proposed optimisation technique consists of four steps:

1. consider an existing, realistic geometry and regard it as a "zero configuration";
2. compute the thermo-fluiddynamic field around it and calculate the entropy generation rates in the passage, recording the value of the overall enthalpy drop of the gas as well as the value and location of the maximum material temperature;
3. keeping all other parameters fixed (in particular, maintaining the same kinematic, dynamic and thermal boundary conditions), slightly perturb the geometry of the passage, by introducing arbitrary modifications in the blade profile, in the tube diameter and pitch, in the number and location of cooling holes, etc.;
4. compute anew the velocity and temperature fields and, if necessary, tune the inlet conditions so that the integral constraints are abided by; recompute the entropy generation rates and compare them to those of the previous configurations.

By iterated application of the very strong heuristics which can be derived from the entropy generation rate, an "optimal" configuration can be found that still satisfies the design specifications but is affected by minimal losses. Some applications of the procedure are discussed, for realistic gas turbine blade profiles, both in 2-D and in 3-D, and for a finned-tube heat exchanger. Such an approach was implicitly already contained in the first formulation of an "entropy minimisation strategy" outlined in [1]: the innovative character of the analysis presented here consists in the fact that *local entropy generation rates are extracted from the flow field and can thus be directly used as objective function in a (discrete) pseudo-optimisation procedure*. The method results in a very powerful tool for the design of "better" turbomachinery components, because the ratio of the *global* value of the entropy generation rate (i.e., of its integral over the entire computational domain) to the total enthalpy drop across the blading or the heat exchanging channel is proportional to the cascade efficiency vz. to the heat exchanger effectiveness, and because the *local* entropy generation values, properly split into their *viscous* and *thermal* components, give a complete and thermodynamically correct indication of the "problem areas" in which design changes ought to be introduced to improve component performance.

2. The Calculation Procedure for the Local Entropy Generation Rates and the Entropy-Based Constrained Minimisation Procedure

The governing equations of the phenomena described here are the Navier-Stokes equations, the continuity- and the energy equation; the formulation applies to the laminar or turbulent, steady, incompressible flow of a fluid of constant properties. The governing equations and the numerical techniques are extensively discussed in [3,14,15,19]. With the exception of the results presented in Section 3.3.1, the entropy generation rates are computed from the converged flow field using the following equations:

$$s_v = \frac{1}{T}\left[\left(\frac{\partial u}{\partial x}\right)^2 + \left(\frac{\partial v}{\partial y}\right)^2 + \left(\frac{\partial w}{\partial z}\right)^2\right] + \frac{1}{2}\left[\left(\frac{\partial u}{\partial x} + \frac{\partial v}{\partial y}\right)^2 + \left(\frac{\partial v}{\partial z} + \frac{\partial w}{\partial y}\right)^2 + \left(\frac{\partial u}{\partial z} + \frac{\partial w}{\partial x}\right)^2\right] \tag{1}$$

$$s_t = \frac{1}{T^2}\left[\left(\frac{\partial T}{\partial x}\right)^2 + \left(\frac{\partial T}{\partial y}\right)^2 + \left(\frac{\partial T}{\partial z}\right)^2\right] \tag{2}$$

In this Lecture these two contributions are handled separately, and indicated by *viscous-* and *thermal entropy generation rate* respectively. Notice that these expressions are *dimensionless*: to compute the dimensional specific entropy generation rates, one must multiply s_v and s_t respectively by the so-called *entropy scales*:

$$S_v = \frac{\mu U^2}{L^2 \Delta T} \quad \text{and} \quad S_t = \frac{k_{gas}}{L^2} \qquad (3), (4)$$

The value of the entropy generation rate in a finite domain can be computed from a global point of view, using the perfect gas approximation:

$$\dot{s}_{1\to2} = \frac{d(s_{1\to2})}{dt} = c_p \, ln\left(\frac{T_2}{T_1}\right) - R \, ln\left(\frac{p_2}{p_1}\right) \tag{5}$$

or can be computed in a *local perspective*, considering the thermal and viscous contributions to the overall dissipation [1]:

$$s_{1\to2} = \int_V \left[\frac{\mu\Phi}{T} + \frac{k_{gas}(\nabla T)^2}{T^2}\right] dV = \int_V (S_v s_v + S_t s_t) dV \tag{6}$$

In the right-hand side of equation 6, the first term in the integrand represents the irreversible entropy production rate due to local shear (Φ being the so-called dissipation function, which includes in this case the contributions of the turbulent stresses), while the second term represents the irreversible entropy production due to local thermal imbalances.

The procedure proposed in this work can be now described and justified on the basis of the above remarks. For a given heat-exchanger or cascade configuration, if a complete quantitative description of the velocity- and temperature fields is available, one can compute s_v and s_t *locally*, i.e., *in every point of the flow field*; their integral over the entire flow domain (equation 6) gives then a value which is numerically very close to the $s_{1\text{-}2}$ obtained by equation 5, and is therefore a measure of the work lost in viscous- and thermal irreversibilities in the flow; notice however that the two contributions in equation 6 can be computed separately, while this would not be possible if one used equation 5; furthermore, the integral in equation 6 explicitly includes local inhomogeneities in the u_i's and T.

Now, if the configuration is modified, for instance by adding a row of cooling holes on the leading edge of the turbine vanes, or by increasing the fin spacing in the heat exchanger, the same calculation can be repeated, and the new values of s_v and s_t, as well as their integral over the entire flow domain, will *allow a direct comparison of the relative performance of the modified configuration with respect to the original one*, and, what is more important, will *show how the map of the viscous- and thermal irreversibility distribution in the flow domain depends explicitly on the design changes*. The information thus gathered is of significance to the designer, who can use it to:

a) choose the configuration which produces the lower total irreversibility (lower entropy generation rate, equation 6);
b) determine -by inspection- where design changes might be desirable to reduce irreversible losses under the applicable design constraints;
c) determine -also by inspection- which design changes are more apt to reduce irreversible losses at off-design conditions.

The following Sections describe several applications of this procedure to representative 2-and 3-D geometries of turbine cascades and heat exchangers.

3. Examples of application of the Optimisation Procedure

3.1. AIR-COOLED GAS TURBINE NOZZLE: CONFIGURATION STUDY

Consider a two-dimensional section of a gas turbine first-row stator blade (figure 1): for the purpose of the present study, the profile can be taken as being representative of the radial section at mid-span of a generic gas turbine vane, internally cooled by a radial flow of air (perpendicular to the plane of the drawing) through three separate internal slots, equipped with cooling holes from which the cooling air is discharged into the mainstream gas. The effectiveness of film cooling on the stator vanes depends on nozzle geometry and on the location and number of the cooling holes (unsteadiness effects have been not addressed here): to choose the most appropriate design configuration, it is very important to fully understand the flow characteristics in the channel, which in this context means the combination of number, type and location of the cooling holes which produces the lowest thermal stresses compatible with the type of materials and operation. But, there is another face to the medal: any discrete holes film cooling configuration will produce losses in the expander, which decrease its performance. [1]

The difference between the configurations analysed in this study consists in the presence (or absence), location and geometry of one or more rows of cooling holes (represented in this 2-D schematization by their longitudinal midsection). In the configuration without cooling holes ("a" here below) the cooling air is assumed to be discharged from the vane tip.

3.1.1. Cascade Data

The relevant geometric parameters of the cascade are shown in Fig. 1. Details on the material properties and on the flow characteristics are given in [14]. The dimensionless characteristic flow parameters are shown in Table 1.

---

[1] A major contribution to these losses is given by the reduced mass flow in the compressor, which requires a portion of the expander's shaft work to be used to pressurise air which then bypasses combustion; but there are also pressure losses in the cooling circuit, which reduce the effectiveness of the convective cooling, and mixing losses at the ejection points (holes), which disturb the boundary layer/mainstream interface and modify the endwall/wake interaction (thereby also reducing the cooling effectiveness). We shall consider only losses related to the blade passage, and neglect all others.

TABLE 1. Dimensionless flow parameters for case study 3.1.1.

$L_{ref} = 0.15$ m	Re = 86000
$U_{ref} = 103$ m/s	Pr = 0.7 (at T = 1300°K)
$T_{ref} = 300$ °K	Pe = 60200
$\Delta T_{ref} = 1000$ °K	Br = 154

3.1.2. Boundary Conditions

The boundary conditions are the same for all four configurations, and are shown in Table 2.

TABLE 2. Boundary conditions for case study 3.1.1.

Inlet velocity	turbulent, symmetric profile with $V_\infty = V_1 = 103$ m/s
Inlet temperature	plug profile, with T_∞=1300 K (this value was varied to achieve constant blade enthalpy drop, see Table 3);
Inlet turbulent quantities	$k_\infty = 0.03\ V_1^2 \quad \varepsilon_\infty = 5 \cdot 10^{-9} \frac{V_1^3}{L_c}$;
Blade inner wall temperature	$T_{inside} = 1118$ K;
Outflow conditions	$\frac{\partial T}{\partial n} = \frac{\partial W}{\partial n} = \frac{\partial k}{\partial n} = \frac{\partial \varepsilon}{\partial n} = 0$;
Wall values (see Section 2)	$u=v=0$, k=0, $\varepsilon = \lim_{y \to 0} \varepsilon_{wl}$

3.1.3. The Numerical Procedure

Since circumferential periodicity is assumed, only a portion of the 2-D cascade (a "cell" containing one blade) is actually computed. All relevant details about the procedure and the numerics of this example are given in [14].

3.1.4. Results and Discussion

The velocity and temperature fields in the fluid and the blade temperature contours are presented respectively in Fig. 2,3,4. The general features of the profiles are in qualitative agreement with the flow patterns one would expect in such a cascade (see [14] for a complete discussion). The entropy generation rates are shown in Fig. 5 and 6, while their integral values over the entire flow domain are reported in Table 3. The following conclusions can be drawn:

1) both $\dot{s}_v$ and $\dot{s}_t$ are almost entirely produced in the boundary layer around the blade (Fig. 9,10, but see also [14]), and are caused by high local gradients at the wall and in the wake; *a "better" fluiddynamic blade design could reduce the viscous entropy generation rate to a certain extent, but not of a substantial amount* (the turbulent boundary layer and the turbulent wake cannot be simply dispensed with!). *Thermal irreversibilities, on the other hand, can be substantially reduced by avoiding the strong temperature gradients at the blade walls*: from this point of view, and pending numerical validation of this conclusion, transpiration cooling (i.e.,

continuous ejection of the cooling air through permeable vane walls) would be the best solution. At any rate, the thermal gradients are lower (cooler blade) for configuration d1 (Fig. 3).

2) Among the four configurations examined here, the one with the highest $\dot{s}_t$ is the internally cooled blade (without discrete hole film cooling), and this was expected, because this configuration also attains the highest wall temperatures, and therefore the highest thermal gradients. What was *not expected*, however, is that $\dot{s}_v$ *can be lowered by the addition of cooling holes.* A possible *a posteriori* physical explanation is that the velocity gradients at the wall are significantly reduced on a large portion of both the suction and the pressure side by the thin layers of cooling air which remain attached to the wall and act as a sort of "slip layers" between the viscous sublayer and the mainstream flow. The numerical values shown in Table 3 demonstrate that an *acceleration of the trailing edge cooling flow* (case d1) *brings about an improvement in the efficiency* ($\dot{s}_v$ decreases), but that *an excessive acceleration* (case d2) *has just the opposite effect.* Adopting this point of view, transpiration cooling could slightly increase the cascade performance, by decreasing the viscous losses as well. As remarked in [14], these conclusions have to be carefully checked by specific numerical simulations, and must be taken here only as "workable hypotheses".
3) Flattening the leading edge of the blade (case d) increases $\dot{s}_v$ as well, because the upstream section of the blade is affected by higher stagnation point losses.
4) The local Bejan number, Be = $s_t/(s_t + s_v)$, which is clearly a synthetic measure of the relative magnitude of the two irreversibilities, attains a value of about 0.5 at the wall, and of 0.4 - 0.6 around the cooling holes, showing that in these flow regions the thermal and viscous contributions to the irreversible losses are comparable. In the so-called "quasi-inviscid regions", which surround the boundary layer at the wall and the mixing layers around the wake, Be reaches a peak value of about 0.95, showing the diminishing importance of s_v and the effects of strong convective thermal mixing. Upstream of the vane and in the so-called "isothermal core", which spans the central portion of the interblade passage, Be ≈ 0.05.
5) The average or integral Bejan number, $\overline{Be}$, takes values in the vicinity of 0.5 for all configurations (see Table 3).

TABLE 3. Global results of the entropy generation calculations for case study 3.1.1 (Note: $\dot{s}_v$, $\dot{s}_t$ and $\dot{s}_{tot}$ in W/°K, Δh in J/kg, T_0 = 300 °K).

Sub-Case	s_t	s_v	s_{tot}	ζ	η	Δh_i	Δh_{real}	$\overline{Be}$
design ("a")	76	84.6	160.6	0.318	0.871	70.8	61.7	0.47
"b"	59.3	76.3	135.7	0.26	0.849	72.7	61.7	0.44
"c"	65.5	77.3	152.8	0.288	0.888	69.5	61.7	0.46
"d1"	64.5	73.7	148.3	0.27	0.860	71.7	61.7	0.47

3.2. INFLUENCE OF THE ANGLE OF ATTACK ON THE IRREVERSIBLE LOSSES IN A GAS TURBINE ROTOR BLADE

Consider a two-dimensional section of a highly loaded gas turbine first-row rotor blade (Fig. 7): for this configuration, extensive experimental data are available (see [15] for details). Again, the profile is representative of the radial mid-section of the vane, which is internally cooled by a radial flow of air through three separate internal slots. In all cases examined here, the cooling air is assumed to be discharged from the vane tip. The unsteadiness due to rotation effects is not addressed by these steady-state simulations: attention is placed to the response of the rotor to off-design operating conditions, which in this context means to study the effects of a variation of the angle of attack α_1 (angle between the relative velocity and the geometrical tangent to the profile at the leading edge), which takes the values -8.4°, -4.4°, -0.4° (design value), +3.6° and +7.6°.

3.2.1. Cascade Data

The midplane section of the blade on which the calculations were performed is shown in Fig. 7, and the detailed description of the physics of the flow is given in [15]. The values assumed of the dimensionless characteristic flow parameters are the same as those shown in Table 1: Re has now been calculated for the relative flow ($U_{ref} = W_{ref} = W_1 =$ 103 m/s).

3.2.2. Boundary Conditions

The boundary conditions (which are the same for all four configurations, except for the flow direction at the inlet boundary) are identical to those displayed in Table 3.

3.2.3. The Numerical Procedure

The method, the code and the general features of the numerical solution are the same as those of example 3.1: all relevant details are offered in [15].

3.2.4. Results and Discussion

The details of the numerical solution are not shown here: the general large-scale features of the flow fields are similar for different angles of attack, and the differences are barely detectable in the maps of s_v and s_t given in [15]). What is of interest here is to examine in detail the entropy production rates and their distribution, and to predict the performance of a cascade on the basis of its total entropy production. The following remarks apply:

1. a variation of α_1 induces a modification of the local characteristics of the boundary layer in the vicinity of the stagnation point, and the distribution of the viscous entropy generation rate changes accordingly: one would expect that a departure from the design conditions brings about an increase of the irreversible losses, i.e., that the value of the integral of $\dot{s}_v$ over the entire domain increases with $\Delta\alpha$ (where $\Delta\alpha = \alpha_{design} - \alpha_{off\text{-}design}$), no matter whether the sign of $\Delta\alpha$ is positive or negative. Figure 8 shows that this is indeed the case, and that the increase is almost quadratic (the usual loss correlations predict a quadratic dependence of the viscous losses on α_1). The thermal entropy production, on the other hand, is not substantially affected by the

angle of attack, and only minor variations result from the distortion of the boundary layer in the proximity of the stagnation point.

2. Blade performance, measured by the irreversibility coefficient ζ [6], remains almost constant for small variations of α_l about its design value, but decreases substantially for angles of attack far from it. This means that the blade leading edge is capable of accommodating small variations in the flowrate without severe performance deterioration.
3. The cooling effectiveness is only slightly affected by a variation of the angle of attack (the blade surface heat flux profiles do not vary appreciably with α_1, see [15]),: but the thermal entropy production varies of a non-negligible amount, as shown by its global value (integral of $\dot{s}_t$ over the entire domain, figure 8). The conclusion is that the temperature distribution around the blade, though contributing significantly to the overall irreversibility of the expansion process, is nevertheless *not a relevant parameter* in this optimisation.
4. Neither the local Bejan number, Be = $s_t/(s_v + s_t)$, nor its integral counterpart, $\overline{Be}$, show substantial variations with α_1, confirming that the ratio of the two irreversible entropy production rates does not depend on the flowrate coefficient $\phi = V_m/U$, at least within the range covered by the present investigation.

These conclusions -as remarked in [15]- ought to be viewed with some caution for three reasons: first, the simulation being at steady-state, the details of the flow are obviously valid only as "statistical averages", and the time-dependent details of the solution are entirely neglected; second, the turbulence has been modelled by a 2-D k-ε closure, so that the behaviour of the turbulent quantities in the vicinity of the wall has been imposed *a priori* by the choice of model parameters. Now, most of s_v and s_t are produced in the sub- and inner layers, and therefore the results might be strongly model-dependent. Last, a 2-D simulation undoubtedly fails to account for both the cooling jet/mainstream interactions and the "centrifugal" displacement of the streamlines, which are essentially three-dimensional phenomena: so, it may well be that there are features of both the temperature and the entropy profiles in and around the blade that cannot be captured but with 3-D simulations.

3.3. THREE-DIMENSIONAL SIMULATIONS

The objections raised against 2-D simulations have been addressed, as suggested in [14], in two separate and independent investigations [10,20], the first one for steady and the second for unsteady flow in a first stage of a gas turbine. The two methods employed different turbulence modelling techniques, and neither of them took shock-related effects into account.

3.3.1. Steady, Uncooled Gas Turbine Rotor

A synthesis of the results presented in [20] for a blade with the midplane section shown in figures 9 is offered in Fig. 10: the aerodynamic losses in the rotor have been calculated independently (by integrating equation 5 over the relevant areas of the computational domain) so that their relative importance could be assessed. The rotor blade is uncooled, so that the results are in effect of interest for the choice of the profile shape, and can be used to directly test the effects of design changes. The use of equation 5, and its integration over separate areas of the flow, is questionable, because it is difficult to imagine that there is no mutual influence among losses of different types; furthermore, viscous and thermal losses cannot be properly separated. Nevertheless, the study clearly demonstrates the feasibility of the method as a design tool.

3.3.2. Unsteady, Air-Cooled Gas Turbine Stage

It is well known from textbook analyses that unsteady effects linked to relative rotation are important for a series of theoretical and practical reasons. The optimal design of a cooled vane, therefore, is likely to be affected by the fully 3-D flow characteristics in the entire stage (nozzle + rotor): to understand these characteristics, a three-dimensional calculation was conducted on a complete first stage [10]; the entropy generation contours obtained by that simulation are shown in Fig. 11. The entropy generation rate was computed using equation 6. The loss patterns appear clearly from the entropy contours: losses are relatively low in the central portion of the channel, where both s_v and s_t are low; the contribution of the thermal gradients is noticeable around the blade, and the local "high loss" spots correspond to the cold air injection into the main stream. At the leading edge, an additional contribution by s_v increases the irreversible losses; downstream, wake effects (again, mainly due to the s_v contribution) are noticeable. This study is impressive, in that it is a real full-scale demonstration of the applicability of the technique to industrial cases.

3.4. OPTIMAL CONFIGURATION OF A SINGLE FINNED TUBE IN A GAS-TO-LIQUID HEAT EXCHANGER

In the cooling of an air stream in a finned-tube heat exchanger (for which both the tube external diameter and the fin thickness are prescribed), there are two competing effects to be accounted for: the friction losses, which decrease with increasing fin spacing, and the overall heat transfer, which prescribes the ratio of the fin diameter to the fin spacing. In this case, an "optimal" configuration can be found by entropic analysis which is different (and more effective) than the one obtained by separately minimising the pressure losses or maximising the heat transfer per meter of tube.

3.4.1. Description of the Analysed Configurations

The initial configuration is the "optimal" one suggested by current engineering practice, which dictates the choice of the outer fin diameter as a function of the tube external diameter and of the fin spacing (see [19] for details). The configuration is depicted in Fig. 12 and Table 4: in the numerical simulation, the tube diameter d_t as well as the fin thickness t have been kept constant, while both the fin protrusion ℓ and the fin spacing s have been varied. The values assumed for the air- and material properties and the dimensionless characteristic flow parameters are not reported here, but can be found in [3,19].

Four different fin protrusion lengths (ℓ= 10,15,20 and 25 mm) have been considered, and for each one ten different values for the fin spacing have been computed: all in all, forty different configurations have been numerically evaluated. The fluid flowing inside of the tube has been assumed to have an internal heat transfer coefficient much higher than that of the external air, so that the external tube surface could be considered isothermal. The Reynolds number based on the tube diameter, calculated with the fluid and material properties described in detail in [3,19], is Re = 966, and therefore it can be safely assumed that a laminar solution will suffice. The problem is axisymmetrical, so that it is convenient to model only a portion of the domain: we chose a control volume that included a slice of tube of axial length equal to $(s+t)/2$ and of radial width equal to twice the fin external radius (see Fig. 12).

TABLE 4. Configuration data for case study 3.4.

Entity	Design value	Range of variations
tube diameter d_t, mm	30	fixed
fin protrusion ℓ, mm	15	10, 20, 25
fin spacing s, mm	3	1÷ 9.5 (in non-uniform steps)
fin thickness t, mm	0.5	fixed

3.4.2. Boundary Conditions

The boundary conditions (Table 5) are the same for all configurations, the only difference being of course the aspect ratio of the finned tube.

TABLE 5. Boundary conditions for the finned tube case (refer to Fig. 19).

inlet velocity (section "A")	plug profile with $U_\infty = 0.6$ m/s
inlet temperature (section "A")	plug profile with $T_\infty = 293$ K
tube outer wall temperature (section "E")	$T_w = 333$ K
outflow conditions (section "C")	$\partial f / \partial n = 0$ ("f" being the generic variable u,v,w,T)
symmetry conditions	sections "B" and "I"

3.4.3. The Numerical Procedure

The computational domain is shown in Fig. 13: since axial periodicity is assumed, only a portion of the tube (a "cell" containing one fin) is actually computed; in addition, the problem being symmetrical with respect to the longitudinal midplane through the tube axis, only half of the tube/fin arrangement has been modelled. Details on the numerical technique are given in [19].

3.4.4. Results and Discussion

A representative view of the velocity and temperature fields is shown in Fig. 14. These results are in excellent agreement with an unrelated numerical and experimental data on similar configurations [9]: for what the discussion of the flow structure is concerned, the reader is referred to [3]. The results prompt the following remarks:

1. the heat flux from a single fin is shown in Fig. 21: for a fixed fin thickness it increases -as expected- with the fin protrusion ℓ and also increases with fin spacing, tending asymptotically to the "isolated fin" limit. As a consequence, the heat flux per meter of tube, $Q_{meter} = Q_{fin}/(s+t)$, exhibits a maximum (Fig. 17, top) which

corresponds approximately to a spacing of 0.025 to 0.03 m, which is identical with the "optimal" value derived analytically in [4];

2. the viscous dissipation is -due to the low Re- of orders of magnitude lower than its thermal counterpart: considering that the entropy generation rates scale defined by equations 3 and 4, the order of magnitude of their ratio will be:

$$\frac{S_{0,t}}{S_{0,v}} \approx \frac{k_{gas}\Delta T}{\mu U^2} = 155000 \qquad (15)$$

and the ratio between the computed macroscopic values (obtained by integrating the local values over the entire flow domain) is about the same. With increasing fin spacing, the velocity gradients decrease, and so does the viscous entropy generation rate per meter of tube (Fig. 16);

3. The thermal entropy generation (Fig. 17, bottom) attains its maximum in the thermal boundary layer and in the thermal wake of the finned tube. The thermal entropy production per meter, on the other hand, displays a maximum which is attained for much wider fin spacing than the corresponding maximum in the heat flux. This is due to the development of the thermal boundary layer in the interfin space, which gives origin to two distinct temperature profiles in the channel between fins, depending, in a way that we are not going to discuss here, on a sort of entry length parameter [19].

The analysis of this finned tube configuration leads to the following conclusions:

1) for the low Reynolds number considered here, $\dot{s}_t$ is much higher than $\dot{s}_v$ (their ratio actually is of the same order of magnitude as the Brinkman number): thus, it has to be expected that in most laminar flows, as long as the temperature difference is not very small, the dominant loss will be that due to the thermal irreversibilities. In particular, for air between 300 and 800 °K, $\mu/k \approx (0.71 \div 0.65)10^{-3}$, so that $U^2/\Delta T$ ought to be of order 1000 for the two entropy production rates to be of the same order of magnitude (while in the present example $U^2/\Delta T = 0.012$!).

2) The "optimal" fin spacing based on the maximisation of the heat flux per meter of tube, which for the present configuration would be chosen around 2.5 *mm* (figure 17), still gives origin to a high thermal entropy generation rate. From a 2-nd Law viewpoint, it would be better to use a larger fin spacing, thus reducing the irreversible losses, and to accept the corresponding decrease in overall heat flux per meter. In [19] a possible thermo-economic criterion for choosing the fin spacing is presented which suggests to take, for each fin protrusion ζ the spacing s for which the thermal entropy generation rate per meter is half of its peak value: in the cases analysed in this study, this would correspond to $s \sim 5 \div 6$ *mm*.

4. Conclusions

A new procedure to assess the losses in a process or transformation in which a fluid exchanges energy with a thermo-mechanical component is described in this paper. As a performance parameter, the total entropy generation rate is proposed. This quantity offers several advantages in evaluating the effectiveness of an energy conversion process:

1) $\dot{s}_{tot}$ can be computed numerically from known flow fields (where local values for both velocity and temperature are available);
2) proper integral constraints can be imposed (like for example the overall enthalpy drop across the cascade, or the maximum blade temperature) to represent realistic operational conditions;
3) $\dot{s}_{tot}$ can be conveniently split into its viscous- and thermal contributions, thus allowing better insight in the flow phenomena;
4) $\dot{s}_{tot}$ ontains information at two physical levels: at a *local* level, it shows not only where irreversibilities are present, but also to what extent they are sensitive to design changes; at a *global* -or integral- level, it gives a measure of the "degree of irreversibility" of the flow process in the cascade, which enables a designer to assess the relative performance of different configurations;
5) $\dot{s}_{tot}$ can be therefore readily linked to the *real* efficiency of the machine of which the cascade is a component through -for example- the "entropy loss" coefficient [6], or the Gouy-Stodola lost work.

The procedure has been applied to realistic 2-D and 3-D models of turbomachinery components, and for some of the configurations the geometry was varied to verify the claim made in Section 1 that it would be possible to compute the "classical" performance parameters on the basis of the irreversible losses generated in the flow. The results confirm and complement those described in [14,15,19], and the conclusions which were stated in those previous studies can be repeated here:

i- the procedure (which was applied to a stator vane in [14] and to a rotor vane in [10,15,20) is indeed applicable to realistic 3-D analyses of an entire stage;

ii- there are areas which require further investigation. First of all, the turbulence model (especially the wall functions) is likely to affect the absolute numerical value of the computed entropy generation rates; furthermore, unsteady and three-dimensional effects can change the picture, as endwall and wake losses, tip leakage effects, and interactions with upstream flow disturbances affect the relative magnitude of $\dot{s}_v$ and $\dot{s}_t$;

iii- in general, the procedure is only as good as the flow field from which the velocity- and temperature data were extracted: if -however- experimental data about the flow fields are known, the calculation of $\dot{s}_v$ and $\dot{s}_t$ is straightforward.

iv- design decisions always involve some form of economic (sub-)optimisation: it appears indeed possible (see also [19]) to introduce "cost figures" in the entropic analysis and derive "engineering optima".

Since compressible flow simulations results are not yet available for an entropic analysis like the one described here, these remain open issues as well; but it is likely that 2-D compressible simulations ought to be addressed before going into complete 3-D simulations, which are still extremely demanding in terms of computer resources. On the basis of the experience gained in several applications of the method outlined in this paper (and described in more detail in the literature quoted here), the author recommends that more case studies be analysed, to better focus the numerical procedure and to gain deeper insight in the local flow phenomena.

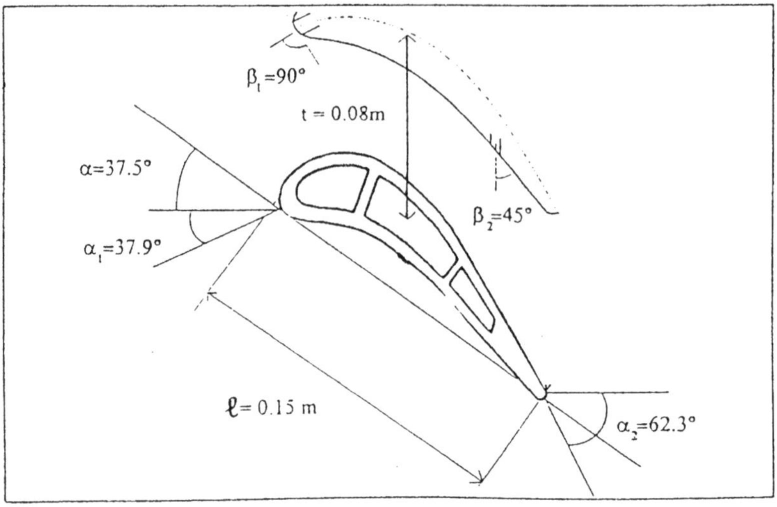

Figure 1. Mid-plane section of an air-cooled first stage nozzle blade: geometry definition.

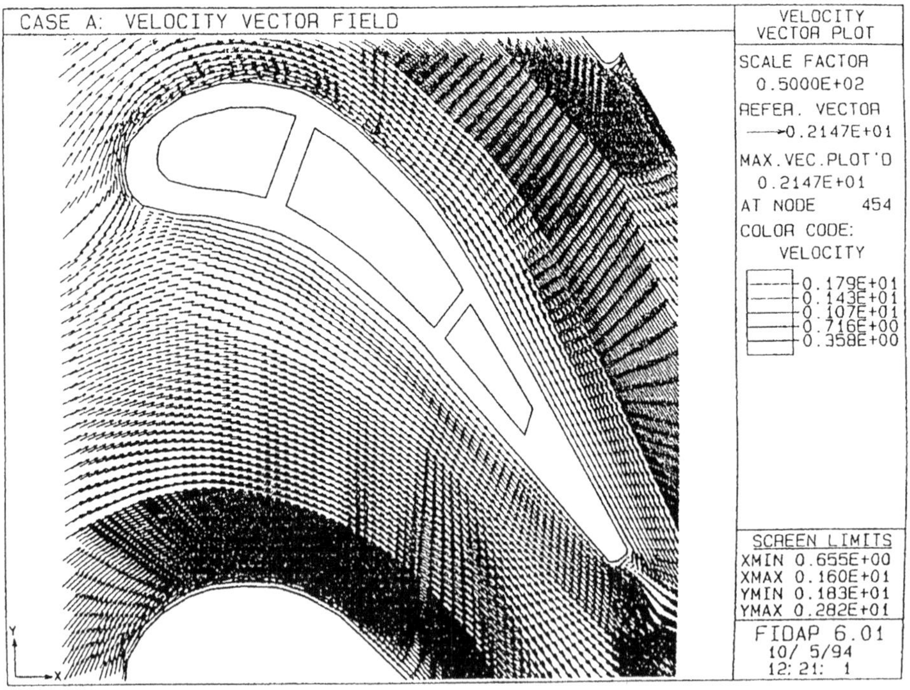

Figure 2. Velocity vector field for Cass Study 3.1, Sub-case "a".

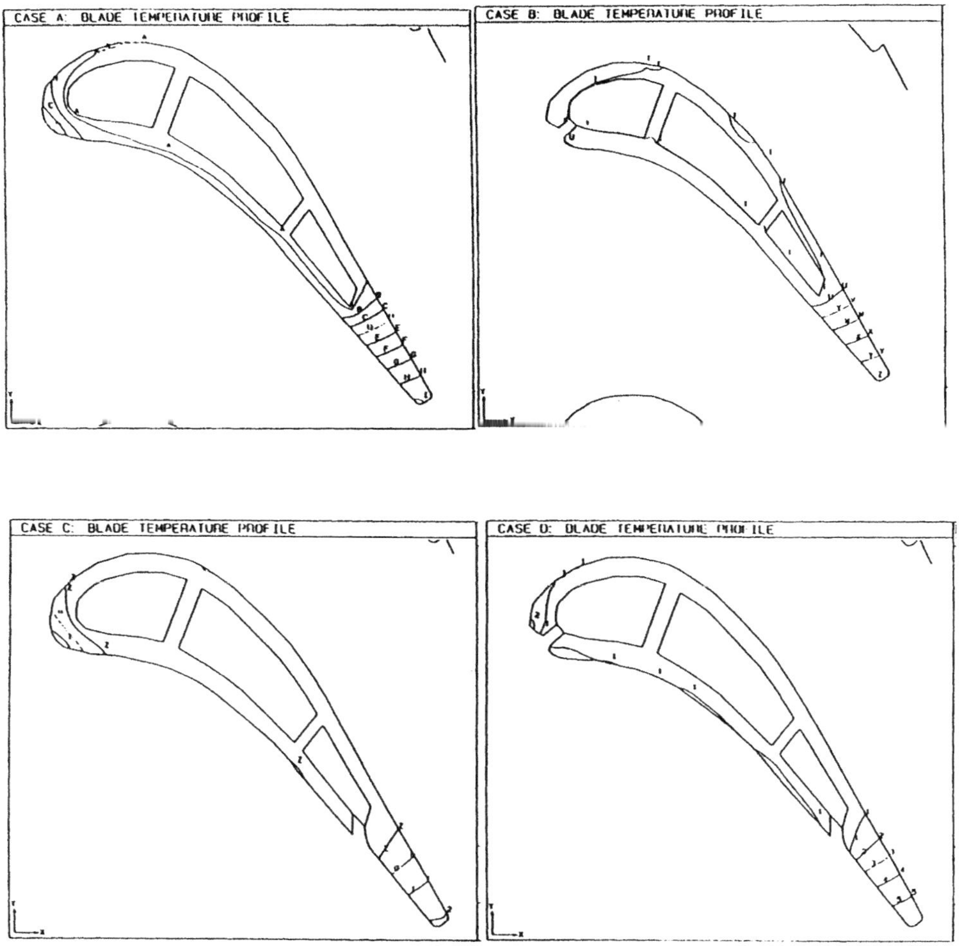

Figure 3. Blade temperature contours for Case 3.1.

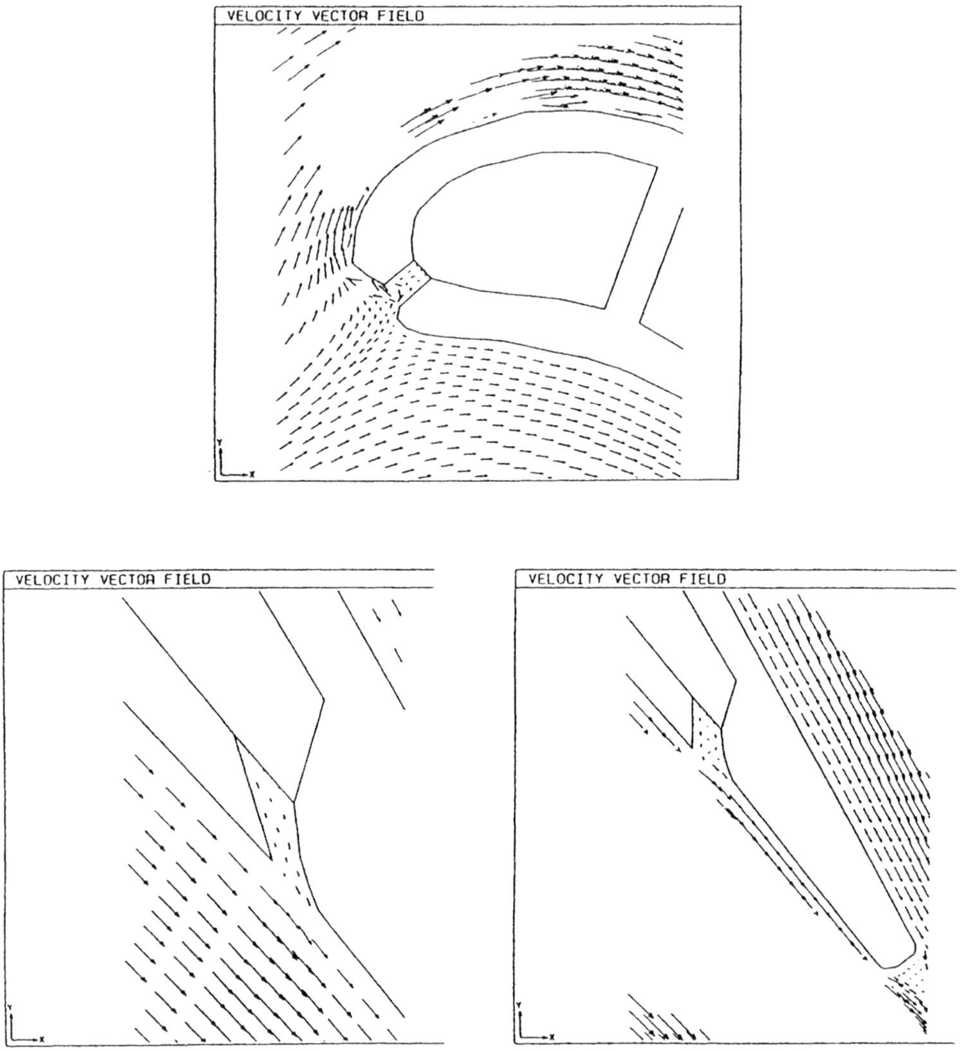

Figure 4. Details of the velocity vector fields for Sub-cases "b", "d_1" and "d_2".

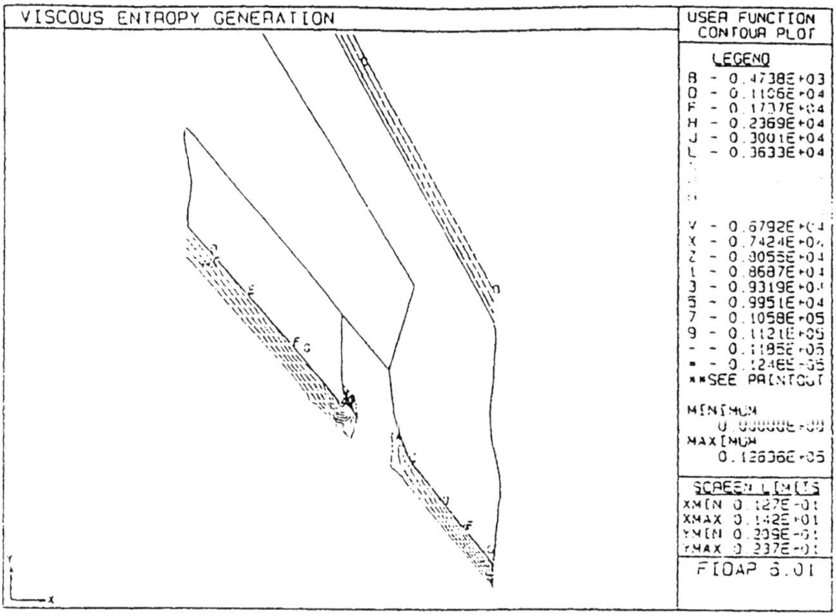

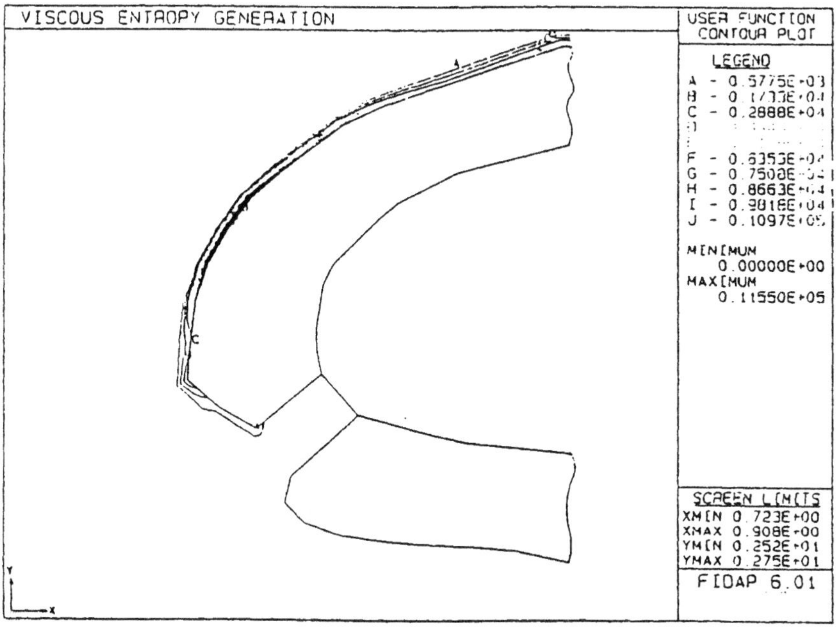

Figure 5. Viscous entropy generation rates for Case 3.1 (details).

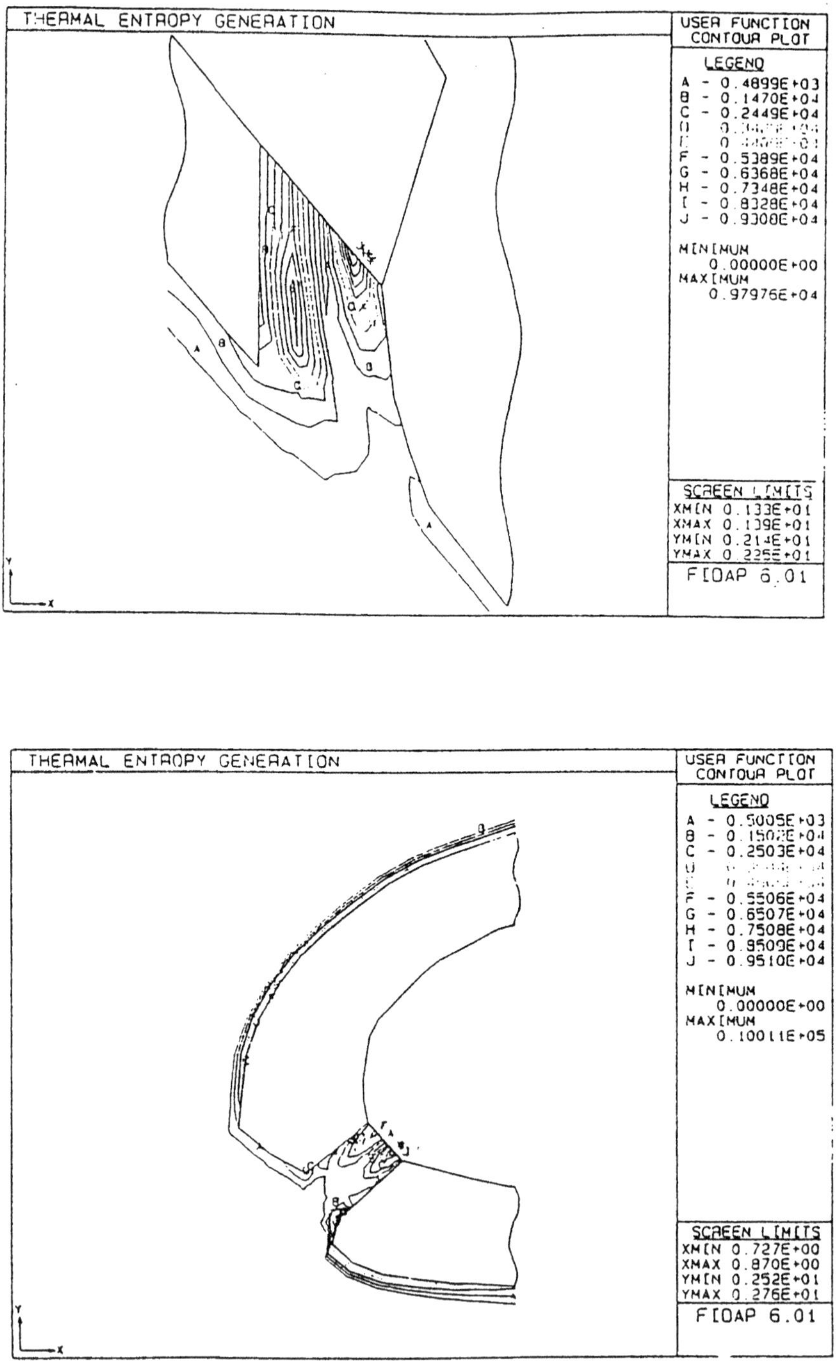

Figure 6. Thermal entropy generation rates for Case 3.1 (details).

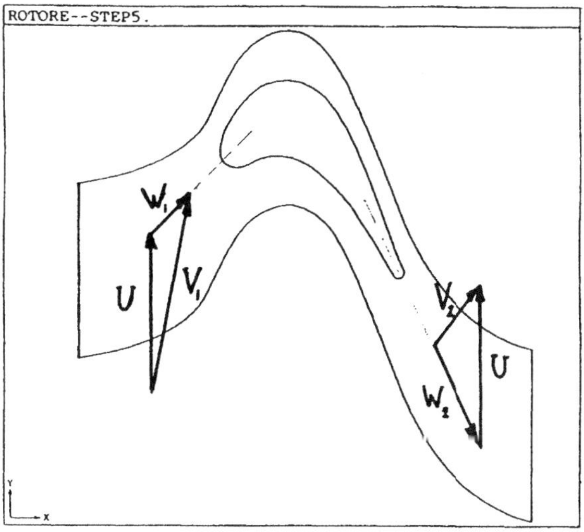

Figure 7. Mid-plane section of an air-cooled first stage rotor blade: geometry definition for Case 3.2.

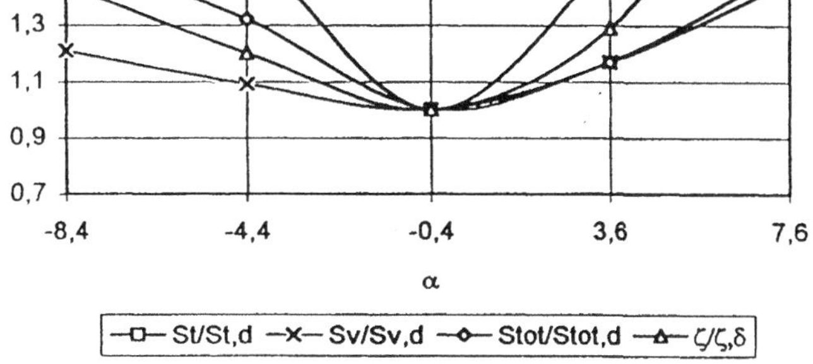

Figure 8. Case 3.2: variation of the global entropy generation rates as a function of the angle of attack.

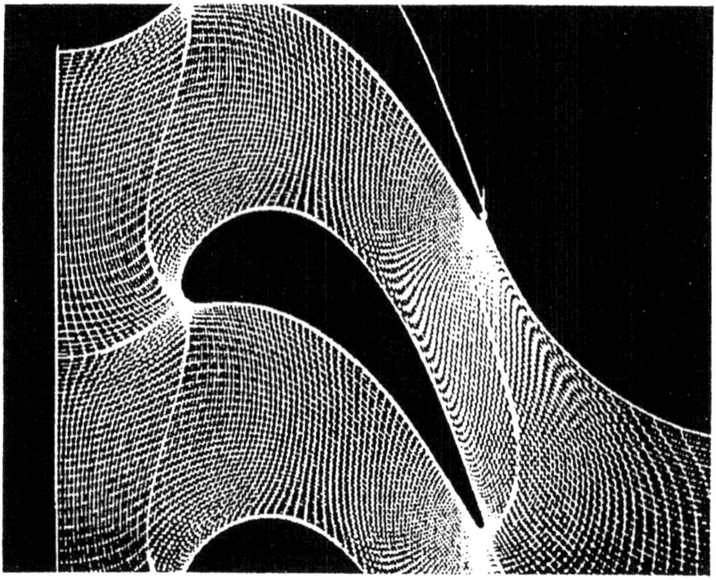

Figure 9. Case Study 3.3.1: geometry definition (midplane section).

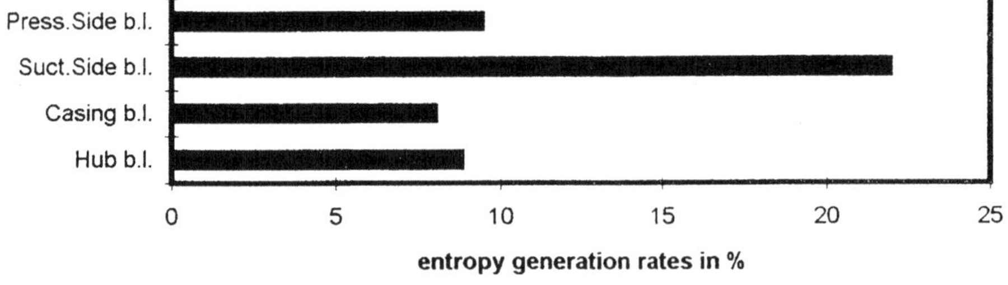

Figure 10. Case Study 3.3.1: entropy generation rates attributed to the single losses (adapted from [20]).

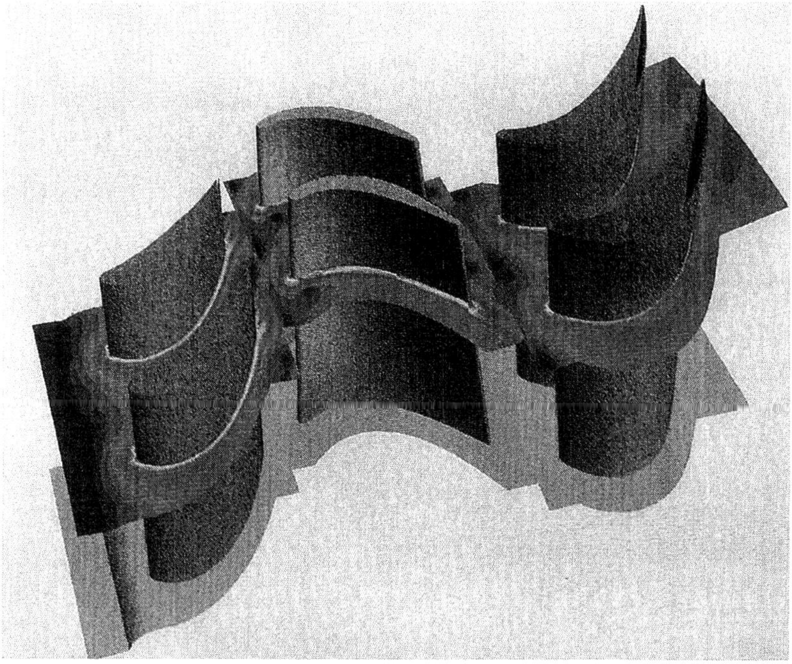

Figure 11. Case Study 3.3.2: entropy generation rates on the midplane section.

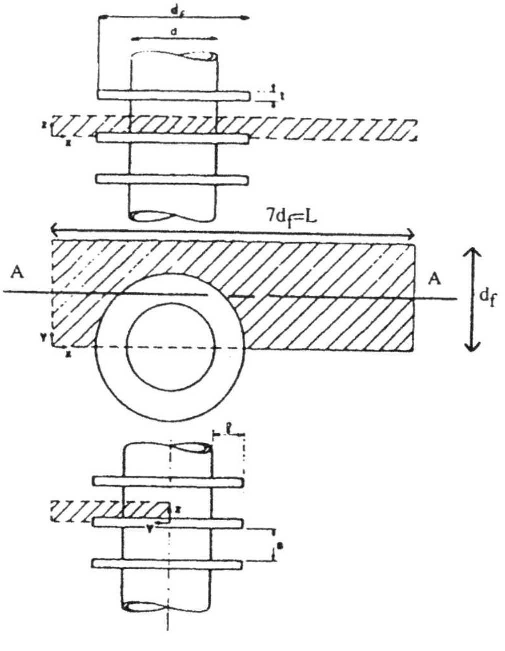

Figure 12. A "unit element" of a water-to-finned tube heat exchanger.

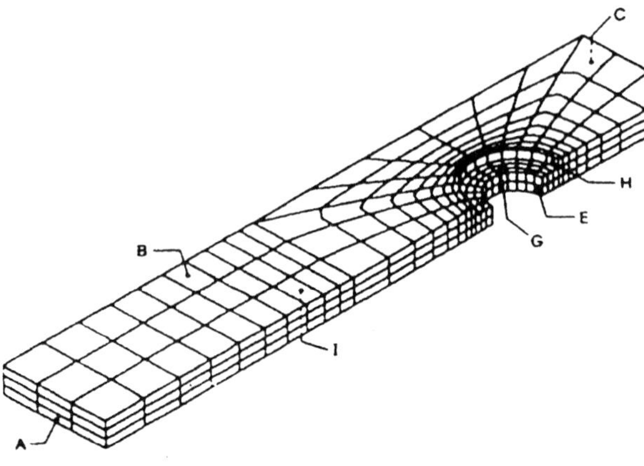

Figure 13. Case Study 3.4: the computational mesh.

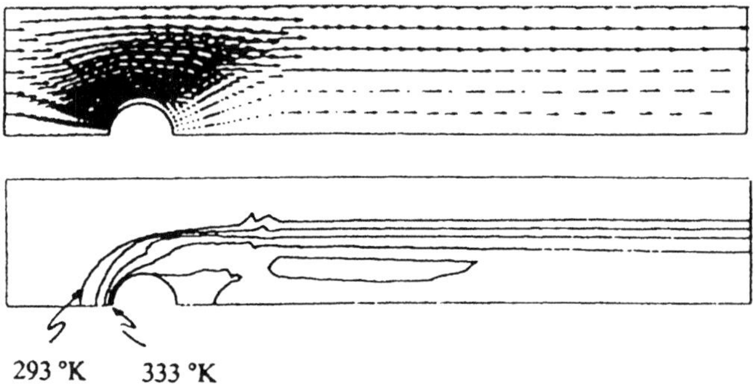

Figure 14. Case Study 3.4: velocity and temperature fields for the original design configuration.

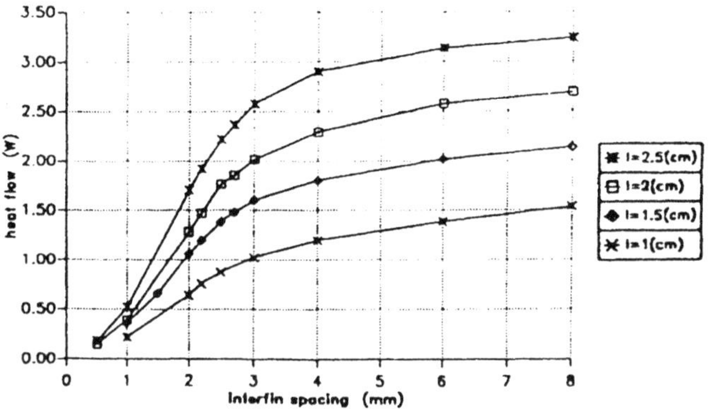

Figure 15. Case Study 3.4: heat flux from one single fin.

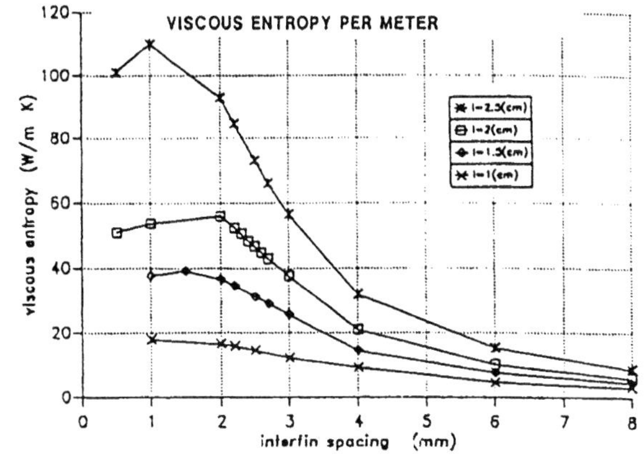

Figure 16. Case Study 3.4: viscous entropy generation rates per meter of tube.

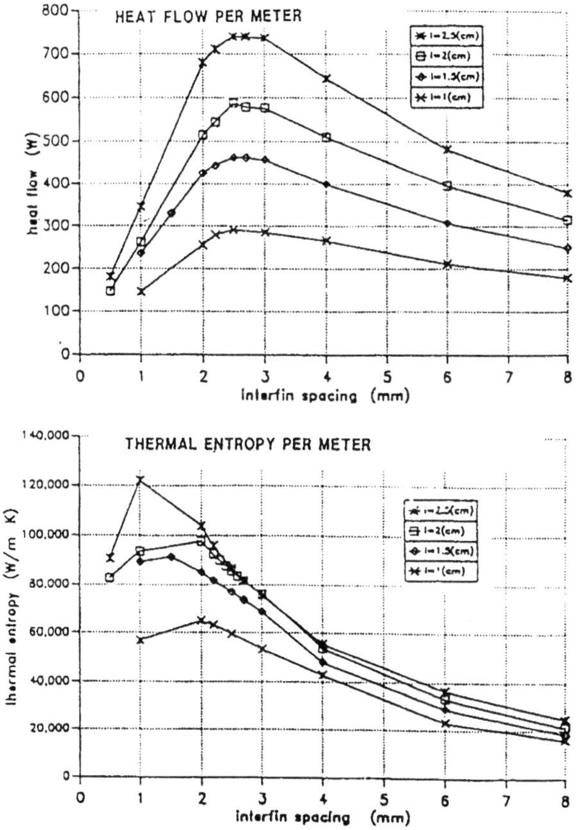

Figure 17. Case Study 3.4: thermal entropy generation rates and heat flux per meter of tube.

5. References

1. A.Bejan: *Entropy Generation through Heat and Fluid Flow*,Wiley, New York, 1982.
2. A.Bejan, E.Sciubba: The optimal spacing of parallel plates cooled by forced convection, *Intl. J. of H. & M. Transf.*, 35, n.12, 1992.
3. P.L.Benedetti, E.Sciubba: *Numerical calculation of the local rate of entropy generation in the flow around a heated finned tube*, in *ASME AES-3*, 1993.
4. A.E.Biermann, H.H.Ellenbrock Jr.: The design of fins for air-cooled cylinders, *NACA Rep. 726*, 1976.
5. R.J.Boyle: Navier-Stokes analysis of turbine blade heat transfer, *J.of Turbomach.*, 113, Jul. 1991.
6. J.D.Denton: Loss mechanisms in turbomachines, *J.of Turbomach.*, 115, Oct. 1993.
7. D.J.Dorney, R.L.Davis: Navier-Stokes analysis of turbine blade heat transfer and performance, *J.of Turbomach.*, 114, Oct. 1992.
8. M.K.Drost, M.D.White: Numerical predictions of local entropy generation in an impinging jet, *ASME J.H.T.*, 113, 1991.
9. A.Haught, M.S.Engelman: Numerical and experimental simulation of air flow heating in a tube fin heat exchanger, in *ASME HTD-103*, 1988.
10. A.Holzwarth: *Private Communication*, April 1997.
11. T.V.Jones, C.M.B.Russell: Local heat transfer coefficients on finned tubes, in *ASME HTD-21*, 1981.
12. S.Mereu, E.Sciubba, A.Bejan: The optimal cooling of a stack of heat generating boards with fixed pressure drop, flowrate or pumping power, *Intl. J. of H.& M. Transf.*, 36, n.15, 1992.
13. J.Moore, J.G.Moore: Entropy production rates from viscous flow calculations, Part 1- a turbulent boundary layer, *ASME paper 83-GT-70*, 1983.
14. G.Natalini, E.Sciubba: Entropy generation rates in air-cooled gas turbine nozzles: a numerical configuration study, *Proc. IGTI-COGEN-TURBO*, 1994.
15. G.Natalini, E.Sciubba: Entropy generation in a 2-D cascade at different angles of attack: a numerical study, *Proc. IGTI-COGEN-TURBO*, 1997.
16. S.Paoletti, F.Rispoli, E.Sciubba: Calculation of exergetic losses in compact heat exchanger passages, in *ASME AES-10/2*, 1989.
17. E.Sciubba: Numerical calculation of local irreversibilities in compact heat exchangers, in *Proc. NATO-TIBTD Workshop on 2nd Law*, Erciyes Univ., Kayseri 1990.
18. E.Sciubba: Entropy generation rates as a true measure of thermal and viscous losses in thermo-mechanical components and systems, *Proc. Bucharest Heat Transfer and Thermodynamics Workshop*, Polytechnic Univ. of Bucharest, 1994.
19. E.Sciubba: A minimum entropy generation procedure for the discrete pseudo-optimisation of finned tube heat exchangers, *Rev.Gen.Therm.*, vol.35, 1996.
20. T.J.Sullivan: Novel aerodynamic loss analysis technique based on CFD predictions of local entropy production, *SAE Paper 951430*, 1995.
21. K.Takeishi, S.Aoki, T.Sato, K.Tsukagoshi: Film cooling on a gas turbine rotor blade, *J.of Turbomach.*, 114, Oct.1992.

AVAILABLE ENERGY VERSUS ENTROPY

A. ÖZTÜRK
İstanbul Technical University
Mechanical Engineering Department
Gümüşsuyu, İstanbul 80191 - Turkey

1. Introduction

As known, the second law of thermodynamics is an entropy law. The foundation of the second law of thermodynamics is usually attributed to Clausius and Kelvin, around 1850, after S.Carnot's principles *"heat cannot be converted completely and continuously into work"* [1]. In Classical Thermodynamics, the first law of thermodynamics is introduced and then followed with a discussion of the definitions of the reversible cycle (Carnot cycle) and --related to Carnot cycle-- absolute temperature. Next, the Kelvin-Planck statement and Clausius statement for the second law of thermodynamics are introduced . Then the Clausius inequality

$$\oint \frac{\delta E_Q}{T} \leq 0 \qquad (1)$$

is proved in different ways [2-5]. Entropy is then defined and the entropy equation is formulated as the second law of thermodynamics *(Fig. 1)*. The application of the second law of thermodynamics to the various processes has been presented in different ways, such as, (i) the net entropy increase (or entropy generation, or entropy production, or entropy creation, etc.), (ii) irreversibility, (iii) lost work, and finally (iv) the loss of available energy (or exergy).

Usually, it is considered that the available energy equation is the combination of the first and the second laws of thermodynamics, in other words, the combination of the energy and entropy equations [6]. Indeed, if those two equations are written for control mass or control volume, the available energy equation can be obtained with some mathematical and thermodynamical procedures [5, 6]. Even though some authors, like Obert [7], have concluded that *"The available energy of the isolated system decreases in all real processes (and is conserved in reversible processes),"* traditionally all scholars introduce the formulations the available energy equation after the entropy equation.

This work introduces a new and conceptually easier approach based on a direct proof of the known statement of inequality. In this approach definitions are provided for concepts like the available energy of heat (maximum work produced from heat) and the available energy of work (useful work). Then the known inequality is proven

A. Bejan and E. Mamut (eds.), Thermodynamic Optimization of Complex Energy Systems, 187–194.

directly in terms of the available energy of heat and the available energy of work.

In the literature, this well-known inequality statement is usually induced after Clausius inequality, in terms of entropy and with long computations based on concepts such as reversible work, maximum reversible useful work, and so on. The statement as developed by this author is proved directly and is thus another version of the second law of thermodynamics. I have termed this statement *"the new inequality"* [8].

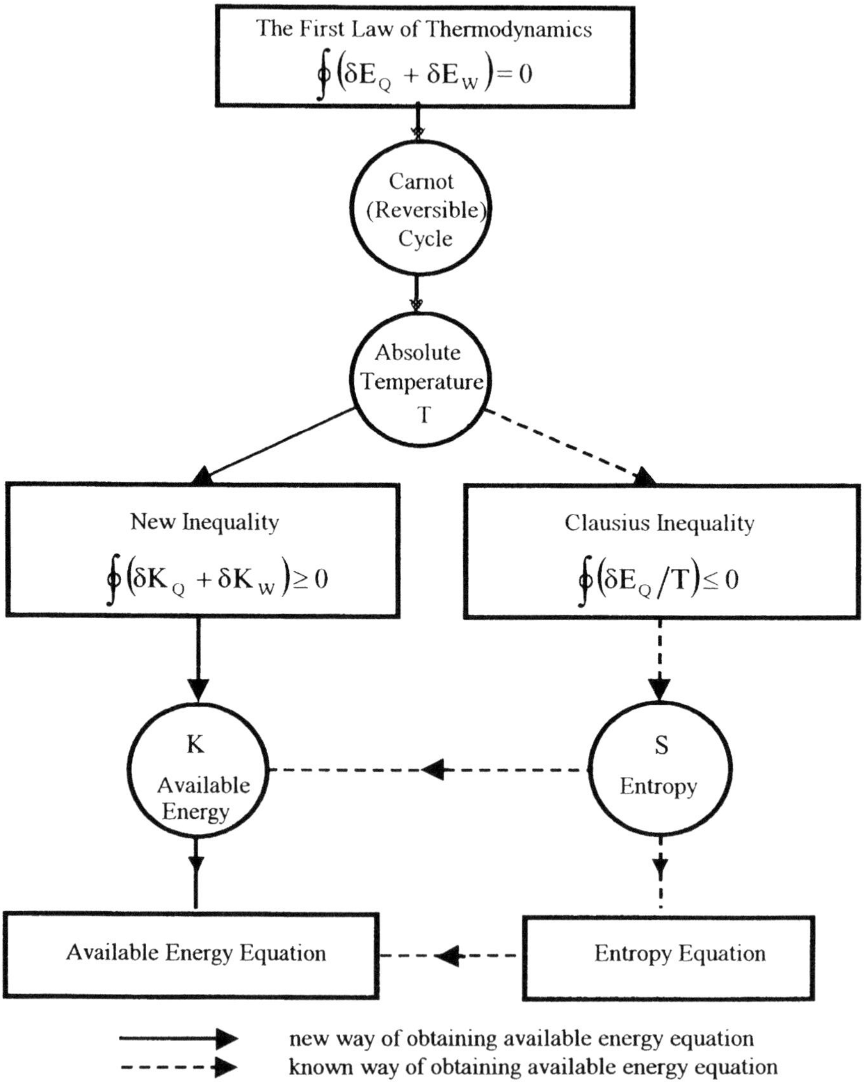

Figure 1. Illustration of available energy versus entropy.

2. New Inequality

Recalling the reversible cycle (Carnot cycle) and the existence of absolute temperature, the maximum work which can be achieved from the amount of heat (E_Q) at the temperature level T with respect to T_0 ambient temperature is called *the available energy of heat*

$$\delta K_Q = \eta_{C0}\, \delta E_Q \tag{2}$$

where, $\eta_{C0} = 1 - T_0 / T$ is the Carnot efficiency. According to the second law of thermodynamics, the actual work obtained from the same amount of heat is less than the available energy of heat and if the process were completely reversible they could be equal ($\delta E_W \leq \delta K_Q$).

The available energy of the work is

$$\delta K_W = \delta E_W + P_0\, dV \tag{3}$$

which is actually the useful work.

Now, let us determine the amount of work done on a control mass during the infinitesimal reversible process shown in *Fig.2.* Let us assume the amount of heat exchange of the control mass is δE_Q during this process. For only internal irreversibilities, the heat exchange can be done with the ambient by means of reversible engine. Applying the first law and the definition of the absolute temperature scale to the reversible engine the following equations can be written, respectively

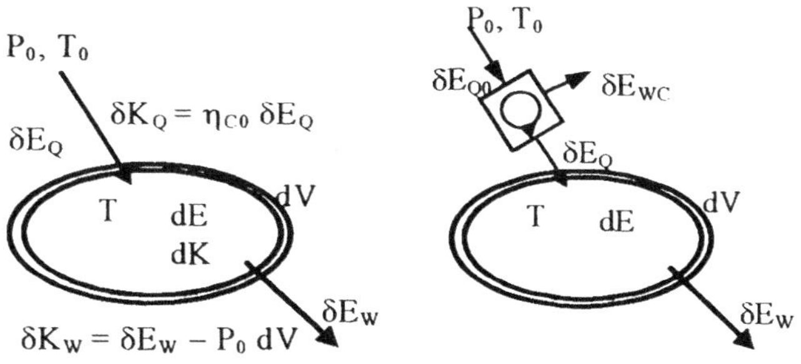

Figure 2. The illustration of the new inequality and Clausius inequality.

$$-\delta E_Q + \delta E_{Q0} + \delta E_{WC} = 0 \tag{4}$$

$$\frac{-\delta E_Q}{T} + \frac{\delta E_{Q0}}{T_0} = 0 \tag{5}$$

where, δE_Q is the heat transfer to the control mass and therefore the minus sign (–) comes for the heat transfer of the reversible engine. The sign convention is used in this work, as both the heat and the work transferred to the control mass is considered positive. From the Eqs.(2), (4) and (5) eliminating δE_{Q0} and δE_Q, the work of the reversible engine can be found as follows,

$$\delta E_{WC} = \left(1 - \frac{T_0}{T}\right) \delta E_Q = \delta K_Q \tag{6}$$

On the other hand, the work of the control mass, from Eq.(2), can be written as follows

$$\delta E_W = \delta K_W - P_0 \, dV \tag{7}$$

Now, in a manner which is conceptually very simple, the work done on the control mass and on the reversible engine can be written in terms of available energy of heat and the available energy of work, from the Eqs.(6) and (7)

$$\delta E_{Wtot} = \delta E_{WC} + \delta E_W = \delta K_Q + \delta K_W - P_0 \, dV \tag{8}$$

Recalling the Kelvin-Planck statement for the second law of thermodynamics *"It is impossible to construct a heat engine using one heat reservoir to produce the work,"* so, work can be converted to heat and the analytical statement is reduced as follows, by means of our sign convention

$$\oint \delta E_W \geq 0 \tag{9}$$

Applying the Kelvin-Planck statement to the Eq.(8), the new inequality for the cyclic process of a control mass

$$\oint \left(\delta K_Q + \delta K_W\right) \geq 0 \tag{10}$$

is developed directly, because V is a property of the control mass and the cyclic integral of a property is zero. Now, from this, the second law of thermodynamics can be stated as follows: *"During the cycle of a control mass, the available energy flow to and/or from the system as heat and/or work can not be negative."* In other words, available energy is destructed because of internal irreversibilities. If the cycle is internally reversible there is no destruction of available energy.

With the combination of the first law of thermodynamics

$$\oint \left(\delta E_Q + \delta E_W\right) = 0 \tag{11}$$

Clausius inequality can be derived from this new inequality using the Eqs. (2) and (3).

$$\oint \left[\left(1 - \frac{T_0}{T}\right) \delta E_Q + \left(\delta E_W + P_0 \, dV\right) \right] \geq 0 \tag{12}$$

Thus both inequalities are equivalent.

3. Available Energy - Availability (Available Internal Energy)

The available energy of a system may be defined as follows: *the maximum possible useful work that could be achieved by bringing the system to a dead state --a state in equilibrium to its environment-- by ideal processes.* If the available energy of a system is denoted by K, the available internal energy (B) which is an extensive property of a substance is similar to internal energy, however it is also associated with environmental properties, such as P_0 and T_0. Then, subtracting available kinetic energy and available potential energy of the system from the available energy, the availability is determined as follows:

$$B = K - \frac{1}{2} mV^2 - mg(z - z_0) \tag{13}$$

where, m is the mass, V is the velocity, z is the elevation of the system, z_0 is the elevation of dead state which can be taken as zero, and g is the gravitational acceleration.

The available energy of a control mass, in this work, will be directly derived from the new inequality in like manner to the realization of energy from the first law of thermodynamics (Eq.11). Consider any two arbitrary reversible processes of a system, 1A2 and 1B2, which are between the initial state 1 and the final state 2. With the application of a third reversible process 2C1, each of these two processes becomes reversible cycles; then the new inequality can be applied *(Fig.3)*. The new inequality will be zero as follows,

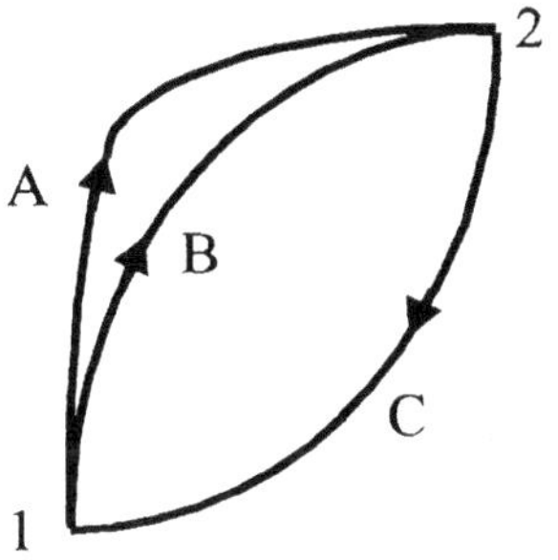

Figure 3. Two reversible cycles used to determine the available energy as a property.

$$\int_{1A}^{2} \left(\delta K_Q + \delta K_W\right) + \int_{2C}^{1} \left(\delta K_Q + \delta K_W\right) = 0 \tag{14a}$$

$$\int_{1B}^{2} \left(\delta K_Q + \delta K_W\right) + \int_{2C}^{1} \left(\delta K_Q + \delta K_W\right) = 0 \tag{14b}$$

Subtracting Eq.(14b) from Eq.(14a) it is concluded that

$$\int_{1A}^{2} \left(\delta K_Q + \delta K_W\right) = \int_{1B}^{2} \left(\delta K_Q + \delta K_W\right) = 0 \tag{15}$$

This shows that the integral of $(\delta K_Q + \delta K_W)_{rev}$ is the same for all reversible paths between state 1 and state 2 since these two processes (A and B) have been chosen arbitrarily. This means the value of this integral depends only on the end states, but not the path. Thus, one can conclude that the quantity of $(\delta K_Q + \delta K_W)_{rev}$ defines the change of a thermodynamic property that is obviously the available energy of the control mass:

$$dK = \left(\delta K_Q + \delta K_W\right)_{rev} = dB + dK_{KE} + dK_{PE} \tag{16}$$

By substituting Eqs.(2) and (3) into Eq.(16) and neglecting both the available kinetic and potential energies, the differential change of available internal energy,

$$dB = \left[\left(1 - \frac{T_0}{T}\right)\delta E_Q + \left(\delta E_W + P_0\, dV\right)\right]_{rev} \tag{17}$$

and eliminating the heat by means of the first law of thermodynamics and with the only moving boundary work $\delta E_W = -P\, dV$, differential of available internal energy per unit mass can be determined as

$$db = T_0\left[\left(\frac{1}{T_0} - \frac{1}{T}\right)du + \left(\frac{P_0}{T_0} - \frac{P}{T}\right)dv\right] \tag{18}$$

and then substituting internal energy equation into the above equation,

$$db = \left(1 - \frac{T_0}{T}\right)c_v\, dT + \left[\left(T - T_0\right)\left(\frac{\partial P}{\partial T}\right)_v - \left(P - P_0\right)\right]dv \tag{19}$$

can be obtained without entropy as well as the other properties [9]. On the other hand, arranging the Eq.(17) by means of the first law of the thermodynamics ($\delta E_Q + \delta E_W = dU$)

$$dB = dU + P_0\, dV - T_0\left(\frac{\delta E_Q}{T}\right)_{rev} \tag{20}$$

and using the definition of entropy the well-known relation can be derived easily

$$dB = dU + P_0\, dV - T_0\, dS \tag{21}$$

4. Available Energy Equation

Consider two arbitrary processes of a system between end states 1 and 2, and that one is reversible (A) and the other irreversible (B). These two cyclic processes are then concluded with an arbitrary reversible process (C) *(Fig.4)*. Because the cycle 1A2C1 is reversible,

$$\int_{1A}^{2}(\delta K_Q + \delta K_W)_{rev} + \int_{2C}^{1}(\delta K_Q + \delta K_W)_{rev} = 0 \tag{22a}$$

can be written similar to Eq.(14a). Applying the new inequality to the irreversible cycle, it gives the result

$$\int_{1\ B}^{2}(\delta K_Q + \delta K_W)_{irr} + \int_{2C}^{1}(\delta K_Q + \delta K_W)_{rev} > 0 \tag{22b}$$

Subtracting Eq.(22b) from Eq.(22a) the following inequality can be written:

$$\int_{1A}^{2}(\delta K_Q + \delta K_W)_{rev} < \int_{1B}^{2}(\delta K_Q + \delta K_W)_{irr} \tag{23}$$

Since the process 1A2 is reversible, by substituting Eq.(16), without using entropy we are able to obtain the *available energy equation* for a system

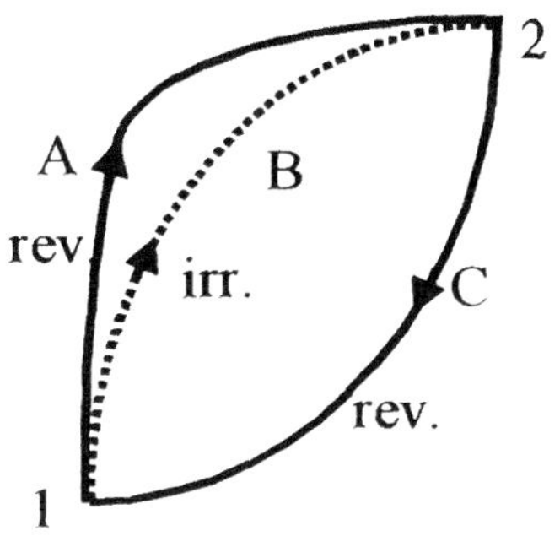

Figure 4. The reversible and irreversible cycles used to develop the available energy equation.

$$dK \le (\delta K_Q + \delta K_W) \tag{24}$$

and the available energy loss for a process.

$$K_{L12,int} = \int_1^2 (\delta K_Q + \delta K_W) + K_1 - K_2 \tag{25}$$

In the Eq. (24), equality assigns for a reversible process and the inequality for an irreversible process. The available energy loss, in these equations, are due to internal irreversibilities. If the control mass contacts more than one reservoir having temperatures of T_{Ri}, the total loss of available energy, including external irreversibilities, is obtained as follows:

$$K_{L12} = \sum_i \left(1 - \frac{T_0}{T_{Ri}}\right) E_{Qi,12} + E_{Wuse,12} + K_1 - K_2 \tag{26}$$

by adding and subtracting the summation of $\Sigma(E_{Qi}/T_{Ri})$ where, E_{Qi} s are the heat transfer interactions with the heat reservoirs T_{Ri}.

5. Conclusion

This is a new approach for obtaining availability which is a property of substances and the available energy equation directly from the new inequality without involving the concept of entropy. This considerably simplifies the traditional procedures and these concepts become much easier to understand, especially for students who had often found the traditional procedure ambiguous. Previously some authors had tried not to use entropy, however that approach had made the discussion even more obscure [10]. The new inequality is much simpler and more meaningful than Clausius inequality because it involves the available energy of heat and work instead of "transformation",

Q/T. The new inequality is also superior in that it resembles the first law of thermodynamics (Eq. 11).

By means of the new inequality, the second law of thermodynamics becomes easily understandable to everyone, not only just to students of the subject.

Of course, at this point, we can not abandon the concept of entropy, but initially we can discuss the available energy equation, and then follow with the entropy equation.

Obviously, the equation

$$dK = \delta K_Q + \delta K_W \quad (27)$$

for a reversible process and

$$dK = \delta K_W \quad (28)$$

are achieved for a reversible adiabatic process from the Eq.(24). For an adiabatic control mass, the second law of thermodynamics results

$$dK \leq \delta K_W \quad (29)$$

and for an isolated system

$$dK \leq 0 \quad (30)$$

the available energy of the system decreases and the change of available energy becomes zero for equilibrium.

Finally, in parallel to Clausius' famous expression, we can say that *"the energy of the universe is constant, but the available energy of the universe decreases continuously."*

References

1. Carnot, S. (1824) Reflections on the motive power of fire, and on machines fitted to develop that power, in J. Kestin, (ed.), *The Second Law of Thermodynamics Benchmark Papers on Energy, Vol.5*. Dowden, Hutchington & Ross. Inc.1976.
2. Van Wylen G.J., and Sonntag, R.E. (1986) *Fundamentals of Classical Thermodynamics*, 3rd Ed., Wiley.
3. Lee, F.L. and Sears, F.W. (1963) *Thermodynamics*, Second Ed., Addison-Wesley.
4. Wark, K. (1983) *Thermodynamics*, 4th ed., McGraw-Hill.
5. Öztürk, A. and Kılıç, A. (1993) *Thermodynamics With Solved Problems*, 3rd Ed., Çağlayan (in Turkish).
6. Bejan, A. (1988) *Advanced Engineering Thermodynamics*, Wiley.
7. Obert, E. F. (1960) *Concepts of Thermodynamics*, McGraw-Hill.
8. Öztürk, A. (1996) Teaching second law without entropy, in G. Svedberg et.al. (eds.) *Efficiency, Costs, Optimization, Simulation and Environmental Aspects of Energy Systems*, ECOS'96 June 22-27, Stockholm-Sweden.
9. Öztürk, A. (1997) Some thermodynamic relations involving availability not entropy, in G. Manfrida et.al. (eds.) *Florence World Energy Research Symposium*, FLOWERS'97, July 30-August 1, Florence-Italy.
10. Lucca, G. (1991) On the opportunity, from the didactic point of view, to derive entropy from exergy, in Reistad, G.M. et.al. (eds.), *Second Law Analysis-Industrial and Environmental Applications, AES-Vol.25/HTD-Vol.191*, Winter Annual Meeting of ASME, the Heat Transfer Division, Dec.1-6, Atlanta, Georgia.

EXERGY ANALYSIS IN THE PROCESS INDUSTRY

R.L. CORNELISSEN and G.G. HIRS
University of Twente, Dept. of Mechanical Eng.,
Chair of Energy Technology
P.O. Box 217, 7500 AE Enschede, The Netherlands

1. Introduction

The method of exergy analysis, as it is used nowadays, has been developed by Szargut [1]. He and his co-workers have analysed several plants systematically. In his method the complete production chain for the product has been analysed. Another systematically analysis of process plants is performed by Hedman [2] in the USA. His work is limited to the analysis of plants only, without adding new elements to the method of exergy analysis.

Nowadays, the use of exergy analysis in the process industry in the Netherlands is increasing steadily. Most big companies in this industry have experienced, often with help of the universities, sometimes without, the use of exergy analysis. The reactions of the companies are positive. Shell, Akzo Nobel and DSM, three main chemical companies have performed exergy analysis of several plants. On basis of this analysis, several improvements are suggested. However, the financial pay back period of these improvements is in many cases too long, because investments for energy saving are often high in existing plants. So, it is suggested to implement the use of exergy analysis in the design process of new plants.

Hoogovens, the Dutch steel maker, has used exergy analysis to analyse a new production process of steel making [3]. With exergy analysis a comparison is made with the blast furnace process. Furthermore, the locations of the losses are found in the two processes. On basis of this knowledge, the existing process can be optimised, while for the new process more insight is gained on which components major improvements are possible. Unilever has used exergy analysis for analysing its products. The complete life cycle of the washing process has been carried out [4]. The insights of this analysis are used to develop better washing powders. New products designs have been tested on basis of exergy analysis. The results have been used to improve the products and even to stop a product development.

In this article a cryogenic air separation plant is analysed. The analysis is performed on a plant of Hoek Loos in the Rotterdam harbour area.

A. Bejan and E. Mamut (eds.), Thermodynamic Optimization of Complex Energy Systems, 195–208.

2. Method

The exergy analysis consist of the following steps:

1. Setting up of the mass balance. This balance has to be closed
2. Setting up of the energy balance. This balance has also to be closed. Quite often, (parts of) the energy balance is used to set up the material balance.
3. Carrying out of the exergy analysis.

For performing an exergy analysis a reference state has to be chosen. In the Netherlands the reference system of Szargut [1] is mostly used. The standard environmental conditions are a reference temperature of 298.15 K and a pressure of 1 atm. However, sometimes it is necessary to change these conditions. For example, in the case of a power plant, which use cooling water of 15°C. Problems arise when two reference conditions are required. For example, as in the just mentioned power plant the temperature of the inlet air is 10°C. A choice between the reference temperature of 10 or 15°C has to be made. In both cases one of the inputs, water or air, will have an exergy content. No systematic approach for this situation has been developed yet.

Most exergy analyses for process plants are performed with flowsheeters. The most common flowsheeter is ASPEN PLUS. A special exergy calculating routine, EXERCOM, has been developed for ASPEN PLUS by STORK ENGINEERING [5]. The routine facilitates the performing of a exergy analysis greatly.

On basis of the results of the exergy analysis improvements are suggested. Suggestions for improvements are based on two criteria, the exergy losses and the efficiency of the components. On basis of the exergy losses the maximum amount of exergy saving is known, while the efficiency of the components gives an indication of the technical possibilities for improvements. No systematic method for improvements has been developed yet, as is the case for pinch technology.

3. Cryogenic Air Separation Plant

The analysed plant uses the principle of two-column separation. This is based on a low and a high pressure distillation column. The condenser of the high pressure lower column (LC) is cooled by the reboiler of the low pressure upper column (UC). This principle is widely used for cryogenic air separation plants. The flowsheet of the plant is shown in Figure 1. The incoming air is compressed to 6.2 bar and after cooling in the spray cooler (SC) water and carbon dioxide are removed by adsorption in the molecular filters. Subsequently, the air is cooled in the main heat exchanger (MHE) to -172 °C and enters the lower column (LC), which has 54 stages. Here, the air is separated into the top product nitrogen, which is partly liquefied in the condenser and the bottom product (B), which is a mixture of 38 vol.% oxygen, 60 vol. % nitrogen and 1.6 vol. % argon. The gaseous part of the nitrogen (TN) is liquefied in a special section. Here, the nitrogen is compressed to 46 bar with the aid of compressor 1, 2 and 3. Compressor 3 is driven by the hot expansion turbine (HET) and compressor 2 by the cold expansion turbine (CET). The liquefaction of the stream takes place in heat exchanger 2 (HE 2).

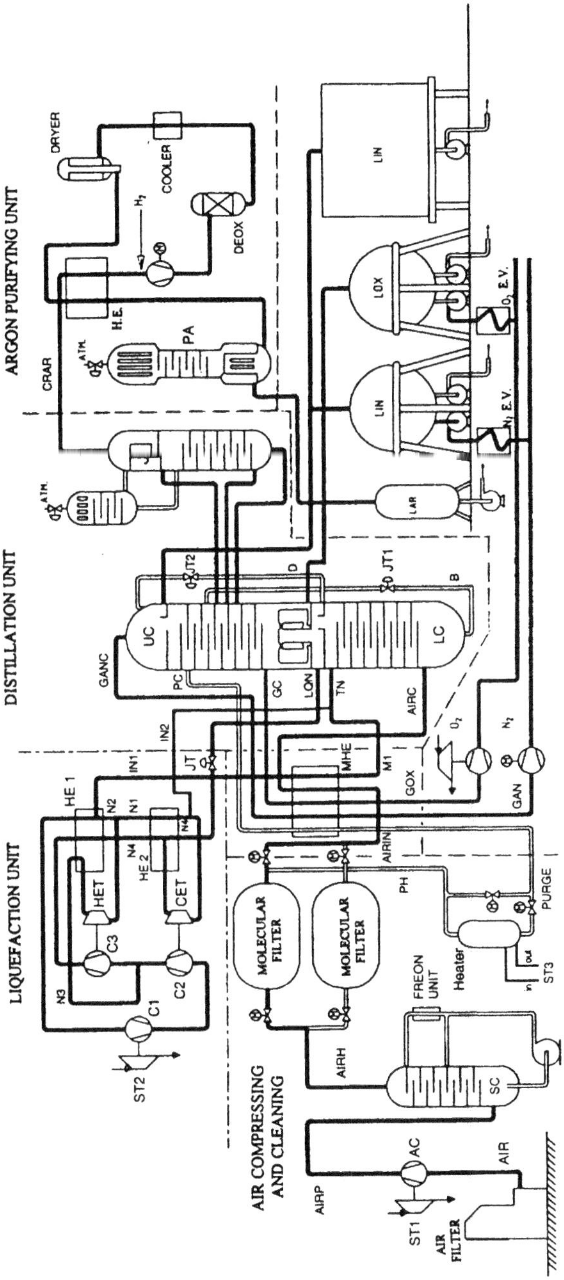

Figure 1. Flowsheet of the analysed cryogenic air separation plant

Further cooling of the liquid nitrogen takes place in the throttling valve (TV). The nitrogen stays almost completely liquefied after passing this valve and enters the lower column again. The bottom product and the liquid nitrogen leaving the lower column are throttled to ca. 1.5 bar in valve 1(JT 1) and valve 2 (JT2) and enter the upper column (UC), which has 96 stages. Of this column the top product is gaseous (GAN) and liquid nitrogen (LIN) and the bottom product is gaseous (GOX) and liquid oxygen (LOX). One of the side products is unpurified argon, consisting of argon with oxygen and nitrogen. Most of this oxygen and nitrogen returns from the crude argon distillation column to the upper column. The other side product is purge gas (PC), which is used to clean the molecular filters. It consists mainly of nitrogen and the gas is heated to 170°C before cleaning the filters. The top product of the crude argon distillation column (CRAR) enters the argon purifying unit, where it is compressed to 3 bar and deoxygenated by burning the oxygen with hydrogen. After cooling, separating the water and further cooling to -180°C the argon is further purified in the pure argon distillation column (PA). Compression of the products to the desired pressure is left out in this analysis. The flow and product specification are displayed in Table 1. Because of the flexibility of the plant variations of the flow are possible. The analysis has been performed with the flow given in Table 1.

TABLE 1. Flow and product specification.

Components	Flow (in kg/s)	Purity (in vol. %)
Air	16.39	--
Gaseous Oxygen (GOX)	2.82	99.5
Liquid Oxygen (LOX)	0.70	99.6
Gaseous Nitrogen (GAN)	6.18	99.95
Liquid Nitrogen (LIN)	2.86	99.999
Liquid Argon (LAR)	0.13	99

4. Rational efficiency of the different units

The general form of the rational efficiency is defined as the ratio of the desired exergy output to the exergy used [6]:

$$\psi = \frac{\dot{E}_{desired\ output}}{\dot{E}_{used}} \tag{1}$$

where

$$\dot{E}_{used} = \dot{E}_{desired\ output} + \dot{I} \tag{2}$$

and with help of formula (2) can (1) by rewritten in

$$\psi = 1 - \frac{\dot{I}}{\dot{E}_{used}} \tag{3}$$

For the compressors the following formula is used

$$\psi_{compressor} = \left(\dot{E}_{out} - \dot{E}_{in}\right) / \dot{W}_{in} \tag{4}$$

where W is the work input of the compressor. For turbines the rational efficiency is defined as the work output divided by the exergy decrease of the incoming and outing flows, in formula form

$$\psi_{turbines} = \frac{\dot{W}_{out}}{\dot{E}_{out} - \dot{E}_{in}} \tag{5}$$

For the sake of simplicity no distinction is made in the rational efficiency of the heat exchanger for the thermal and mechanical component. The rational efficiency is then defined as

$$\psi_{\mathrm{HE,heating}} = \frac{\dot{E}_{\mathrm{cold,out}} - \dot{E}_{\mathrm{cold,in}}}{\dot{E}_{\mathrm{hot,in}} - \dot{E}_{\mathrm{hot,out}}} \tag{6a}$$

for operating above T_0 and

$$\psi_{\mathrm{HE,heating}} = \frac{\dot{E}_{\mathrm{cold,out}} - \dot{E}_{\mathrm{cold,in}}}{\dot{E}_{\mathrm{hot,in}} - \dot{E}_{\mathrm{hot,out}}} \tag{6b}$$

for operating below T_0.

No chemical reactions occur in the heat exchangers, so the difference in exergy of the streams is taken, because the chemical component will have no influence on the rational efficiency. Because no distinction is made between the thermal and mechanical component in the case of the throttling valve it was not possible to define a desired product, so no efficiency could be calculated. No desired product can be defined in the case of mixing and also no rational efficiency can be obtained.

For the air separation and cleaning unit the exergy balance[1] is

---

[1] Streams $ST1_{,in}$ and $ST1_{,out}$ refer to the incoming and outgoing streams of steam turbine 1 (ST1). This notation is used further on for more components displayed in Fig. 1.

$$\dot{E}_{\mathrm{AIR}} + \dot{E}_{\mathrm{ST1,in}} + \dot{W}_{\mathrm{freon}} + \dot{E}_{\mathrm{PURGE}} + \dot{E}_{\mathrm{ST3,in}} = \dot{E}_{\mathrm{AIRIN}} + \dot{E}_{\mathrm{ST1,out}} + \dot{E}_{\mathrm{ST3,out}} + \dot{I}_{\mathrm{ACC}} \qquad (7)$$

The desired product is

$$\dot{E}_{desired\,output} = \dot{E}_{\mathrm{AIRIN}} - \dot{E}_{\mathrm{AIR}} \qquad (8)$$

which leads to the following expression for the exergy used:

$$\dot{E}_{used} = \dot{E}_{\mathrm{ST1,in}} - \dot{E}_{\mathrm{ST2,out}} + \dot{E}_{\mathrm{ST3,in}} - \dot{E}_{\mathrm{ST3,out}} + \dot{W}_{\mathrm{freon}} + \dot{E}_{\mathrm{PURGE}} \qquad (9)$$

The rational efficiency is then

$$\psi_{ACC} = \frac{\dot{E}_{\mathrm{AIRIN}} - \dot{E}_{\mathrm{AIR}}}{\dot{E}_{\mathrm{ST1,in}} - \dot{E}_{\mathrm{ST1,out}} + \dot{E}_{\mathrm{ST3,in}} - \dot{E}_{\mathrm{ST3,out}} + \dot{W}_{\mathrm{freon}} + \dot{E}_{\mathrm{PURGE}}} \qquad (10)$$

or in the rewritten form by (3)

$$\psi_{\mathrm{ACC}} = \frac{\dot{E}_{\mathrm{AIRIN}} - \dot{E}_{\mathrm{AIR}}}{\dot{E}_{\mathrm{AIRIN}} - \dot{E}_{\mathrm{AIR}} + \dot{I}_{\mathrm{ACC}}}$$

For the lower distillation column the exergy balance is

$$\dot{E}_{\mathrm{AIRC}} + \dot{E}_{\mathrm{LQN}} + \dot{E}^{Q}_{\mathrm{cond}} = \dot{E}_{\mathrm{B}} + \dot{E}_{\mathrm{D}} + \dot{E}_{\mathrm{TN}} + \dot{I}_{\mathrm{LC}} \qquad (11)$$

The desired product is the increase of the chemical component

$$\dot{E}_{desired\,output} = (\sum_{fr} \dot{E}^{\mathrm{o}}_{\mathrm{i}} - \dot{E}^{\mathrm{o}}_{\mathrm{mix}}) \qquad (12)$$

$$= \dot{E}^{\mathrm{o}}_{\mathrm{B}} + \dot{E}^{\mathrm{o}}_{\mathrm{D}} + \dot{E}^{\mathrm{o}}_{\mathrm{TN}} - \dot{E}^{\mathrm{o}}_{\mathrm{AIRC}} - \dot{E}^{\mathrm{o}}_{\mathrm{LQN}}$$

where the mixture is the incoming stream(s) and the fractions are the leaving streams. $\dot{E}^{\mathrm{o}}_{\mathrm{LQN}}$ and $\dot{E}^{\mathrm{o}}_{\mathrm{TN}}$ are equal and can be left out for further analysis. This leads to the following expression for the exergy used:

$$\dot{E}_{used} = (\dot{E}^{\mathrm{ph}}_{\mathrm{mix}} - \sum_{fr} \dot{E}^{\mathrm{ph}}_{\mathrm{i}}) + \dot{E}^{Q}_{\mathrm{cond}} \qquad (13)$$

$$= (\dot{E}^{\mathrm{ph}}_{\mathrm{AIRC}} + \dot{E}^{\mathrm{ph}}_{\mathrm{LQN}} - \dot{E}^{\mathrm{ph}}_{\mathrm{B}} - \dot{E}^{\mathrm{ph}}_{\mathrm{D}} - \dot{E}^{\mathrm{ph}}_{\mathrm{TN}}) + \dot{E}^{Q}_{\mathrm{cond}}$$

The rational efficiency for the lower distillation column is then

$$\psi_{LC} = \frac{\dot{E}_B^o + \dot{E}_D^o + \dot{E}_{TN}^o - \dot{E}_{AIRC}^o - \dot{E}_{LQN}^o}{(\dot{E}_{AIRC}^{ph} + \dot{E}_{LQN}^{ph} - \dot{E}_B^{ph} - \dot{E}_D^{ph} - \dot{E}_{TN}^{ph}) + \dot{E}_{cond}^Q} \tag{14}$$

For the upper distillation column the exergy balance is

$$\dot{E}_{\mathrm{JT1,in}} + \dot{E}_{\mathrm{JT2,out}} = \sum_{fr} \dot{E}_{\mathrm{i}} + \dot{E}_{\mathrm{cond}}^Q + \dot{I}_{\mathrm{LC}} \tag{15}$$

where the fractions (fr) are GANC, LIN, CRAR, PC, GC and LOX. For the upper column, including the crude argon distillation column, the desired product is

$$\dot{E}_{desired\ output} = (\sum_{fr} \dot{E}_{\mathrm{i}}^{\mathrm{o}} - \dot{E}_{\mathrm{mix}}^{\mathrm{o}}) + \dot{E}_{\mathrm{cond}}^Q \tag{16}$$

where mix is JT1, in and JT2, out and this results to the following expression for the exergy used:

$$\dot{E}_{used} = (\dot{E}_{\mathrm{mix}}^{\mathrm{ph}} - \sum_{fr} \dot{E}_{\mathrm{i}}^{\mathrm{ph}}) \tag{17}$$

The rational efficiency for the upper distillation column is then

$$\psi_{\mathrm{UC}} = \frac{(\sum_{fr} \dot{E}_{\mathrm{i}}^{\mathrm{o}} - \dot{E}_{\mathrm{mix}}^{\mathrm{o}}) + \dot{E}_{\mathrm{cond}}^Q}{(\dot{E}_{\mathrm{mix}}^{\mathrm{ph}} - \sum_{fr} \dot{E}_{\mathrm{i}}^{\mathrm{ph}})} \tag{18}$$

Because there is no heat exchange with the environment, the rational efficiency for the distillation section is

$$\psi_{\mathrm{DIST}} = \frac{(\sum_{fr} \dot{E}_{\mathrm{i}}^{\mathrm{o}} - \dot{E}_{\mathrm{AIRC}}^{\mathrm{o}})}{(\dot{E}_{\mathrm{AIRC}}^{\mathrm{ph}} - \sum_{fr} \dot{E}_{\mathrm{i}}^{\mathrm{ph}})} \tag{19}$$

as can be derived by setting up the exergy balance. The fractions are the flows leaving the section. The same formula can be applied for the argon purifying unit by replacing AIRC by CRAR in (19). The outgoing fraction is then only the LAR. The top product is released and dissipated in the environment.

The desired output of liquefaction unit is seen as the exergy increase of the outgoing flow compared to the incoming flow because of the cooling and liquefying of the nitrogen. By setting up the exergy balance and defining the desired exergy output and thus the exergy used as just described the following formula is obtained

$$\psi_{\mathrm{LIQ}} = \frac{\dot{E}_{\mathrm{LQN}} - (\dot{E}_{\mathrm{IN1}} + \dot{E}_{\mathrm{IN2}})}{(\dot{E}_{\mathrm{ST2,in}} - \dot{E}_{\mathrm{ST2,out}})} \tag{20}$$

The exergy balance for the whole air separation plant is given by:

$$\dot{E}_{\mathrm{air}} + \sum_{all} \dot{E}_{\mathrm{Steam,in}} + \dot{W}_{\mathrm{freon}} = \sum_{frall} (\dot{E}^{\mathrm{ph}}_{\mathrm{out,i}} + \dot{E}^{\mathrm{o}}_{\mathrm{i}}) + \sum_{all} \dot{E}_{\mathrm{Steam,out}} + \sum_{all} \dot{I}_{\mathrm{i}} \tag{21}$$

where $\sum_{all} \dot{E}_{\mathrm{Steam,in}}$ and $\sum_{all} \dot{E}_{\mathrm{Steam,out}}$ are all steam in- and outputs of the plant, respectively, $\dot{E}_{\mathrm{air}}$ is the exergy of the incoming air and $\sum_{frall} \dot{E}^{\mathrm{ph}}_{\mathrm{out,i}}$ means the physical exergy of the fractions leaving the distillation plant, which are GAN, LIN, GOX, LOX and LAR as displayed in Figure 1. This physical exergy can be used in other processes. From (9) it is clear that the exergy used is:

$$\dot{E}_{used} = \sum_{all} \dot{E}_{Steam,in} - \sum_{all} \dot{E}_{Steam,out} + \dot{W}_{freon} \tag{22}$$

This leads with reference to (2) and with the aid of expressions (21) and (22) to the following expression for the desired exergy output:

$$\dot{E}_{desired\,output} = \sum_{frall} (\dot{E}^{\mathrm{ph}}_{\mathrm{out,i}} + \dot{E}^{\mathrm{o}}_{\mathrm{i}}) - \dot{E}_{\mathrm{air}} \tag{23}$$

Using (1), the rational efficiency for the whole air separation plant is given by:

$$\psi_{\mathrm{plant}} = \frac{\sum_{frall} (\dot{E}^{\mathrm{ph}}_{\mathrm{out,i}} + \dot{E}^{\mathrm{o}}_{i}) - \dot{E}_{air}}{\sum_{all} \dot{E}_{\mathrm{Steam,in}} - \sum_{all} \dot{E}_{\mathrm{Steam,out}} + \dot{W}_{\mathrm{freon}}} \tag{24}$$

or from (3)

$$\psi_{\mathrm{plant}} = \frac{\sum_{frall} (\dot{E}^{\mathrm{ph}}_{\mathrm{out,i}} + \dot{E}^{\mathrm{o}}_{\mathrm{i}}) - \dot{E}_{\mathrm{air}}}{\sum_{frall} (\dot{E}^{\mathrm{ph}}_{\mathrm{out,i}} + \dot{E}^{\mathrm{o}}_{\mathrm{i}}) - \dot{E}_{\mathrm{air}} + \sum_{all} \dot{I}_{\mathrm{i}}} \tag{25}$$

It could be argued that the desired product of the plant is the chemical exergy increase only. This would lead to the following formula for the whole plant

$$\psi_{plant} = \frac{\sum_{frall} \dot{E}_i^o - \dot{E}_{air}^o}{\sum_{frall} \dot{E}_{out,i}^{ph} - \dot{E}_{air}^{ph} + \sum_{all} \dot{E}_{Steam,in} - \sum_{all} \dot{E}_{Steam,out} + \dot{W}_{freon}} \qquad (26)$$

5. Exergy Analysis

5.1. AIR COMPRESSOR AND CLEANING UNIT

The main exergy loss, 1708 kW, is caused by the compression of the air to 6.2 bar as can be seen in Table 2. The exergy loss of steam turbine 1 (ST1), which delivers the required 4.3 MW of power for the air compressor (AC), is 708 kW (see Table 3). The consumption of steam of 400°C and 40 bar is 15.5 ton a hour. The rational efficiency of steam turbine 1 as calculated with (5) is 0.86. For cooling the air to 7°C the electricity use for the freon unit is 110 kW $(\dot{W}_{freon})$. Most of the exergy increase due to the cooling of the air is lost in the molecular filter, because the air is here heated to just below the environmental temperature. The molecular filters are cleaned by the hot purge gas. By the heating of the purge gas to 170°C with steam of 210°C and 11 bar and dumping the gas into the environment after the cleaning of the filters 225 kW of exergy is lost. The rational efficiency of the air compressor and cleaning unit is calculated to be 0.48 with formula 10.

TABLE 2. The in- and outgoing flows of the air compressor (AC).

Components	Temp. (in °C)	Press. (in bar)	Flow[1)] (kg/sec)	Enthalpy (in kW)	Exergy (in kW)
AIR	20.0	1.01	16.39	-2638.9	0.7
Work input				4334.9	4334.9
Heat transfer to envir.				-3362.6	0
AIRP	79.4	6.20	-16.39	1666.7	-2627.8
Total				0.1	1707.8

[1)] The minus sign means the stream is leaving the AC

TABLE 3. The in- and outgoing flows of steam turbine 1 (ST1).

Components	Temp. (in °C)	Press. (in bar)	Flow[1)] (kg/sec)	Enthalpy (in kW)	Exergy (in Kw)
Steam in (ST1in)	400.0	40.00	4.31	-54921	5497.3
Steam out (Stout)	29.6	0.04	-4.31	59256	-454.1
Work output				-4335	-4334.9
Total				0	708.3

[1)] The minus sign means the stream is leaving the ST1

5.2. MAIN HEAT EXCHANGER (MHE)

The exergy losses are caused by temperature difference between the hot and cold streams and pressure losses. The mean temperature difference is 4.2 K. The rational efficiency is 0.86 as calculated by (6b), where AIRIN is the hot stream. In Figure 1 INA and INB are displayed as one stream (IN). In reality M1 is split in the main stream INA and a small stream INB in the main heat exchanger (MHE) and combined at constant temperature after heat exchanger 1 (HE 1). The exergy calculations take into account this split as can be seen in Table 4.

TABLE 4. The in- and outgoing flows of the main heat exchanger.

Components	Temp. (in °C)	Press. (in bar)	Flow[1)] (kg/sec)	Enthalpy (in kW)	Exergy (in kW)
AIRIN	20.0	6.20	16.20	-110.1	2567.6
GOXC	-178.1	1.67	2.82	-527.4	81.3
Purge cold (PC)	-180.6	1.43	3.49	-743.8	665.9
M1	-177.5	5.85	3.60	-795.1	1195.1
GANC (GC)	-182.0	1.40	6.20	-1349.6	1292.2
AIRC	-172.6	5.85	-16.20	3373.5	-4688.0
GOX	17.0	1.47	-2.82	21.9	-426.5
INA (IN)	17.0	5.75	-3.50	34.9	-626.3
INB (IN)	-84.0	5.75	-0.10	11.7	-20.7
GAN	17.0	1.30	-6.20	53.9	-293.4
PURGE	17.0	1.43	-3.49	30.1	-145.6
Total				-0.3	333.2

[1)] The minus sign means the stream is leaving the MHE.

5.3. DISTILLATION UNIT

Because of its complex nature no complete flow scheme, including all stream data, has been given in this article. In Table 5 it can be seen how the different components operate by showing the exergy losses and rational efficiencies. The heat transfer between the reboiler of the upper column and the condenser of the lower column cause an exergy loss of 82 kW due to the temperature difference of 1.1 K. This results according to (6b) in a rational efficiency of 0.98. The throttling valves (JT1 and JT2) between the columns causes an exergy loss of 125 kW. The exergy loss in the lower column (LC) is 62 kW and the exergy loss in the upper column (UC) and crude argon column is 487 kW. Of the exergy loss in the crude argon column 108 kW is caused in the condenser by the temperature difference of 2.8 K between the coolant and the top of the column. The chemical exergy increase of the products of the lower and upper distillation column is 242.7 kJ and 428.5 kJ, respectively. Using (14) and (18) the rational efficiency of the lower and upper column were found to be 0.80 and 0.92, respectively. The high efficiency of the upper column is caused by the high exergy transfer from its reboiler to the condenser of the lower column. An exergy loss of 32 kW takes place in the heat exchanger (not displayed in Figure 1), which cools the products of the lower column, B and D, before going into the valves which lead to the upper column. Using (6b) a rational efficiency of 0.88 is obtained for this heat exchanger. The rational efficiency of the distillation section is 0.46 as calculated by (19) with a chemical exergy increase of 671.2 kJ.

TABLE 5. Results of the exergy analysis in the distillation unit.

Unit	exergy losses (in kW)	rational efficiency
heat exchange reboiler/condenser	82	0.98
Throttling valves columns	125	--
Lower distillation column	62	0.80
Upper distillation column	487	0.91
Heat exchanger	32	0.88
Total	788	0.46

5.4. LIQUEFACTION UNIT

Because of its complex nature the incoming and outgoing flows of the different components are not given. In Table 6 it can be seen how the different components operate by showing the exergy losses and rational efficiencies.

TABLE 6. Results of the exergy analysis in the liquefaction unit.

Unit	exergy losses (in kW)	Rational efficiency
Five stages compressor (C1)	1965	0.67
Steam turbine	1021	0.86
Compressor 2 (C2)	205	0.70
Compressor 3 (C3)	141	0.69
Capacity mismatch	344	-----
Hot expansion turbine (HET)	241	0.77
Cold expansion turbine (CET)	428	0.61
Heat exchanger 1 (HE 1)	256	0.72
Heat exchanger 2 (HE 2)	173	0.90
Mixers	5	----
Throttling valve (TV)	74	----
Total	4853	0.25

For the compression of the nitrogen to 30 bar a five stage compressor with intercooling (C1) is used. Steam turbine 2 (ST2), which delivers the required 6.0 MW for the compressor, uses 20 ton steam a hour of 420°C and 40 bar. A capacity mismatch between compressor 3 (C3) , which compresses to 46 bar, and the hot expansion turbine (HET) results in a exergy loss of 344 kW. The relative low efficiency of the turbines is caused by the fact that the expansion of the flows takes place below the environmental temperature. The temperature difference between the isentropic temperature and the real outlet temperature leads to a great exergy loss due to the low temperature. Especially, in the case of the cold expansion turbine (CET), where the outlet temperature is -177 °C.

The throttling process in the valve (TV) causes a temperature decrease of 3.4 K. The low exergy loss in the valve is achieved because the flow enters the valve in the liquid phase and leaves the valve still almost completely liquid.

5.5. ARGON PURIFYING UNIT

The exergy loss in the heat exchanger (H.E.) and compressor is 1.2 kW and 2.9 kW, respectively. The deoxygenation of the stream with hydrogen with an exergy of 65 kW causes an irreversibility of 27 kW. The exergy loss taking place due to the cooling of argon flow from 950°C to 15°C leads to an exergy loss of 43 kW. The heat is not used somewhere else. The exergy loss in the pure argon distillation column is 10 kW. A purge stream of 0.8 kW leaves the column at the top. The rational efficiency of this unit is 0.017 according to (19). No rational efficiencies are calculated for the separate components due to the small exergy losses of these components.

5.6. TOTAL

The overall exergy loss is 8810 kW as can be seen in Table 7. The total exergy of the products is 3514 kW, resulting in an overall rational efficiency of 0.28 according to (25). The efficiency is so high due to the great physical exergy of the products of 2765 kW, mainly the physical exergy of the liquid products, which is 2602 kW. If only the chemical exergy of the products is seen as the desired output, the rational efficiency becomes 0.071 by using (26).

TABLE 7. Results of the exergy analysis.

Unit	exergy losses (in kW)	rational efficiency
Air compressor and cleaning	2751	0.48
Main heat exchanger	333	0.86
Distillation unit	788	0.52
Liquefaction unit	4853	0.25
Argon purifying unit	85	0.02
Total	8810	0.28

6. Results

More than half of the exergy loss is taking place in the liquefaction unit, while almost one third is lost in the air compression unit. Minor exergy losses are taking place in the distillation unit and heat exchangers. The greatest exergy losses are caused by the compressors and to a lesser extent by the turbines. The relatively low rational efficiency of the turbines operating in the cryogenic region is striking, while the efficiency of the heat exchangers is very high.

7. Improvements

By using the excess work in the case of the capacity mismatch 344 kW can be saved. The increase of the polytropic efficiency of the air compressor (AC) from 0.70 to 0.85 will decrease the power use by 880 kW. This will give an associated exergy saving in steam turbine 1 (ST1) of 139 kW. The increase of the polytropic efficiency of the nitrogen compressor (C1) from 0.75 to 0.85 leads to a decrease of the power use by 759 kW. It gives also an associated exergy saving of 108 kW in steam turbine 2 (ST2). The rational efficiencies of both compressors (AC and C1) becomes 0.77. The increase of the polytropic efficiency of 0.83 of the cold expansion turbine (CET) to the same value as the hot expansion turbine (HET), will give an exergy saving of 75 kW and results in a rational efficiency of 0.66 for the cold expansion turbine.

The rational efficiency of the heat exchangers is high to very high. So the possibilities for exergy saving are very limited. Only the exergy loss in heat exchanger 1 (HE 1) of the liquefaction unit could be halved be improving the heat integration. This would lead to an exergy saving of about 125 kW.

8. Conclusions for the Cryogenic Air Separation

The exergy analysis of the cryogenic air distillation plant has pinpointed and quantified the exergy loss in the different plant sections. The biggest exergy loss is caused by the compressors. This exergy loss can be reduced by almost one half by using better compressors. Including other suggested improvements the exergy loss can be reduced by 25%. However, the cryogenic air separation unit itself is well designed from an exergetic point of view and for a further improvement of the rational efficiency of air separation alternative processes have to be used or developed.

References

1. Szargut, J., Morris, D.R. and Steward, F.R. (1988) *Exergy Analysis of Thermal, Chemical, and Metallurgical Processes*, Hemispere Publishing Corporation.
2. Hedman, B. (1981), *Application of the Second Law of Thermodynamics to Industrial Processes*, Ph.D. thesis, Drexel University,
3. Griffioen, A. (1996), Exergy Analysis of the Blast Furnace, Corex and CCF process, Master thesis, University of Twente.
4. Mozes, E, Cornelissen, R.L., Hirs, G.G. and Boom, R.M. (1997) Exergy Analysis of the Conventional Washing Machine. *Proc. of Flowers '97*, (Manfrida, Eds), Florence, 931-937.
5. Hinderink, A.P., Kerkhof, P.J.M., Lie, A.B.K. (1996) De Swaan Arons, J. and Van der Kooi, H.J., Exergy Analysis with a Flowsheeting Simulator-I. Theory; Calculating Exergies of Material Streams, *Chemical Engineering Science*, Vol. 51, No 20, 4693-4700.
6. Kotas, T.J. (1995) *The Exergy Method of Thermal Plant Analysis*, 2-nd edition, Krieger Publishing Company, USA.

9. Nomenclature

E, ε	= exergy, molar exergy
GAN	= gaseous nitrogen
GOX	= gaseous oxygen
LAR	= liquid argon
LIN	= liquid nitrogen
I	= irreversibility
P	= pressure
R	= universal gas constant
T	= temperature
ψ	= rational efficiency

Subscripts

AC	= air compressor
ACC	= air compressing and cleaning unit
CET	= cold expansion turbine
cold	= cold stream
CRAR	= crude argon
cold	= cold stream
cond	= condenser
DIST	= distillation section
HE	= heat exchanger
HEN	= heat exchanger network
HET	= hot expansion turbine
hot	= hot stream
fr	= fractions
hot	= hot stream
i	= i-th item
in	= incoming
JT	= Joule-Thomson valve
LC	= lower distillation column
LIQ	= liquefaction unit
min	= minimum
mix	= mixture
out	= outgoing
PA	= pure argon distillation column
PC	= purge cold
PH	= purge hot
S	= STEAM
Scf	= spray cooler
ST	= steam turbine
TN	= top nitrogen
UC	= upper distillation column
VDU	= vacuum distillation unit

Superscripts

0	= environmental state, chemical component
ph	= physical component
Q	= heat
ΔT	= thermal component
·	= rate with respect to time

EXERGETIC LIFE CYCLE ANALYSIS OF COMPONENTS IN A SYSTEM

R.L. CORNELISSEN and G.G. HIRS
University of Twente, Dept. of Mechanical Eng.,
Chair of Energy Technology
P.O. Box 217, 7500 AE Enschede, The Netherlands

1. Introduction

Exergy analysis and life cycle analysis have been developed separately. Exergy analysis has been described extensively in the books of Kotas [1] and Szargut [2]. Life cycle analysis (LCA) has been described by Consoli et al. [3] and Heijungs et al. [4]. The latter one gives a detailed methodology to use in a LCA. The methodology in the LCA includes the effects of all the phases of the production, use and recycling on the environment. In this paper the methodology has been performed using only one criterion which is the minimisation of the life cycle irreversibility associated with the delivery of domestic hot water. The complete LCA involves other factors e.g. pollution of air and water, noise, etc., which were not considered here. The concept of cumulative exergy, introduced by Szargut, uses the method of accumulation of the exergy consumption to a defined point in the life cycle analysis (Szargut et al. [2]. The cumulative exergy consumption of a product takes into account all the exergy destruction for the manufacture of the product. However, in this method the exergy destruction associated with the disposal of the product and the influence of recycling which cause changes in the exergy destruction are not taken into account.

The ELCA method is described by Cornelissen [5]. A comparison between the LCA and the ELCA is performed for the porcelain mug and the disposable polystyrene cup. It is shown that the life cycle irreversibility, the exergy loss during the life cycle, is the parameter for the depletion of natural resources. Furthermore, the general results of the ELCA and the LCA were similar. More research has to be performed to show in which cases this takes place. It is expected that in energy intensive processes or components operating in these processes the results of the ELCA and the LCA are similar [6]. However, the ELCA is less time consuming and gives more insight in the location of the losses. On basis of this knowledge better improvements can be suggested.

Because of its widespread use the heat exchanger has been selected as an example. Bejan [7] studied extensively the optimisation of a heat exchanger, excluding exergy destruction associated with use of materials and cumulative exergy of generation of heat and power. His approach uses the concept of

A. Bejan and E. Mamut (eds.), Thermodynamic Optimization of Complex Energy Systems, 209–219.

entropy generation minimisation. An extension to his approach to include material use has been made by Aceves-Saborio et al. [8]. They took into account the cumulative exergy of the material, but did not include the irreversibility due to the pressure drops.

Tondeur and Kvaalen [9] have shown that in the case of heat exchangers or separation devices involving a given heat transfer and achieving a specified transfer duty, the total entropy produced is minimal when the local rate of entropy production is uniformly distributed along space variables and time.

In this article the optimal design of a component, a heat exchanger, has been obtained on basis of the concepts of ELCA. This means the minimisation of the life cycle irreversibility of the heat exchanger.

Optimisation of a Heat Exchanger

1.1. THE HEAT EXCHANGER

The heat exchanger analysed is a balanced counter flow heat exchanger, which is used in a district heating system to heat the domestic tap water. The inner tube carries the cold stream and the surrounding outer tube carries the hot stream. An equal mass flow of the hot and cold water has been assumed. The inner and outer tube have been constructed from copper and steel, respectively. The combined annular heat exchanger is helically wound as shown in Figure 1. The influence of winding on the pressure drop has been neglected. The thermal insulation of the heat exchanger has been assumed to be perfect.

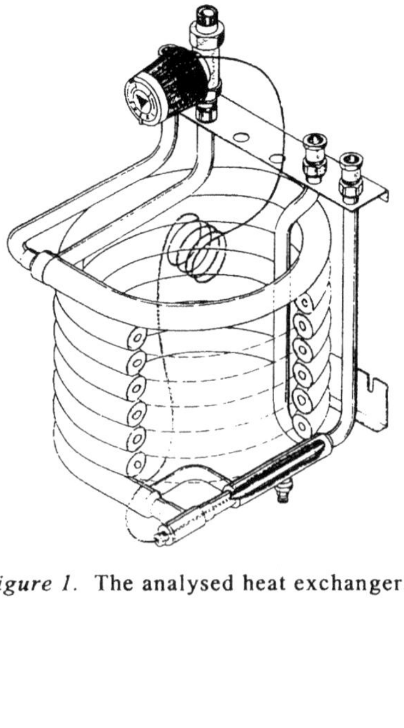

Figure 1. The analysed heat exchanger.

1.1.1. Theory

The following formula can be derived for the irreversibility in the heat exchanger due to the stream to stream heat transfer and pressure drops:

$$\dot{I} = \dot{I}^{\Delta T} + \dot{I}^{\Delta P} \tag{1}$$

with

$$\dot{I}^{\Delta T} = T_0 \left[\dot{m}\, c_p \ln \frac{T_{1,out}}{T_{1,in}} + \dot{m}\, c_p \ln \frac{T_{2,out}}{T_{2,in}} \right] \tag{1a}$$

and

$$\dot{I}^{\Delta P} = \frac{\dot{m}}{\rho}(P_{1,in} - P_{1,out}) + \frac{\dot{m}}{\rho}(P_{2,in} - P_{2,out}) \tag{1b}$$

The heat exchanger effectiveness is:

$$e = \frac{T_{1,in} - T_{1,out}}{T_{1,in} - T_{2,in}} = \frac{T_{2,in} - T_{2,out}}{T_{1,in} - T_{2,in}} \tag{2}$$

In a nearly ideal heat exchanger limit $N_{tu} \gg 1$ and therefore $1 - e \approx N_{tu}^{-1}$, where $N_{tu} = \alpha A / (\dot{m} c_p)$. Neglecting the heat resistance of the tube wall we have according to Bejan [7]

$$\dot{I}^{\Delta T} = T_0 \left[\frac{\dot{m}^2 c_p^2 \tau^2}{\alpha_1 A} + \frac{\dot{m}^2 c_p^2 \tau^2}{\alpha_2 A} \right] = T_0 \cdot \dot{m} c_p \left[\frac{\tau^2}{N_{tu,1}} + \frac{\tau^2}{N_{tu,2}} \right] \tag{3}$$

$$\text{with } \tau = \frac{|T_{2,in} - T_{1,in}|}{\sqrt{T_{1,in} T_{2,in}}}.$$

By using the force balance inside the tube(s) we can write for the pressure drops:

$$\Delta P_1 = P_{1,in} - P_{1,out} = 2 f_1(\mathrm{Re}) \cdot \rho \cdot \bar{u}_1^2 \frac{L}{D_1} \tag{4a}$$

$$\Delta P_2 = P_{2,in} - P_{2,out} = 2 f_2(\mathrm{Re}) \cdot \rho \cdot \bar{u}_2^2 \frac{L \cdot (D_1 + 2d_1 + D_2)}{(D_2^2 - (D_1 + 2d_1)^2)} \tag{4b}$$

Using $\bar{u}_1 = \dfrac{4\dot{m}}{\rho \pi D_1^2}$ and $\bar{u}_2 = \dfrac{4\dot{m}}{\rho \pi \left[D_2^2 - (D_1 + 2d_1)^2 \right]}$ and substituting (4a) and (4b) in (1b) yields

$$\dot{I}^{\Delta P} = \frac{32}{\pi^2}\left[f_1(\text{Re})\frac{\dot{m}^3}{\rho^2}\frac{L}{D_1^5} + f_2(\text{Re})\frac{\dot{m}^3}{\rho^2}\frac{L\cdot(D_1+2d_1+D_2)}{\left[D_2^2-(D_1+2d_1)^2\right]^3}\right] \tag{5}$$

or in the rewritten form

$$\dot{I}^{\Delta P} = 2\pi\mu\left[f_1(\text{Re})\cdot \text{Re}\,\bar{u}_1^2 L + f_2(\text{Re})\cdot \text{Re}\,\bar{u}_2^2 \frac{L\cdot(D_1+2d_1+D_2)}{D_{h,2}}\right]$$

with $D_{h,2} = D_2 - D_1 - 2d_1$

In the turbulent flow region in tubes and annular spaces with a limited temperature difference of 5 K for liquids between the bulk fluid and pipe surface temperature we have according to Chapman [10]

$$Nu = \frac{\alpha D_h}{\lambda} = 0.023\,\text{Re}^{0.8}\,\text{Pr}^n = 0.023\cdot\left(\frac{\rho\bar{u}D_h}{\mu}\right)^{0.8}\cdot\left(\frac{c_p\mu}{\lambda}\right)^n \tag{6}$$

with n= 0.3 or 0.4 for cooling or heating, respectively. Experimental data of Kays and London [11] gives a similar relation.

The friction factor in tubes according to the friction law of Blasius is given in Rogers and Mayhew [12] as

$$f(\text{Re}) = \frac{0.0791}{\text{Re}^{0.25}} = 0.0791\cdot\left(\frac{\mu}{\rho\bar{u}D_h}\right)^{0.25} \tag{7}$$

Substituting (6) and (7) in (5) and using $A \approx \pi\cdot L\cdot D_1$ we obtain

$$\dot{I}^{\Delta T} = T_0\frac{14.3\,\dot{m}^{1.2}}{\pi^{0.2}}\left[\frac{c_p^{1.6}\tau^2\,\mu^{0.4}}{\lambda^{0.6}}\frac{D_1{}^{0.8}}{L} + \frac{c_p^{1.7}\tau^2\,\mu^{0.5}}{\lambda^{0.7}}\frac{(D_2+D_1+2d_1)^{0.8}\,D_{h,2}}{L\cdot D_1}\right] \tag{8a}$$

with $D_{h,2} = D_2 - D_1 - 2d_1$ and

$$\dot{I}^{\Delta P} = \frac{1.79}{\pi^{1.75}}\frac{\dot{m}^{2.75}\,\mu^{0.25}}{\rho^2}\left[\frac{L}{D_1^{4.75}} + \frac{L\cdot(D_1+2d_1+D_2)^{1.25}}{\left[D_2^2-(D_1+2d_1)^2\right]^3}\right] \tag{8b}$$

Cumulative losses due to heat and power generation. Irreversibilities in heat exchangers are associated with external exergy destruction, for example, to overcome the frictional losses in the heat exchanger a pressure difference is needed. The generation of heat and power is associated with exergy destruction, which has to be taken into account in the optimisation, according to the LCA. We assume the following situation, based on the district heating system in Enschede, The Netherlands. In this system a cogeneration heat and power plant, a steam and gas turbine plant (STAG), is used, which has an exergetic efficiency of 50%, when there is no useful heat production. The exergetic efficiency of the extraction type of plant stays nearly constant when a part of the steam is used for the district heating system. Irreversibilities associated with the building of the combined heat and power plant and the transport of the fuel, natural gas, are neglected.

The exergetic efficiency of the heat transport to the houses is estimated to be 0.5 for a widespread distribution net[1] (wide net) and 0.75 for a very dense distribution net[1] (dense net). The exergy destruction in the wide net is greater because of more power needed to overcome the frictional pressure drops and more heat transfer to the environment. A great part of the exergy destruction in the heat transport takes place because of the temperature difference in the heat exchanger between the main transport tube and the local distribution net. It has been assumed that the exergy destruction associated with the heat transport is independent of the district heating water temperature. The exergetic efficiency of the pumps is assumed to be 0.7.

Hence the exergetic cost for the analysed heat exchanger of pressure rise, k_P, and heat, k_T, can be calculated to be 2.85 and 4 for the wide net and 2.85 and 2.66 for the dense net, respectively. The exergetic cost of a product is the amount of exergy which is needed for the production of one exergy unit of the product.

For the exergy destruction associated with the operation of the heat exchangers we have

$$\dot{I}_{oper} = k_T \dot{I}^{\Delta T} + k_P \dot{I}^{\Delta P} \tag{9}$$

Irreversibilities associated with the use of the material. The life cycle flow diagram of the heat exchanger is displayed in Figure 2. The exergy destruction in each process is shown in Table 1.

TABLE 1. Exergy destruction associated with the production of material and manufacturing of tubes.

Process	I_S (MJ/kg)	I_{Cu} (MJ/kg)
Primary process	10.5	60
Secondary process	4.4	20
Manufacturing process	5.7	15

The data listed in Table 1 is obtained from Cornelissen [5].

[1] This is a hypothetical situation. In reality the exergy losses are higher because the peak in heat demand is supplied by auxiliary heating boilers using natural gas. Their exergetic efficiency is very poor.

The heat exchanger is located in an insulated box made of polyurethane foam (PUR) with a thickness of 0.10 m. The cumulative energy consumption for the production of PUR with a density of 30 kg/m^3 is 98 MJ/kg according to Kindler et al. [13]. The exergy content is estimated to be 27 MJ/kg on basis of the lower heating value according to Kindler et al. [13]. So the cumulative exergy destruction is 71 MJ/kg (C_{PUR}). The box containing the heat exchanger has two outer sides of 0.70 meter. The height of the box is calculated from the length of the heat exchanger. The heat exchanger has been helical wound in three coaxial tubular layers. The mean diameter of the tube windings is 0.45 meter. No recycling of the PUR has been assumed.

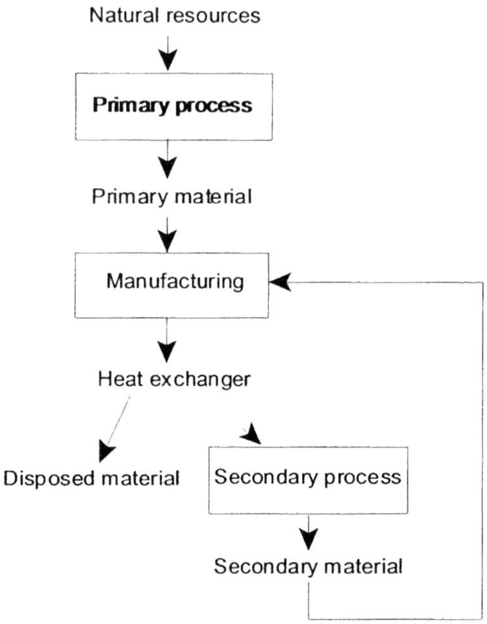

Figure 2. Life cycle of the heat exchanger.

The exergy destruction associated with the manufacture of the heat exchangers is due to the production of copper tube and steel tube, welding and the production of the PUR-foam box.

$$\dot{I}_{man} = \frac{M_{Cu}C_{Cu} + M_{Fe}\,C_S + LC_{w,S} + M_{PUR}C_{PUR}}{t} =$$

$$\frac{\pi L}{t}\left[(D_1 + 0.5d_1)d_1\rho_{Cu}(C_{man,Cu} + x_{Cu}C_{\sec,Cu} + (1 - x_{Cu})C_{pri,Cu})\right]$$

$$+\frac{\pi L}{t}\left[(D_2+0.5d_2)d_2\,\rho_S(C_{man,S}+x_S C_{\sec,S}+(1-x_S)C_{pri,S})+C_{w,S}\right]$$
$$+\frac{1}{t}\left[\left(4\cdot\frac{L\cdot(0.7-0.1)}{3\cdot\pi\cdot 0.45}\cdot(D_2+2\cdot d2)+2\cdot 0.49\right)\cdot 0.10\,\rho_{PUR}C_{PUR}\right] \qquad (10)$$

in which t is the operating time of the heat exchanger during its life cycle and x is the recycling ratio which is the proportion of secondary material, i.e. material which is recycled.

2. Results

From the above considerations we obtain an expression for the total life cycle irreversibility, which has to be minimised.

$$\dot{I}_{LC}=\dot{I}_{oper}+\dot{I}_{man} \qquad (11)$$

Where $\dot{I}_{oper}$ and $\dot{I}_{man}$ are stated in (9) and (10), respectively.

The following operating parameters have been assumed for the heat exchanger. The incoming temperature of the cold domestic tap water is 15°C. The domestic tap water is heated to 65°C. The temperature of the incoming district heating water is variable. The environmental temperature, T_0, is 25°C. The operating time for the heat exchanger is 30 minutes a day on full load for 10 years. The mass flow of the district heating water as the domestic water is 0.1 kg/s. The mean temperature of the inlet and outlet streams is used for the heat capacity, viscosity[2] and thermal conductivity of water. The wall thickness of the inner and outer tube are 0.8 and 2 mm, respectively. The recycling ratio is set to be 0.9 for the copper and steel parts of the tube.

2.1. REFERENCE CONFIGURATION

As a reference situation a domestic water heat exchanger is taken with the hypothetical fixed length of 30 meter. The optimised inner and outer tube diameter are $8.56\cdot 10^{-3}$ m and $1.29\cdot 10^{-2}$ m for the wide net and $9.04\cdot 10^{-3}$ m and $1.36\cdot 10^{-2}$ m for the dense net. The life cycle irreversibility of the heat exchanger in full operation is $1.82\cdot 10^{3}$ W for the wide net and $1.34\cdot 10^{3}$ W for the dense net. Resulting from a ΔT, ΔP_1 and ΔP_2 of 5.24 K, 1.34 bar and 7.57 bar for the wide net and 5.57 K, 1.03 and 5.48 bar for the dense net, respectively. Where ΔT is the temperature difference between the hot and cold stream, which is constant, because we have a balanced counter flow heat

[2] The value of the viscosity is strongly temperature dependent. The viscosity at 15°C, 70 °C and 80°C is $1.14\cdot 10^{-3}$ Pa·s, $0.406\cdot 10^{-3}$ Pa·s and $0.357\cdot 10^{-3}$ Pa·s, respectively. So the assumption of the mean temperature will cause a deviation from the real situation.

exchanger. $T_{1,out}$, which can be calculated by $T_{2,in}$ - ΔT, is set to 65°C by use of iterative calculations.

2.2. RESULTS OF OPTIMISATION

The minimisation of $\dot{I}_{LC}$ for the 3 variables, D_1, D_2 and L gives the minimum value of life cycle irreversibility per heat exchanger in the wide net of the district heating system of $1.06 \cdot 10^3$ W. The optimum geometrical parameters were found to be $D_1 = 1.07 \cdot 10^{-2}$ m, $D_2 = 1.69 \cdot 10^{-2}$ m and L = 105.8 m. The ΔT, ΔP_1 and ΔP_2 of this optimised heat exchanger are 1.95 K, 1.62 bar and 3.65 bar, respectively.

The dense net life cycle irreversibility is $8.76 \cdot 10^2$ W for $D_1 = 1.07 \cdot 10^{-2}$ m, $D_2 = 1.69 \cdot 10^{-2}$ m and L = 87.0 m. The ΔT, ΔP_1 and ΔP_2 of the optimised heat exchanger are 2.93 K, 1.34 bar and 3.00 bar, respectively.

We see that for optimal geometrical parameters the tube diameters are independent of the efficiency of the distribution net, the dense or wide net. However, the length of the tube is strongly dependent on the type of distribution net. The components of the life cycle irreversibility in the heat exchanger are displayed in Table 2.

TABLE 2. Components of life cycle irreversibility in Watts.

Component	Wide net (L=30m)	dense net (L=30m)	Wide net	dense net
Thermal	1451	1033	514	422
Mechanical	255	186	151	124
Manufacture	114	118	395	330
Total	1820	1337	1060	876

The contribution of the use of copper, steel and PUR-foam to the irreversibility associated with the manufacture is 39%, 42% and 19% for the wide and dense net optimisation and 32%, 33% and 35% for the both heat exchangers with a length of 30 meter, respectively.

2.3. DISCUSSION

The effect of the tube diameters and the length on life cycle irreversibility for the optimal geometrical parameters is shown in Figure 3. The cross sections of flow areas of the inner and outer passages are set fixed in the ratio of 1: 0.863 to get a 3 dimensional figure. This ratio has been obtained for the optimisation of the wide net. In Figure 3 can be seen that the life cycle irreversibility rate rises for smaller tube length and inner tube diameter. At zero length or at zero tube diameter the life cycle irreversibility rate becomes of course infinite. The effects of the diameters and the length on $\dot{I}_{LC}$ is greatest at their smallest values. The optimal inner tube diameter varies from around $0.8 \cdot 10^{-2}$ m for smaller lengths to $1.1 \cdot 10^{-2}$ for longer lengths of the tube.

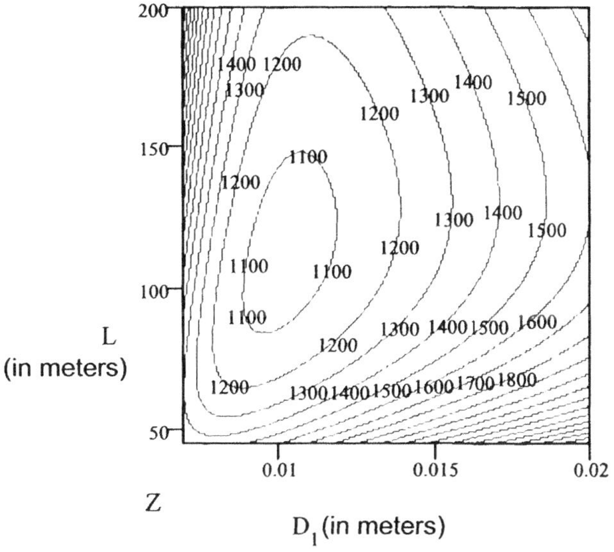

Figure 3. Life cycle irreversibility (in Watts) against length and diameter of the heat exchanger for the wide net

The analysis carried out in this work is only applicable to the situation when both the heat and the power are supplied from a cogeneration plant. If the heat is provided by a source which produces only heat from combustion of fossil fuel the exergy saved in the heat exchanger will be lost in the heater. There is no similar trade-off between exergy saving during operation and exergy use during construction of the heat exchanger as in the case when heat and power are generated separately. The saving in exergy resulting from the use of a longer heat exchanger requiring lower heating water temperature can not in this case be utilised elsewhere e.g. for generating more electricity.

It is possible that a heat exchanger with more than one heated water tube, arranged in parallel inside the larger outer tube carrying the heating, may lead to a more efficient heat exchanger process. However, this matter must be left for a further investigation.

The Reynolds numbers of the inner and outer flow are $19.8 \cdot 10^3$ and $7.15 \cdot 10^3$ for the optimal configuration of the heat exchanger, respectively, when the mean temperature of the inlet and outlet streams is taken for the viscosity of the water. For the fixed length optimisation the Reynolds numbers are between $7.72 \cdot 10^3$ and $24.4 \cdot 10^3$ for the inner and outer flow, so the condition for the turbulent flow region is fulfilled.

Heat transfer inside the heat exchanger box, which leads to exergy destruction, is neglected. If this is taken into account an insulation directly wound around the tubes would probably be more effective than the assumed box. However, if this is the case the analysis shown would not changed drastically, because the exergy use during construction would still increase when the tubes become longer.

3. Conclusions

With the Exergetic Life Cycle Analysis, the combination of exergy analysis and life cycle analysis, the optimal design of a heat exchanger can be obtained. For all energy systems where there is a trade-off between exergy saving during operation and exergy use during construction of the energy system this method should be adopted to get the true optimum from the point of view of conservation of exergy reservoirs of natural resources.

In the case under study the optimal design parameters of the heat exchangers are obtained under the specified conditions. The dense net, which is a more energy efficient heat supply system than the wide net, has the same inner tubes diameters as the wide net whilst the length of the heat exchanger is smaller for the former than for the latter. The dense net has lower life cycle irreversibility due to the manufacture of the heat exchanger compared to the wide net, because less exergy is saved by the same increase of exergy use due to the manufacture. In the optimised situation the life cycle irreversibility is more uniformly distributed between the component irreversibilities than in the fixed length of 30 meters situation.

In general we can conclude that the thermodynamic optimisation of the design parameters of a subsystem is dependent of the thermodynamic efficiency of the whole system and that the different components of the life cycle irreversibility of heat exchangers are more uniformly distributed when there are less restrictions on the design parameters for the optimisation.

4. Nomenclature

A = heat transfer area
c_p = heat capacity
C = cumulative exergy destruction
D = inner diameter of the tube
d = thickness of the tube wall
I = irreversibility or exergy destruction
L = length of the tubes
M = mass
$\dot{m}$ = mass flow
Nu = Nusselt number
N_{tu} = number of heat transfer units
P = pressure
Pr = Prandtl number
Re = Reynolds' number
T = temperature
T_1 = inner tube temperature
T_2 = outer tube temperature
$\overline{u}$ = mean velocity of the fluid in the tube
x = recycling ratio of the material

Greek letters

α = heat transfer coefficient
e = effectiveness
ρ = density
λ = coefficient of thermal conduction
μ = dynamic viscosity

Subscripts

0 = environmental
1 = inner tube of the heat exchanger

2 = outer tube of the heat exchanger
Cu = copper
h = hydraulic
S = steel
in = inlet
LC = life cycle
man = manufacturing
mat = material
oper = operating
out = outlet
pri = primary
PUR = polyurethane foam
sec = secondary
tot = total
w = welding

Superscripts

ΔP = mechanical component
ΔT = thermal component

5. References

1. Kotas, T.J., (1995) *The Exergy Method of Thermal Plant Analysis*, Krieger Publishing Company, Melbourne, Florida, USA .
2. Szargut, J., Morris, D.R. and Steward, F.R., (1988) *Exergy Analysis of Thermal, Chemical and Metallurgical Processes*, Hemispere Publishing Corporation.
3. Consoli, F., et al., (1993), Guidelines for Life Cycle Assessment of Products: A 'Code of Practice', SETAC, Brussels, Belgium.
4. Heijungs, R. et al., (1992) *Environmental Life cycle Analysis of Products*, Center for environmental studies, University of Leiden, The Netherlands.
5. Cornelissen, R.L., (1997) *Thermodynamics and Sustainable development, the Use of Exergy Analysis and Reduction of Irreversibility*, Ph.D. thesis, University of Twente.
6. Cornelissen, R.L., Bourdri, J.C. and Kalf, M.A., (1998) The value of the ELCA above the LCA, KEMA/University of Twente/CCS, in Dutch, in press.
7. Bejan, A., (1982) *Entropy Generation through Heat and Fluid Flow*, John Wiley & Sons.
8. Aceves-Saborio, S., Ranasinghe, J. and Reistad, G.M., (1989) *ASME Journal of Heat Transfer* **111**, 29 - 36.
9. Tondeur, D. and Kvaalen, E., (1987) *Industrial & Engineering Chemistry Research* **26**, 50– 56.
10. Chapman, A.J., (1974) *Heat Transfer*, Macmillan, New York .
11. Kays, W.M. and London, A.L., *Compact Heat Exchangers*, third edition, McGraw-Hill Book Company .
12. Rogers, G.F.C. and Mayhew, Y.R., (1967) *Engineering Thermodynamics, Work and Heat Transfer*, Longmans, London.
13. Kindler H. and Nikles A., (1980), *Energy for thePproduction of Materials - Method of Calculation and Energy use of Plastics (in German)*, Kunststoffe **70**, 802-807.

THE INTIMATE CONNECTION BETWEEN EXERGY AND THE ENVIRONMENT

I. DINCER[1] and M.A. ROSEN[2]
[1] *Department of Mechanical Engineering, KFUPM, Dhahran 31261, Saudi Arabia.*
[2] *Department of Mechanical Engineering, Ryerson Polytechnic University, Toronto, Ontario M5B 2K3, Canada.*

1. Introduction

It is well known that there is always an environmental cost associated with the thermal, chemical, and/or nuclear emissions which are a necessary consequence of carrying out the processes that give benefits to mankind. The environmental impact of emissions is reduced by increasing the efficiency of resource utilization. Sometimes, in practice, this is referred to as "*energy conservation.*" However, increasing efficiency generally entails greater use of materials, labor, and more complex devices. The additional cost may be justified by the added security associated with a decreased dependence on energy resources and by the social peace obtained through increased productive employment.

In 1970s, there was a primary focus on the relationship between energy and economics. At that time, the linkage between energy and the environment did not receive as much attention. As environmental concerns, such as acid rain, ozone depletion and global climate change, became major issues in the 1980s, interest in the link between energy utilization and the environment became more pronounced (especially in the late 1980s and early 1990s). Since then, there has been increasing attention on this connection. Despite many studies concerning the close relationship between energy and the environment, there have been limited works on the link between the exergy and environment concepts [e.g., 1-5].

Many suggest that the impact of energy resource utilization on the environment is best addressed by considering exergy. The exergy of a quantity of energy or a substance is a measure of its usefulness, quality or potential to cause change. Exergy appears to be an effective measure of the potential of a substance to impact the environment. In practice, the authors feel that a thorough understanding of what exergy is, and how it provides insights into the efficiency and performance of energy systems, are required for the engineer or scientist working in the area of energy systems and the environment. Based on Bejan [6] approach combining thermodynamics, heat transfer and fluid mechanics as a distinct discipline, *entropy generation*, here, we make a similar approach for exergy, because of its interdisciplinary character as the confluence of energy, environment and sustainable development (Fig. 1).

A. Bejan and E. Mamut (eds.), Thermodynamic Optimization of Complex Energy Systems, 221–230.

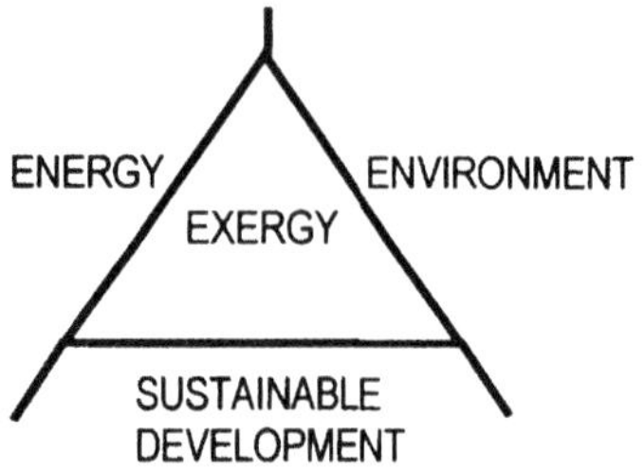

Figure 1. The interdisciplinary triangle covered by the field of exergy analysis.

Energy production, transformation, transport and use have important impacts on the earth's environment. Energy policies increasingly play an important central role in relating to a broad range of local, regional and global environmental concerns. Given the complexity of these problems, there has been a growing need to understand the link between exergy utilization and environmental impacts. The main objective of this paper is to present such a study, and to provide useful insights and direction to those involved in energy and environment studies for analyzing and solving environmental problems using the exergy concept.

2. Environmental Problems

Environmental problems span a continuously growing range of pollutants, hazards and eco-system degradation factors that affect areas ranging from local through regional to global. Some of these problems may arise from observable, chronic effects on, for instance, human health, while others may stem from the perceived risk of a possible accidental release of hazardous materials. A significant number of these environmental issues are caused by or relate to the production, transformation and end-use of energy. For example, eleven major areas of environmental concern in which energy plays a significant role have been identified [7, 8] and include major environmental accidents, water pollution, maritime pollution, land use and siting impact, radiation and radioactivity, solid waste disposal, hazardous air pollutants, ambient air quality, acid deposition, stratospheric ozone depletion, and global climate change.

The interface between energy and the environment is complex and constantly evolving. Generally, the ability of science to identify and quantify the production and effects of potentially harmful substances has greatly advanced. Throughout the 1970s most environmental analyses and legal instruments of control focused on conventional pollutants (i.e., SO_x, NO_x, CO, and particulates). Recently, concern has been extended to the control of (i) hazardous air pollutants, which are usually toxic chemical substances that are harmful in small doses and (ii) globally significant pollutants such as CO_2. Developments in industrial processes and structures have led to new environmental problems. For instance, in the energy sector, major increases in the transport of industrial goods and people by car has led to an increase in road traffic which has led to an increase in the attention paid to the effects and sources of NO_x and volatile organic compound (VOC) emissions. Consequently, while concern was given to energy policy and economic considerations in the 1970s and 1980s, environmental control and energy efficiency have received increasing attention over the last decade.

An important aspect of environmental impact is that the aesthetics of man and his machines on the planet. The low cost of power from fossil fuels has made man increasingly dependent on them, allowed significant pollution increases, and thus engendered a reduction on the planet's ecological diversity. The topic of planet's evolution is a subject for researchers in this field, and they can play a vital role in guiding industrial society toward the use of exergy as an efficient tool to reduce the energy consumption.

3. Energy Use and Environmental Impact

Energy resources are required to supply the basic human needs of food, water and shelter, and for improving the quality of life. It was indicated by the United Nations [9] that it is crucial to consider the energy sector in any broad atmosphere strategy. The need was identified for two major areas of activity: programs to increase energy efficiency and programs to encourage the transition to environmentally sound energy systems. The major program areas accepted for consideration promote: (i) the energy transition; (ii) increased energy efficiency and consequently, increased exergy efficiency; (iii) renewable energy sources; and (iv) sustainable transportation systems. Additionally, it was reported that (i) a major energy efficiency program would provide an important means of reducing CO_2 emissions, and (ii) these activities should be accompanied by strong new measures to reduce the fossil fuel components of the energy mix and develop alternative energy sources. These ideas have been reflected in many recent energy-related studies, which have concentrated on the provision of the energy service needed for an activity, with the lowest environmental impact and cost and with the maximum of energy security possible.

4. Thermodynamics and the Environment

As indicated by several researchers [4, 10, 11], the second law of thermodynamics is instrumental in providing insights into environmental impact. It is well known that heat transfer from industrial devices to the environment is of major concern to engineers and scientists working on energy and environment issues. Irresponsible management of waste energy can significantly increase the temperature of portions of the environment, resulting in thermal pollution. If not carefully controlled, thermal pollution can seriously disrupt marine life in lakes and rivers. However, by careful design and management, the waste energy dumped into large bodies of water can sometimes be used to improve significantly the quality of marine life by keeping the local temperature increases within safe and desirable levels.

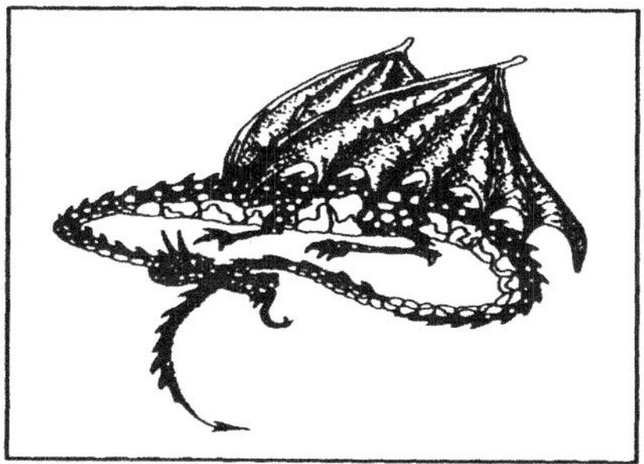

Figure 2. The mythological creature Ouroboros, which survived and regenerated itself by eating only its own tail [20].

Mankind has long been intrigued by the implications of the laws of thermodynamics on the environment. One myth speaks of Ouroboros, a serpent-like creature which survived and regenerated itself by eating only its own tail (Fig. 2). By neither taking from nor adding to its environment, this creature was said to be completely environmentally benign and self-sufficient.

It is useful to examine this creature in light of the thermodynamic principles recognized today. Ouroboros' existence would not have violated the conservation law for mass. However, it would have violated the first law of thermodynamics (which states energy is conserved), since Ouroboros would have had to expend energy in eating, and therefore would have required an external energy input to compensate for this expenditure. This is true, even if it is assumed that Ouroboros was an isolated system (i.e., it received no energy from the sun, and emitted no energy during any process). More importantly, Ouroboros' existence would have violated the second law (which states that exergy is reduced for all real processes), since Ouroboros would have had to obtain exergy to regenerate the tail it ate into an equally ordered part of its body (or it would ultimately have dissipated itself to an unordered lump of mass), and therefore would have required an external exergy input. Thus, Ouroboros would have to have had an impact on its environment. Besides showing that within the limits imposed by the laws of thermodynamics, all real processes must have some impact on the environment, this example is intended to illustrate the following key point: the second law is instrumental in providing insights into environmental impact. It is of course true that the principles, demonstrated here through a historical example, are relevant today. For example, technologies are still striven for today that have the characteristics of Ouroboros (i.e., that are environmentally benign and self-sufficient). In fact, researchers at the University of Minnesota have built an "energy-conserving" house, called Ouroboros [12].

5. Exergy and the Environment

Energy and environment studies which lead to increased energy efficiency can reduce environmental impact by reducing energy losses. Within the scope of exergy methods, such activities lead to increased exergy efficiency. In practice, the efficiency improvements can be identified by means of modeling and computer simulation studies. Increased efficiency can often contribute in a major way to achieving energy security in an environmentally acceptable way by the direct reduction of emissions that might otherwise have occurred. Increased efficiency also reduces the requirement for new facilities for the production, transportation, transformation and distribution of the various energy forms; these additional facilities all carry some environmental impacts. To control environmental pollution, efficiency improvement actions often need to be supported by pollution control technologies or fuel substitution. It is through regional or national actions, rather than through individual projects, that improved exergy efficiency can have a major impact on environmental protection.

Exergy is defined as the maximum amount of work which can be produced by a stream of matter, heat or work as it comes to equilibrium with a reference environment. Exergy is a measure of the potential of a stream to cause change, as a consequence of not being completely stable relative to the reference environment. For exergy analysis, the state of the reference environment, or the reference state, must be specified completely. This is commonly done by specifying the temperature, pressure and chemical composition of the reference environment. Exergy is not subject to a conservation law. Rather exergy is consumed or destroyed, due to irreversibilities in any process.

Exergy analysis is a method that uses the conservation of mass and conservation of energy principles together with the second law of thermodynamics for the design and analysis of energy systems. The exergy method can be suitable for furthering the goal of more efficient energy-resource use, for it enables the locations, types, and true magnitudes of wastes and losses to be determined. Therefore, exergy analysis can reveal whether or not and by how much it is possible to design more efficient energy systems by reducing the inefficiencies in existing systems.

Most recently, Cornelissen [13] has conducted an excellent study on the thermodynamics and sustainable development and pointed out that one of the keystones for obtaining sustainable development is the use of exergy analysis. Energy can never be lost as stated in the first law of thermodynamics. Exergy can be lost and this loss, called irreversibility, created during the use of non-renewables has to be minimized to obtain sustainable development. He also showed that all environmental effects associated with emissions and the environmental effect of depletion can be expressed in terms of one indicator, which is based on physical principles.

Several researchers have suggested that the best way to link the second law and environmental impact is through exergy because it is a measure of the departure of the state of a system from that of the environment [e.g., 2, 11]. The magnitude of the exergy of a system depends on the states of both the system and the environment. This departure is zero only when the system is in equilibrium with its environment. The concept of the environment as it applies to exergy analysis is discussed in detail elsewhere [1, 3, 14]. Detailed discussions of exergy analysis for many processes and systems are also given elsewhere [14-17]. Tribus and McIrvine [18] suggest that performing exergy analyses of the natural processes occurring on the earth could form a foundation for ecologically sound planning because it would indicate the disturbance caused by large-scale changes.

An understanding of the relations between exergy and the environment may reveal the underlying fundamental patterns and forces affecting changes in the environment, and help researchers to deal better with environmental damage. Three relationships between exergy and environmental impact, which have been introduced previously [4, 5], are discussed in the following subsections.

5.1. ORDER DESTRUCTION-CHAOS CREATION

The destruction of order, or the creation of chaos, is a form of environmental damage. Fundamentally, entropy is a measure of chaos. A system of high entropy is more chaotic or disordered than one of low entropy. For example, a field with papers scattered about has higher entropy than the field with the papers neatly piled. Conversely, negentropy and exergy are measures of order. Relative to the same environment, the exergy of an ordered system is greater than that of a chaotic one. The difference between the exergy values of the two systems containing papers described above is a measure of the minimum work required to convert the chaotic system to the ordered one (i.e., in collecting the scattered papers). In reality, more than this minimum work, which only applies if a reversible clean-up process is employed, is required. The exergy destroyed when the wind scatters a stack of papers is a measure of the order destroyed during the process. The ideas relating exergy and order in the environment become more abstract if human values are considered [10]. People are distressed when they see a landscape polluted with papers chaotically scattered about, while they value the order of a clean field with the papers neatly piled at the side. Perhaps human values are related to exergy and order.

5.2. RESOURCE DEGRADATION

The degradation of resources found in nature is a form of environmental damage. Kestin [19] defines a resource as a material, found in nature or created artificially, which is in a state of disequilibrium with the environment. Resources have exergy as a consequence of this disequilibrium. For some resources (e.g., metal ores), it is their composition that is valued. Many processes exist to increase the value of such resources by purifying them (i.e., by increasing their exergy). This is done at the expense of consuming at least an equivalent amount of exergy elsewhere (e.g., burning coal to produce process heat for metal ore refining). For other resources (e.g., fuels), it is normally their reactivity that is valued (i.e., their potential to cause change, or "drive" a task or process). By preserving exergy through increased efficiency (i.e., using as little exergy as necessary for a process), environmental damage is reduced. Increased efficiency also has the effect of reducing exergy emissions which, as discussed in the next section, also play a role in environmental damage. The earth is an open system subject to a net influx of exergy from the sun. It is the exergy (or order states) delivered with solar radiation that is valued; all the energy received from the sun is ultimately radiated out to the universe. Environmental damage can be reduced by taking advantage of the openness of the earth and utilizing solar radiation (instead of degrading resources found in nature to supply exergy demands). This would not be possible if the earth was a closed system, for it would eventually become more and more degraded, or "entropic."

5.3. WASTE EXERGY EMISSIONS

The exergy associated with process wastes emitted to the environment can be viewed as a potential for environmental damage. Typical process wastes have exergy, a potential to cause change, as a consequence of not being in stable equilibrium with the environment. When emitted to the environment, this exergy represents a potential to change the environment. In some cases, this exergy may cause a change perceived to be beneficial (e.g., the increased rate of growth of fish and plants near the cooling-water outlets from thermal power plants). More often, however, emitted exergy causes a change which is damaging to the environment (e.g., the deaths of fish and plants in some lakes due to the release of specific substances in stack gases as they react and come to equilibrium with the environment). Emissions of exergy to the environment can also interfere with the net input of exergy via solar radiation to the earth. The carbon dioxide emitted in stack gases from many processes changes the atmospheric CO_2 content, affecting the receiving and re-radiating of solar radiation by the earth. The relation between waste exergy emissions and environmental damage has been recognized by several researchers. By considering the economic value of exergy in fuels, Reistad [1] developed an air-pollution rating in which the air-pollution cost for a fuel was estimated as either the cost to remove the pollutant, or the cost to society of the pollution (i.e., the tax which should be levied if pollutants are not removed from effluent streams). Reistad claimed the rating was preferable to the mainly empirical ratings then in use.

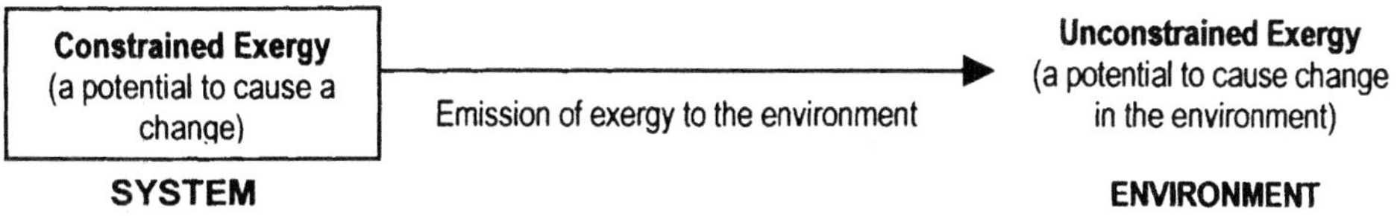

Figure 3. The comparison of constrained and unconstrained exergy. Exergy constrained in a system represents a resource. When emitted to the environment, exergy becomes unconstrained and represents a driving potential for environmental damage.

The reader may at this point detect a paradox. It was pointed out in the second last section that exergy in the environment, in the form of resources, is of value; while exergy in the environment, in the form of emissions, is stated in the last section to have a "negative" value, in that it has a potential to cause environmental damage. This apparent paradox can be resolved if the word constrained is included in the discussions (see Fig. 3). Constrained sources of exergy in the environment are of value. Most resources found in nature are constrained. Unconstrained emissions of exergy to the environment have negative value. If emissions to the environment are constrained (e.g., separating out sulfur from stack gases), then there are two potential benefits: (i) the potential for environmental damage is blocked from entering the environment, and (ii) the now-constrained emission potentially becomes a valued commodity, a source of exergy.

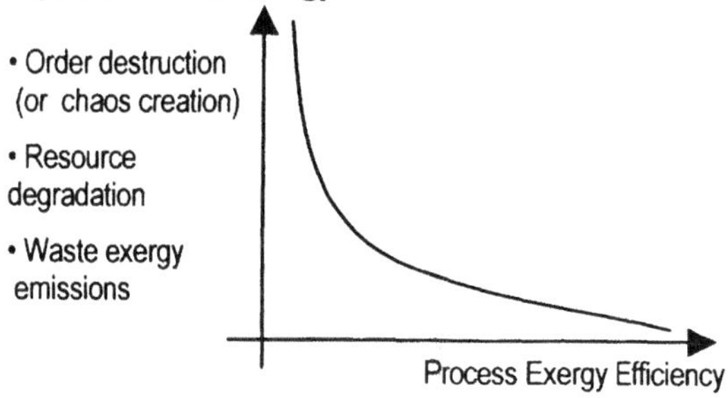

Figure 4. Qualitative illustration (in relative units) of how decreasing the exergy efficiency of a process causes an increase in the related environmental impact associated with the process, through either order destruction-chaos creation, or resource degradation, or waste exergy emissions.

5.4. QUALITATIVE ILLUSTRATION

The manner by which decreasing the exergy efficiency of a process causes the related environmental impact (regardless of which of the three measures discussed in Sections 5.1-5.3 is considered) to increase is illustrated a qualitatively and approximately in Fig. 4.

6. Illustrative Example

The ideas discussed in this paper are demonstrated by the authors for the process of electricity generation, by considering the coal-fired Nanticoke Generating Station (NGS). The individual units of NGS each have net outputs of approximately 500 MWe. A substantial base of data has been obtained for NGS, which has been operating since 1981. Overall balances of exergy, energy and mass for NGS are illustrated in Fig. 5, where the rectangle in the center of each diagram represents NGS. The overall system consists of four main sections [4]:

a) Steam Generators: Eight pulverized-coal-fired natural circulation steam generators each produce 453.6 kg/s steam at 16.89 MPa and 538°C, and 411.3 kg/s of reheat steam at 4.00 MPa and 538°C. Coal is consumed at full load at a rate of 47.9 kg/s. Air is supplied to the furnace by two 1080 kW, 600-rpm motor-driven forced draft fans. Regenerative air preheaters are used. The flue gas passes through an electrostatic precipitator rated at 99.5% collection efficiency, and exits the plant at 120°C through two multi-flued, 198 m high chimneys.

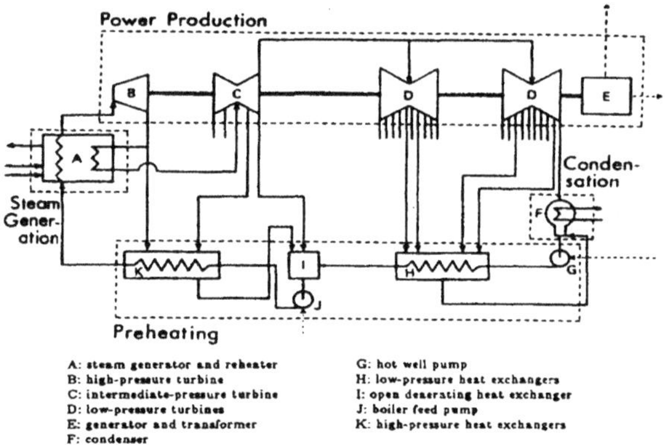

Figure 5. Breakdown of a single unit in the NGS electrical generating station into four main sections.

b) Turbine Generators and Transformers: The steam produced in the steam generators is passed through a series of turbine generators and a transformer. The net power output is 505 MW. Each unit has a 3600-rpm, tandem-compound, impulse-reaction turbine generator containing one single-flow high-pressure cylinder, one double-flow intermediate-pressure cylinder and two double-flow low-pressure cylinders. Steam exhausted from the high-pressure cylinder is reheated in the combustor. Extraction steam from several points on the turbines preheats feed water in several low- and high-pressure heat exchangers and one spray-type open deaerating heat exchanger. The low-pressure turbines exhaust to the condenser at 5 kPa.

c) Condensers: Cooling water condenses the steam exhausted from the turbines. A flow rate of cooling water of 18.9 m^3/s is used to achieve a temperature rise of 8.3°C in the cooling water across the condenser.

d) Preheating Heat Exchangers and Pumps: The temperature and pressure of the condensed steam are increased to 253°C and 16.89 MPa in a series of pumps and heat exchangers.

In this example of a conventional coal-fired electrical generating station, each of the relationships between exergy and environmental impact described in the last is demonstrated, and described below:

- Waste exergy emissions from the plant occur with (i) the gaseous wastes exiting through the stack, (ii) the solid wastes exiting the combustor, (iii) and the waste heat released both to the lake, which supplies condenser cooling water, and to the atmosphere. The exergy associated with each of these emissions represents a potential to impact on the environment. The societial consensus regarding emissions of harmful chemical constituents in stack gases and "thermal pollution" in local bodies of water indicates that the potential for impact of these emissions is already generally recognized, but not from an exergy-related perspective.

- During the electricity generation process, a finite resource, coal, is degraded as it is used to drive the process. Although a degree of degradation is necessary for any real process, increased process efficiency can reduce this degradation for the same services/products. In the extreme, if the process considered in the example were made thermodynamically ideal, the exergy efficiency would increase from 37% to 100%, and coal use as well as related emissions would decrease by over 60%. These insights are provided by the exergy, not energy, analysis results.
- The more abstract concepts of order destruction and chaos creation are also illustrated. Certainly the degradation of coal fuel to stack gases and solid wastes represents a destruction of order, especially since less exergy is present in the products of the process. The emissions of wastes to the environment increases the chaos in the world by allowing the constrained material and heat products of combustion to be released without constraints into environment.

The authors feel that this example (i) provides some practical illustrations of the more abstract concepts discussed throughout this paper, and (ii) highlights the importance of understanding and considering the relations between exergy, rather than energy, and environmental impact, when addressing environmental challenges and problems. It is clear that an enhanced understanding of the environmental problems relating to energy presents a high-priority need and urgent challenge, both to allow the problems to be addressed and to ensure the solutions are beneficial for the economy and the energy and energy systems themselves.

7. Conclusions

This paper discusses the relations between environmental impact and thermodynamics, in general, and the thermodynamic property exergy, in particular. Historical and modern examples demonstrate these concepts. Three main relations between exergy and environmental impact are identified: (i) order destruction-chaos creation, (ii) resource degradation, and (iii) waste-exergy emissions. An apparent paradox is described between the usefulness of a high-exergy quantity, and the simultaneous potential for environmental disruption it possesses. It is concluded that the potential usefulness of exergy analysis in addressing and solving environmental problems is substantial, but that further work is required before this potential can be properly and fully tapped.

8. Acknowledgements

The support for this work provided by King Fahd University of Petroleum and Minerals, Ryerson Polytechnic University and the Natural Sciences and Engineering Research Council of Canada is gratefully acknowledged.

9. References

1. Reistad, G.M. (1970) *Availability: Concepts and Applications*, Ph.D. Thesis, University of Wisconsin, Madison.
2. Szargut, J. (1980) International progress in second law analysis, *Energy* 5, 709-718.

3. Wepfer, W.J. and Gaggioli, R.A. (1980) Reference datums for available energy, in *Thermodynamics: Second Law Analysis*, ed. R.A. Gaggioli, ACS Symposium Series 122, Washington, DC, 77-92.
4. Rosen, M.A. (1986) *The Development and Application of a Process Analysis Methodology and Code Based on Exergy, Cost, Energy and Mass*, Ph.D. Thesis, University of Toronto, Toronto.
5. Rosen, M.A. and Dincer, I. (1997) On Exergy and environmental impact, *Int. J. Energy Research* 21, 643-654.
6. Bejan, A. (1996) *Entropy Generation Minimization*, CRC Press, Boca Raton, FL.
7. Anon. (1989) *Energy and the Environment: Policy Overview*, OECD-International Energy Agency, Paris.
8. Dincer, I. (1998) Energy and environmental impacts: present and future perspectives, *Energy Sources* 20(4) (in press).
9. Strong, M.F. (1992) Energy, environment and development, *Energy Policy* 20, 490-494.
10. Hafele, W. (1981) *Energy in a Finite World: A Global Systems Analysis*, Ballinger, Toronto.
11. Edgerton, R.H. (1982) *Available Energy and Environmental Economics*, D.C. Heath, Toronto.
12. Markovich, S.J. (1978) Autonomous living in the ouroboros house, in Solar Energy Handbook, *Popular Science*, 46-48.
13. Cornelissen, R.L. (1997) *Thermodyanmics and Sustainable Development*, Ph.D. Thesis, University of Twente, the Netherlands.
14. Kotas, T.J. (1985) *The Exergy Method of Thermal Plant Analysis*, Butterworths, Toronto.
15. Ahrendts, J. (1980) Reference states, *Energy* **5**, 667-668.
16. Moran, M.J. (1989) *Availability Analysis: A Guide to Efficient Energy Use*, ASME, New York.
17. Moran, M.J. and Sciubba, E. (1994) Exergy Analysis: Principles and Practice, *J. Engineering for Gas Turbines and Power* 116, 285-290
18. Tribus, M. and McIrivne, E.C. (1971) Energy and information, *Sci. American* 225(3), 179-188.
19. Kestin, J.(1980) Availability: the concept and associated terminology, *Energy* 5, 679-692.
20. Warchol, K. (1983) *Computer News*, Computing Services, University of Toronto 207 (cover page).

EFFECT OF VARIATION OF ENVIRONMENTAL CONDITIONS ON EXERGY AND ON POWER CONVERSION

Y.A. GÖĞÜŞ[1] and Ö.E. ATAER[2]
[1]*Aeronautical Engineering Dept.*
Middle East Technical University,
06531 Ankara, Turkey

[2] *Mechanical Engineering Dept.*
Gazi University, 06570 Ankara, Turkey

1. Introduction

Most thermal engineering problems are solved at sufficient accuracy by second law analysis of either steady state or control mass models. However problems with transient or periodic changing conditions pose also challenges. Changes in environmental conditions may influence thermal performance of energy conversion systems positively or negatively [1-4]. Shifting greater load of system to times with favourable environmental conditions, e.g. in power plants to times with low environment temperature, can increase daily average thermal efficiency. This lecture concentrates on the effects of variation of environmental conditions and of demand, utilising recently obtained results [1, 2].

2. Momentary Magnitude of Exergy and Exergy Balance

Change of environmental conditions during the process, which is being optimised, will effect on one hand the momentary value of exergy, on the other hand the magnitude of exergy destruction. Definition of exergy concept and Gouy-Stodala theorem base upon existence of a reservoir (environment) so large that its state is not effected by the process. However the condition of the environment may change due to other reason p_o (t), T_o (t) and this necessitates a modified exergy balance. Exergy change of a system consists of 5 groups of terms; three of them are due to heat, work, matter interactions, fourth due to exergy destruction effect of entropy generation and the last one (ΔE_o) due to change of environmental conditions [3, 5]:

$$E(t_2) - E(t_1) = I_1 - (W - I_2) + (E_{f,in} - E_{f,ex}) - I_3 + \Delta E_o \tag{1}$$

$$I_1 = \int_{t_1}^{t_2} (1 - \frac{T_o(t)}{T_b(t)})\dot{q}dt \ ; \quad I_2 = \int_{t_1}^{t_2} p_o(t)\frac{dV(t)}{dt}dt \qquad (2), (3)$$

where $T_b(t)$ is boundary temperature at which heat is transferred to system,

A. Bejan and E. Mamut (eds.), Thermodynamic Optimization of Complex Energy Systems, 231–240.

$$E_f = \int_{t_1}^{t_2} \dot{m}(t)\{[h(t) - T_o(t)s(t)] - [h_o(t) - T_o(t)s_o(t)]\}dt\,; \quad I_3 = \int_{t_1}^{t_2} T_o(t)\dot{S}_{gen}\,dt \qquad (4), (5)$$

Each group of terms of Eq. (1), except the last one (ΔE_o), can be expressed in terms of their values for constant environment conditions (index r) and corrections. E.g.:

$$E(t_2) - E(t_1) = E_2 - E_1 = \Delta E = \Delta E_r - \Delta E_{or} - \Delta E_{cs} + \Delta E_{cv} \qquad (6a)$$

$$\Delta E_r = \Delta U - T_{or}\Delta S + p_{or}\Delta V\,; \qquad \Delta E_{or} = \Delta U_o - T_{or}\Delta S_o + p_{or}\Delta V_o \qquad (6b,c)$$

$$\Delta E_{cs} = (T_{o2} - T_{or})(S_2 - S_{o2}) - (T_{o1} - T_{or})(S_1 - S_{o1}) \qquad (6d)$$

$$\Delta E_{cv} = (p_{o2} - p_{or})(V_2 - V_{o2}) - (p_{o1} - p_{or})(V_1 - V_{o1}) \qquad (6e)$$

For other terms, similar operations can be made. Comparing the new form of Eq. (1) with exergy balance with reference to fixed environmental conditions:

$$\Delta E_r = I_{1r} - [W - I_{2r}] - I_{3r} - \Delta E_{f,r} \qquad (7)$$

one obtains a general expression for the correction term ΔE_o:

$$\Delta E_o \cong -\Delta E_{or} + I_{1,3or} + I_{2or} - \Delta E_{f,or} \qquad (8)$$

where the first three terms are for control volume and the last for flow substance:

$$\Delta E_{or} = [U_o(t_2) - T_{or}S_o(t_2) + p_{or}V_o(t_2)] - [U_o(t_1) - T_{or}S_o(t_1) + p_{or}V_o(t_1)] \qquad (9a)$$

$$I_{1,3or} = \int_{t_1}^{t_2} (T_o - T_{or})\frac{dS}{dt}dt\,; \quad I_{2or} = \int (p_o - P_{or})\frac{dV}{dt}dt \qquad (9b,c)$$

$$\Delta E_{f,or} = \int_{t_1}^{t_2} [\dot{m}_{ex}(t) - \dot{m}_{in}(t)][h_o(t) - T_{or}s_o(t)]dt - \int_{t_1}^{t_2} [\dot{m}_{ex}(t) - \dot{m}_{in}(t)](h_{or} - T_{or}s_{or})dt$$

(9d)

Special cases for control mass and steady state can be obtained easily from these general expressions.

3. Effect of Periodic Change of Source Temperature and General Characteristics of the Method

Environment temperature and source temperature will be assumed to change harmonically, the later with a phase difference ϕ ahead of the former;

$$T_o = T_{om} + T_{oa}\sin\tau\,; \quad T_s = T_{sm} + T_{sa}\sin(\tau + w) \qquad (10), (11)$$

where the independent variable τ is dimensionless time: $\tau = 2\pi ft = \omega t$, with t: time, ω: angular speed and f: frequency, w is phase difference between source and environment temperature. Fig. 1 shows the schema of the engine and temperature variation.

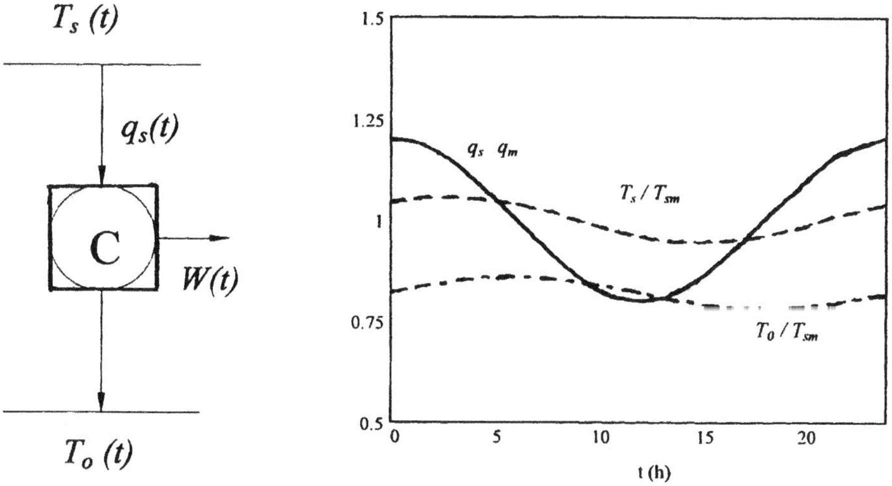

(a) (b)

Figure 1. (a) Schema of a reversible engine running between source temperature T_s and environment temperature T_o ; (b) Temperatures and heat supply q_s by time for an example case (Table 1, first line).

With reference to average source temperature, T_{sm}, these equations can be written in dimensionless form introducing $x = T_{oa}/T_{sm}$, $y = T_{sa}/T_{sm}$ and $z = T_{om}/T_{sm}$. Similarly the instantaneous heat supply, q_s, to the reversible harmonic engine can be expressed in dimensionless form utilising the average heat supply, q_{sm}:

$$T_o / T_{sm} = z + x\sin\tau \; ; \quad T_s / T_{sm} = 1 + y\sin(\tau + w) \qquad (12), (13)$$

$$q_s / q_{sm} = 1 + a\sin(\tau + \phi) \qquad (14)$$

Here $a = q_{sa}/q_{sm}$ is the relative magnitude of the amplitude of heat supply. The phase shift, ϕ , is the unknown variable which is adjusted to maximise the role played by changing source temperature. Substituting Eq.s (12) and (13) into Carnot efficiency Eq. (15) and retaining only the first two terms of the binomial expansion of $[1 + y\sin(\tau + w)]^{-1}$ and rearranging the product of multiplication by neglecting the second degree terms, the instantaneous thermal efficiency of the heat engine is then:

$$\eta_C = 1 - T_o / T_s \; ; \quad \eta_C = 1 - z + yz\sin(\tau + w) - x\sin\tau \qquad (15), (16)$$

where (1-z) can be interpreted as steady efficiency. The error of these approximations is less than 1 % for $T_{sa} < 0.3T_{sm}$. The instantaneous power of the engine is:

$$W' = [1 + a\sin(\tau + \phi)][(1 - z + yz\sin(\tau + w) - x\sin\tau] \qquad (17)$$

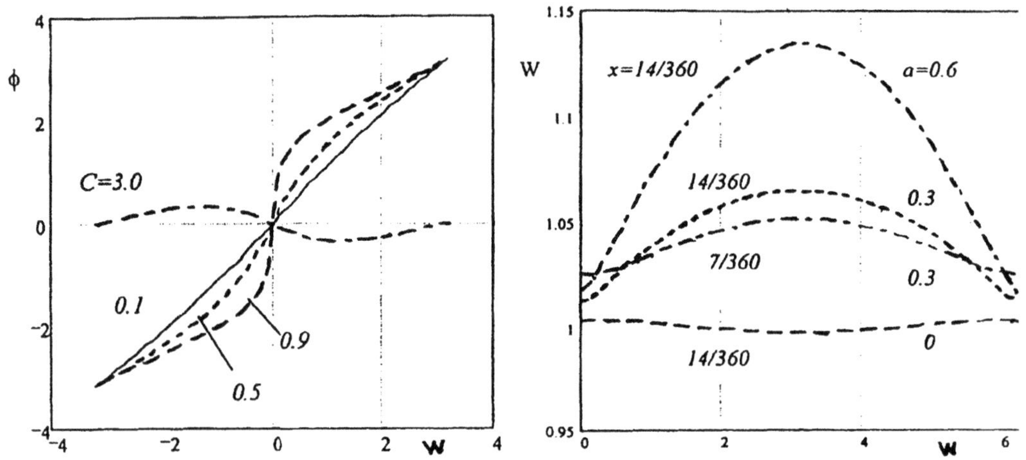

Figure 2. (a) Variation of ϕ along the phase shift of source temperature w for $c = x/(yz)$ values 3.0; 0.9; 0.5 and 0.1. (b) Variation of dimensionless work W along w for z = 295/360, y = 20/360 and different a, x values.

Integrating Eq. (17) with respect to time over one period and bringing in dimensionless form, by means of the work of reversible engine running at constant average heat supply q_{sm}, between T_{sm} and T_{om}, results:

$$W = \frac{2\pi - 2z\pi - ax\pi\cos\phi + ayz\pi\cos(\phi - w)}{2\pi(1-z)} \tag{18}$$

To find the value of ϕ which maximises this work, its derivative with respect to ϕ will be taken and solved for ϕ:

$$\phi = \arctan\left[\frac{yz\sin w}{-x + yz\cos w}\right] + n\pi \tag{19}$$

This is the optimum value of ϕ that maximises the work. The value of n is determined by the second derivative of Eq. 18 with respect to ϕ. For yz.cos(w)- x = 0 and for $\phi = \pm\pi/2$, physically meaningful solution should be chosen. As an example, the data of the first line of Table 1 have been used and variations on Fig. 1b are obtained. Fig. 2a shows the variation of ϕ along w in the range -π to π for different values of x/(yz) = c: 3.0; 0.9; 0,5; 0.1.

Figure 2b shows how the dimensionless work of one period is affected by the amplitudes x and a. As expected, increase of ‘a’ increases the importance of choosing proper phase shift ϕ for operation. Fig. 3 shows the effects of z and y on dimensionless work for two different w.

TABLE 1. Standard and auxiliary values of the parameters.

x	y	z	w	d	a	p
14/360	20/360	295/360	π/4	0	0.3	0.4
7/360	0	295/700	π	π/6	0.5	0.6

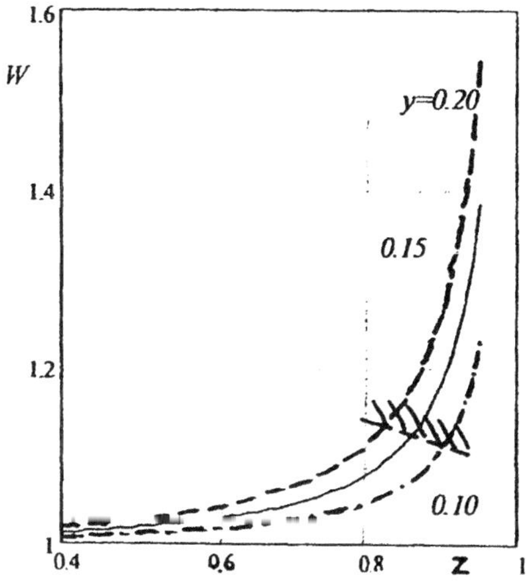

(a)

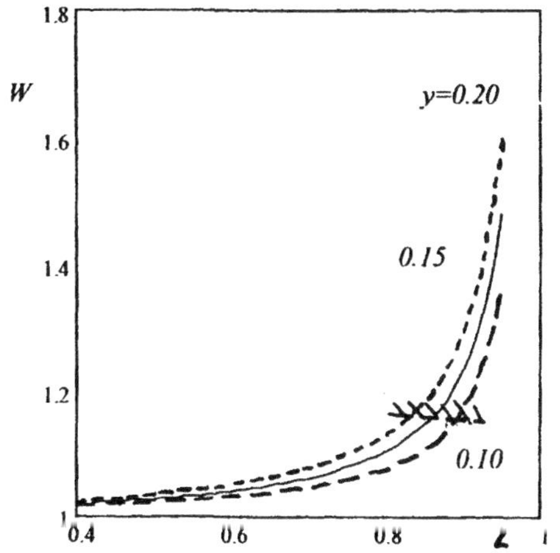

(b)

Figure 3. Effect of z and y on dimensionless work W at a = 0.3, x = 14/360 and (a) w = π/6, (b) w = 5π/6.

4. Periodic Change of Price and Its Coupling with Source Temperature

Hourly changes in power demand result that utility companies change price P of the produced power hourly. As a first approximation this change can be expressed by a harmonic function in dimensionless form:

$$P / P_m = 1 + p\sin(\tau + d) \tag{20}$$

Here P_m is the average price and p is the dimensionless amplitude of price variation. The total monetary value of the produced power over a period, V, can be obtained by multiplying the produced work, Eq. (18) with price P of Eq. (20). Utilising average power W and average price P_m over dimensionless period 2π one can express dimensionless value of the work produced over one period as:

$$\frac{V}{q_m(1-z)2\pi \cdot P_m} = \int_0^{2\pi} (1 + a\sin(\tau + \phi))(1 - z - x\sin\tau + yz\sin(\tau + w))(1 + p\sin(\tau + d))d\tau \tag{21}$$

Derivation of the result of integration with respect to ϕ and solving it gives

$$\phi = \arctan\left(\frac{zy\sin w + p(1-z)\sin d}{zy\cos w + p(1-z)\cos d - x}\right) + n\pi \tag{22}$$

This is the optimum phase shift of operation. For the simple case of constant source temperature (y = 0) the effect of price variation on optimum phase shift is similar to the effect of variation of source temperature as explained in Section 3; the only difference is that w will be replaced by d and yz will be replaced by p (1-z):

$$\phi = \arctan\left(\frac{(1-z)p\sin d}{p(1-z)\cos d - x}\right) + n\pi \tag{23}$$

- Observations made on Fig. 2 on the relation $\phi(w)$ and W(w) with given values of 'a' and y are valid for this case also, however for the latter instead of yz = 0.0455 with y = 0 and p (1-z) = 0.0455 or z = 295/360 and p = 0.25.

5. Finite Time Effect

The problem of endoreversible engine with external resistances as shown on Fig. 4a. Instantaneous Novikov-Curzon-Ahlborn relation may be applied [6]:

$$\eta = 1 - \left(T_o(t) / T_s(t) \right)^{1/2} \tag{24}$$

Substituting Eq. (10, 11) and expanding firstly the denominator in binomial series for power (-1), multiplying by nominator and expanding in binomial series for power (1/2) results in:

$$\eta = 1 - z^{1/2}\left[1 - \frac{1}{2}(y\sin(\tau + w) - \frac{y}{z}\sin\tau + \frac{xy}{z}\sin\tau\sin(\tau + w))\right] \tag{25}$$

The average of this efficiency over one period, assuming that the rate of operation is approximately constant, results:

$$\eta = 1 - z^{1/2}\left[1 + \frac{\pi}{2}\frac{xy}{z}\cos(w)\right] \tag{26}$$

In other words Novikov-Curzon-Ahlborn efficiency needs a correction factor near 1.0 in case temperatures of sink and source oscillate.

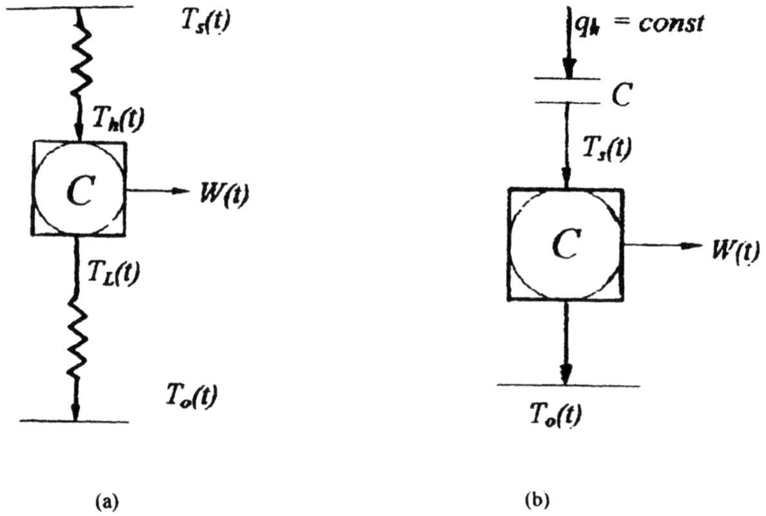

Figure 4. Special cases of reversible engine (a) with external irreversibilities, (b) with a storage on source side where heat flow is uniform.

6. Effect of Storage and Coupling with Price

Utilisation of change of environmental temperature by means of thermal storage at heat source, which has an average temperature T_{sm} , base on harmonic operation of the engine. Energy balance of thermal storage and heat supply expressions are:

$$C_s dT_s / dt = q_h(t) - q_s(t) \;; \qquad q_h = q_{hm} + q_{ha}\sin(\omega t + g) \qquad (27), (28)$$

The integral over a period gives: $q_{hm} = q_{sm}$. Solving this equation one can obtain

$$T_s = T_{sm} + \frac{q_{ha}}{C_s\omega}\sin(-\omega t - g') + \frac{q_{sa}}{C_s\omega}\sin(\omega t - \phi') \qquad (29)$$

where $g' = g + \pi/2$ and $\phi' = \phi + \pi/2$ and average temperature of the storage T_{sm} is known. Considering this problem as a special case of harmonic source temperature variation, one can obtain optimum ϕ for given g. An important special case is constant heat supply, i.e. $q_{ha} = 0$. For this case one obtains

$$T_s = T_{sm} + y\sin(\omega t + \phi'); \qquad y = \frac{q_{sa}}{C_s\omega T_{sm}} \qquad (30a,b)$$

As expected, allowed largest q_{sa} improves production, limited by T_{smax} ;

$$\frac{q_{sa}}{C_s\omega} = T_{sa} \;; \qquad T_{sa} \le T_{s\max} - T_{sm} \qquad (31a,b)$$

From the optimum value of ϕ (Eq. 19)

$$\phi = \arctan\left(\frac{\sin w}{\cos w - c}\right) + n\pi \qquad (32)$$

With $w = \phi' = \phi + \pi/2$, one obtains a value for ϕ, as solution of the equation:

$$\phi = \arctan\left(\frac{\sin(\phi + \pi/2)}{\cos(\phi + \pi/2) - c}\right) + n\pi \qquad (33)$$

The solution of this equation for optimum ϕ is

$$\phi = -\arcsin(1/c) \quad \text{and} \quad \phi = \pi + \arcsin(1/c) \qquad (34a,b)$$

For $c < 1$ there is no horizontal slope, i.e. the function is changing monotonically. Substituting $T_{sa} / T_{sm} = y = (q_{sa} / \omega C_s)(1/T_{sm})$ one obtains $c = (x/z).(\omega T_{sm} C_s / q_{sa})$.

In case of coupling thermal storage effect with the effect of change of price, optimum phase angle will be:

$$\tan\phi = \frac{\sin(\phi + \pi/2) + p'\sin d}{\cos(\phi + \pi/2) + p'\cos d - c}; \quad p' = p(1-z) \tag{35a,b}$$

However, for frequently encountered situations either $d \cong 0$ or $p = 0$. For $d = 0$

$$\phi = -\arcsin(1/(c - p')) \tag{36}$$

7. A General Thermal Module and Its Applications

Above mentioned specific results about external resistance (Novikov-Curson-Ahlborn problem) and heat storage cases can be combined in a general thermal module which consists of : First a series of resistances and storages between a source and a reversible engine, second a reversible engine and third a series of resistances and storages between the reversible engine and a sink. Such a set of thermal resistances, storages and a reversible engine is called thermal module [1].

For perodic change of source temperature (T_s) and environment temperature (T_o), the rate of operation of the reversible engine ($W' = q_h - q_l$), working between its supply temperature T_h ($< T_s$) and discharge temperature T_l ($> T_o$) may be changed periodically at such a suitable phase (ϕ) and amplitude (q_{ha}) that work output will be maximized, with constraints of given T_s , T_o and $q_{h,max}$.

Equations for ϕ and q_{ha} which maximise work output of a period are obtained by combining Eq. (19) (between temperatures T_h and T_l) and relations between: At source side T_s and T_h ; at sink side T_l and T_o respectively. The second group of relations are obtained using complex variable and matrix method explained in [7] for one dimensional periodic heat conduction through composite layers.

8. Upper Harmonics of Temperature Variation

Any non harmonic variation of source or environment temperature can be expanded in Fourier series with parameters T_{sa1} , T_{sa2} , ...; ϕ_{s1} , ϕ_{s2}, ...; $T_{\infty a1}, T_{\infty a2}, \ldots$ The procedure given above can be applied to upper harmonic functions, just in the same way to find $\phi_{q1}, \phi_{q2}, \phi_{q3}, \ldots, \phi_{qi}$. It was shown that $q_{sa} = q_{lim} - q_{sm}$. The amplitudes q_{sa2} , q_{sa3} ,...have to satisfy the condition that at their own maximum where $t = t_{imax}$, $i = 1,2,3,\ldots$(with $2\pi \cdot t_{imax} + \phi_{qi} = \pi/2 + 2n\pi$, etc.) same condition will be satisfied.

$$q_{lim} \geq q_{sm} + \sum_{i=1}^{j-1} q_{sai} \sin(\omega_i t_{jmax} + \phi_{qi}) + q_{saj}, \quad j = 2,3,\ldots \tag{37}$$

9. Other Types of Engines

The extension of the method of finding optimum operation from a single harmonic variation of source and environment temperatures to Fourier series expansion of an arbitrary temperature variation , as described in the previous section is only intuitive. It needs mathematical proof whether the result is a true optimum.

The general model assumes a work output (or input , which will be indicated by a negative value). However, this work can be in form of exergies. The engine may increase exergy , if required also energy of a fluid which flows through. This difference in type of work exchanged does not need any change in the above theory. To improve efficiency and economy of solar driven absorption system, one may use weak solution storage and ammonia receiver as exergy storage. The high pressure side should run according to variations of source and environment temperatures. The low pressure side should run according to variations of refrigeration demand and environment temperature. The only constraint between two engines is the capacity of matter-exergy storages. Finding the optimum design parameter (characterizing the interface, i.e. storage capacity) one can optimise the complex system .

The following constraints will be discussed: First, same thermal source will be used by the engines and the constraint is the given total heat supply to these; i.e. parallel running. Second, the heat output of one engine may be completely or partly the heat supply of the other, through storage and resistance; series running. Third, the exergy output of one engine may be consumed totally or partially by another engine, having an exergy transfer function in between; i.e. exergy series running. The study of such complex systems will be in three stages: First, determination of mean values (q_m , T_m ,...) ignoring the fluctuating components. Second, determination of optimum phase of working for each engine (ϕ_{q1} , ϕ_{q2} ,...). Third, Determination of optimum amplitude of fluctuating component of heat supplies each engine.

The first step is straight forward; constraints or design values of component will suffice to find mean values which can be considered also steady state values. The second and then third stages require application of one of the various methods like circuit theory, linear programming or sectioning. In sectioning, the interfaces of subsystems will be characterised by parameters of which the optimum values will be searched for. The optimum phase and amplitude of working of each engine will be expressed in terms of these interface values.

10. Applications

Low temperature power production with solar energy driven Stirling engine may be considered as an example of application of Section 3. Relatively large value of y ($=T_{as}/S_M$) eg. 0.2 and z ($=T_{om}/S_M$) e.g. 0.8 is expected to be observed. These values of parameters with $w = \pi/4$, $x = 14/360$ can be used to obtain optimum phase shift ϕ from Fig. 2a and dimensionless work per cycle from Fig. 2b, 3a, 3b.

Another field of application is geothermal power production. Although steady operation has various advantages, on one hand change of environment temperature on the other hand change of demand, i.e. price, makes hourly change of power production rate through natural or engineered thermal storage attractive.

For approximate coincidence of demand function with environment temperature function, d = 0, for dimensionless thermal storage e.g. $C_s\omega T_{sm}/(q_{sm}\,a) = [(10^5\ \text{kJ / K}\ 360\ \text{K})\ 2\pi/(86400\ \text{s})]/(333\ \text{kW}\ 0.3) = 26.2$ and $x = 14/360$, $z = 295/360$, $p = 0.4$ one obtains $c = (x/z)26.2 = 1.24$; $c - p' = 1.17$. This results in $\phi = -58.7^{\circ}$ and $238.7\,^{\circ}$ (i.e. -1.0245 and

4.166). w = 31.3 and 328.7 (i.e. 0.5462 and 5.737). For these solutions from Fig. 2b; W= 1.045 i.e. 4.53 % more work due to usage of thermal storage and periodic operation.

11. Conclusions

To calculate exergy destruction or find exergy of control mass system or exit exergy of control volume with timewise variation of environmental condition exergy balance equation can be written in terms of fixed reference condition and correction terms which can be calculated easily.

Day and night temperature difference is relatively large for energy systems with low source temperature. This effect may coincide with variation of source temperature, variation of price, existence of thermal storage at source or at sink and effect of external heat transfer resistance. In each of these cases, harmonically changing rate of operation; with an optimum phase shift with respect to environment temperature improves overall efficiency. Solar and geothermal energy applications can utilise the results given above.

Increasing amplitude of the operation brings advantage if it does not disturb source temperature. In case of heat storage, temperature fluctuation depends strongly on rate of operation and puts limits on it. Optimum phase shift for a practically important case: heat storage at source and constant heat supply, is an especially simple expression:

$$\phi = -\arcsin(yz/x) \quad \text{and} \quad \phi = \pi + \arcsin(yz/x) \tag{38a,b}$$

Further study on different applications of the derived relations is recommended.

Acknowledgement

First author thanks sincerely Alexander von Humboldt Stiftung, German F.R. for financial support during one month visit at Institiut für Thermodynamik und Warmetechnik, Universitaet Stuttgart and Director of Institut Prof. E. Hahne for his hospitality.

References

1. Göğüş, Y., Ataer, Ö. E., İleri, A. (1993) A thermal module to study exergetic utilization of environmental temperature variation in J.S Szargut (ed.), *Proceedings of Energy Systems and Ecology Conference*, Cracow, 857-866.
2. Göğüş, Y. A., Ataer, Ö. E. (1998) Thermal power generation considering periodic change of environment temperature in M. Feidt (ed), *ECOS'98, Efficiency, Costs, Optimization, Simulation and Environmental Aspects of Energy, Energy Systems and Processes*, Nancy
3. Szargut, J., Morris, D. R. and Steward, F.R. (1988) *Exergy Analysis of Thermal, Chemical, and Metallurgical Processes*, Hemisphere Publishing Corporation, New York.
4. Göğüş, Y., Ataer, Ö. E., (1998) *Time-Dependence Effects on Thermodynamic Optimization by Second Law*, METU, Ankara.
5. Bejan, A., Tsatsaronis, G., Moran, M., (1996) *Thermal Design and Optimization*, John Wiley and Sons Inc., Toronto.
6. Bejan, A. (1996) *Entropy Generation Minimization*, CRC Press, New York.
7. Carslaw, H.S., Jaeger, J.C. (1959) *Conduction of Heat in Solids*, Oxford University Press, London.

BOND GRAPHS AND INFLUENCE COEFFICIENTS APPLICATIONS

Y.A. GÖĞÜŞ
Aeronautical Engineering Dept.
Middle East Technical University
06531 Ankara, Turkey

1. Introduction

A complex thermodynamic system needs as a first step clarification of components and their interaction. In many cases the succeeding step which completes the study is steady state analysis. However when the first step is accomplished i.e. interrelation and internal behaviour of components are clarified, the unsteady behaviour of the system also can be studied easily. The aim of this lecture is first to review bond graphs, give examples and compare with linear graps, second to review the method of influence coefficients, which is used for steady state processes and show how to apply it to unsteady processes.

2. Bond Graph

2.1. ELEMENTS, INTERACTIONS AND VARIABLES

Bond graphs have interactions as lines and elements at end points of lines, in comparison to linear graphs which indicate elements by lines and points of interaction of these elements with each other by nodes at the end of these lines. Arrows on the lines show the positive reference direction of flow and energy. Fig. 1 shows an example. Elements, which undergo only one type of interaction with the rest of system, are called one-port element although they have in general two terminals. Two basic types of one-port elements are storage (C) and dissipator (R).

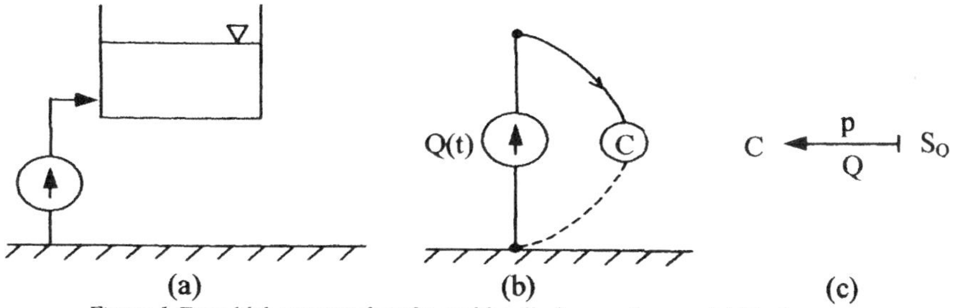

Figure 1. For a high water tank and a positive displacement pump (a) Hardware (b) Linear graph (c) Bond graph.

A. Bejan and E. Mamut (eds.), Thermodynamic Optimization of Complex Energy Systems, 241–250.

In both types of graphs each one-port has two relevant properties, such that during an interaction the product of these gives the energy transferred. One of them is called "flow variable", the other "effort varible". Table 1 shows these variables for different energy domains.

In complex graphs, each bond has a number and if more than one energy domains exist, symbols of effort and flow variables are attached to bonds. For heat interaction the flow variable is taken as heat rate in simple treatment, which is called "pseudo bond graph". Proper treatment leads to entropy rate.

TABLE 1. Variables and their category for different energy domains in bond graphs.

General	Flow (f)	Effort (e)
Linear movement	Velocity	Force
Rotational movement	Angular velocity	Torque
Hydraulic system	Volume flow rate	Pressure
Electrical system	Current	Voltage
Thermal system	Entropy flow rate	Temperature

2.2. CAUSALITY

There is a short perpendicular line on one end of each bond, which is called "causal stroke" and indicates the element on which effort (e.g. voltage, pressure, temperature, force, torque) is imposed by the other element. A source element is starting point of determining the causal strokes in a system. S_e is a source for effort hence always S_e ——| and S_f |—— , i.e. for a flow source the element with which it interacts, determines the effort.

2.3. TWO-PORTS AND JOINTS

Some components of engineering systems have two ports, each interacting with a different part of the graph, even in different energy domains. Energy conserving linear two-port elements has constitutive equation set of following form:

$$\begin{vmatrix} V_1 \\ f_1 \end{vmatrix} = \begin{vmatrix} C_{11} & C_{12} \\ C_{21} & C_{22} \end{vmatrix} \cdot \begin{vmatrix} V_2 \\ f_2 \end{vmatrix} \quad (1)$$

If $C_{12} = C_{21} = 0$ and $C_{11} = -1/C_{22}$, it has properties of an electrical transformer; it is called transformer and C_{11} = TF. On the other hand if $C_{11} = C_{22} = 0$ and $C_{12} = -1/C_{21}$, the two-port element has properties of a gyroscope, it is called "gyrator", C_{12} = GY.

The factors TF and GY are determined by the physical characteristics of these elements. In some devices TF or GY is a variable of time, the device is called "modulated".

For convenience the nodes at which two or more bonds coincide are treated as "joints" and depending on the connection of bonds such a joint is called zero-junction or one-junction:

0-junction: parallel ;	$e_1 = e_2 = e_3$	$f_1 + f_2 + f_3 = 0$
1-junction: series ;	$f_1 = f_2 = f_3$	$e_1 + e_2 + e_3 = 0$

Possible flow and causal directions and causal relations of two-port elements and junctions are shown on Table 2. If energy storage elements are capacitors and causal relations are integral type (i.e. flow is imposed on capacitor, e is function of $\int_0^t f \cdot dt$, starting with sources like in thermal systems one can determine all causal strokes without any problem. For special cases references [1] and [2] should be consulted.

TABLE 2. Possible reference flow and causal directions of two-ports and junctions.

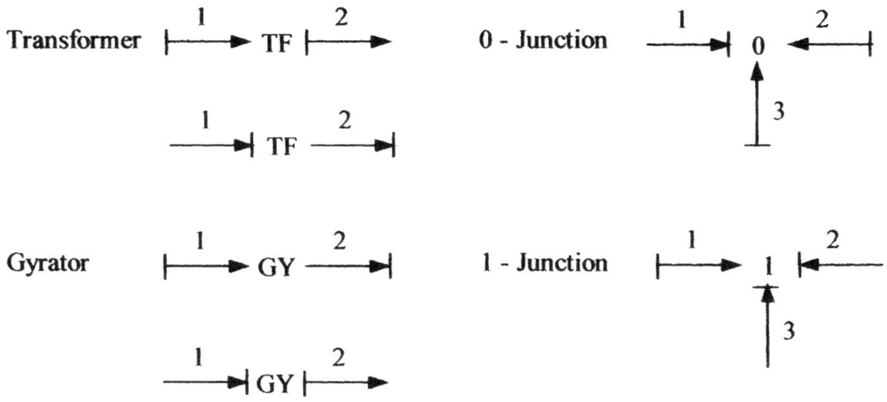

2.4. CAPACITY AND RESISTANCE FIELDS

The lumped components of a system can have simultaneous interactions with various energy domains and at the same time energy storage possibility or dissipative function; so that they are neither pure transformer nor pure capacitance. Such components are needed especially in thermodynamics. They are called "field" and shown with C or R with more than one bonds. In such a treatment entropy flow rate and temperature are flow and effort of thermal energy domains.

2.5. STATE SPACE EQUATIONS

The bond graph is completed when:

a) Nodes (components) are identified (constitutive equations written)

b) Bonds established (numbers and effort and flow variables are indicated)

c) Reference flow directions are chosen
d) Causal directions and relations are determined

A completed bond graph is of great help even for steady state analysis of complex thermal systems. However bond or linear graphs are indispensable for dynamic analysis.

Because of various advantages, the following form is preferred as the set of equations, which determine the state of a system.

$$\dot{\vec{x}} = f(\vec{x},\vec{u},t) \tag{2}$$

$\vec{x} = x_1, x_2, \dots, x_n$ and n is the number of independent (primary) variables, i.e. order of system. $\vec{u}$ Consists of $u_1(t)$, $u_2(t)$, ... , $u_m(t)$ input variables.The order of the system is equal to the number of state variables.

Three succeeding steps to establish state-space equations are:

i) Select input, energy and co-energy variables. Input variables have been shown as sources. Energy variables are; $h = \int_0^t f \cdot dt$ for capacities with integral causality. Co-energy (or dual) variables are effort for capacities and flow for impedances. They are secondary variables, which will be eliminated.

ii) Formulate the initial set of system equations

iii) Reduce the initial equations to state-space form

3. Influence Coefficients

3.1. STRUCTURAL BONDS AND SINGLE SERIES OF COMPONENTS

Like in investigation of other properties of complex systems, exergy analysis depends greatly upon behaviour of components and on the relation between component performance and system performance. Quantitative study of these relations allows concentrating our efforts mainly on the important components to improve system in efficient manner. Below explanation is based to a great extend upon the explanation of the subject by Kotas [4].

Coefficient of Structural Bonds (CSB) are shown by $\sigma_{k,i}$ and defined as follows, by indicating irreversibility rate by I:

$$\sigma_{k,i} = \frac{\left(\dfrac{\partial I_T}{\partial x_i}\right)}{\left(\dfrac{\partial I_k}{\partial x_i}\right)} = \left(\frac{\partial I_T}{\partial I_k}\right)\Bigg|_{x_i = var} \tag{3}$$

i.e. change of total irreversibility of system (subscript T) per change of irreversibility of component k; both due to change of a parameter of system, x_i.

It is assumed that system is in steady state and input parameters are not changed during the change of the parameter x_i at an amount dx_i , except that the exergy input

E_{in} needs to be modified to keep the exergy of product E_{out} unchanged. Since E_{out} is kept constant, change of system irreversibility is balanced by change of inflow exergy, all as rates: $\Delta I_T = \Delta E_{in}$.

Testing system components k and system parameters x_i with respect to $\sigma_{k,i}$ will indicate largest $\sigma_{k,i}$, i.e. will show which component's irreversibility should be reduced more strongly to achieve greatest reduction of system irreversibility. If some $\sigma_{k,i}$ are smaller than 1, reasons should be found and remedies implemented.

For a special system consisting of n components connected in series (Fig. 2) and each with exergy ratio w_k :

$$w_k = \frac{E_{k\,out}}{E_{k\,in}} < 1 \tag{4}$$

The σ_k increases monotonically with increasing k along exergy flow direction. By expressing I_T, and I_j in terms of the fixed E_{out} and $\prod^{j-1} w_k$, taking derivative of these with respect to w_j one can show that

$$\sigma_j = 1/\prod^{j-1} w_k \tag{5}$$

i.e. if due to change of a parameter x_i the irreversibility of a component j changes, this change will have greater effect on in-flowing exergy (ΔE_{in} ↑) as the component is nearer to exit (j↑).

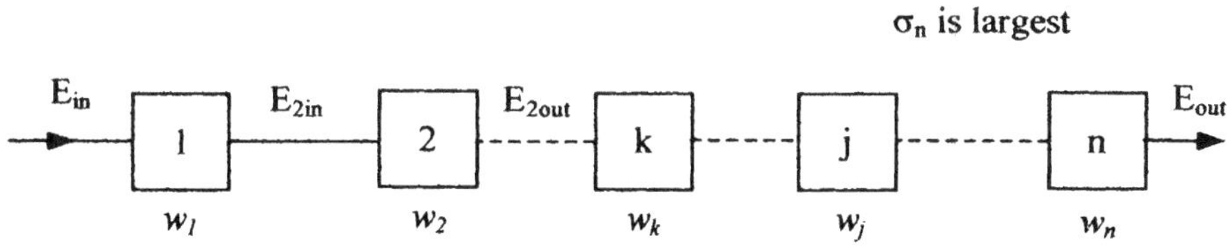

Figure 2. A system consisting of n components connected in series in exergy flow direction.

3.2. COEFFICIENT OF EXTERNAL BONDS

To study a large system through a study of its subsystems is convenient. However during optimisation exergy inflow and outflow from subsystems should be variables, in contrast with the whole system for which E_{out} is kept constant. Since subsystem has many components with different irreversibilities, the ratio of change of exergy input of complete system (ΔE_{in}) to change of an exergy (ΔE_j) on the boundary of subsystem (both due to change of x_i) is a suitable parameter, which is called coefficient of external bonds (CEB):

$$X_{j,i} = \left(\frac{\partial E_{in}}{\partial E_j} \right) \Bigg|_{x_i = var} \tag{6}$$

Here E_{in} is for the complete system and E_j is an exergy flow rate on the boundary of subsystem. This concept is useful because one can show by partial derivation that

$$\sigma_{k,i}^{s} = X_{j,i} \cdot \sigma_{k,i}^{ss} \tag{7}$$

Here superscript s means whole system, ss the subsystem under consideration.

Considering irreversibility in a component as difference of incoming and outgoing exergies, and taking derivative of both sides with respect to variable x_i, comparing the result with the derivative of total irreversibility of system, which is equal to derivative of exergy inflow:

$$\frac{\left(\frac{\partial I_k}{\partial x_i} \right)}{\left(\frac{\partial I_T}{\partial x_i} \right)} = \frac{\left[\sum_{in} \frac{\partial E_j}{\partial x_i} - \sum_{out} \frac{\partial E_j}{\partial x_i} \right]}{\left[\sum \frac{\partial E_{in}}{\partial x_i} \right]} \tag{8}$$

One obtains:

$$\sigma_{k,i} = \sum_{in} \frac{1}{X_{j,i}} - \sum_{out} \frac{1}{X_{j,i}} \tag{9}$$

Values of these coefficients for components and subsystems should be found from numerical simulation at few points of the interesting range.

4. Applications

4.1. STEADY STATE

To model steady state thermal systems, material flow diagrams with indication of energy transfer, which may be considered as simplified linear graphs, will be sufficient. Fig. 3a shows [5] the schema of the hardware and Fig. 3b a graph of an Absorption Heat Transformer (AHT).

The AHT is a device, which can deliver heat at a higher temperature than the temperature of the fluid by which it is fed, by consuming negligible amount of work. As shown in figure, the AHT consists of: an evaporator which is at moderately high temperature, to utilise exergy of the source; a Low Temperature heat Exchanger (LTE) to gain more energy from the source; a High Temperature heat Exchanger (HTE), to bring a part of the supply fluid to desired high temperature, to close the thermodynamic cycle of the actual working substance; and the auxiliary devices like pumps, valves to maintain steady state.

To identify the independent parameters of the AHT it is necessary first of all to find the number of parameters and the total number of constraints; the degrees of freedom of the AHT is determined as the difference of these numbers. The parameters are pressure, temperature, mass fraction and mass flow rate at the connection points of the components, as well as energy and two mass interactions at the components; altogether 73. The constraints are conservation equations for energy and species in the components, pressure drop and the phase equilibrium assumptions, altogether 68. The degrees of freedom of the AHT is determined as 5.

4.2. UNSTEADY ELECTROCHEMICAL ENERGY STORAGE

As mentioned above, simultaneous compression, heat interaction and matter or electron through-flow necessitates to consider thermodynamic systems as an energy capacitance with more than one port, i.e. capacitance field. In following bond graph modelling of electrochemical cell is summarised along the approach by Karnopp [6]. Gibbs equation [7]:

$$dU = T\,dS - p\,dV + \mu_1\,dm_1 + \ldots + \mu_m\,dm_m \tag{10}$$

shows that temperature, pressure and chemical potential are effort variables, entropy flux, volume change rate and matter flow rates are flows.

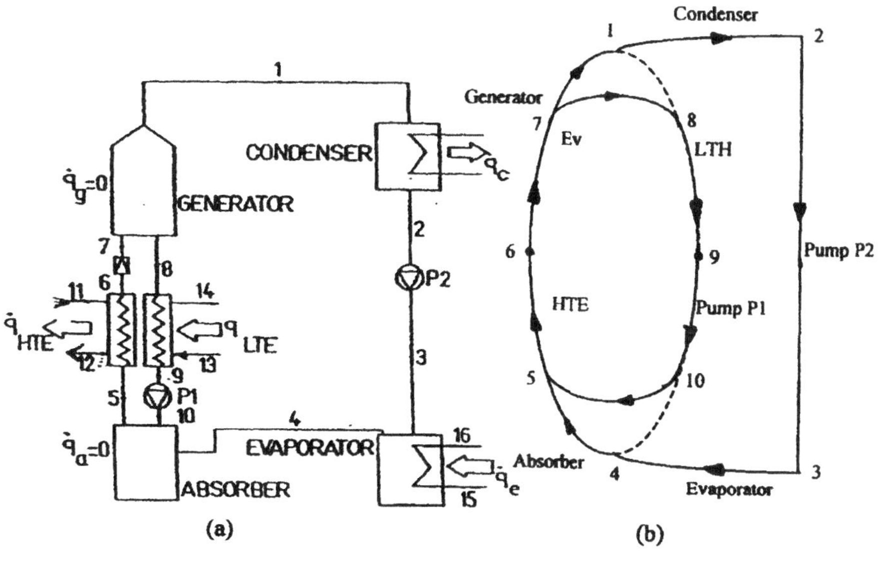

Figure 3. Absorption heat transformer (a) Hardware (b) Simplified linear graph

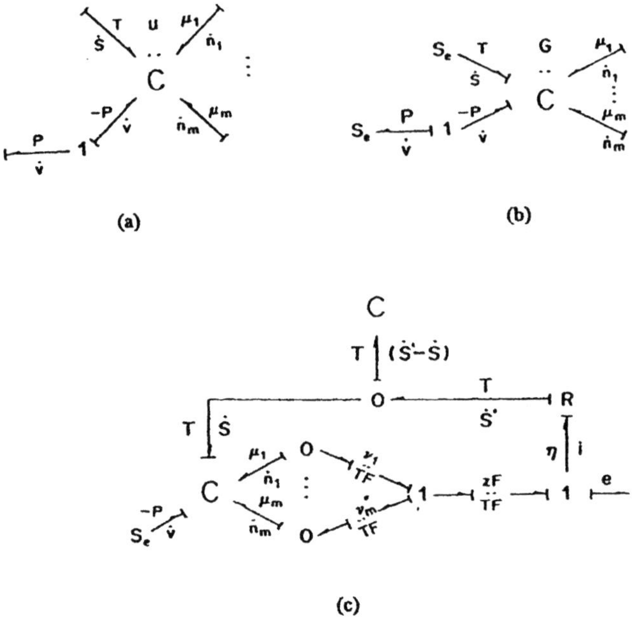

Figure 4. Bond graph of an electrochemical cell [6] (a) Basic case Eq. (10), (b) With transformed Gibbs function, Eq. (11), (c) With chemical reaction.

Fig. 4a shows bond graph of such a system with causal strokes, indicating that S, V, m_i are imposed by the environment. However, Legandre transformation gives Gibbs function with T and p independent variables, (Fig.4b)

$$dG = -S\,dT + V\,dp + \mu_1\, m_1 + \ldots + \mu_m\, m_m \tag{11}$$

(To change the sign of pressure to positive (+) a 1-juction i.e. series combination with source of pressure is used). Affinity A and rate of reaction J can represent the chemical reaction:

$$A = \sum \pm \nu_i \mu_i \qquad\qquad \dot{n}_i = \pm \nu_i J \tag{12a,b}$$

The reaction will be indicated by transformers of species. Because of its finite rate also a resistance R is needed. However this resistance of reaction rate can be considered as function of overvoltage η instead of affinity. Because of reaction rate-temperature relation it is connected to the heat capacity of system. When all these effects are combined and for this purpose 0- and 1-juctions are used, one obtains for one side of an electrochemical cell the bond graph given in Fig.4c.

Such complex components like electrochemical cell require bond graph analysis. However very frequently they are used in other complex and large systems, for which linear graph analysis [8] is better developed.

Koenig et al. [3] describe foundations of the method to transfer information of subsystem analysis to a complete system. However their book uses linear graph techniques. Contributions to make the connection of subsystem analysis with complete system analysis easier are needed, although the basic knowledge already exists.

4.3. INFLUENCE COEFFICIENTS FOR TRANSIENT CASE

Above structural bond coefficient $\sigma_{k,i}$ was defined for steady state. However in some cases the thermal system functions in unsteady way. E.g. an aircraft engine always starts to run from a stopped condition, undergoes a series of changes during the flight and after landing it stops again.

A thermal system, like aircraft engine, which runs during one flight under successive different load conditions, should be considered for a complete mission and instead of structural bond coefficient, structural-functional bond coefficient should be introduced:

$$\sigma_{k,j,i} = \frac{\partial I_{k,j} / \partial x_i}{\partial \left[\int^{\tau} I_T dt / \tau \right] \Big/ \partial x_i} \tag{13}$$

Here $I_{k,j}$ is the rate of irreversibility of component k at time interval j; τ is complete time interval of a mission and I_T is total irreversibility of system.

In case of a simple exergy flow system consisting of series connected components, the irreversibility near the product end has greater importance (Fig. 2). Similarly the irreversibility a component at a time interval near the end of flight has greater importance.

5. Conclusions

Bond graphs and influence coefficients are valuable tools for analysis and exergetic improvement of large systems respectively. However procedures for their application are not simple. Further studies on these subjects will lead to simpler techniques of their usage.

Acknowledgment

The author thanks Prof.Ö.E. Ataer and Prof.Y. Ercan for valuable discussions on these subjects.

References

1. Rowell, D., Wormley, D.N. (1997) *System Dynamics: An Introduction*, Prentice Hall, New Jersey.
2. Karnopp, N.C., Margolis, D.L., Rosenberg, R.C., (1990) *System Dynamics: A Unified Approach*, John Wiley Sons, New York.
3. Koening, H.E., Tokad, Y., Kesavan, K.H., Hedges, G.H. (1980) *Analysis of Discrete Physical Systems*, Mc Graw Hill, New York.
4. Kotas, T.J. (1985) *The Exergy Method of Thermal Plant Analysis*, Butterworths, London.
5. Ataer, O.E., Göğüs, Y.A. (1995) Simulation of Closed Absorption Heat transformer Utilizing Bond Graph, in Y.A. Göğüs, A. Öztürk, G. Tsatsaronis (eds.) ECOS '95 Efficiency, Costs, Optimisation, Simulation and Environmental Impact of Energy Systems, İstanbul.
6. Karnopp, D. (1990) Bond Graph Models for Electromechanical Energy Storage, Electrical Chemical and Thermal Effects, J. the Franklin Institute, **327**, 983-992.
7. Bejan A. (1997) *Advanced Engineering Thermodynamics*, John Wiley Sons, New York.
8. Chandrashekar, M., Wong, F.C. (1982) Thermodynamic Systems Analysis-I. A Graph Theoretic Approach, Energy, **7**, 539-566.

REPOWERING OPTIONS FOR EXISTING POWER PLANTS

P.F. MATHIEU
University of Liège – Institut de Mécanique
Department of Nuclear Engineering and Power Plants
Rue Ernest Solvay, 21 – Bât. C3
4000 Liège, Belgium

1. Why Repower an Existing Power Plant ?

There are mainly two repowering options which can be summarized as follows :

- The « **Substitute Heat Recovery Steam Generator(HRSG)** » option which consists in the replacement of the existing combustion boiler by a heat recovery steam generator behind a Gas Turbine (GT), making so a combined cycle (CCGT) «Fig. 1».
- The « **Hot Windbox** » option which consists in the addition of a GT to the existing boiler in 3 possible ways :

- the air forced draft fan and the air heater are replaced by a GT whose exhaust gas feeds the existing burners «Fig. 1».
- the GT exhaust gas is only used to reheat partly or totally the feedwater as well as producing low pressure superheated steam in a separate heat recovery boiler; in this case, it is a parallel powered CC configuration.
- When the repowered plant is coal fired, both the GT and the air forced draft fan feed the boiler's burners ; the boiler exhaust gas is then used both to reheat partially the feedwater.

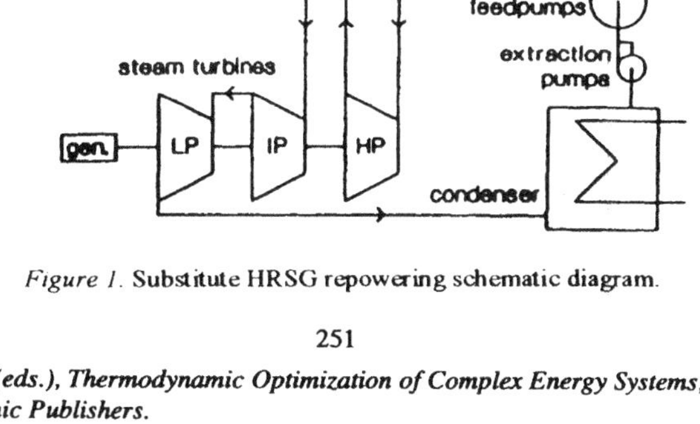

Figure 1. Substitute HRSG repowering schematic diagram.

A. Bejan and E. Mamut (eds.), Thermodynamic Optimization of Complex Energy Systems, 251–260.

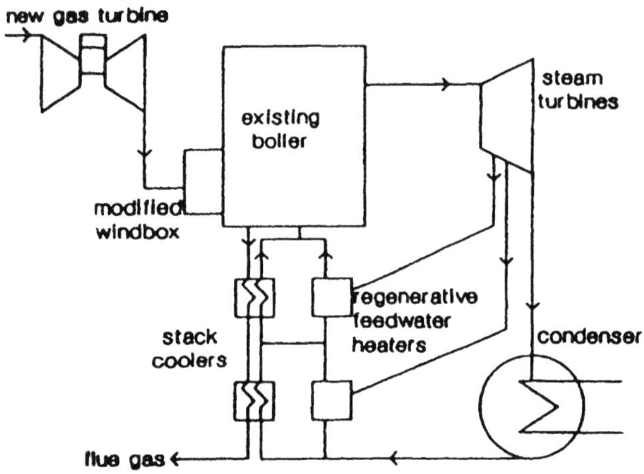

Figure 2. Hot windbox repowering schematic diagram.

1.1. THE MAIN REASONS FOR A REPOWERING ARE :

1.1.1. Energy/Environment

- Increase of the installed power by addition of the GT power output to the existing steam cycle.
- Increase of the global plant efficiency with or without an increase of the total power output. In the Netherlands, in particular, the hot windbox repowering is considered as a means of burning high grade fuels, like natural gas, at high efficiency.
- Decrease of the emissions, mainly of NO_x and CO_2. The GT and boiler are in series and the system looks like a staged combustion.
- Increase of the share of of natural gas in the fuel basket and so increase of the flexibility of the repowered plant with respect to the fuel.
- Increase of the operational flexibility, the combination of a GT with a Rankine cycle allowing the load following transients, the covering of the peak and intermediate loads, the operation at part load.
- The possibility of phasing the construction, the GT being able to be installed first with a by-pass stack.

1.1.2. Site related reasons

- Lack of new available sites in the country
- Lack of heat sinks (rivers, sea,...) and recuperation of the existing cold source (condenser, cooling tower, pipes, pumps, ...)
- No place for the erection of a new cooling tower
- The installation of an additional GT is easy and rapid, not requiring heavy civil engineering works.

1.1.3. Economy

- Recovering of all the existing components with a minimum of adaptations
- Low investment and O&M costs of the GT and so decrease of the generated kWh cost
- Industrial maturity of all the components
- Short pay-back time and low financial risks

In order to give orders of magnitude in a repowering, a typical rise of the installed power is 30%, a relative improvement of the efficiency is 10% (4-5% pts) and the reduction of NO_x emissions can reach 70%. As a general rule, a repowering, like a cogeneration, is a **case by case study** ; even if there are some common guidelines, it is very dependent on the state of the exisiting installations and of the site.

1.2. COAL VERSUS GAS

The repowering of a given power plant with a given GT (hot windbox option) efficiency is about 2% pts lower when the existing boiler is fueled with coal instead of natural gas due to some significant differences of the plant when it is fired with coal or with gas.

- The fuel LHV are different (42-45MJ/kg for natural gas and 20-25MJ/kg for coal) and the needed excess air is about 15% higher with coal. The sizes of the burners windboxes and of the air pipes and of exhaust gas pipes are larger. Hence the size of the boiler is 15 to 20% larger.
- The distribution of radiation and convection heat transfer is different so that the steam properties at the superheaters outlet are different
- A coal plant needs more auxiliaries for the coal transport, grinding, for the flue gas desulfuration and for the electrostatic extraction of the particles. The electricity consumption is consequently higher.

2. Comparison of Two Repowering Options

Amongst the many ways in which a reheat steam cycle may be repowered using one or more GTs, two basically different repowering schemes are considered here.

2.1. THE SUBSTITUTE HEAT RECOVERY BOILER REPOWERING OPTION

As above mentioned, in this scheme, the existing boiler is replaced with one or more HRSGs. Feedwater heaters are usually removed, and the feedwater is used to cool the stack down to the lowest temperature which is compatible with the fuel (sulfur content, dew point). The steam mass flow requirement of the steam set determines the number and size of the GTs. Figure 1 shows a schematic diagram for such a repowered configuration. This repowering method is the only one available if the existing boiler is close to the end of its lifetime.

2.2. THE HOT WINDBOX REPOWERING OPTION

In this scheme (fig. 2), the existing boiler is kept in service. It should therefore have a proper remaining life and a good availability. The GT replaces the forced draft fan and exhausts directly into the boiler windbox. This latter needs to be modified because the GT exhaust temperature is higher than the preheated air temperature. However, the oxygen content of this exhaust gas is still 16% and is sufficient for the combustion to take place. Of course, the combustion process (adiabatic flame temperature, radiation and convection share in the flame zone) is affected by the change in the oxydiser temperature and composition. Also the evaporator and superheater surfaces could be no longer adapted to the new temperature profile. The question is : is it necessary or not to modify the heat exchangers surfaces and if yes, how ? Generally, only the economizer surface is changed and is increased to extract the required total heat from the flue gas. The oxygen requirement for the boiler combustion provides the necessary guidance for the GTs to be installed. Figure 2 shows a schematic diagram for such a repowered cycle. It can be seen that some of the feedwater heaters are kept in service and operate in parallel with the new stack cooler.

3. Specific Technical Issues

The choice between the substitute HRSG and the hot windbox options requires the analysis of the following technical issues which are specific to repowering.

3.1. STEAM BLEEDS

In the substitute HRSG scheme, the best design point efficiency is always achieved when all the steam bleeds from the steam turbine set are closed. In the hot windbox scheme, the maximum design point efficiency is obtained with an *optimum* bleeding flow from the existing steam turbines. This latter flow is usually much lower than the original one. Hence, most of the steam bleeds, but not all of them must be closed. In the case of a hot windbox configuration, the global efficiency exhibits an optimum with respect to the total bleeding flow.

3.2. BACK-END LOAD AND DERATING

When the steam bleeds are closed, the steam flow within the LP stages of the steam turbine set is increased. Typically, closing all the bleeds would give a 25 to 30% increase of the low pressure stages mass flow. This is mechanically unacceptable for the bearings and for some of the blades themselves. It would be very difficult and costly to modify the LP stages and bearings to accommodate this increased flow and the associated mechanical load. Moreover, the condenser and boiler feedpumps would need some modifications as well as the heat load is increased.

The solution is then to *derate* the steam turbine so that its power output is left unchanged when all the bleeds are closed. This only asks for minor changes to the inlet valves to the HP turbine section. If the utility is not looking for additional capacity but only for fuel economy (as in the Netherlands), the steam set may be further derated but with an associated decrease of its effectiveness. Using this technique, the steam turbine derating may compensate for the GT additional output leaving the total plant output nearly unchanged after repowering.

3.3. FUEL FLEXIBILITY

After the first oil crisis, Western countries have converted some of their steam plants in order to burn all the fossil fuels, namely, natural gas, oil and pulverized coal. Clearly, if the boiler is replaced with a HRSG, coal cannot be used anymore. The GT itself must be running on natural gas or on a light distillate. If the HRSG is supplementary fired, this additonal combustion has to run on natural gas or liquid fuel as well.

On the other hand, if the hot windbox option is used, the GT must be running on gas or on distillate as well. When a coal-fired boiler is repowered, the forced draft fans are kept in service. The coal mills can then be operated with the required flows either of preheated air or of a mixture of GT exhaust gas and preheated air. In this latter case, a small air-air preheater has to be used. The efficiency penalty, mainly due to the increased auxiliary power requirement, may typically amount to two % pts in that case. However, because of the increased windbox temperature, gas velocities in the boiler are increased. This is thought to be unacceptable if coal is used, because these higher velocities would lead to erosion in the superheater and reheater sections of the boiler. Hence, it is a great concern that the ability to run on coal could be lost when repowering. In general, the existing boiler and the GT operate on the same fuel.

3.4. REMAINING LIFE AND AVAILABILITY

The repowering always associates new equipments with existing ones. The remaining life of existing equipments and the associated availability must be taken into account. If the boilers are old or if they are used for load following, this mode of operation leading to frequent temperature changes which shorten the lifetime of the hottest components, the existing boiler has to be replaced by a new HRSG. It appears to be the most suitable solution in such a context as it is the case in Belgium where the base load is covered by nuclear power plants.

3.5. EMISSIONS

Emissions characteristics are different and must be addressed in a different way depending on the selected repowering option.

3.5.1. Substitute Heat Recovery Steam Generator option

Without any supplementary firing, pollutants of the repowered cycle are those of the GT which is selected. Only NO_x could prove unacceptable. NO_x emissions are reduced by well-known techniques like water or steam injection or dry low NO_x combustors. If some supplementary firing is used, CO, unburned hydrocarbons and NO_x emissions now depend on the supplementary firing combustion. It is impossible to generalize, because many factors (burner design, flow patterns,...) must be taken into account. In many cases, CO_2 and SO_2 emissions depend only on the fuel that is burnt and on the efficiency of the plant. Because the substitue HRSG repowering increases the efficiency by about 20%, CO_2 and SO_2 emissions are cut by about 20% as well.

3.5.2. Hot windbox repowering

In this case, the combustion in the GT is followed by a combustion in the existing boiler. Because some of the work is extracted by the GT before the boiler combustion is conducted, the final burnt gas temperature is less than for a classical steam cycle running with the same overall fuel/air ratio. This may lead to large reductions in NO_x levels. A staged combustion could further reduce the NO_x content of the GT exhaust. For CO_2 and SO_2 , the same applies as for the subsitute HRSG option, except that the cut in emissions is about 10% for a given fuel, because of the smaller efficiency improvement.

4. Design Point Performance Analysis

4.1. DATA

The design point performance analysis was carried out using a home made computer simulation program. The GT size is determined by the program itself, but its characteristics are those of a 150MW-class machine (see Table 1).

TABLE 1. GT characteristics.

Compressor (expander) isentropic effectiveness : 0.91 (0.86)
TIT : 1110°C
Mechanical effectiveness : 0.99
Inlet ΔP (outlet ΔP) : 2000 Pa (4000 Pa)
Inlet conditions : dry air, 20°C, 1 bar

The pressure ratio and the machine output are optimized to provide the best possible repowered cycle efficiency. Exhaust mass flow and temperature are dependent on the above mentioned parameters.

The selected steam cycle, which is running on coal, represents what state of the art was some twenty years ago, and hence could be a good candidate for repowering. Its main characteristics are summarized in Table 2.

TABLE 2. Steam cycle characteristics.

Drum pressure : 111 bar
Superheater and reheater outlet temperature : 540°C
Superheater discharge flow : 138 kg/s
Reheater inlet pressure : 36 bar
HP (LP) steam turbine isentropic effectiveness : 0.81 (0.83)
Design point power output : 167MW
Design point efficiency : 35.7% (LHV – natural gas)
Design condenser pressure : 0.07 bar
Condenser cooling water inlet temperature : 20°C
Pump effectivenesses : 0.75
Economizer inlet pressure : 116.6 bar

2 LP and 2 HP steam extractions (bleeding flows are optimized when repowering). Other characteristics cannot be detailed here but include realistic values of pressure losses, approach and pinch point temperature differences, auxiliaries, generator effectiveness, ...

4.2. HOT WINDBOX REPOWERING

The temperature of the exhaust gases entering the windbox is 580°C. Both the GT and the boiler are running on natural gas and the stack temperature cannot fall below 120°C. The GT that fulfils the boiler O_2 requirement is found to have a power output of 56.7 MW with an exhaust flow of 193.6 kg/s.

For a given GT size, the most interesting problem is the feedwater heating. It can be calculated that, starting with a total bleed flow fraction Y=0.125, characterising the original steam cycle, closing some of the bleeds until this parameter is 0.08 multiplies the repowered plant efficiency by 1.01. The feedwater heating then needs to be done in parallel by the new stack gas coolers and the steam feedwater heaters (see Fig. 2). With the optimised bleeding flows, the net repowered plant efficiency (natural gas – LHV) reaches 40.6%, which is 4.9 % pts above the original steam cycle. Given the necessary steam turbine derating, a 21% increase in output is associated with this repowering option. The utility could decide to further derate the steam turbine set if no additional power is required or if there is already an excess electric power in the country.

GT optimisation. When considering a range of GTs, an optimum pressure ratio was found at 12.5. The influence of the pressure ratio on the plant efficiency is however quite small, and using a GT with a slightly different pressure ratio does not give any measurable efficiency penalty. Not surprisingly, the GT TIT has a much stronger effect. Figure 3 shows that the efficiency increases almost linearly with the GT TIT ; i.e. the efficiency is increased by 1.1% for each 100°C in TIT.

4.3. SUBSTITUTE HEAT RECOVERY STEAM GENERATOR REPOWERING

The substitute HRSG repowering is computed using the same GT as for the hot windbox option. Once again, the size is not predetermined and is computed by the program. The exhaust temperature (580°C) allows to fulfil the boiler requirement even without supplementary firing.

The assumptions specific to the HRSG are mentioned in Table 3. For a given GT, the bleeding level and the supplementary firing in the GT exhaust are the more interesting parameters.

TABLE 3. HRSG assumptions

Single pressure level – natural circulation
Economizer approach point : 20°C
Superheater pressure loss : 3% of the drum pressure
Reheater pressure loss : 2% of its inlet pressure
Economizer pressure loss : 2% of the drum pressure
Gas side pressure loss : 4000 Pa

Bleeding level. The best efficiency is achieved with all the bleeds closed, that is with all the feedwater heating done by the HRSG economizer. This situation is similar to that of a genuine combined cycle. If all the bleeds of the original steam cycle were kept in operation, this would turn into a 1.2 % point (i.e. 2.7%) efficiency penalty.

Supplementary firing. Because the steam cycle is a reheated one, it has a quite important heat requirement at high temperature. The required heat may be provided by a supplementary firing. The share of the latter in the total heat provided to the cycle is

represented by the ratio $\dot{Q}_{GT}/\dot{Q}_{tot}$ which is the ratio of the heat input to the GT cycle over the overall heat input (GT combustor + supplementary firing). This ratio may therically vary between 0 (steam cycle) and 1 (unfired HRSG). Starting with 1, a HRSG can cope with values of this parameter down to about 0.7, which corresponds to temperatures of the gas leaving the supplementary firing around 800°C. Quite fortunately, the calculations show that the efficiency peaks at 48% when the considered ratio is 0.75, that is when the GT fuel mass flow is ¾ of the overall fuel flow.
The required GT has a power output of 208.8 MW and an exhaust flow of 713.27 kg/s.

GT optimisation. The same conclusions as for the hot windbox repowering apply. The efficiency peaks for a GT pressure ratio of 12.5, with a very flat optimum. The influence of the GT TIT is very strong (even stronger than for the hot windbox option). Figure 6 shows that the repowered plant efficiency goes up by 3% for each 100°C increase in GT TIT. Line number two on the same figure is the unfired case which has a slight efficiency penalty when compared with the optimum supplementary firing case because the supplementary firing achieves a lower stack temperature with the reheated non-optimized steam cycle under consideration. In the heat transfer diagram, the gas cooling curve rotates around the pinch point (see Fig. 4).

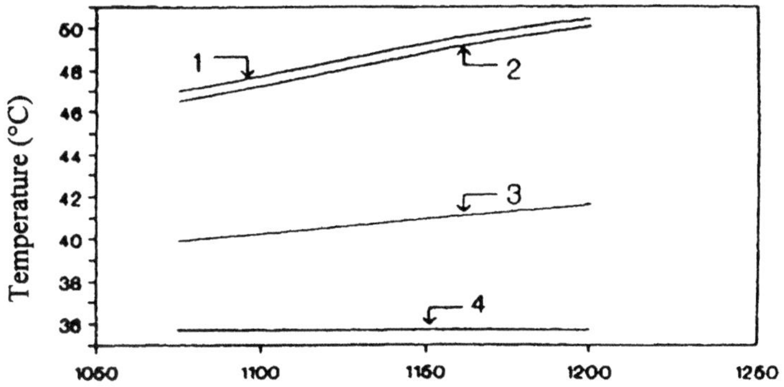

Figure 3. Net cycle efficiency (LHV) versus gas turbine TIT.

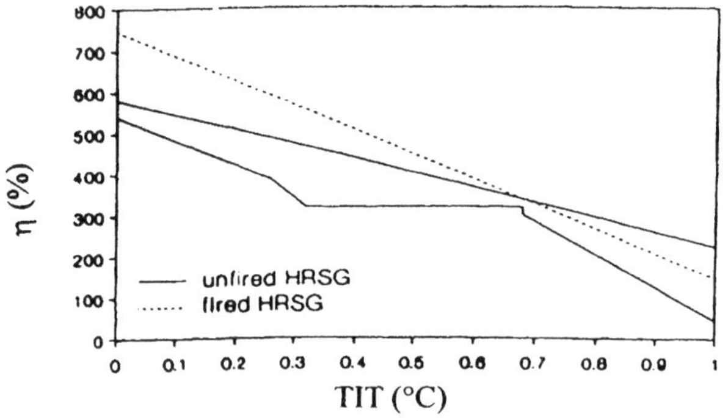

1. Fired HRSG option
2. Unfired HRSG option
3. Hot windbox option
4. Original steam cycle

Figure 4. Heat transfer diagrams.

5. Discussion of the Results

The repowering options illustrated above are now compared against each other.

5.1. UNFIRED AGAINST FIRED HRSG

The unfired HRSG case is first compared with the fired HRSG case, which proves to be a better solution for reheat steam cycles. This fired HRSG case is then compared with the hot windbox option. For decision making, the selected repowering option then needs to be economically assessed with reference to a new combined cycle. Because the steam cycle under consideration is a reheat one, the best efficiency is achieved with some supplementary firing. Figure 3 shows that the efficiency advantage of the fired case is about 0.5 % point over the unfired one. This difference is due to the lower stack temperature achievable through the HRSG heat exchange when supplementary firing is used. Both heat exchange diagrams are superimposed on figure 4 for comparison purposes.

Figures 5 and 6 compare the energy fluxes for these two solutions. These figures are based on the assumption that the heat input to the system is constant, to help the comparison. This points out the difference in repartition between the heat rejections to the environment through the stack and the condenser, depending on the option which is used. For example, the lower stack temperature achieved when the HRSG is supplementary fired reduces the stack losses from 22 to 8% of the energy input. For a given GT size, some more powerful steam cycles may be repowered using the supplementary fired option. Steam plants up to 150 MW may be repowered using the unfired scheme with two 150 MW-class GTs. A 300 MW steam plant may be repowered using two 150 MW-class GTs exhausting in a HRSG which is supplementary fired up to 750°C. For the unfired HRSG option, the ratio of the GT output to the steam turbine output is 2 ; this ratio becomes 1 in the optimum fired case. Supplementary firing drawbacks are complexity, reliability and emissions. The supplementary firing is an additional system in itself which can be a source of unavailability. The supplementary combustion increases the emissions from the GT combustor.

5.2. HOT WINDBOX OPTION VERSUS THE FIRED HRSG

The fired HRSG option achieves an efficiency improvement much more important than that of the hot windbox. Figures 6 and 7 compare the energy fluxes for these two options. However, off design considerations tend to decrease this hot windbox design point disadvantage. The substitute HRSG with the optimum supplementary firing roughly doubles the output, while the hot windbox multiplies it by a factor between 1 and 1.30, depending on the chosen steam turbine derating.

When the repowered cycle has to operate on a load following basis, the off design performances of supplementary fired HRSG cycles are very suitable for this. In particular, the upper range of the part load operation covered with the supplementary firing as the controlling parameter exhibits very good efficiencies. When a country is faced with the need to install additional power, repowering with a supplementary fired HRSG is very attractive as it roughly doubles the output.

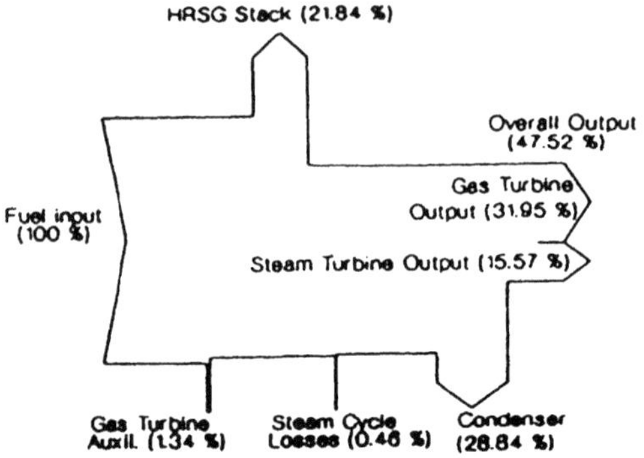

Figure 5. Unfired HRSG option - energy flux diagram.

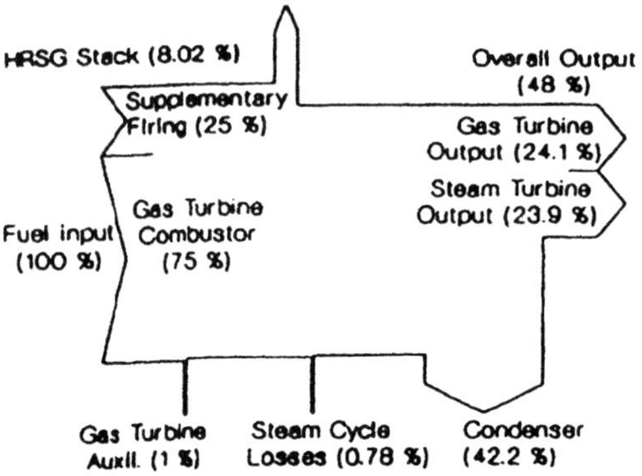

Figure 6. Fired HRSG option - energy flux diagram.

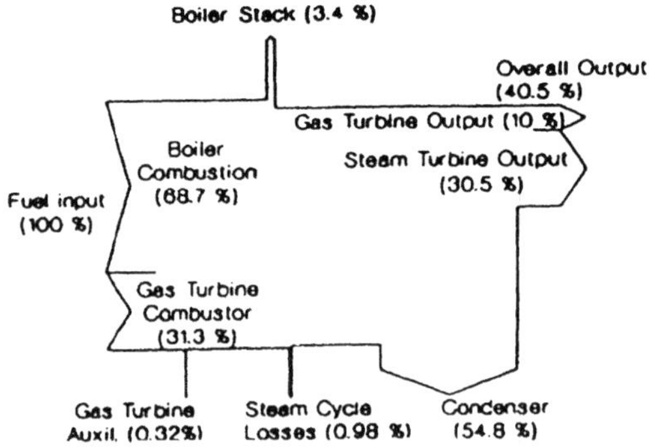

Figure 7. Hot windbox option - energy flux diagram.

6. Conclusions

This study shows the optimal repowering option. This is not a reason why not repowering a coal or lignite or other low grade fuels fired power plants; however the efficiency and output will be less than when natural gas is the only fuel. Also some new technical issues will have to be solved.

COGENERATION BASED ON GAS TURBINES, GAS ENGINES AND FUEL CELLS

P.F. MATHIEU
University of Liège – Institut de Mécanique
Department of Nuclear Engineering and Power Plants
Rue Ernest Solvay, 21 – Bât. C3
4000 Liège, Belgium

1. Main Features of Cogeneration

Cogeneration means simultaneous or combined production of heat and electrical power (CHP). The heat can be used to produce hot water or steam for a system or a process, or/and to feed a refrigeration system. In the case of production of electricity, heating and chilling, one speaks of trigeneration. Usually cogeneration leads to a better utilisation of the fuel than a separate generation of the same amounts of electricity and heat. When using a Rankine cycle either in a fossil fuel fired plant or in a combined cycle with gas turbine (CCGT), there are two main ways to cogenerate electricity and heat :

- with a back-pressure steam turbine (the condenser operates at the pressure of the process heat or steam water).
- through bleeding from the steam turbine at the process pressure.

  We will not examine here theses two options, usually used in large-scale cogeneration plants (for instance, for district heating) and concentrate on small cogeneration (less than let's say 40MW).

The assets of cogeneration are :

- Cut of primary energy use and preservation of fuel reserves thanks to high energy efficiency (85-90 %), that means an extraction of up to 2.5 times the amount of useful heat and power from raw fuels in conventional electricity generation (35 to 40 % for a coal-fired power plant).
- Lower emissions-for a small size gas fired CHP, for example :
  - the emissions of CO_2 are cut by more than 50 % due to the high efficiency
  - the emissions of SO_2, the most responsible of acid rains, are practically zero due to the use of natural gas, the cleanest of the fossil fuels
  - the primary energy use is cut by some 35% ; the fuel saving when compared to separate production is typically 10 to 15% but can be higher
  - the NO_x emissions are controlled and brought at levels not higher than with conventional generation.
- Production of electricity and heat close to the users, most of the time on the same site. This is another way to consider power generation, with decentralized small-scale power plants delivering either alternating or direct current. Until recently, we

A. Bejan and E. Mamut (eds.), Thermodynamic Optimization of Complex Energy Systems, 261–270.

were used to power generation by ever larger power plants located far from the users and delivering synchronized alternating current to high-voltage transmission lines and to local networks at lower voltages. This makes the industrial society vulnerable to any significant disruption of electricity. Nowadays, electricity generation plants move into the liberalization of the market and as a consequence, the investors turn their money to smaller stations, that can be built more rapidly and sited more easily, reducing so the risks on investments and giving shorter pay-back times (less than 5 years against 20 to 30 years to recover the investment).

The cost of CHP plants is a leading factor for the decision-making.

- The interest of a CHP is not only based on performance and environmental impact considerations but mainly on its cost and this latter depends a lot on strategy of the distributor regarding the price paid on the one hand for the purchase of the surplus generated electricity sent to the network and on the other hand for the selling of electricity taken from the network.
- Like the CCGT, on-site cogeneration systems use most of the time natural gas as the primary fuel but unlike the CCGT, they remain available even if the network is unavailable for one or another reason. At this stage the question of the availability of natural gas is raised. The combustion of this latter is clean when compared to coal or oil for the same power output, with no sulphur oxides, controllable NO_x formation by primary (control of flame temperature, use of staged combustion) and secondary measures (use of selective catalytic reactors) and smaller emissions of CO_2 (higher efficiencies and higher hydrogen to carbon ratio). If the natural gas is available in the country, it can be transported continuously to the site, eliminating the need for storage and handling.
- Operation and maintenance of on-site generating systems will not require a staff on the site itself. A remote monitoring will become a standard practice based on remote real-time data acquisition, analysis and diagnosis. Candidates for cogeneration will be, on top of major industrial installations, hospitals, universities, schools, public buildings, hotels, office blocks, airports, shopping centers, ...

1.1. SMALL- SCALE COGENERATION

The best candidates for on-site cogeneration at present and for the near future are :

- very small gas turbines, industrial or aeroderived, with or without supplementary firing (SF).
- advanced diesel, gas and Stirling engines.
- fuel cells : like batteries and photovoltaic cells, they generate direct current used in particular in electronics, computers, TV and videos.

Due to a high utilisation of the fuel and the use of natural gas, cogeneration is one of the most recommended option to decrease the CO_2 emissions. After the Kyoto summit in December '97, most of the western countries have decided to increase the share of CHP in their electricity generation network. Moreover, systems like fuel cells do not generate CO_2 in operation, however they are still very costly. In a sustainable energy development, this is a proper way to supplement the increase of the share of renewables like photovoltaics, windmills, micro hydrostations and the use of biomass.

2. Cogeneration Based on the Use of a Gas Turbine

In the simplest case, this system is similar to a CCGT and electricity is generated by an industrial or an aeroderived gas turbine (GT). The flue gas at the GT exhaust is at 450-550°C depending on the GT type and its sensible heat is recovered in a heat recovery boiler with one (or possibly several) pressure level(s) to generate steam. Unlike in a CCGT, this steam is used for a process and not in a Rankine cycle. The heat in the flue gas is recovered free of charge, except that investment is needed for the recovery boiler. The electrical efficiency is the same as that of the GT but the design of the recovery boiler is simpler. This provides high pressure and temperature steam and is well adapted to the steam demand of industrial processes, like in refinery, sugar production, paper mills, breweries,... When the steam demand is not steady, the heat generation must be adapted by a regulation. There are 3 ways for regulating the heat produced :

- The opening or closing of the variable inlet guide vanes (VIGV) of the GT compressor, increasing or decreasing the inlet air mass flow rate and consequently the flue gas mass flow and hence its heat content. However, this regulating mode has a negative impact on the performance of the GT which operates at part load with a lower effciency than at the design point (over a range of about 20% of the nominal power).
- The use of a supplementary firing (SF). This system consists in a battery of burners fueled with gas, installed either in the GT exhaust gas or inside the heat recovery boiler itself. In this arrangement, the GT is designed on the basis of the minimum heat demand and the associated investment cost on the GT side is the lowest. The heat surplus is provided by the SF whose role is to rise the gas temperature at the boiler inlet, only to provide the additional heat demand without producing electricity. Consequently, the resulting save in fuel consumption is decreased with respect to a separate combustion (see the example in § 4). In addition, the possible conventional steam boilers existing on the site can be used in back up to cover the peaks of the heat demand.
- If the GT is designed to provide an intermediate heat demand, the 2 types of regulation may be used. However, the feasibility of this is limited by the restricted choice in the GT available off-the-shelf on the market. Consequently the selection of the regulation mode is the result of a techno-economic evaluation taking into account the investment costs on both the GT and the boiler sides, the heat demand curve, the fuel cost, the sale price of the generated kWh.

3. Calculation of the Fuel Saving Compared to Separate Production

A CHP plant is characterised by the 3 following parameters : *E/H*, the cogeneration and exergetic efficiencies η_{co} and η_{ex}.

3.1. COGENERATION ENERGY AND EXERGY EFFICIENCIES

The cogeneration efficiency is given by :

$$\eta_{CO} = \left(W + Q\right) / \left(M_f .LHV\right) = E + H = H\left[1 + E / H\right] \tag{1}$$

where $E = W/\left(\dot{M}_f.LHV\right)\%$ and $H = \dot{Q}/\left(\dot{M}_f.LHV\right)\%$ are the electricity and useful heat fractions, LHV is the low heating value (MJ/kg), $\dot{M}_f$ represents the fuel mass flow rate (kg/s), $\dot{Q}$ the useful heat power (W) given up to the process and $\dot{W}$ the net electrical power (W).

By definition, the exergetic efficiency is :

$$\eta_{ex} = \left(\dot{W}.FQ(E) + \dot{Q}.FQ(H)\right) \Big/ \left(\dot{M}_f.LHV\right) = E \times FQ(E) + H \times FQ(H) \text{ (in \%)} \quad (2)$$

where the quality factor is given by : $FQ(H) = 1 - T_o/T_{mi}$, T_{mi} being the mean integrated or the average temperature of the closed loop of the 2 flows, the steam flowing at the boiler outlet (point 1) and the water flowing at the return (point 2) in the boiler. The amount of heat given up by the hot gases to the steam/water loop is $\dot{Q} = \dot{m}\,(h_1 - h_2)$ where $\dot{m}$ is the steam and the return water mass flow rate in the closed loop. The exergy flow rate is :

$$\dot{m}\,\Delta e_{12} = \dot{m}\left[(h_1 - h_2) - T_o(s_1 - s_2)\right] = \dot{Q}\left[1 - T_o/T_{mi}\right] = \dot{Q} \times FQ \quad (3)$$

where $T_{mi} = (h_1 - h_2)/(s_1 - s_2)$.
In the particular case when the water return flow is at the reference temperature T_o, as in an open circuit, $FQ(H)$ is the usual quality factor of the steam flow at temperature T, namely : $FQ\ (H) = 1 - T_o/T$.

3.2. FUEL SAVING

The efficiency of the separate production is given by :

$$\eta_c = \left(\dot{W} + \dot{Q}\right) / \left(\dot{Q}_e + \dot{Q}_b\right) \quad (4)$$

where $\dot{Q}_e$ and $\dot{Q}_b$ represent respectively the amounts of primary energy, namely $\dot{M}_f \times LHV$, for the generation of the amounts of electricity $\dot{W}$ and of heat $\dot{Q}$ in the separate production. After some algebraic transformations, η_c writes :

$$\eta_c = \left(\frac{e}{\eta_e} + \frac{1-e}{\eta_b}\right)^{-1} \quad (5)$$

where $e = E/(E+H)$ and $1-e = H/(E+H)$ whereas $\eta_e = \dot{W}/\dot{Q}_e$ and $\eta_b = \dot{Q}/\dot{Q}_b$ are respectively the efficiencies of the reference electricity power plant (here 55%) and of the reference boiler (here 90%) in separate production.
The cogeneration is interesting compared to separate production if $\eta_{co} > \eta_c$

Fig. 1 shows a set of curves representing the cogeneration efficiency η_{CO} versus the generated electricity fraction e with the heat quality factor *FQ* as a parameter. The lower curve represents the decrease of the separate production efficiency η_C when the electricity fraction varies from 0 to 100% and provides the efficiencies for the reference cases respectively of the boiler (point A, where e=0%, η_b=90% whilst 10% of the fuel energy are the heat losses) and of the power plant (point B, where e=100%, η_e=55%, that is the efficiency of a CCGT operating on natural gas). The curve $\eta_{CO}(e)$ moves upwards in the diagram when *FQ* decreases, with A and B as fixed points. For each curve *FQ*, there is an optimum of fuel saving, that is a function of *e*. This optimum is the point of the curve the most remote from the curve η_C, hence where the difference $\eta_{CO}(FQ) - \eta_C$ is the largest for a given FQ and not the maximum of η_{CO}. When *FQ* is low, this optimum is reached for a low value of e. This region corresponds to the use of small GTs and gas engines with a recovery boiler. When *FQ* increases, that is when the process steam is at higher pressure and temperature, the optimum moves to the right towards higher electricity fractions *e*. We cross then regions where more complex systems are used like successively GT + recovery boiler + supplementary firing, then GT + recovery boiler and finally, for high values of *e*, combined cycles with steam extraction. It is also observed that the optimal zone is fairly flat showing that a rather large range of GTs size is possible without leading to big efficiency penalties. Of course, in the final choice of a configuration, a techno-economic compromise solution has to be found. This illustrates quite clearly that each design of a cogeneration plant is a ***case by case study***.

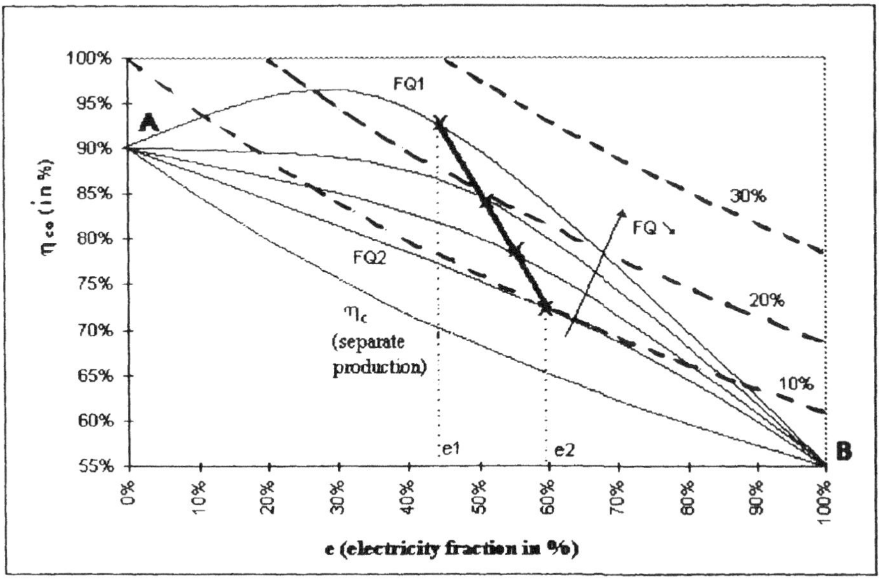

Figure 1. Cogeneration efficiency η_{CO} as a function of the electricity fraction *e*, the quality factor *FQ* being the parameter. In dotted lines, the iso fuel saving curves.

The fuel saving is the difference between the fuel (or primary energy in GJ) consumption in separate systems and in the cogeneration plant. In percentage of the fuel consumption (or heat input) in the cogeneration process ,it is given by :

$$\text{Fuel saving (\%)} = E/\eta_e + H/\eta_b - 1 \tag{6}$$

where E/η_e and H/η_b represent the relative fuel consumptions (or the heat input) in a CCGT with a η_e=55% and in a boiler with a thermal efficiency η_b=90%. On Fig. 1, the curves of iso-fuel saving are parallel to the curve η_c and tangent to the curves corresponding to a given FQ at the optimal point. For example, the curve corresponding to a fuel saving of 10% compared to the efficiency η_c of separate production with the same fuel represents all the points where the same amount of $E+H$ is obtained with only 90% of fuel consumption instead of 100% and the corresponding η_{co} is η_c/0.9. Let us consider the curve with a constant FQ_2, a saving of 10% is obtained for e_2=60%. η_c is there equal to 65% and hence η_{co} (optimum on the curve FQ_2) is 72%. A saving of 20% can only be obtained with lower values of FQ. So, for e_1=45%, η_c=70% and η_{co} =95% (optimum and not maximum of the curve with a constant FQ_1). This represents a fuel saving of [(100-70)/95]=27%. The line joining the optima for all the curves FQ (crosses on Fig.1) can be used as a first guide for the best cogeneration system. The lower the value of FQ, the lower the value of e corresponding to a maximum of η_{co} and to the optimum. In practice, the leading parameter is generally the heat demand at a given temperature and pressure (FQ is given). From the corresponding FQ curve on Fig.1, the optimal electricity generation e (%) can be derived and the fuel saving as well and the GT can be **selected to match at the best** the heat demand.

4. Applications of a Small Scale Cogeneration System

4.1. COGENERATION BASED ON THE USE OF A GAS TURBINE AND A FIRED RECOVERY BOILER

Fig.2 shows the cogeneration unit ordered by the oil company ESSO and built in partnership with the private electricity company ELECTRABEL, installed on the site of a refinery next to the harbour of Antwerpen in Belgium. The refinery asks for a steam flow at 42 bar and 385°C ranging from 70 to 170 ton/hour. The water is returned at 42 bar, 95°C. What are the energy and exergy efficiencies ? Here, the GT has been selected and is rated at 36MW with 32% efficiency (E=32%). Hence the heat power input in the combustion chamber is :

$$\dot{Q}_{in} = W / \eta_e = 36 / 0.32 = 112.5 MW \tag{7}$$

and the fuel mass flow is :

$$\dot{M}_f = \dot{Q}_{in} / LHV = 112.5 / 45 = 2.5 kg/s \tag{8}$$

the fuel being natural gas with a LHV=45MJ/kg.

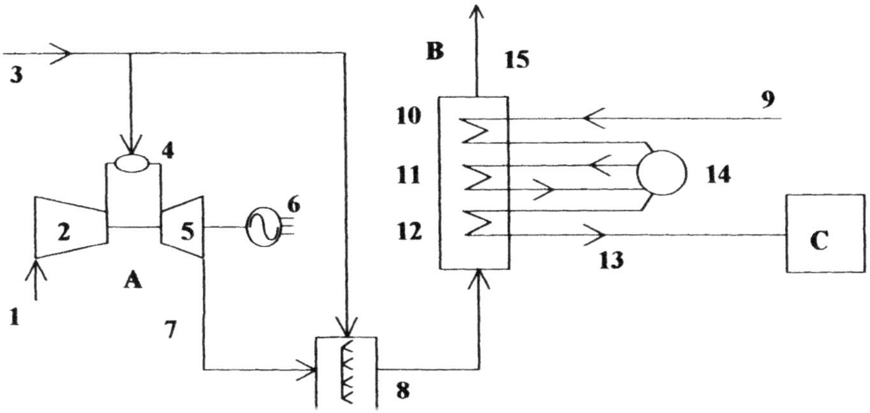

A. Gas turbine	1. Air intake	6. Generator	11. Evaporator
B. Recovery boiler	2. Compressor	7. Exhaust gases	12. Superheater
C. Oil refinery	3. Gas supply	8. Supplementary firing	13. Steam pipe
	4. Combustion chamber	9. Feed water	14. Steam drum
	5. Expander	10. Economizer	15. Stack

Figure 2. Combined Heat and Power unit (CHP) in a refinery

4.1.1. *Production of the minimum steam demand* : 70t/h (=19.44kg/s)

The enthalpies and entropies of the steam and water are :

- at 42 bar, 385°C : h_1=3173.8kJ/kg; s_1=6.69kJ/kgK
- at 42 bar, 95°C : h_2=401.1 kJ/kg; s_2=1.25kJ/kgK

The heat power given up to the refinery steam/water closed circuit from the one-pressure level unfired recovery boiler is :

$$\dot{Q}_{th} = 19.44 \times (3173.8 - 401.1) = 53914\ kW = 53.9MW \tag{9}$$

- the thermal efficiency is the heat power fraction transfered to the process/heat input in the hot source : $H = 53.9/112.5 = 48\%$
- the energy cogeneration efficiency is :

$$\eta_{co} = \left(\dot{W}_e + \dot{Q}_{th}\right)/\dot{Q}_{in} = (36 + 53.9)/112.5 = 80\%(= 32\% + 48\%) \tag{10}$$

- the exergy efficiency which is the useful energy you can extract from the fuel chemical energy taking into account the quality of the heat is given by :

$$\eta_{ex} = E \times 1 + H \times FQ \tag{11}$$

with $FQ = 1 - T_o / T_{mi} = 0.43; T_{mi} = (h_1 - h_2)/(s_1 - s_2) = 509.6K$ and $T_o = 288K$ (12)

Here the feedwater is returned at 95°C, and not at the reference temperature as it is often the case in an open circuit. This is taken into account in T_{mi} through the enthalpies and the entropies of the steam and water flows. Then one has :

$$\eta_{ex} = 0.32 \times 1 + 0.48 \times 0.43 = 52.8\% \quad (13)$$

$$E / H = 0.32 / 0.48 = 0.66 \quad (14)$$

$$\text{Fuel saving} = E / 0.55 + H / 0.9 - 1 = 0.584 + 0.53 - 1 = 11.4\% \quad (15)$$

For a sake of accuracy and correctness, a complete heat transfer balance from the gas turbine exhaust gas to the water/steam process has to be calculated taking into account the temperature of the feedwater entering the recovery boiler. The effectiveness of this heat transfer is generally less than the reference case (90%). Hence, when the feedwater comes back at a higher temperature than T_o, the fuel saving is lower.

4.1.2. *Production of the maximum steam demand* : 170t/h (=47.22kg/s)
Now the supplementary firing (SF) is needed in the GT exhaust gas. The total power given up to the refinery is : 47.22 × (3173.8 – 401.1) = 130.93MW.
From the previous point, the GT flue gas contribution is 53.9 MW ; as a result, the SF provides 77.03MW, that is 1.7kg/s of natural gas consumption. Consequently, the total heat input in the system is 112.5MW from the GT and 77.03MW from the SF, namely 189.5 MW and the total fuel consumption is 2.5 +1.7=4.2kg/s (=189.5/45). Then one has :

$$E = 36/189.5 = 19\%,\ H = 130.9/189.5 = 69\%,\ E/H = 0.27 \text{ and } \eta_{co} = 88\% \quad (16)$$

$$\text{Exergy of the heat} = m\left[(h_1 - h_2) - T_O(s_1 - s_2)\right] = 56.9MW \quad (17)$$

$$\eta_{ex} = (36 + 56.9)/189.5 = 49\% \quad (18)$$

$$\text{Fuel saving} = 0.343 + 0.77 - 1 = 11.3\% \quad (19)$$

When the SF is in operation, H and η_{co} increase from 48 to 69% and from 80 to 88% respectively whereas at the same time E decreases from 32 to 19% because the cogeneration system operates closer to a boiler. Fig.1 shows that, when e decreases from 40% (32/80) down to 21% (19/88), FQ remaining constant (0.43) since the temperatures of the process are unchanged (the SF only increases the steam mass flow), η_{co} increases by 8% points whilst the fuel saving hardly decreases by 0.1%point. This is due to the shape of the iso-fuel saving curves which are very flat around the optimum on the considered FQ curve where the design point without SF is normally located. On the other hand, η_{ex} decreases because the fuel used in the SF does not take part to the electricity production.

5. The Next Generation of Small-Scale CHP : the Fuel Cell Power Plant ?

Fuel cells are under an increasing development as *efficient* (high energy conversion efficiency), *clean* (low emissions of NO_x, SO_x, CO, unburnt hydrocarbons and no CO_2)

and *quiet* (very low noise and vibration levels) generation primemovers, for both stationary and automotive applications. Their *modular* nature provides the potential to cover a range of unit size from a few kW_e to several MW_e. They are fueled with hydrogen or a syngas ($CO+H_2$) and produce water. Four main types are under consideration for CHP applications and are characterised by their operating temperature, their electrolyte, the materials :

Phosphoric Acid Fuel Cell (PAFC) : operating temperature around 190°C, η_{co}=82% with E=40%, could operate on steam reformed natural gas (CO_2+H_2). Available on the market as small scale packaged cogeneration system in the range 50 to $200kWh_e$.

Solid Polymer Fuel Cell (SPFC) : operates at low temperature (80°C), solid polymeric membrane as electrolyte, high power density and good transient response, η_e=32 to 34%, low O&M costs. It is very well suited for the transportation sector. Available in 2000.

Molten Carbonate Fuel Cell (MCFC) : operates at 650°C, accepts a syngas ($CO+H_2$), possibility of internal reforming (η_e=55% instead of 45% with external reforming). It is suited to large scale applications (50kW and more).

Solid Oxide Fuel Cell (SOFC) : operates in the range 800 to 1000°C, with a ceramic electrolyte. When incorporated in combined cycles, η_e is 65 to 70%. Power output : a few kW_e to several MW_e. It will not be commercially available prior to 2010.

6. CHP Criteria

6.1. TEMPERATURE

The PAFC is able to supply low pressure hot water at around 90°C or low pressure steam at around 120°C. Therefore, it is already used in commercial/building related installations. The SPFC (80°C) is suited to produce low pressure hot water.

6.2. HEAT TO POWER RATIO

For an overall η_{co}=90%, η_e of MCFC and SOFC exceeds 50% giving H/E less than 1. They are not adapted to cogen applications. As to the PAFC and SPFC, their η_e is in the range 30 to 40% providing H/E around 1.8. This is applicable in the buildings sector.

6.3. CYCLIC OPERATION

SOFC, MCFC and PAFC are inherently unsuitable for stop/start cyclic operation. In conclusion, for many CHP applications, the high electric efficiency, which is usually considered as a target, has a significantly reduced advantage, unless full credits can be obtained for electricity exports. Consequently, the fuel cells cogeneration plants will probably not enter the market of large scale, industrial cogeneration sector, characterised by large H/E. The 2 low temperature fuel cell systems, PAFC and SPFC are potentially usable in the commercial and building sectors provided the heat demand is continuous (for the PAFC) and the temperature required in the host sites are low (< 80°C for a SPFC). The micro-scale market (less than 50kWe) represents a large potential market for fuel cell based CHP units, but it is far from being a certainty. In conclusion, the cost and the inability of the fuel cells to follow the heat demand transients are challenges for their future development and application in cogeneration.

7. Conclusions

The matching of the heat demand and of a GT raises the question of how to design a small scale cogeneration plant. If steam is wanted, it is recommended to use a GT and a simple recovery boiler with one-pressure level and possibly a supplementary firing. On the other hand, a gas engine is better suited to hot water production, using the heat of both the exhaust gas and the cooling circuit of the engine. In a project of partnership with an electricity distributor, the *MINIMUM* heat demand on the site is the basis of the design, namely the steam mass flow rate and the exhaust gas temperature. The next stage is to design a one-pressure boiler and carry out a calculation of the heat transfer from the gas to the water/steam circuit. The data are :

On the water/steam side : the feedwater temperature, the steam pressure and temperature, the MINIMUM steam mass flow rate, the recovery boiler type (1 pressure level), the approach temperature.

On the GT side : the power output, the open cycle efficiency, the pressure ratio, the turbine inlet temperature (TIT), the exhaust gas temperature and mass flow rate. From heat balances on the heat transfers above and under the pinch point, the stack temperature and the pinch point difference temperature are derived. Now the heat transfer surfaces can be calculated, namely of the economizer, the evaporator and of the superheater. Consequently, the cost of the boiler can be estimated. It has now to be checked, from the global heat balance, whether or not the selected GT is fitting the required steam mass flow $\dot{M}_{st}$. The same procedure is remade with different available GTs and the one giving the closest $\dot{M}_{st}$ to the required one is selected. If the $\dot{M}_{st}$ provided by this GT is higher than the required value, then the GT has to be operated at partial load by closing the variable inlet guide vanes (VIGV) accordingly.

Roughly speaking, $\dot{M}_{st}$ and $\dot{M}_{air}$ are more or less proportional. For example, if the ratio of the calculated and required steam mass flows is 0.8, the corresponding ratio of air mass flows at the compressor inlet is also roughly 0.8 and the VIGVs have to be opened at 80% of the full opening. Of course, running at partial load, both the power output and efficiency of the GT decrease, increasing consequently the fuel consumption and decreasing the fuel saving compared to a separate production. However, the range of action of the VIVGs is limited, typically to a reduction of the air mass flow by 20% from its design value when they are fully closed. Consequently, the appropriate GT has to be found. In the opposite situation, when the $\dot{M}_{st}$ provided by the GT is lower than the minimum required value, a supplementary firing has to be installed and designed to follow the heat demand.

Amongst the different options obtained according to this procedure, one selects the most flexible one (generally a GT with supplementary firing) giving the highest E and η_{co} for a given H.

GAS DYNAMICS CYCLES OF THERMAL AND REFRIGERATING MACHINES

A.I. LEONTIEV
Academician of the Russian Academy of Sciences, Professor
Moscow State Technical University named after N.E. Bauman
2-nd Baumanskaya, Moscow 107005 Russia

1. Introduction

On 26 July 1824 year, at a session of section on general physics of French Academy of sciences, the academician P. Gerard has stated the contents of the book of the young French engineer Sadie Carnot "Reflections about driving force of a fire and machines, capable to develop this force". On this session there were practically all scientific stars of France: Navier, Laplace, Fourier, Gay-Lussac, Poisson and other, which, unfortunately, could not estimate ingenious ideas, incorporated in work Sadie Carnot. The scientific and engineering world of France has not displayed in any form of interest to work S. Carnot. And only through a few tens of years, due to work Clapeyron, Thomson, Clausius it has become clear, that Carnot has for the first time formulated the basic laws of thermodynamics and under the right it is considered founder of this classical discipline. The great physicist Thomson (lord Kelvin) has named the book Carnot "as the largest achievement of a science XIX of century".

S. Carnot has for the first time entered concept about convertible thermal processes, that has allowed him to offer and to analyse a cycle of the ideal thermal machine of a maximum (peak) efficiency.

On *Figure 1* the Carnot cycles are represented in *T-S* and *P-V* coordinates. S. Carnot has shown, what the convertible cycle of the ideal thermal machine, consisting from two isotherms with heat addition and heat rejection (1-2 and 3-4) and two adiabats of expansion and compression (2-3 and 4-1) is the most effective of all possible and efficiency of such cycle depends only on the relation of the maximum and minimum temperatures ($\eta_C = 1 - T_2 / T_1$). The similar conclusions were made and for a return cycle of the refrigerating machine.

2. Analysis of Offered Ideal Cycles

As is known, one of lacks of a Carnot cycle is a significant difference of specific volumes of a working body, that results in the large size of a flowing part of blade machines or length of the cylinder in a piston machines. This lack can be removed by replacement of adiabatic processes of expansion and compression on isochoric (cycle of Stirling) or isobaric processes (cycle of Ericsson) with intermediate regeneration of a

A. Bejan and E. Mamut (eds.), Thermodynamic Optimization of Complex Energy Systems, 271–278.

heat. In this case, at identical with a Carnot cycle profitability, it is possible essentially to reduce the sizes of a working part of the thermal machine.

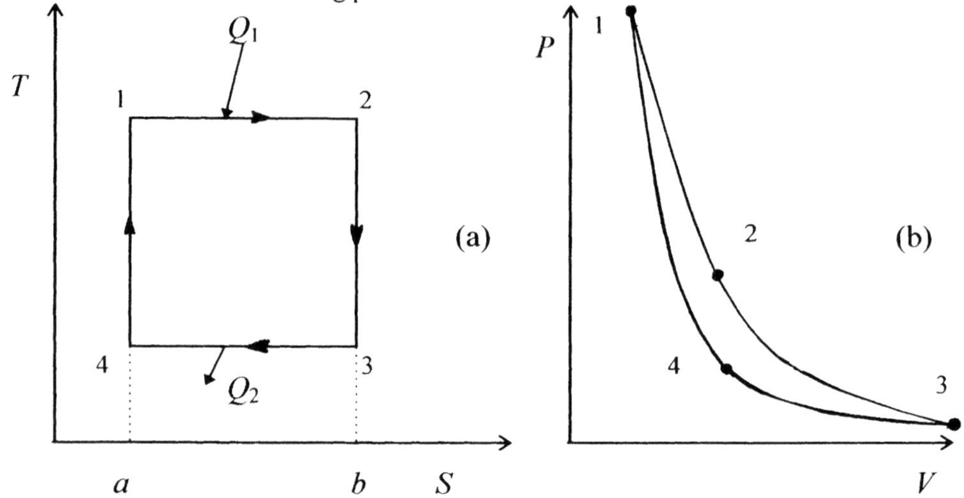

Figure 1. Carnot Cycle in *T-S* and *P-V* coordinates.

In the proposed *Figure 1* a gas dynamics cycles of the thermal and refrigerating machine idea of internal regeneration of a heat is in essence used.

2.1. CYCLE OF THE THERMAL MACHINE

We shall consider a cycle of the ideal thermal machine in *T-S* coordinates (*Figure 2, a*).

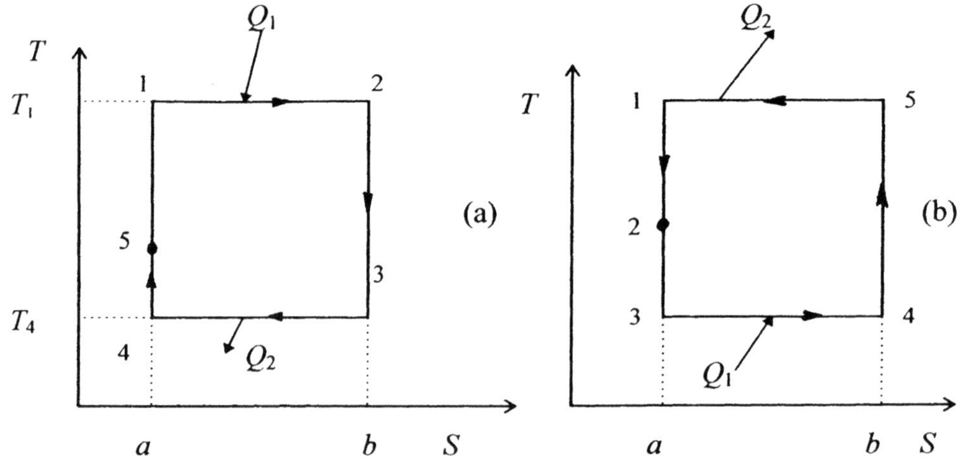

Figure 2. Gas dynamics cycles thermal (*a*) and refrigerating (*b*) machines

During (1-2) comes true isothermal addition of a heat to a working body and the whole added heat Q_1 will be transformed to useful work. Further, as against a Carnot cycle, we carry out adiabatic expansion of gas in supersonic nozzle from pressure P_2 up to pressure P_3. During 3-4 isothermal rejection of a heat Q_2 from a supersonic gas flow comes true. The isothermal rejection of a heat can be carried out changing the area of the channel on length. In this case, as it follows from the theory ideal thermal nozzle, there will be braking a flow of gas with growth of static and full pressure. During (4-5) comes true adiabatic braking of a supersonic flow in diffuser up to pressure P_5. Stagnation temperature in a point 5 (T_5^*) will be less than stagnation temperature in a point 1 (T_1^*) as during (3-4) we have rejected from gas quantity of a heat, determined by the area *a*-4-3-*b*-*a*. Hence, to close a cycle, it is necessary from a point 5 up to a point 1 compress gas in the compressor, by spending capacity

$$L_C = C_p\left(T_1^* - T_5^*\right) = Q_2. \tag{1}$$

Thus, efficiency of a considered convertible ideal cycle will be equal to efficiency of a Carnot cycle:

$$\eta_C = \frac{Q_1 - Q_2}{Q_1} = 1 - \frac{T_3}{T_2}. \tag{2}$$

As it is visible from *Figure 2, a* the sizes of the thermal machine, and also the sizes of the compressor in an offered cycle turn out much less, than in a Carnot cycle, at the same efficiency. As it can see from *Figure 2, a* and with increase of a degree of expansion in nozzle, i.e. with reduction of temperature T_3, the work, spent on the compressor, decreases.

On *Figure 3, a* is shown ideal gas dynamics cycle of the thermal machine without the compressor. After isothermal expansion of a working body in the cylinder (or blade of the gas turbine) (1-2), where the whole added heat Q_1 will be transformed to useful

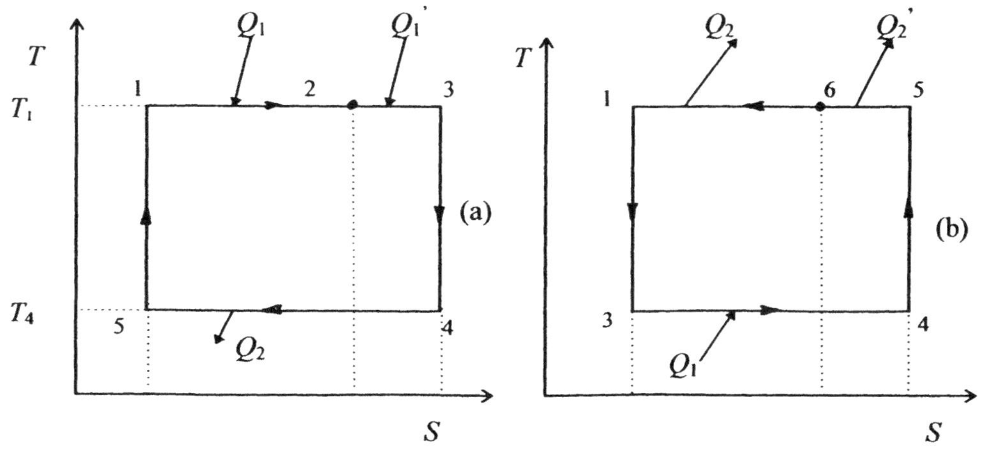

Figure 3. Gas dynamics cycles thermal (*a*) and refrigerating (*b*) machines without compressor and expander

work, the gas acts in the subsonic channel, where comes true isothermal addition of a heat Q_1'. In this case, according to the theory thermal nozzle, static pressure and stagnation pressure fall, and stagnation temperature is increased by Q_1'/C_p. The gas is accelerated up to speed of a sound in a point 3 and further extends on adiabat (3-4) in supersonic nozzle. The rejection of a heat from a supersonic gas flow comes true in isothermal process 4-5. For closing of a cycle without the compressor it is necessary to maintain a condition $Q_2 = Q_1'$. Then stagnation temperature of gas in a point 5 will be equal to temperature a point 1 and the cycle becomes closed by adiabat of compression (5-1) in supersonic diffuser. It is possible to show, that stagnation pressure in a point 5 under accepted conditions will be equal stagnation pressure in a point 1.

In isothermal process 1-2 available work is equal

$$L = \int_1^2 \frac{dP}{\rho} = RT_1^* \ln \frac{P_2}{P_1}. \tag{3}$$

As $P_1 = P_1^*$ and $P_2 = P_2^*$ and $L = Q_1'$: $\frac{P_2^*}{P_1^*} = 1/\exp(Q_1'/RT_1^*)$.

During process 2-3

$$\frac{\Delta S_{2-3}}{R} = -\ln\frac{P_3^*}{P_2^*} + \frac{K}{K-1}\ln\frac{T_3^*}{T_1^*}. \tag{4}$$

From here: $\ln\frac{P_3^*}{P_2^*} = -\frac{\Delta S_{2-3}}{R} + \frac{K}{K-1}\ln\frac{T_3^*}{T_1^*}$.

On the other hand for process 4-5 we have $\ln\frac{P_5^*}{P_3^*} = \frac{\Delta S_{4-5}}{R} + \frac{K}{K-1}\ln\frac{T_5^*}{T_3^*}$.

As $T_5^* = T_1^*$, and $\Delta S_{4-5} = \Delta S_{1-2} + \Delta S_{2-3}$,

$$\frac{P_5^*}{P_2^*} = \exp(\Delta S_{1-2}/R) = \exp(Q_1'/RT_1^*). \tag{5}$$

Hence $P_5^* = P_1^*$.

2.2. CYCLE OF THE REFRIGERATING MACHINE

Similarly it is possible to present gas dynamics cycle of the ideal refrigerating machine (*Figure 2, b*). The cycle begins with adiabatic expansion of gas in the thermal machine (1-2) with fulfilment of work. In process (2-3) of adiabatic expansion of gas in supersonic nozzle there is the fall of static pressure and static temperature at the constant stagnation parameters (T^* and P^*). In isothermal process (3-4) comes true addition of a heat, in result stagnation temperature (up to initial meaning $T_4^* = T_1^*$) is increased, for this purpose it is necessary, that $Q_1 / C_p = T_1^* - T_2^*$, and stagnation pressure $P_4^* <$ P_3^* falls. The process of adiabatic braking of gas (4-5) comes true in supersonic ideal diffuser. The cycle becomes closed by isothermal compression of gas (6-1) with rejection of a heat Q_2 and expense of capacity on the compressor, determined area (1-*a*-*b*-5-1). The part of this capacity takes from the turbine.

As it is visible from *Figure 2, b* refrigerating factor of an offered ideal cycle is equal to refrigerating factor of a Carnot cycle:

$$\varepsilon = Q_1 / (Q_2 - Q_1) = 1 / (T_1 / T_3 - 1). \qquad (6)$$

By analogy to a cycle of the thermal machine it is possible to offer a cycle of the refrigerating machine without expander (*Figure3, b*).

In this case, after adiabatic expansion of gas in nozzle (1-2), isothermal addition of a heat (2-3) and adiabatic compression of gas in diffuser (3-4) isothermal rejection of a heat in the subsonic channel (4-5) in quantity $Q_2^{'} = Q_1$ comes true. The cycle becomes closed by isothermal compression of gas in the compressor (5-1), with rejection of a heat Q_2. It is visible, that refrigerating factor of such cycle is equal to refrigerating factor of a Carnot cycle:

$$\varepsilon = \frac{Q_1}{\left(Q_2^{'} + Q_2^{''}\right) - Q_1} = \frac{T_2\left(\Delta S_{2-3}\right)}{T_1 \Delta S_{2-3} - T_2\left(\Delta S_{2-3}\right)} = 1 / (T_1 / T_2 - 1). \qquad (7)$$

From *Figure 3, b* follows, that in an offered cycle the sizes of the compressor of a refrigerating machinery in comparison with a Carnot cycle are much reduced and is eliminated expander. Does not represent complexities to carry out all thermodynamic calculations offered ideal gas dynamics cycles, using the theory ideal thermal nozzle.

In an ideal cycle of the thermal engine the processes (2-3) and (4-5) proceed without irreversible losses of energy because of work of friction forces. Actually dissipation effects render essential influence to gas flow, especially at supersonic speeds.

It is known, that at supersonic speeds of gas flow temperature thermal isolated wall (equilibrium temperature) is defined (determined) from the formula:

$$T_W^* = T_o + r \frac{W_o^2}{2C_p} = T_o\left(1 + r \cdot \frac{k-1}{2} \mathrm{M}^2\right), \qquad (8)$$

where r - recovery factor, in a general case dependent from a longitudinal gradient of pressure and cross flow of substance. For a flow of a flat impenetrable plate $r = \mathrm{Pr}^{1/2}$ - for a laminar boundary layer and $r = \mathrm{Pr}^{1/3}$ - for a turbulent boundary layer.

For the majority of gases the criterion Prandtl (Pr) is close to unit and equilibrium temperature poorly differs from temperature of adiabatic braking

$$T_o^* = T_o\left(1 - \frac{k-1}{2}\mathrm{M}_o^2\right). \tag{9}$$

It is known, that the Reynolds analogy, establishing connection between heat transfer coefficient and friction coefficient, is fair and for supersonic speeds of gas flow, if heat transfer coefficient to define as

$$\alpha = q_W / \left(T_W^* - T_W\right). \tag{10}$$

3. Analysis of Offered Real Cycles

Thus, in a general case of rejection of a heat in process (4-5) or addition of a heat in process (2-3) (*Figure 3, b*) temperature of a wall will differ from thermodynamic temperature of gas, that will result in irreversible losses of energy in comparison with ideal convertible processes. For reduction of these irreversible losses it is necessary to reduce a difference of temperatures between gas and environment. It is possible in a limiting cycle at the decreasing of recovery factors. As already was marked, by use as a working body of gases, a recovery factor poorly differs from unit, however for some mixes of gases (He-Xe or H_2 - Xe) the Prandtl number, and consequently and r, it can be essential less unit.

Other opportunity of reduction of recovery factor as the result of creation of a positive gradient of pressure along a flow of gas. As is known, [3] in area of separation of a boundary layer, at a cross flow of the cylinder very low, even negative meanings of recovery factors were measured. During (4-5) supersonic gas flow comes true with a positive gradient of pressure and it is possible to expect favourable influence of this gradient to recovery factor. However it is necessary to note, that in a cycle of the refrigerating machine the process of addition of a heat (2-3, *Figure 3, b*) comes true with a negative gradient of pressure. In work [1,2] influence of a cross flow of substance to recovery factor is shown, however use of this effect at realisation of thermodynamics cycles is represented inconvenient. The interesting opportunity of reduction of recovery factor opens at organisation of jump of condensation on an exit from supersonic nozzle. In this case the condensate can drop out on a wall of the channel and recovery factor aspires to zero, since temperature of liquid drops is close to thermodynamic temperature of gas. Especially perspective use of this effect in process (2-3, *Figure 3, b*) cycle of the refrigerating machine is represented.

The basic question, arising at practical realisation of offered cycles, consists in an opportunity of restoration of stagnation pressure in cooled channels (process (4-5, *Figure 3, a*) in a cycle of the thermal machine and process (4-5, *Figure 3, b*) in a cycle of the refrigerating machine). In work Gukhman [4] was shown, that at flow of ideal gas on a cylindrical pipe it is impossible to realise thermal supersonic nozzle, if the heat transfer comes true through a wall of a pipe.

From generalisation of Reynolds analogy follows

$$\frac{2\mathrm{St}}{C_f} = \frac{f}{\mathrm{Pr}^{2/3}}, \tag{11}$$

where the function f takes into account influence of a longitudinal gradient of pressure to Reynolds analogy.

For isothermal flow of gas in the cooled channel

$$\frac{\Delta S_R}{\Delta S_{FR}} = -\frac{C_p \cdot f}{W_o^2\,\mathrm{Pr}^{2/3}}\left(T_W^* - T_W\right). \tag{12}$$

For a case interesting us $T_W = T_0$ and from the equation (12) we receive

$$\frac{\Delta S_R}{\Delta S_{FR}} = -\frac{r \cdot f}{2\,\mathrm{Pr}^{2/3}}. \tag{13}$$

From the formula (13) follows, that at $f=1$ and $\mathrm{Pr}=1$, $\Delta S_R = -0.5\Delta S_{FR}$, i.e. it is impossible to realise process of cooling of gas in the channel with reduction of entropy, since always $\Delta S_{FR} > \Delta S_R$.

The change of stagnation pressure of gas is connected to change of entropy by the formula

$$\frac{dP^*}{P^*} = \frac{k}{k-1}\cdot\frac{dT^*}{T^*} - \frac{dS}{R}. \tag{14}$$

As

$$dS = dS_{FR} + dS_R = dS_R\left(1 - \frac{2\,\mathrm{Pr}^{2/3}}{r \cdot f}\right), \tag{15}$$

and we have

$$dS_R = \frac{dQ_R}{T_o} = \frac{C_p dT^*}{T_o}, \tag{16}$$

hence

$$\frac{dP^*}{P^*} = \frac{k}{k-1}\cdot\frac{dT^*}{T_o}\left[\frac{T_o}{T^*} - 1 + \frac{2\,\mathrm{Pr}^{2/3}}{r \cdot f}\right] \tag{17}$$

or

$$\frac{dP^*}{P^*} = \frac{k}{k-1}\cdot\frac{dT^*}{T_o}\left[\frac{2\,\mathrm{Pr}^{2/3}}{r \cdot f} - \frac{k-1}{k+1}\lambda^2\right]. \tag{18}$$

For the cooled channel $dT^* < 0$, hence a condition of increase of stagnation pressure ($dP^* > 0$) is following

$$\frac{2\,\mathrm{Pr}^{2/3}}{r \cdot f} < \frac{k-1}{k+1}\lambda^2. \tag{19}$$

From the formula (19) it is visible, that at Pr =1 and f=1 this condition is impracticable, as the maximum meaning of the right part of an inequality (19) $\lambda^2(k-1)/(k+1)=1$. However if to take into account influence of Prandtl number and longitudinal gradient of pressure on Reynolds analogy and recovery factor, the inequality (19) can be maintained, that will result in growth of stagnation pressure in the cooled channel. In work [5] it is underlined an opportunity of cooling of gas by injection of water. In this case the Reynolds analogy is not restriction for growth of stagnation pressure, though there are the additional losses, connected to interaction of a gas flow to drops. Resulted in work [5] calculations of parameters of a supersonic gas flow, cooled by an injection of a liquid, specify essential increase of stagnation pressure (up to 40 %) at the rather small additives of water. The essential irreversible losses can take place at braking a flow in supersonic diffuser. It is known [6], that in this case practically it is impossible to carry out braking of a flow without shocks.

The analysis of cycles of thermal machines, realising a cycle offered is brought in work [7].

4. Conclusion

In this paper ideal cycles of thermal and refrigerating machines are analysed, in which for closing are used gas dynamics processes of gas flow in channels with heat addition and heat rejection . Is shown, that for ideal processes the profitability of such cycles is equal to profitability of Carnot cycles. At the same time basic defect of Carnot cycles - large degrees of expansion in the turbine is eliminated. Moreover in an ideal cycle the process of compression of gas comes true in the cooled channel.

There are open the questions of start of machines, working on these cycles, and also problem of adjustment of capacity. It is a lot of unsolved problems, connected to use of phase of transitions (evaporation and condensation).

However for practical realisation of offered cycles more thorough theoretical and experimental researches is necessary.

References

1. Kutateladze, S.S., Leontiev, A.I. (1990) *Heat Transfer, Mass Transfer and Friction in Turbulent Boundary Layuers,* Hemisphere Publishing Corporation, New-York.
2. Hartnett, J.P. (1985) *Mass Transfer Cooling,* Handbook of Heat Transfer Application, McGraw-Hill Book Company, New-York.
3. Eckert E, Weise W. (1942) *Forsch. Gebiete Ingenieurw,* **13**, 246 (in German)
4. Gukhman A.A. (1939) To Theory of Limited States of Gas Flow, *Journal Tekhnicheskoy Fisiki,* **9,** №6. (in Russian).
5. Vulis L.A. (1950) *Termodynamics of Gas Flow,* Gosenergoizdat, Moscow (in Russian).
6. Abramovich G. N. (1976) *Applied Gas Dynamics*, Nauka, Moscow (in Russian).
7. Leontiev A.I., Shmidt K.L. (1997) Ideal Cycle of Closed Gas Turbine Installation without Compressor, *Izvestiya Akademie Nauk, Seria Energetica*, **3**, 132-140 (in Russian).

ENERGY SAVING TECHNIQUES IN DISTILLATION
Thermodynamic Efficiency and Energy Conservation

Z. FONYO and E. REV
Technical University of Budapest,
Department of Chemical Engineering,
H-1521 Budapest, Hungary

1. Introduction

When considering a separation process there are two entirely different interpretations of the word „efficiency". The first refers to product purity (i.e. separation efficiency) and can be regarded as a pointer to the theoretical possible improvements of separation in the given process, considering the separation unit as a set of theoretical (equilibrium) stages. The second refers to energy consumption (i.e. thermodynamic efficiency) and can be regarded an indicator to possible reductions of energy consumption in the industrial separation. The first aim of the present paper is to interpret the thermodynamic efficiency theoretically and expound a comprehensive view on heat degradation, losses and the efficiency of thermodynamic coupled processes. After the theoretical foundation, the various representations of thermodynamic efficiency (TE) are examined by application to several examples.

Various alternative definitions of thermodynamic efficiency were previously suggested in literature and recommended in precedence to others. However, these definitions are open to different interpretations as far as separation is concerned [1]. In addition, most seem to share the features that different phenomena such as heat losses, heat degradation and coupling of the heat and component transports are not treated separately in the theoretical explanation and the different (energy, entropy and exergy) representations are not distinguished unambiguously at the practical investigations.

The industrial separation processes (such as distillation or gas absorption with stripping or extraction with solvent recovery) can be treated as steady system of the first order, where the coupling of heat and component transports can be regarded as the essence of the processes. By separating the entropy consuming and entropy producing processes, the coupling thermodynamic

A. Bejan and E. Mamut (eds.), Thermodynamic Optimization of Complex Energy Systems, 279–296.

efficiency can be defined, which expresses the effectiveness of the coupling. From thermodynamic point of view, in the case of separation our goal is to utilize the entropy produced by heat transport as completely as possible by the entropy consumption of component separation. In addition to this coupling thermodynamic efficiency (CTE), so called overall thermodynamic efficiency (OTE) was also defined taking the additional losses and entropy producing effects into account, inherent with all thermal processes.

The uniform energetic interpretation of distillation and other separations based on the thermodynamics will be of assistance, in the evaluation and practical implementation of energy conservation techniques of these processes.

2. Separation Processes as First Order Steady Systems

Although the linearity of the integral expressions for the transfer processes is questionable in the irreversible thermodynamics, the linearity of the phenomenological equations can be assumed and proved to be appropriate in the mathematical treatment. The entropy production Φ of a system in this theory is given by:

$$\Phi=\sum_k J_k X_k \geq 0 \tag{1}$$

where J_k is the thermodynamic flux k, and X_k is the conjugate thermodynamic force or driving force. If each flux depends only on the respecting conjugate driving force, i.e.:

$$J_k=L_{kk} X_k\,,\ (k=1,2,...) \tag{2}$$

only diagonal diagonal L_{kk} phenomenological coefficients occur in the expression (1):

$$\Phi=\sum_k J_k X_k=\sum_k L_{kk} X_k^2=\sum_k \Phi_k\,. \tag{3}$$

In this case each process occurs spontaneously and independently inherent with certain entropy production.

In general each flux depends on all driving forces and so cross-coefficients also occur in the entropy production:

$$\Phi=\sum_k \sum_n L_{nk} X_n X_k=\sum_k \Phi_k\,, \tag{4}$$

thus entropy consuming processes can also proceed in addition to entropy producing processes. When considering, e.g. two fluxes the entropy production is given by:

$$\Phi=\Phi_1+\Phi_2=J_1 X_1+J_2 X_2=L_{11} X_1^2+(L_{12}+L_{21}) X_1 X_2+L_{22} X_2^2\,. \tag{5}$$

In this case the positive sign of each Φ_k is not required, the only restriction may be formulated as:

$$L_{11}X_1^2+(L_{12}+L_{21})X_1X_2+L_{22}X_2^2>0\ , \tag{6}$$

an one of the two fluxes can be an entropy consuming one forced by the other flux owing to cross-effects. Such a process is called a *coupled process* thermodynamically when entropy consuming subprocess utilizes the entropy production of the other subprocess to a certain extent.

Every system if left to itself tends to change toward a final state of rest or equilibrium. In the equilibrium state, the entropy production is zero, because macroscopic fluxes do not exist. In a system left to itself, the driving forces and drived fluxes decrease gradually and the system strives after zero entropy production.

In steady (or stationary) states, the state parameters are independent of time, but the entropy production is not zero, because finite driving forces and fluxes do exist. Stationary non-equilibrium states have the important property that under certain conditions (of which the most important are the supposition of constancy of phenomenological coefficients and the validity of the Onsager reciprocal relations), they are characterized by *a minimum of entropy production*, compatible with the external constraints imposed on the system. This principle is called the Glansdorf-Prigogine theorem in literature [2,3].

According to the number of external constraints imposed on the system, the steady systems with coupled processes can be categorized as follows [4]:

- *zero order*, if not a single variable is fixed, all properties are time independent and the system is in a state of rest or state of equilibrium;
- *first order*, if one single driving force fixes one property (e.g. temperature and composition) to a constant value;
- *second order*, if two driving forces fix two properties (e.g. temperature and composition) to constant values; etc.

The criterion of minimum entropy production for a first order steady system with two fluxes, when X_1 is fixed, can be formulated as:

$$\frac{\partial\Phi}{\partial X_2}=(L_{12}+L_{21})X_1+2L_{22}X_2=0 \tag{7}$$

From this expression it follows that:

$$X_2=-\frac{L_{12}+L_{21}}{2L_{22}}X_1 \tag{8}$$

and substituting this formula for X_2 into (5), the entropy production may be rewritten in the form:

$$\Phi = L_{11} X_1^2 \left[1 - \frac{(L_{12}+L_{21})^2}{4L_{11}L_{22}} \right] = L_{11} X_1^2 (1-p^2) \tag{9}$$

where p is the so called *coupling factor*, i.e. the measure of the coupling between entropy producing and entropy consuming processes.

In the case of thermal separation processes, the heat flow is flux 1, maintained by an X_1 driving force of constant value due to an external constraint, and flux 2 is the resulting component flow. Thus during the separation process, one part of the entropy produced by heat transport can be utilized by the entropy consuming process of the component separation. The range of the coupling factor is:

$$0 \leq p \leq 1 \tag{10}$$

It is clear that at $p = 0$ the system is completely irreversible, since $\Phi = L_{11} X_1^2$ i.e. only an entropy producing process takes place (coupling does not exist). If $p = 1$, $\Phi = 0$ i.e. the system is reversible, the entropy produced by heat flow is completely utilized by the entropy consuming component separation (total is completely utilized by the entropy consuming component separation (total coupling exists).

3. Coupling Thermodynamic Efficiency

In general it can be said that the entropy production may be calculated for the first order steady systems as the difference of the diagonal elements, without considering the off-diagonal elements:

$$\Phi = L_{11} X_1^2 - L_{22} X_2^2 \quad . \tag{11}$$

In the case of separation, the first term would be the entropy production of the compatible independent heat flux and the second term would be the entropy consumption of the compatible independent component flux. The ratio of these consumed and produced quantities of entropy was defined as thermodynamic efficiency η by Odum and Pinkerton [5]:

$$\eta_s^c = \frac{L_{22} X_2^2}{L_{11} X_1^2} \tag{12}$$

(Index S refers to entropy representation; superscript C refers to coupling.) It may be remarked that this kind of efficiency is equal to the square of the coupling factor defined by Kedem and Caplan [6], i.e. using definition (12) and Eq.(8) we obtain:

$$\eta_s^c = \frac{L_{22} X_2^2}{L_{11} X_1^2} = \frac{L_{22}(L_{12}+L_{21})^2 X_1^2}{L_{11} X_1^2 4L_{22}^2} = \frac{(L_{12}+L_{21})^2}{4L_{11} L_{22}} = p^2 \tag{13}$$

This ratio can be called the entropy representation of the coupling thermodynamic efficiency (CTE) and serves as a measure of coupling between heat and component transports or a measure of the utilization of the entropy produced. The efficiency reflects the essence of the process, which is the coupling, but does not take the other entropy producing effects into account. Since the measure of coupling is expressed by the ratio of utilized and produced entropy, this CTE value is defined here in entropy representation. Later on, energy and exergy representation of TE will also be distinguished.

4. Illustrative Separation Processes

Some of the most common separation processes can be interpreted as first order steady systems. This interpretation is rather plausible for the processes where heat is used as the separating agent (e.g. distillation), but the comprehension is somewhat more complicated for the processes when the separating agent is a stream of mass (e.g. sorption and extraction). The later ones should be considered with the recovery systems of the separating agent together. The energy supplied to a separation process may be in the form of thermal, mechanical, electrical or chemical energy, depending upon the type of process.

4.1. DISTILLATION

An ordinary distillation column is a good example of a separation process driven by heat input. From an energy point of view our goal is to utilize the entropy produced by heat transport as completely as possible by the entropy consumption of the component separation.

The minimum isothermal „work of separation" ($W_{min,T}$) represents a lower bound on energy that must be consumed by a separation process [7]. For a multicomponent liquid mixture being separated isothermally into two products we have:

$$W_{\min,T} = -RT_0\left(D\sum_i x_{iD} \ln\gamma_{iD} x_{iD} + W\sum_i x_{iw} \ln\gamma_i x_{iW} - F\sum_i x_{iF} \ln\gamma_i x_{iF} \right) \tag{14}$$

which corresponds to the utilized entropy consumption of the compatible independent component flux in the numerator of the CTE:

$$L_{22} X_2^2 = -R\left(D\sum_i x_{iD} \ln\gamma_{iD} x_{iD} + W\sum_i x_{iw} \ln\gamma_i x_{iF} - F\sum_i x_{iF} \ln\gamma_i x_{iF} \right) \tag{15}$$

(R is the gas constant; T is temperature; index o refers to ambient state; F, D, and W are feed, distillate, and weight product flow rates, respectively; these letter as indices refer to these streams, x is component mole fraction, γ is activity coefficient.)

The „net work consumption" W_n of the process is defined as the difference between the work that could have been obtained from the heat Q entering the system and the work that could be obtained from the heat leaving the system. If no mechanical work is involved and the enthalpy difference between products and feed is negligible compared to the heat input, the net work consumption is:

$$W_n = QT_0\left(\frac{1}{T_D} - \frac{1}{T_W}\right) \tag{16}$$

which corresponds to the entropy production of the compatible independent heat flux in the denominator of the CTE:

$$L_{11}X_1^2 = Q\left(\frac{1}{T_D} - \frac{1}{T_W}\right) . \tag{17}$$

Thus the expression of CTE may be given as:

$$\eta_s^c = \frac{-R\left(D\sum_i x_{iD}\ln\gamma_i x_{iD} + W\sum_i x_{iw}\ln\gamma_i x_{iw} - F\sum_i x_{iF}\ln\gamma_i x_{iF}\right)}{Q\left(\frac{1}{T_D} - \frac{1}{T_W}\right)} \tag{18}$$

when distillation is treated as a first order steady system.

4.2. ABSORPTION AND STRIPPING

In simple gas absorption, a solvent is used to separate a soluble from an insoluble gas by passing the mixture through a multistage contractor countercurrent to a solvent. Accordingly, an absorption column is a typical example of a separation process, where a stream of mass is used as the separating agent instead of heat [8]. In such cases, however, the stripping, the recovery of the separating agent, should also be considered as part of the system.

Stripping is the inverse of absorption and involves the removal of a volatile component from a liquid into a gas phase. The energy supplied to the stripper may be in the form of thermal or mechanical energy, but generally heat is used. Consequently, the absorber-stripper combination can also be considered an example of a separation process driven by heat input.

The minimum isothermal work of separation and the net work consumption of the process can be defined and interpreted identically with distillation. Thus the coupling thermodynamic efficiency can be defined as follows:

$$\eta_s^c = \frac{-R\left(G\sum_i x_{iG}\ln\gamma_{iG} + V\sum_i x_{iV}\ln\gamma_i x_{iV} - F\sum_i x_{iF}\ln\gamma_i x_{iF}\right)}{Q\left(\frac{1}{T_T}-\frac{1}{T_B}\right)} \tag{19}$$

when the absorption-stripper combination is considered as a first order steady system.

4.3. EXTRACTION AND SOLVENT REMOVAL

Liquid-liquid extraction also employs liquid separating agent or solvent (s), but this is used to separate a liquid instead of a gaseous mixture. Solvent extraction processes may be classified as broadly single and double solvent operations, according to the number of solvents used to effect the separation. In the case of liquid-liquid extraction, such as in gas absorption, the first order steady system includes the solvent removal units, in addition to the extractor. In solvent removal units, heat is generally used to recover the solvents.

The minimum isothermal work of separation is the same as previously, but the net work consumption is the sum of the work consumed in two solvent removal units. Accordingly the expression of CTE may be formulated as (*A* and *B* are the extructs):

$$\eta_s^c = \frac{R\left(A\sum_i x_{iA}\ln\gamma_i x_{iA} + B\sum_i x_{iB}\ln\gamma_i x_{iB} - F\sum_i x_{iF}\ln\gamma_i x_{iF}\right)}{Q_1\left(\frac{1}{T_2}-\frac{1}{T_1}\right)+Q_2\left(\frac{1}{T_4}-\frac{1}{T_3}\right)} \tag{20}$$

when the extraction-solvent removal combination is treated as a first order steady system.

4.4. OTHER SEPARATIONS

The previous two examples are suitable to demonstrate that the separating agent performs a function similar to that of heat in the case of distillation and the recovery of the separating agent requires energy supplied to the system. Certain further separation processes employing a liquid separating agent, such as solid-liquid extraction or extractive and azeotropic distillation, and other ones employing solid separating agent, such as gas adsorption and ion exchange, can all be interpreted as first order steady systems in a thermodynamic sense. In certain processes, however, the energy required by the removal units of the separating agents should be supplied to the process in other forms than heat. E.g. the production of reflux in the ion exchange process is effected by means of chemical energy or in vapor recompression distillation a heat pump is used, requiring mechanical energy, to transfer the heat removed from the condenser to the reboiler by raising its temperature level. Mechanical energy is also used in the case of heatless adsorption, with regeneration accomplished by means of frequent lowering of the pressure on the adsorbent beds.

Although the calculation of the minimum isothermal work of separation is similar in all cases, the calculation of the net work consumption should be replaced with appropriate formulas when using other forms of energy than heat.

5. Heat Losses and Heat Degradation of Uncoupled Process

In any uncoupled stationary process without losses, the energy added to the system is equal to the energy withdrawn from the system. In the case of any kind of losses, the efficiency of energy conservation in a system can be measured by:

$$\nu_E = \frac{\sum E_{out}}{\sum E_{in}} \tag{21}$$

where $\sum E_{in}$ is the sum of energy added to, and $\sum E_{out}$ is the sum of energy withdrawn from the system. This kind of efficiency is defined on the First Law basis and may be called *First Law efficiency*.

However, this definition of efficiency does not take into account the temperature at which heat is added to or withdrawn from the system, so the efficiency on the Second Law basis has to be defined additionally. The maximum portion of heat, which can be converted into other kinds of energy, can be determined by the Carnot efficiency:

$$\eta_{Ct} = \frac{T_2 - T_1}{T_2} = 1 - \frac{T_1}{T_2} \tag{22}$$

when heat enters the system at a temperature T_2 and leaves at a temperature T_1. Generally the available energy or exergy B of an amount of heat Q at a temperature T is limited by:

$$B = \left(1 - \frac{T_0}{T}\right) Q \tag{23}$$

where T_o is the ambient temperature and heat losses are assumed to be zero. The change in available energy of a system without heat losses is a measure of the heat degradation. The exergy conservation of a system can be expressed by:

$$\nu_B = \frac{\sum B_{out}}{\sum B_{in}} \tag{24}$$

where $\sum B_{in}$ is the sum of exergy added to, $\sum B_{out}$ is the sum of exergy withdrawn from the system. This kind of efficiency is defined on the Second Law basis and may be called the Second Law efficiency.

In the case of real stationary heat transportation processes, the heat losses and heat degradation occur simultaneously, thus the definition (24) can serve as a combination of the First and Second Law efficiencies by measuring the overall exergy changes of the streams through the system. It should be pointed out that this definition of efficiency cannot be applied to coupled processes where the exergy conservation is not an indicator of effectiveness.

6. Overall Thermodynamic Efficiency of Coupled Processes

In the case of thermal separation processes, in addition to coupling the other entropy producing effects should be considered. The goal of these coupled processes is to utilize the entropy produced as completely as possible by the entropy consumption of the separation. This is why the entropy consumption of the separation should be retained in the numerator of the overall thermodynamic efficiency for coupled processes. But in this case, beside the main entropy production caused by heat transport, the other entropy producing effects, such as hydraulic resistance and complementary heat exchanges of the system should also be added to the denominator of the overall efficiency:

$$\eta_s = \frac{L_{22} X_2^2}{L_{11} X_1^2 + (\Delta S)_L} \tag{25}$$

where $L_{22}X_2^2$ is the entropy consumption of the compatible independent component separation $L_{11}X_1^2$ is the entropy production of the compatible independent heat transport, and $(\Delta S)_L$ is the additional entropy production caused by hydraulic resistances, and supplementary heat exchanges, etc.

In addition to the entropy representation above, exergy representation of OTE can be defined when the exergy increased by separation is compared to the sum of the exergy decrease:

$$\eta_B = \frac{B_{SEP}}{B_H + B_L} \tag{26}$$

where B_{SEP} is the exergy increase of the separation B_H is the exergy decrease of heat transport, B_L is the sum of exergy decrease of the other entropy producing effects. This definition of OTE can be rewritten by using the exergy terms of the Eq. (24), i.e. the exergy changes of the output and input streams.

$$\eta_B = \frac{B_{SEP}}{\left(\sum B_{in} - \sum B_{out}\right) + B_{SEP}} . \tag{27}$$

The term B_{SEP} is added to the denominator, because the total exergy expense should be considered here and the difference $\left(\sum B_{in} - \sum B_{out}\right)$ contains only the resulted exergy decrease reduced by the exergy increase of the separations.

7. The Energy Representation of the Thermodynamic Efficiency

The definition of the energy representation of the thermodynamic efficiency for separation processes is somewhat more complicated than entropy or exergy representation. In the case of an ideal mixture, namely when no energy change exists, but entropy (or exergy) consumption and production does exist. The energy representation f the thermodynamic efficiency can be defined, however, by the thermal efficiency of a corresponding heat engine with the same temperature levels and coupling efficiency.

The thermal efficiency of a heat engine receiving heat Q_2 at temperature T_2 releasing heat Q_1 at temperature T_1 and produced shaft work:

$$W = Q_2 - Q_1 , \tag{28}$$

is defined by:

$$\eta_E = \frac{W}{Q_2} = \frac{W}{Q_1 + W} . \tag{29}$$

The shaft work can be obtained from (29):

$$W = Q_1 \frac{\eta_E}{1 - \eta_E} . \tag{30}$$

This work is originated from heat at temperature T_2, so the consumed entropy is:

$$S_{cons} = \frac{W}{T_2} = Q_1 \left(\frac{1}{T_2}\right) \frac{\eta_E}{1 - \eta_E} . \tag{31}$$

The entropy production of heat transfer Q_1 from T_2 to T_1 can be expressed by:

$$S_{prod} = Q_1 \left(\frac{1}{T_1} - \frac{1}{T_2} \right) , \tag{32}$$

consequently the coupling efficiency is:

$$\eta_s^c = \frac{S_{cons}}{S_{prod}} = \frac{\eta_E}{1-\eta_E} \left(\frac{1}{T_2} \right) \frac{1}{\left(T_1^{-1} - T_2^{-1} \right)} = \left(\frac{T_1}{T_2} \right) \frac{\eta_E \eta_{Ct}}{1-\eta_E} \tag{33}$$

where η_{Ct} is the Carnot efficiency (see Eg. 22). The energy representation of TE from Eq.(33) can be written as:

$$\eta_E = \frac{\eta_S^C \eta_{Ct}}{T_1 / T_2 + \eta_S^C \eta_{Ct}} \tag{34}$$

so the energy representation of TE can be calculated from the coupling efficiency, Carnot efficiency and temperatures. Eq. (34) enables us to compare the effectiveness of a separation process and a heat engine.

Finally it may be remarked that the coupling efficiency theoretically, in extreme reversible case, can be equal to unity:

$$\eta_{S,\max}^C = 1 \tag{35}$$

but the energy representation of the efficiency is limited by the value of the Carnot efficiency:

$$\eta_{E,\max} = \frac{\eta_{Ct}}{T_1 / T_2 + \eta_{Ct}} = \eta_{Ct} . \tag{36}$$

8. Systematization of the Various Definitions

The systematized tabulation of the various definitions of thermodynamic efficiency is shown in Table 1. Two kinds of processes can be distinguished: uncoupled and coupled ones. From a thermodynamic point of view, the goal of uncoupled processes is the energy, or more generally, exergy conservation, while in case of coupled processes, like separation, the growth of relative exergy production or that of the entropy consumption of the forced process (here separation) can be regarded as the object.

To drive this forced process, exergy consumption or entropy production is required and the thermodynamic efficiency must reflect the ratio of the appropriate thermodynamic properties (entropy or exergy) of the forced and driving processes. The definitions of the thermodynamic efficiency for coupled processes can be divided further into coupling and overall efficiencies according to the system investigated. The former expresses the effectiveness of the coupling alone, the latter, in addition to coupling, also takes the supplementary entropy producing effects into account. On the other hand, three different representations of TE are distinguished, i.e. energy, entropy and exergy representations, which can indicate the effectiveness of the same process from different points of view.

For uncoupled processes, the energy representation of TE reflects the energy conservation [Eq. (21)], the exergy representation reflects the conservation of available energy [Eq. (24)]. The former is a measure of heat losses alone, the latter takes both the heat degradation and losses into account. The entropy representation of TE for uncoupled processes cannot be defined because the loss of entropy is meaningless.

The fundamental definition of TE for coupled processes is based on entropy consideration. The coupling thermodynamic efficiency expresses the effectiveness of the coupling alone (Eq. 12, 13), while the supplementary entropy producing effects are also considered by the overall thermodynamic efficiency (Eq. 25).

TABLE 1. Definitions of thermodynamic efficiency

TE type:	uncoupled processes	Coupled processes	
Represent-ation		Coupling TE (CTE) η^C	Overall TE (OTE) η^0
Entropy (S)	-	$\eta_S^C = \frac{S_{SEP}}{S_H}$	$\eta_S^O = \frac{S_{SEP}}{S_H + S_L}$
Exergy (B)	$\nu_B = \frac{\sum B_{out}}{\sum B_{in}}$	$\eta_B^C = \frac{B_{SEP}}{B_H}$	$\eta_B^O = \frac{B_{SEP}}{B_H + B_L} = \frac{B_{SEP}}{(\sum B_{in} - \sum B_{out}) + B_{SEP}}$
Energy (E)	$\nu_E = \frac{\sum E_{out}}{\sum E_{in}}$	$\eta_E^C = \frac{\eta_S^C \eta_{Ct}^C}{T_1^C / T_2^C + \eta_S^C \eta_{Ct}^C}$	$\eta_E^O = \frac{\eta_S^O \eta_{Ct}^O}{T_1^O / T_2^O + \eta_S^O \eta_{Ct}^O}$

The exergy representation of the overall TE is defined by Eq. 26 and 27 and the coupling TE can be defined similarly (neglecting B_L from Eq. 26):

$$\eta_B^C = \frac{B_{SEP}}{B_H} \tag{37}$$

where B_{SEP} is the exergy increase of the separation, B_H is the exergy decrease of the heat transport. It should be remarked that the value of the TE in exergy representation is the same as the value of entropy representation, but the former is usually more useful for practical investigations.

The energy representation of the TE is interpreted by the thermal efficiency of the corresponding heat engine with the same coupling efficiency and temperatures (Eq. 34). The coupling and overall definitions differ only in the temperature levels and the Carnot efficiency. Accordingly the thermal efficiency can be regarded as the energy representation of TE of a heat engine, and by using Eq. (33), its value can be converted into entropy (or exergy) representation.

9. Empirical Definitions

A particular advantage often claimed for exergy is that it allows the definition of efficiencies in Second Law terms. This topic has been discussed extensively in German literature, especially in the work of Fratzscher [9], when he defined two alternative empirical definitions in addition to the fundamental indicator of exergy conservation for uncoupled processes (Eq. 24). The first definition of efficiency is given by:

$$\xi_1 = \frac{\left(\sum B_{out} - \sum B_{in}\right)_{purpose}}{\left(\sum B_{in} - \sum B_{out}\right)_{driving}} \tag{38}$$

Here the indices *purpuse* and *driving* refer to the (exergy) increments or decrements happen in the purpose and the driving media. If all the purpose streams are in reference state and the heat of mixing can be neglected we can write:

$$\left(\sum B_{out} - \sum B_{in}\right)_{purpose} = B_{SEP} \,. \tag{39}$$

In this particular case, definition (38) is identical with Eq. (27). Substituting relation (39) into definition (38) it follows namely that:

$$\xi_1=\frac{\left(\sum B_{out}-\sum B_{in}\right)_{purpose}}{\left(\sum B_{in}-\sum B_{out}\right)_{driving}}=$$

$$=\frac{B_{SEP}}{\left(\sum B_{in}-\sum B_{out}\right)_{driving}+B_{SEP}-\left(\sum B_{out}-\sum B_{in}\right)_{purpose}}= \tag{40}$$

$$=\frac{B_{SEP}}{\left(\sum B_{in}-\sum B_{out}\right)+B_{SEP}}\equiv\eta_B\,.$$

This proof also shows that our definition of OTE [Eq. (27)] is in accordance with the so called „exergetic efficiency of distillation" interpreted by Linnhoff and Smith [1].

The second empirical definition of Fratzscher:

$$\xi_2=\frac{\left(\sum B_{out}-\sum B_{in}\right)_{purpose}}{\sum B_{in}}= \tag{41}$$

as well as its modification defined by Linnhoff and Smith:

$$\xi_3=\frac{B_{SEP}}{\sum B_{in}}, \tag{42}$$

can be regarded as approximations of OTE constructed by neglecting the exergy values of the output streams in the denominator. However, definitions (41) and (42) cannot be proposed for indicating the thermodynamic efficiency of separation processes, because the exergy of purpose streams is included in the term $\sum B_{in}$, in spite of the fact that it is not consumed by the process.

10. Localization of Entropy Production

Before categorizing the energy saving techniques let us analyze the origins of the entropy production (Fig.1). According to the reversible model concept the minimum isothermal „work of separation" represents a lower bound on energy that must be consumed by a separation process [10]. The corresponding net entropy consumption is marked as AB. Additional entropy is consumed by the irreversible mixing effects (e.g. non-equilibrium feed conditions) inside the column (BC). Thus instead of the net requirement a greater overall entropy consumption (AC) must be compensated by the entropy producing subprocesses (heat transport).

Because of finite driving forces (adiabatic distillation) and hydraulic resistance the internal entropy production is greater than the entropy consumption to be compensated (CD). In addition to the internal requirements (AD), external entropy production also occur (DE) caused by heat exchangers and imperfect heat conservation.

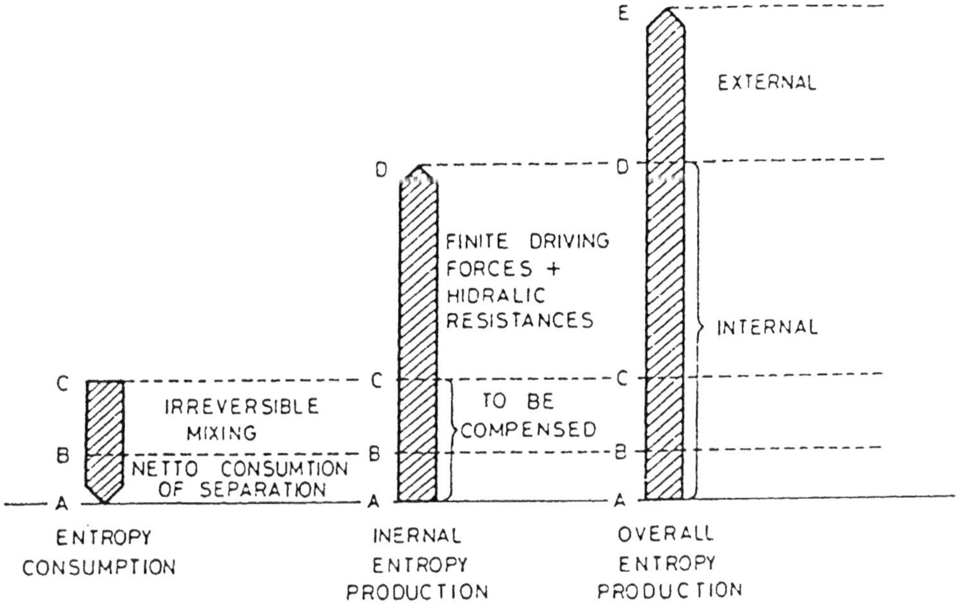

Figure 1. Origins of entropy production

11. Energy Saving Policies

The energy saving concepts can be categorized into three basic policies according to the deciding origins of the entropy production.

The methods of the *first group* aim to reduce the irreversible mixing effects (BC) inside the column, e.g.:

- energy conservation by means of tray retrofit (optimization of the changes in heat duty associated with changes in number of trays)
- optimal feed location

- energy conservation by means of tray retrofit (optimization of the changes in heat duty associated with changes in number of trays)
- optimal feed location
- pressure reduction to decrease heat duty
- thermal coupling between two towers
- application of several feeds and several sidestreams
- use of divided columns
- use of preflash tower.

The techniques of the *second group* strive after less irreversible heat turnover or more uniform driving force distribution to reduce the internal entropy production (CD), e.g.:

- use of interreboilers and intercondensers
- use of intermediate vapor recompression
- non-adiabatic distillation with special countercurrent devices.

The methods of the *third group* aim to improve energy conservation and reduce external entropy production (DE), e.g.:

- use of energy integration concept by multiple-effect distillation, (heat cascading the columns by using the condensation of the overhead vapors from the second column to provide boilup for the first column)
- optimization of feed preheat networks to capture pumparound heat or product heat wasted to air/water
- direct feed to downstream units
- use of direct or indirect vapor recompression.

It is to be noted that certain methods are multifunctional. E.g. some of the first class methods, like pressure reduction, has a positive effect, the thermal coupling between two towers has an adverse effect from the second point of view. Similarly the second class intermediate vapor recompression method has a positive effect from the third point of view etc.

12. Effective Concepts for Energy Efficient Process Design

In this section some chemical engineering techniques are discussed which are being applied to distillation sequences of existing plants in refineries and light-hydrocarbon plants to reduce energy consumption, while maintaining high levels of flexibility and operability [12-14].

Based on our theoretical and economic considerations, computer aided design and plant measurements, the superiority of *generating interreboilers and intercondenders* as well as the *heat integration and heat pumping concept* can be stated among the energy conservation methods. In the industrial distillation plants the largest internal irreversibilities occur due to the nongradual addition and withdrawal of heat and the largest external losses occur due to insufficient energy conservation.

By applying interreboilers and intercondensers the gain in reversibility is no attested as a reduced total heat duty but instead as a lesser degradation of the heat energy passing through the column. In a single reboiler system, the entire heat load must be applied to the base of the column at a high (most expensive) temperature. When the same amount of energy is divided up and added to several intermediate points the temperature levels are progressively lower (cheaper) as the feed tray is approached. The same concepts can be applied to the rectification section, where heat can be removed at successively warmer heat sinks (cheaper) as the feed tray is approached [10].

A new project is underway which will develop and test a design for the optimal thermal control of a distillation process. The control in question minimizes the total costs and is implemented by thermostatted trays maintained by heat pumps or Brayton refrigeration cycles. The main objective of the project is to demonstrate experimentally the potential energy savings in a cost effective design for a distillation column based on the concept of equal thermodynamic distance [15, 16].

According to our experience the energy integration concept and generally the methods of the third policy offers the highest affectivity to reduce energy consumption. By applying the multiple effect distillation, rearrangement of feed preheat networks and direct feed to downstream units can yield 20 to 60 % energy saving in some distillation systems.

This lecture provides also orientation for process engineers on the current application of heat integration and heat pumping techniques for large industrial separation plants. Certain well designed heat pump options can show better economic figures than the base case separation system which uses only steam and cooling water. Thus, in order to realize enhanced integration at the process design the consideration of the different heat pumping techniques is highly recommended. The presented results show that

(1) larger heat load and smaller process temperature difference provide shorter pay back time for heat pumping,

(2) the absorption heat transformation cycles have even chance for implementation at larger process ΔT, when the other configurations are to be discarded [17].

References

1. Linnhoff, B. and Smith, R. (1972) The thermodynamic efficiency of distillation, *I.Chem. E. Symposium Series* **56,** 2.1/47-2.1/73.
2. Glansdorf, P. and Prigogine, I.(1954) Sur les propriétés différentielles de la production d'entropie, *Phisica* **20,** 773-780.
3. De Groot, S.R. and Mazur, P.(1962) *Non-Equilibrium Thermodynamics*, North-Holland Publ. Co., Amsterdam.
4. Szolcsányi, P.(1969) Über die Entropie-Wirkungsgrade von Verfahrenseinheiten, *Wissenschaftliche Zeitschrift der TH für Chemie „Carl Schorlemmer"*, Leune-Merseburg **11**, 351-354.
5. Odum, H.T. and Pinkerton, R.C.(1955), *American Scientist*, **43**, 331. Referred to in [6].
6. Kedem, O. and Caplan, S.R. (1965) Degree of coupling and its relation to efficiency of energy conversion, *Trans. Faraday Soc.* **61**, 1897-1911.
7. King, C.J.(1971) *Separation Processes*, McGraw-Hill Book Co, New York.
8. Pratt, H.R.C.(1967) *Countercurrent Separation Processes*, Elsevier Publ. Co., Amsterdam.
9. Fratzscher, W.(1961) Zum Begriff des exergetischen Wirkungsgrads, *Brennstoff-Waerme-Kraft* **13**, 486-493.
10. Fonyo, Z. (1974) Thermodynamic Analysis of Rectification I and II. Reversible Model of Rectification, Finite Cascade Models, *Int. Chemical Engng.* **14**, (1) 18-27, and (2) 203-210.
11. Fonyo, Z. and Földes, P. (1974) Optimization of a stepwise realization of heat turnover in distillation, *Acta Chimica Acad. Sci.* Hung. **81**, 103-126.
12. Fonyo, Z. and Rev, E. (1982) General Interpretation of the Thermodynamic Efficiency for Separation Process, *Hung. J. Industrial Chemistry* **10**, 89-106.
13. Mizsey, P. and Fonyo, Z. (1990) Toward a More Realistic Overall Process Synthesis: The Combined Approach, *Comp. and Chem. Engng.* **14**. 1213-1236.
14. Rev, E. and Fonyo, Z. (1991) Diverse Pinch Concept for Heat Exchange Network Synthesis: The Case of Different Heat Transfer Conditions, *Chem. Eng. Sci.* **46** 1623-1634.
15. Salamon, P. and Berry, R.S. (1983) Thermodynamic Length and Dissipated Availability, *Physical Review Letters*, **51** 1127-1132.
16. Salamon, P. Nulton, J. and Ihrig, E. (1984) On the Relation between Energy and Entropy Versions of Thermodynamic Length, *Journal of Chemical Physics* **80**, 436-437.
17. Fonyo, Z. and Benkõ, N. (1998) Comparison of Various Heat Pump Assisted Distillation Configurations, *Chemical Engineering Research and Design*, **76**, 348-360.

SECOND LAW BASED OPTIMIZATION OF SYSTEMS WITH THERMOMECHANICAL DISSIPATIVE PROCESSES

E. MAMUT
Ovidius University of Constanta
Department of Mechanical Engineering
124, Mamaia Avenue
8700 Constanta, ROMANIA

1. Dissipative Processes in Mechanical Systems

A mechanical system is designed to perform static, kinematic or dynamic functions. The evolution of a system that involves the change of the value of at least one mechanical parameter is called mechanical process. Every mechanical process involves an exchange of mechanical energy. Because of friction and damping, energy is dissipated in all real mechanical processes. Sometimes the dissipated part of energy is insignificant, and sometimes it can be very important, as in damping systems.

Having as function the damping of mechanical vibrations, the damping system is a good candidate for the study of the influence of thermomechanical processes on dissipative systems. Usually called vibration isolation systems (VIS), these systems are divided into active (AVIS) and passive (PVIS) systems. An AVIS is a complicated system connecting mechanical and electronic sub-systems in so-called mecatronic systems. Passive vibration isolation systems (also called mounting systems) are common, and it is difficult to find an industrial or domestic machine, equipment or installation that is not mounted on, or contains a mounting system. More than 60% of all mounting systems contain rubber mounts. The most usual rubber mount consists of two steel plates bounded by a bulk mass of vibration isolation rubber. There is a wide diversity of shapes and forms of rubber mounts.

The problem that has stimulated the present study was the demand for the improvement of the reliability of rubber mounts. The aim of the study was to analyze the physical aspects of the failure process. Experimental work has been performed at ICECON research institute in Bucharest. Finite-elements (FEM) analysis has been done at Arcelik R&D Center in Istanbul, with the collaboration of Bosforus University and Istanbul Technical University.

2. Influence of Dissipative Processes on the Behavior of Rubber Mounts

The behavior of rubber mounts has been studied on an experimental stand, which allows quasistatic and dynamic loadings. The quasistatic loading is a deformation based on loading with a speed of 5 mm/minute and incremental steps of 1 mm. The related values of the load are indicated on a dynamometer. In order to plot the dependence between load and deformation, the maximum deformation of the mount should be 15% of its height. The stand has been designed to work with sets of 8

A. Bejan and E. Mamut (eds.), Thermodynamic Optimization of Complex Energy Systems, 297–312.

mounts during each testing cycle. The dynamic loading of the mounts is done using an unidirectional variable load with 30,000 N amplitude, and 50 Hz frequency.

The experimental research carried out at ICECON under the guidance of Prof. Polidor Bratu has confirmed the results obtained by other authors, e.g., Poturaev, Dirda and Krush [1, 2]. When the tests were started, the first observation was the heating undergone by the rubber mounts. For this reason the stand was equipped with a temperature measurement system. Continuing the experimental tests, it became obvious that there had to be a relation between the heating process and the failure mechanism of rubber mounts. Comparing the results obtained with different types of rubber (e.g., natural rubber, styrene - butadiene, isoprene rubber) we can identify three types of failure mechanisms of rubber mounts differentiated by the dependence of internal damages on local temperature [1, 2]:

Thermomechanic Failure Mechanism—characterized by a nonsteady regime of temperatures with the tendency of continuous temperature increase until the complete failure of the rubber mount; usual for vibration insulation materials based on natural rubbers, highly crosslinked, and with low contents of zinc oxide.

Fatigue Failure Mechanism—characterized by a steady regime of temperatures, and damage being produced by a fatigue development mechanisms of cracks starting from microscopic structure defaults; could be observed in natural rubber based mixtures with high contents of zinc oxide.

Combined Failure Mechanism—the most common, characterized by a complex influence of temperature on crack development kinetics. Low crosslinked rubber mixtures with low contents of zinc oxide.

In conclusion, we consider the failure process of rubber mounts as a complex process coupling a thermal component, a micro and macrostructure mechanical component, and also some physical and chemical components. In the present study we take into account only the coupled thermal and mechanical components of the process.

The experimental observations show that during the damping process the rubber mount failure starts with a heating process. The distribution of temperature is usually as shown in Fig. 1. The peak temperature could be stabilized after a number of cycles, or could continue to rise. If the temperature is stabilized at low values (e.g., 40 - 50°C), the failure process will be mainly a fatigue crack

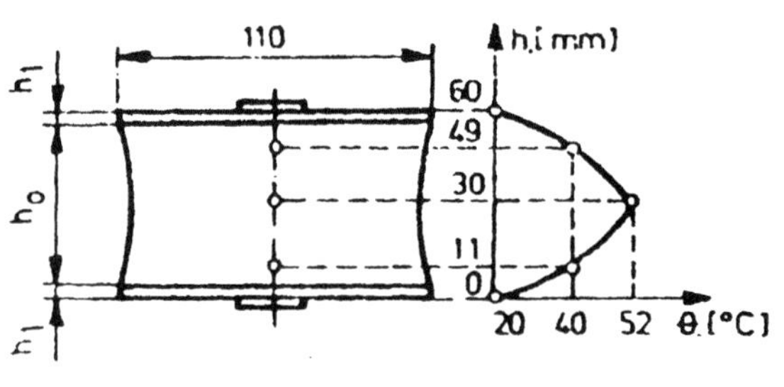

Figure 1. Temperature distribution in a rubber mount.

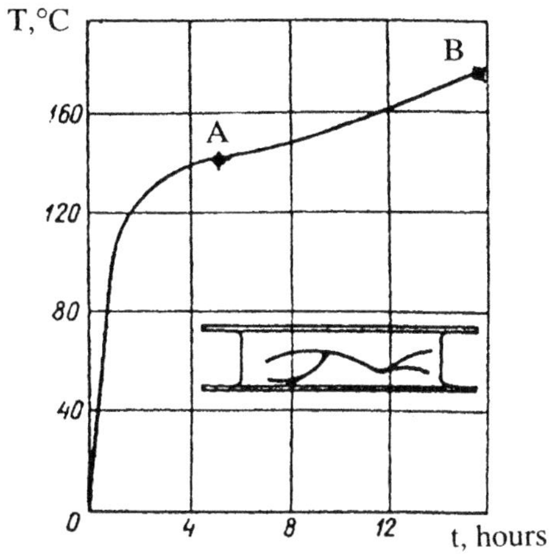

Figure 2. Evolution of temperature and damage in a rubber mount.

occurrence and development type process. Consequently, when the temperature field is steady, the damping process will cause an accumulation of micro-faults until the occurrence of cracks: after that it will continue with a micro-mechanical crack development.

If the temperature rises to high values (e.g., 120 - 160°C), the failure process will be caused by physical and chemical reactions characterized by mixture softening, plastic deformation or rubber-metal bond damage. A generic evolution of temperature and damage in a rubber mount is shown in Fig. 2.

After this initial stage of experimental research the study continued with the analysis of rubber properties, theoretical modeling and the development of a comprehensive theory to explain the failure process of rubber mounts. It was clear that the complexity of the phenomenon could not allow only a mechanical approach followed by an after thought of thermodynamic considerations. As noted by Truesdell [3], it was necessary to develop a new approach.

3. Rubber Properties

The notion "rubber" covers a group of materials, which can be identified by the property of large deformations and almost complete and instantaneous recovery in the absence of deformation load. The large elasticity property (also called hyperelasticity [4]) is not a consequence of a special chemical composition, but a special disposal of the molecular structure components. The concept of "rubber-like" material, covers a set of materials called polymers, which contain long macromolecular chains obtained from repeated units of atoms called monomers. For this reason, rubber-like materials are also called elastomers. The characteristics of an elastomer are [5]:

- the macromolecular chains have to be long and capable of free rotation around the links between the neighboring molecular units;

- the macromolecules have to be liked in a number of positions, in order to constitute three-dimensional networks, either by chemical links or by mechanical crosslinks;
- even with the above mentioned links, the molecules have to be capable of executing with ease relative motion: in other words, the intermolecular attractions have to be weak.

From the mechanical energy dissipation point of view, the main thermomechanical properties of elastomers are [6]: creep, stress relaxation, mechanical hysteresis, heat transfer properties, and temperature dependence.

Creep After Sudden Stress—the property of a material to respond to a sudden stress by a time dependent deformation. Figure 3 shows the time profile of a shear creep experiment [6].

Stress Relaxation After Sudden Strain—the property of a material to respond to a sudden strain, which is subsequently maintained constant by a relaxation in time of the resistive stress. Figure 4 illustrates a generic geometry and time profile of a simple shear stress relaxation experiment following sudden strain [6].

Mechanical Hysteresis—a consequence of the above mentioned properties, which characterize nonlinear materials. Suppose that we strain very slowly a sample of material and then we recover, with the same speed, the initial shape of the sample. If the strain-stress curve in the loading process is different than the strain-stress curve in the unloading process, we obtain a static hysteresis loop (Fig. 5a). If the strain is harmonic, the shape of the loop will be as in Fig. 5b, and is called dynamic hysteresis loop. From the mechanical energy dissipation point of view, the hysteresis loop measures the energy dissipated per cycle.

Heat Transfer Properties—the most significant properties for vulcanizing bulky elastomers are [7]: thermal conductivity, specific heat at constant pressure and

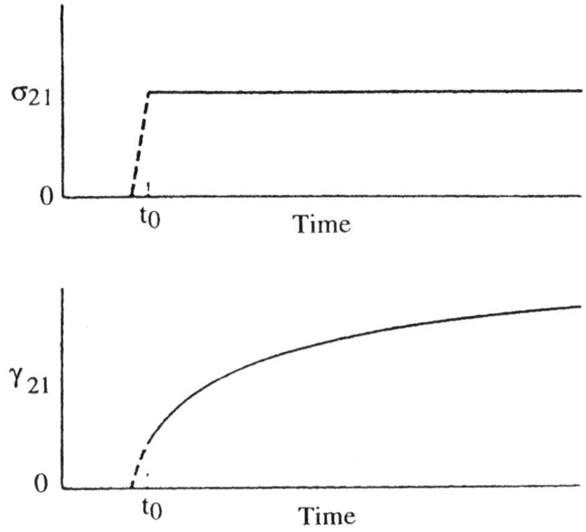

Figure 3. Time profile of a shear creep experiment.

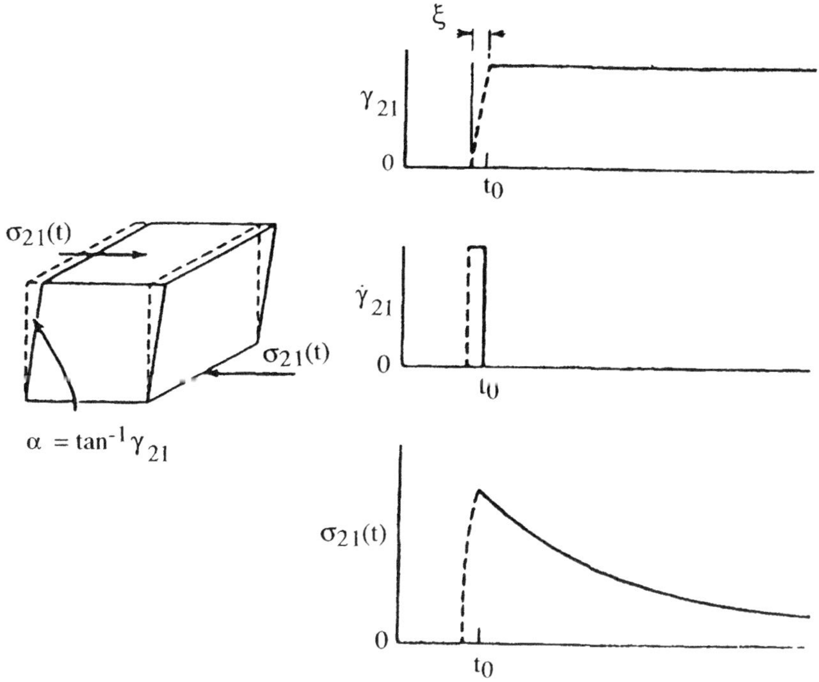

Figure 4. Geometry and time profile of a simple shear stress relaxation experiment following sudden strain.

coefficient of thermal expansion. The thermal conductivity takes values between 0.21 W/mK and 0.3 W/mK. The specific heat at constant pressure takes values between 1.4 kJ/kgK and 1.8 kJ/kgK. The coefficient of thermal expansion is of the order of 5×10^{-4} K^{-1}. The specific heat depends on temperature [7].

Temperature Dependence of the Properties of Elastomers—as in different other materials, between the properties of elastomers there are many and complex relations. For the purpose of the present study, we shall take into account only the dependence of shear modulus on temperature.

4. Finite Element Formulation

A theoretical approach of an engineering problem starts with the modeling phase. Starting from the purpose of the study, several hypotheses are made. The model is conceived as a system of equations and constraints that result from the studied problem.

4.1. CONSTITUTIVE EQUATIONS FOR RUBBER-LIKE THERMO-RHEOLOGIC MATERIALS

The modeling of the thermomechanical processes from metal-rubber vibration

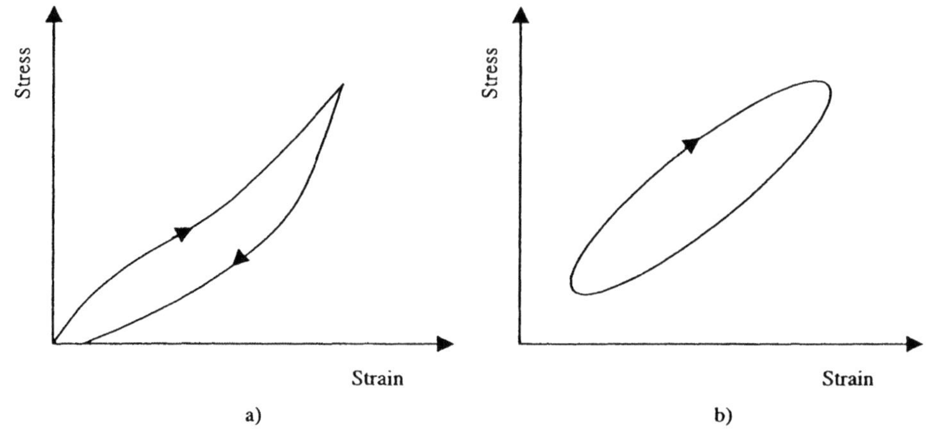

Figure 5. (a) Static hysteresis loop; (b) Dynamic hysteresis loop.

insulators can be developed by two methods, or by a combination of the two. The first method is the statistical or molecular method which starts with a hypothesis about the molecular chains, and describes their behavior based on statistical laws. The second method is the phenomenologic or macroscopic method, which describes the influence of macroscopic parameters on thermoviscoelastic continua. Both methods set up an equation that describes the dependence between the stress tensor, characteristic parameters of the studied continuum thermal field, strain tensor and time (if there exists a time dependence). This equation is called constitutive equation.

The theory of continua introduces a number of axioms that have to be applied to constitutive equations [8]. A number of constitutive equations have been established based on the statistical method, and are known by the names of their authors. The most used in the literature are Flory and Rehner (1943), Treloar (1946), Wang and Guth (1952) and Flory and Erman (1982). Based on the phenomenological method are the relations of Mooney (1940), Rivlin (1948), Valanis and Landel (1967) and Ogden (1972). These equations have been determined at thermodynamic equilibrium or for quasistatic transformations, based on the principles of mechanics and sometimes of thermodynamics. For nonsteady or time-dependent processes, some empirical corrections are usually made.

Given the Mooney-Rivlin equations

$$W = C_{10}(I_1 - 3) + C_{01}(I_2 - 3) \tag{1}$$

where C_{10} and C_{01} are parameters of the constitutive equation, the terms I_1 and I_2 have the following form

$$I_1 = \xi_1^2 + \xi_2^2 + \xi_3^2 \qquad I_2 = \xi_1^{-2} + \xi_2^{-2} + \xi_3^{-2} \tag{2}$$

In these expressions ξ_1, ξ_2 and ξ_3 are the principal specific deformations. For example, in the case of simple strain, we have the following forms

$$\xi_1 = \xi \qquad \xi_2 = \xi_3 = \xi^{-1/2} \tag{3}$$

where ξ is the specific deformation measured in the strain direction. Using the relations (3) we obtain the stress expression for simple strain

$$\sigma - \frac{\partial W}{\partial \xi} = 2C_{10}\left(\xi - \xi^{-2}\right) + 2C_{01}\left(1 - \xi^{-3}\right) \tag{4}$$

and using the reduced form for stress, we have

$$\left[f^*\right] = \sigma / \left(\xi - \xi^{-2}\right) = 2C_{10} + 2C_{01}\,\xi^{-1} \tag{5}$$

Taking into account only the equilibrium states, the temperature dependence is linear [9].

The behavior of several other elastomers is similar. For quasistatic processes the temperature dependence of rubber-like materials is related to $C_{10,\infty}$ and $C_{01,\infty}$ as follows

$$C_{10,\infty} = K_1 T \qquad C_{01,\infty} = K_2 \tag{6}$$

where K_1 and K_2 are constant parameters depending on the type of material.

In order to complete the model we have to add to the obtained equations the laws that describe the evolution of the process, and the appropriate mechanics and thermodynamics principles. In the end we obtain a system of partial differential equations, which has to be solved using numerical methods.

4.2. THE FINITE ELEMENT METHOD

Between several numerical methods, the Finite Element Method (FEM) proved that it can offer good approximation of the results. In variational formulation [10], the method consists of the maximization or minimization in a given domain of a scalar quantity depending on one or more unknown functions. This quantity is named functional. The problem is stated as follows: determine the approximation of a function such that the functional

$$\Pi = \int_A f\left(\{u\}, \frac{\partial}{\partial x}\{u\}, \ldots\right) dA + \int_\Gamma g\left(\{u\}, \frac{\partial}{\partial x}\{u\}, \ldots\right) d\Gamma, \tag{7}$$

defined in the domain A and on its boundary Γ, reaches an extremum. In order to obtain the extreme approximate value of the functional Π, the domain A is divided into sub-domains of finite dimensions, which are called elements. The contribution of each element in setting up the functional is estimated based on a simplified model of the unknown function u. The approximate value of the functional is given by the relation

$$\Pi = \sum \Pi^{(e)} \tag{8}$$

where the index (e) refers to an undefined element. The unknown function u is expressed approximately for each element by a relation of type

$$\{u\}^{(e)} = [N]\{U\}^{(e)} \tag{9}$$

where $\{U\}^{(e)}$ is a column vector containing the values of the function in the characteristic points of the element, and [N] is the shape function that depends on the finite element type and on the coordinates of the internal point of the element where the value of the function u is calculated. The functional Π depends on the all parameters {U} of the domain A, and the extreme condition of the functional has the following form of the system of equations:

$$\partial\Pi / \partial\{U\} = \begin{Bmatrix} \partial\Pi / \partial U_1 \\ \partial\Pi / \partial U_2 \\ \cdot \\ \cdot \\ \cdot \end{Bmatrix} \tag{10}$$

In this way one arrives at the construction rule for the general system of equations that result from each element,

$$\frac{\partial\Pi}{\partial\{U\}} = \sum \frac{\partial\Pi}{\partial\{U\}^{(e)}} = 0 \tag{11}$$

The solution to the system (11) consists of the values of the parameters U of all the finite elements of domain A, which offer the possibility of calculating an approximate value of function u. The numerical method for this solution depends on the shape of expression (9).

The problem of the thermomechanical processes in rubber mounts is a second order problem of thermoviscoelasticity. To solve this problem and to determine the stress, strain and temperature distribution, in the first phase, we start with the uncoupled problem and finally we take into account the couplings.

4.3. FORMULATION OF THE MECHANICAL PROBLEM

There is an extensive literature on the modeling of rubber-like materials. For the purpose of the present study we shall present some considerations, following the demonstrations of Ogden [11], and Jankovich et al. [12]. Given the displacement vector **u** of a point in a given body [11] in R^N in a deformation process which has as components u_i (i = 1, N; N = 2, 3) and $\nabla \mathbf{u} = \delta_{ij} + u_{ij}$ displacement gradients. Furthermore δ_{ij} are Kronecker's coefficient, and $u_{ij} = \partial u_i / \partial x^j$. The notation x_j indicates the local coordinates of the deformed body. The deformation tensor Green-Lagrange is defined by the following relations

$$\varepsilon_{ij} = \frac{1}{2}\left(u_{,j} + u_{,i} + u_{\alpha,i}\, u_{\alpha,j}\right) \tag{12}$$

ε_{ij} is divided into linear and nonlinear components [12]:

$$\varepsilon_{ij} = e_{ij} + \eta_{ij} \qquad e_{ij} = \left(u_{,j} + u_{,i}\right)/2 \tag{13}$$

where $\varepsilon_{ij} = 0.5\ (u_j + u_i)$ is the linear component, and η_{ij} is the nonlinear component. The stress tensor is defined in the sense of Piola Kirchhoff,

$$S^{ij} = \partial W / \partial \varepsilon_{ij} \tag{14}$$

where $W = W(I_1, I_2)$ is the deformation energy function.

We are looking for a field of displacements u which minimizes the total potential energy functional,

$$\Pi(\mathbf{u}) = \int_{\Omega} W(I_1, I_2) d\chi - g(\mathbf{u}) \tag{15}$$

The field of displacements has to satisfy the following incompatibility constraints

$$\det \nabla \mathbf{u} = I_3^{1/2} = 1 \tag{16}$$

or $I_3 = 1$. Given a function f(.) continuous such that $f(1) = 0;\ f(\chi) \neq 0\ \forall \chi \neq 1$, we can write

$$f(I_3) = 0 \tag{17}$$

In this way we can have for the deformation energy function the following form:

$$\overline{W}(\mathbf{u}, p) = W(I_1, I_2) + pf(I_3) \tag{18}$$

where p is a Lagrange multiplier. The problem is reduced to determining the stationary point (u, p) of the potential energy functional,

$$\Pi(\mathbf{u}, p) = \int_{\Omega} \overline{W}(\mathbf{u}, p) d\chi - g(\mathbf{u}) \tag{19}$$

The associated variational principle is

$$\delta W^{ex} = \int_{\Omega} S^{ij} \delta\varepsilon_{ij} d\Omega \qquad 0 = \int_{\Omega} f(I_3) \delta p \tag{20, 21}$$

where δW^{ex} is the virtual work of the external forces and S^{ij} is the stress tensor,

$$S^{ij} = \frac{\partial W(I_1, I_2)}{\partial \varepsilon_{ij}} + p \frac{\partial f(I_3)}{\partial \varepsilon_{ij}} \tag{22}$$

Equation (20) is a nonlinear functional equation, and is solved by the Newton-Ralphson method. For external conservative forces one derives the second-order variation of the deformation energy functional, by taking into account a given displacement u^0 and the p^0, ε^0 and S^0 values for the Lagrange multiplier, deformation and related stress. Fixing $\varepsilon_{ij} = \varepsilon_{ij} + \Delta\varepsilon_{ij}$, and considering in Eq. (20) the Taylor series limited to the second-order terms, we obtain finally:

$$\delta W^{ex} = \int_{\Omega} S^{ij} \delta\Delta\varepsilon_{ij} d\Omega + \int_{\Omega} \Delta\varepsilon_{rs} C^{ijrs} \delta\Delta\varepsilon_{ij} d\Omega + \int_{\Omega} \Delta p \frac{\partial f(I_3)}{\partial \varepsilon_{ij}} \delta\Delta\varepsilon_{ij} d\Omega \tag{23}$$

$$0 = \int_{\Omega} \delta\Delta p \left[\frac{\partial f(I_3)}{\partial \varepsilon_{ij}} \Delta\varepsilon_{ij} + f(I_3) \right] d\Omega \tag{24}$$

where

$$C^{ijrs} = \frac{\partial^2 W}{\partial\varepsilon_{ij}\partial\varepsilon_{rs}} + p^0 \frac{\partial^2 f(I_3)}{\partial\varepsilon_{ij}\partial\varepsilon_{rs}} \tag{25}$$

We consider the case of a bilinear isoparametric element for displacement, and define the following matrices

$$\Delta\mathbf{u}^T\mathbf{K}_L\Delta\mathbf{u} = \int_{V_\varepsilon} \Delta\varepsilon_{rs}C^{ijrs}\Delta\varepsilon_{ij}dV \qquad \Delta\mathbf{u}^T\mathbf{K}_{NL}\Delta\mathbf{u} = \int_{V_\varepsilon} \bar{S}^{ij}\Delta\eta_{ij}dV \tag{26, 27}$$

We also define the vectors

$$\Delta\mathbf{u}^T\mathbf{1} = \int_{V_\varepsilon} \bar{S}^{ij}\Delta\varepsilon_{ij}dV \tag{28}$$

$$\Delta\mathbf{u}^T\mathbf{K}_I\Delta p = \int_{V_\varepsilon} \Delta p \frac{\partial f(I_3)}{\partial\varepsilon_{ij}} \Delta\varepsilon_{ij}dV \tag{29}$$

For elastomers we define

$$i = -\int_{V_\varepsilon} f(I_3)dV \tag{30}$$

Given the load vector at element level, **g**, the relations (23) and (24) take the form

$$\begin{bmatrix} \mathbf{K}_L + \mathbf{K}_{NL} & \mathbf{K}_I \\ \mathbf{K}_I^T & 0 \end{bmatrix} \begin{Bmatrix} \Delta\mathbf{u} \\ \Delta p \end{Bmatrix} = \begin{Bmatrix} \mathbf{g} - \mathbf{1} \\ i \end{Bmatrix} \tag{31}$$

Following [12], a perturbation of order ε is introduced to obtain

$$\begin{bmatrix} \mathbf{K}_L + \mathbf{K}_{NL} & \mathbf{K}_I \\ \mathbf{K}_I^T & -\varepsilon\mathbf{M} \end{bmatrix} \begin{Bmatrix} \Delta\mathbf{u} \\ \Delta p \end{Bmatrix} = \begin{Bmatrix} \mathbf{g} - \mathbf{1} \\ i \end{Bmatrix} \tag{32}$$

where **M** is the matrix of volumes. Finally, the relation (32) becomes

$$\left[\mathbf{K}_L + \mathbf{K}_{NL} + \frac{\mathbf{M}^{-1}}{\varepsilon} \mathbf{K}_I\mathbf{K}_I^T \right] \Delta\mathbf{u} = \left\{ \mathbf{g} - \mathbf{l} - \frac{i}{\mathbf{M}\varepsilon} \mathbf{K}_I \right\} \tag{33a}$$

$$\Delta p = \frac{\mathbf{M}^{-1}}{\varepsilon} \left\{ \mathbf{K}_I^T \Delta\mathbf{u} - i \right\} \tag{33b}$$

In the modeling of viscoelastic damage [24] the deformation energy has the form

$$W(t) = W^{\infty} + \sum_{n=1}^{\infty} W^{n} \exp\left(-t/\lambda^{n}\right) \tag{18'}$$

where W^{∞} is the deformation energy in the static process, W^{n} is the part added for dynamic processes, and λ is a time constant. Solving Eqs. (33a, b) we obtain, u, p, ε_{ij}, and S^{ij} for each element. With these data we can calculate the dissipated energy on each element, per cycle:

$$q_V = \int_{cycle} S^{ij} d\varepsilon_{ij} \tag{34}$$

4.4. HEAT TRANSFER PROBLEM FORMULATION

The conductive heat transfer equation has the following form [13]

$$\mathrm{div}(\lambda \nabla T) - \rho c_v \dot{T} + q_V + r = 0 \tag{35}$$

where $\dot{T}$ is the time derivative of temperature, ∇T is the temperature gradient, ρ is the density, c_v is the specific heat at constant volume, q_V is the dissipated energy per cycle, and r are the internal heat sources. The conditions on the boundary Γ are defined in two sub-domains, Γ_T and Γ_q: on Γ_T we have Dirichlet type conditions, $T = T_0$, and on Γ_q Neumann type conditions, $\lambda(\partial T / \partial n) = q_0$. Using the weighted residual method [14] we can write:

$$-\int_{\Omega} q\delta\nabla T d\Omega - \int_{\Omega} \rho c_V T\delta T d\Omega + \int_{\Omega} (q_V + r)\delta T d\Omega + \int_{\Gamma} q_0 \delta T \delta\Gamma + 0 \tag{36}$$

For each element we have

$$T(x,t) = \mathbf{N}_I(x)\mathbf{T}_I^{(e)}(t) \tag{37}$$

where $N_I(x)$ is the shape function and $\mathbf{T}_I^{(e)}(t)$ is the nodal temp-literature. From Eqs. (36, 37) we obtain the system of ordinary differential equations,

$$\mathbf{MT} + \mathbf{F}^{in} - \mathbf{F}^{ex} = 0 \tag{38}$$

For each element we have

$$\mathbf{M}^{(e)} = \int_{\Omega} \rho c_v \mathbf{N}^T \mathbf{N} d\Omega$$

$$\mathbf{F}^{(e)in} = -\int_{\Omega} \frac{\partial \mathbf{N}}{\partial x} q d\Omega - \int_{\Omega} \mathbf{N} q_v d\Omega - \int_{\Gamma} \mathbf{N} q_0 d T_q \tag{39}$$

$$\mathbf{F}^{(e)ex} = \int_{\Omega} \mathbf{N} r d\Omega$$

and, by their assembly,

$$\mathbf{M} = \mathbf{A}^{(e)}\mathbf{M}^{(e)} \qquad \mathbf{F}^{in} = \mathbf{A}^{(e)}\mathbf{F}^{(e)in} \qquad \mathbf{F}^{ex} = \mathbf{A}^{(e)}\mathbf{F}^{(e)ex} \tag{40}$$

In the steady state $\Delta t = ct$, where c is a constant, therefore

$$\mathbf{T}_{n+1} = \mathbf{T}_n + \Delta t\,\mathbf{T}_n \tag{41}$$

And finally, from Eq. (38) we obtain the system for the temperature distribution:

$$\mathbf{T}_n = \mathbf{M}^{-1}\left(\mathbf{F}_n^{ex} - \mathbf{F}_n^{in}\right) \qquad \mathbf{T}_{n+1} = \mathbf{T}_n + \Delta t\,\mathbf{M}^{-1}\left(\mathbf{F}_n^{ex} - \mathbf{F}_n^{in}\right) \tag{42}$$

The algorithm presented in section 4.3 determines the stress-strain field. For each cycle it is possible to calculate the dissipated energy per element. These data are used in the algorithm presented in section 4.4, in order to obtain the potential field of temperature. Based on the considerations presented in section 4.1, and using the temperature field, we updated the values of C_{01}. Figure 6 shows the iterative cycle of calculations needed to obtain the temperature field after a number of cycles. The cycle was simulated using the Mark K6 code, in order to validate the experimental results and constants used in several formulas. Using the coefficients for the model C_{10} = 550 kPa and C_{01} = 380 kPa the differences between the experimental and calculated hysteresis loops are less than 2%.

5. Thermodynamic Model

In the classical engineering approach reliability analysis is usually a test that follows the series manufacturing phase. Modern mechanical engineering [15] requires the reliability principles to be taken into account in the conception phase of the products. A concurrent engineering project, as most of the present engineering projects are, requires a reliability oriented design. Studying the literature we have seen that the reliability analysis results reported by numerous authors can be summarized as follows:

- the strength of rubber mounts does not depend particularly on one or several specific mechanical parameters, but on the combined result of a group of parameters [2];
- the type of loading is very important in the failure analysis, and each type of mount will have a different behavior a different type of load;
- in the tearing process, more appropriate for describing the rubber behavior seems to be the mechanical work (i.e., an energetic parameter) rather than the stress [1].

The reliability oriented design concept introduces in the design phase of a product or system an optimization problem, in order to increase the lifetime and to reduce the maintenance costs. For this purpose a reliability model is required. The most appropriate type of reliability model for rubber mounts is the model derived from the physics of the failure mechanism [17]. With regard to the physics of the rubber mounts failure mechanism, the mechanical approach does not satisfy the

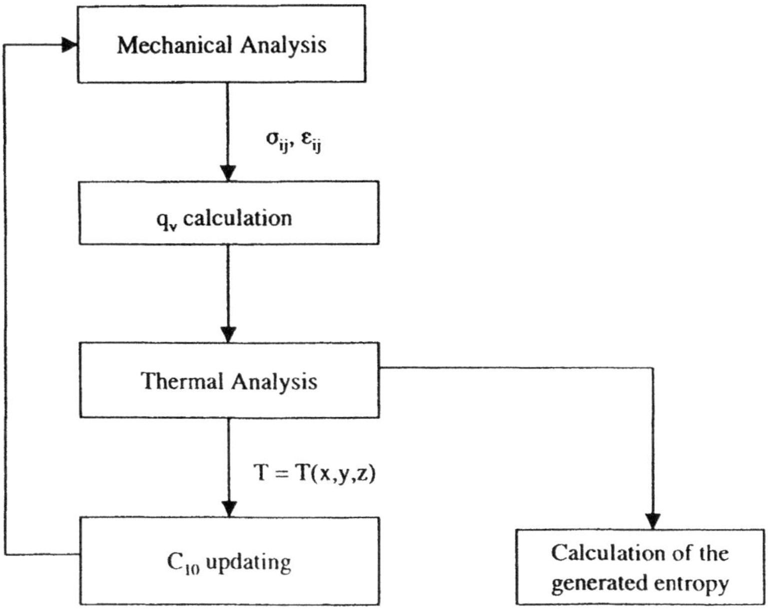

Figure 6. The iterative cycle for calculating the temperature after a number of loading cycles.

requirements of a reliability model. As shown on the left side of Fig. 7, in a mechanical approach we introduce in the system mechanical energy as mechanical work (W_{10}) the system experiences a change in internal energy (ΔU), and the output is mechanical work (W_{out}) and heat (Q_{out}). Every relation obtained in this manner follows from the energy conservation law. In this approach we can define the mechanical response characteristics of the system, but it is not possible to draw any conclusions about the lifetime of the system.

Thermodynamics introduces a new dimension in the analysis of processes: entropy. Introduced by R. Clausius and developed by Maxwell, Boltzmann and Gibbs in a statistical approach, entropy is a concept that continues to be studied. The motivation for the studies is to clarify the concept, as in the case of Carnap [18] and others, or to redefine the foundation of thermodynamics in a quantum mechanics approach, as in the case of Gyftopoulos and his school [19-21].In the case of the rubber mount, the thermodynamic model is conceived as an nonisolated closed system. The boundary allows the exchange of work and heat. The metal plates are made out of mild steel, and the rubber is a Mooney-Rivlin temperature-dependent material. For the purpose of the present study the load is assumed compressive and harmonic. The connection between the thermomechanical process and the reliability model is entropy (Fig. 7, right side). The generated entropy per cycle is given by the following equation [22]:

$$\Delta s = \frac{(\mathbf{q} \cdot \nabla T)}{T^2} + \frac{\Delta u}{T} \tag{43}$$

where Δu is the work lost per cycle (the area of the hysteresis loop):

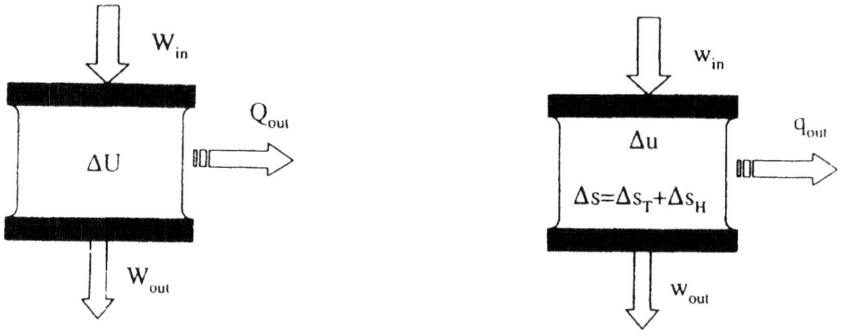

Figure 7. Thermodynamic model of the damping process.

$$\Delta u = \frac{\omega}{2\pi} \int_0^{2\pi/\omega} \sigma_{ij} \dot{\varepsilon}_{ij} d\tau \tag{44}$$

In Eq. (43) the generated entropy has two components: the first is from heat transfer, Δs_T, and the second is from hysteresis work loss, Δs_H. The generated entropy is a criterion for the failure accumulation in a rubber mount [1]. The maximum value of Δs depends on the type of rubber mixture. It can be determined from samples. The optimization steps are:

1. Static and dynamic calculation of the system that should be mounted.
2. Selection of the mounting system architecture and types of mounts.
3. Finite element analysis of every type of mount, or of mounts with every type of loading. Identification of the areas with highest values of Δs per cycle, and concentration of the analysis on those areas.
4. Forecasting of the lifetime in number of cycles.
5. If the forecasting is shorter than the required lifetime, usually we are looking for other types of mounts. If there are constrains, three types of changes are available: the shape of the mount, the improvement of the

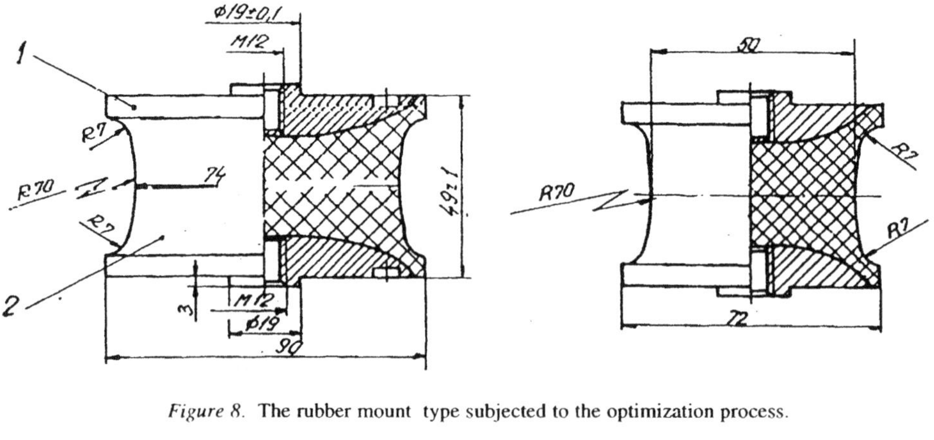

Figure 8. The rubber mount type subjected to the optimization process.

heat transfer properties, and the reduction of the hysteresis loop area.
6. The analysis of the entire system is repeated.
7. Validation of the results by experimental measurements.

The rubber mount shown in Fig. 8 was optimized using the algorithm presented above. A significant reduction of the temperature level was obtained by improving the heat transfer properties of the rubber mixture. Before the optimization, we had a mixture of rubber SAB4a; the optimized mixture was SAB31 (Fig. 9). The mounts have been tested for several types of loadings: $\alpha = 0^\circ$, 30°, 45° and 90°.

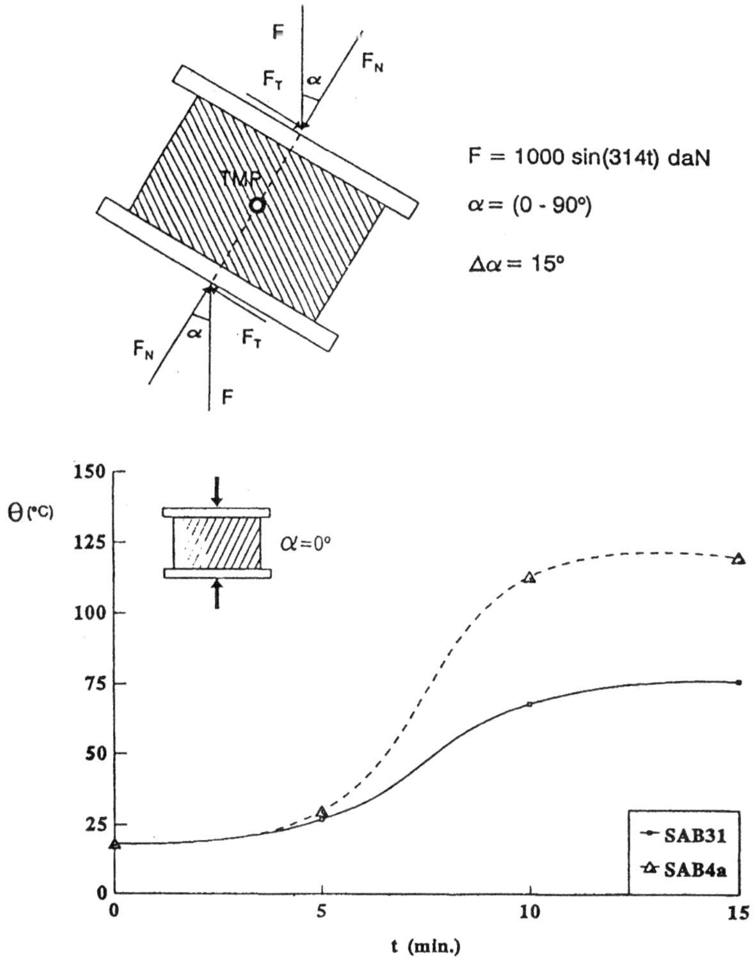

Figure 9. Experimental results obtained after heat transfer optimization: SAB4a rubber before optimization, versus SAB31 rubber after optimization.

6. Conclusions

Entropy generation minimization [23] is a method coming from thermal engineering. The present study shows that it is worth extending the method toward nonthermal areas of engineering. The entropy concept could be a powerful tool in the physical reliability modeling of nonlinear behavior materials. Some research has been conducted in the early sixties, but he modern CAD tools open new possibilities in this area.

7. References

1. Poturaev, V.N. and Dirda, V.I. (1977) *Rezinovie Detalii Mashin*, Mashinostroenie, Moscow.
2. Poturaev, V.I., Dirda, V.I. and Krush, I.I. (1975) *Prikladnaia Mehanika Rezinii*, Nauka Dumka, Kiev.
3. Truesdell, C. (1969) *Rational Thermodynamics*, McGraw-Hill Inc. New York.
4. Naunton, W.J.S. (1961) *The Applied Science of Rubber*, Edward Arnold Publishers, London.
5. Hepburn, C. and Reynolds, R.J.W. (1975) *Elastomers. Criteria for Engineering Design*, Applied Science Publishers, London.
6. Ferry, J.D. (1970) *Viscoelastic Properties of Polymers*, Wiley, New York.
7. Wood, L.A. (1973) *Physical Constants of Different Rubbers*, National Bureau of Standards, Washington D.C.
8. Eringen, C. (1980) *Mechanics of Continua*, Krieger, Malabar, FL.
9. Becker, R.H., Chin, U.Y. and Mark, J.E. (1975) Thermoelastic studies of diene polymers in elongation and compression, *Polymer J.* **7-1**, 234-240.
10. Popa, B., Madarashan, T., Batzaga, N. and Adameshteanu, I. (1978) *Thermal Stresses in Mechanical Engineering*, Editura Tehnica, Bucharest.
11. Ogden, J.T. (1972) *Finite Elements of Nonlinear Continua*, McGraw-Hill, New York.
12. Jankovich, E., Leblanc, F., Durand, M. and Bercovier, M. (1981) A finite element method for the analysis of rubber parts, experimental and analytical assessment, *Computers & Structures* **14-(5-6)**, 385-391.
13. Zhu, Y.Y. and Cescotto, S. (1994) Transient thermal and thermomechanical analysis by mixed FEM, *Computers & Structures* **53-2**, 275-304.
14. Zienkiewich, O.C. (1977) *The Finite Element Method*, McGraw-Hill, New York.
15. Kutz, M., ed. (1998) *Mechanical Engineers' Handbook*, Wiley, New York.
16. Freakley, P.K. and Payne, A.R. (1978) *Theory and Practice of Engineering with Rubber*, Applied Science Publishers, London.
17. Ozekici, S., ed. (1996) *Reliability and Maintenance of Complex Systems*, Springer-Verlag, Berlin.
18. Carnap, R. (1977) *Two Essays on Entropy*, University of California Press, Los Angeles.
19. Gyftopoulos, E.P. (1998) Maxwell's and Boltzmann's triumphant contributions to and misconceived interpretations of thermodynamics, *Int. J. Appl. Thermodynamics* **1-(1-4)**, 9-20.
20. Gyftopoulos, E.P. (1997) Fundamentals of analysis of processes, *Energy Convers. Mgmt.* **38-(15-17)**, 1525-1533.
21. Gyftopoulos, E.P. and Cubukcu, E. (1997) Entropy: thermodynamic definition and quantum expression, *Physical Review* **E55-4**, 3851-3858.
22. Arpaci, V.S. (1966) *Conduction Heat Transfer*, Addison-Wesley, Reading, MA.
23. Bejan, A. (1995) *Entropy Generation Minimization*, CRC Press, Boca Raton, FL.
24. Simo, J.C. (1987) On a fully three-dimensional finite-strain viscoelastic damage model: formulation and computational aspects, *Comp. Math. Appl. Mech. Eng.* **60**, 153-173.

SOLAR ENERGY CONVERSION INTO WORK: SIMPLE UPPER BOUND EFFICIENCIES

V. BADESCU
Candida Oancea Institute of Solar Energy,
Faculty of Mechanical Engineering,
Polytechnic University of Bucharest,
Spl. Independentei 313, Bucharest 79590, Romania

1. Introduction

In this work we present simple upper bound formulae for the efficiency of solar energy conversion into mechanical work. They are *upper bounds* as they refer to a large class of systems. They are *simple* as they are functions of two parameters only, namely the temperatures of the energy sources (i.e. the sun and the ambient). The models we develop here can be classified into two categories, as they deal with enclosed or free (transferred) radiation, respectively.

2. Enclosed Radiation

The main components of any solar power system are a solar converter (which transforms solar energy into heat) and a thermal engine (which uses that heat to generate work). The models of this section do not enter the details of solar converter operation but focus on work production. Two different sorts of thermal engines will be analyzed. First, we shall treat the ideal case of a reversible (Carnot) engine. Second, an endoreversible thermal engine will be considered.

2.1. REVERSIBLE THERMAL ENGINE

In a paper concerning the ideal conversion of enclosed radiation Bejan [1] constructed a model which allowed unification of three existing theories. The model was developed in [2]. Shortly. there is a reversible three-part process performed by the isotropic blackbody radiation contained by a deformable reflecting enclosure (Fig. 1). In the first process ($0 \rightarrow 1$) a volume V_1 is filled with radiation at temperature T_1, while in thermal contact

A. Bejan and E. Mamut (eds.), Thermodynamic Optimization of Complex Energy Systems, 313–322.

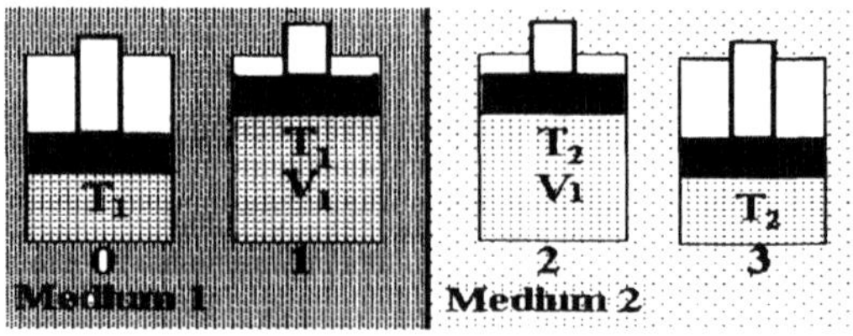

Figure 1. Conversion of enclosed radiation energy into work.

with a hot temperature reservoir consisting in isotropic blackbody radiation (medium 1). Available work is produced during the constant-volume cooling process ($1 \rightarrow 2$), in which the temperature gap between system and the second (cold) temperature reservoir (medium 2) is bridged by a Carnot engine. The ($0 \rightarrow 1$) scenario goes in the reverse direction in the last process ($2 \rightarrow 3$), the system volume decreasing from V_1 to zero. As the volume in states 0 and 3 is zero, the system allows cyclic operation.

In the states 1 and 2 the system parameters are, respectively

$$(V_1, T_1, p_1 = \frac{1}{3}aT_1^4) \text{ and } (V_1, T_2, p_2 = \frac{1}{3}aT_2^4) \qquad (1),(1')$$

where p stands for radiation pressure and $a = 7.565 \times 10^{-16}\ Jm^{-3}K^{-4}$.

During the process ($0 \rightarrow 1$) the volume V_1 is filled reversibly with isotropic blackbody radiation of temperature T_1. The work and heat interactions are, respectively:

$$W_{01} = p_1 V_1 \ , \ Q_{01} = W_{01} + aV_1T_1^4 \qquad (2),(2')$$

During the process ($1 \rightarrow 2$) the radiation is cooled at constant volume down to the temperature T_2. Consequently, the internal energy U drops from $aV_1T_1^4$ to $aV_1T_2^4$. Any infinitesimal energy drop ($-dU$) acts as a positive heat input to the high temperature end of a reversible Carnot cycle, whose instantaneous high temperature is T (this temperature drops gradually from T_1 to T_2). The work extracted is therefore

$$W_{12} = \int_{U=aV_1T_1^4}^{aV_1T_2^4} \eta_{engine}(-dU) \qquad (3)$$

where $\eta_{engine} \equiv 1 - T_2/T$ is the instantaneous efficiency of the Carnot engine, and where $dU = d(aV_1T^4)$. Performing the above integral leads to:

$$W_{12} = aV_1T_1^4[1 - \frac{4}{3}\frac{T_2}{T_1} + \frac{1}{3}(\frac{T_2}{T_1})^4] \qquad (4)$$

Let us assume the second medium consists of blackbody isotropic radiation. In this case, during the process $(2 \to 3)$ the interaction work required to decrease the system volume from V_1 to zero is:

$$W_{23} = -p_2 V_1 \tag{5}$$

The heat engine efficiency of the cycle is defined as:

$$\eta \equiv \frac{\oint \delta W}{Q_{01}} = \frac{W_{01} + W_{12} + W_{23}}{Q_{01}} \tag{6}$$

However, additional assumptions yield different upper bound formulae as follows.

2.1.1. *Jeter's Efficiency*
By using Eqs. (1)-(6) one finds:

$$\eta_J = 1 - T_2/T_1 \tag{7}$$

This is the upper bound proposed by Jeter [3] (see Fig. 2).

2.1.2. *Spanner's Efficiency*
If we replace the reversible process $(0 \to 1)$ with a spontaneous (irreversible) process in which the system does not deliver any work, we have:

$$W_{01irrev} = 0 \tag{8}$$

By using Eqs. (1′),(2′), (3)-(5), (6) and (8) one obtains:

$$\eta_S = 1 - \frac{4}{3}\frac{T_2}{T_1} \tag{9}$$

This is Spanner's efficiency [4] shown in Fig. 2, too.

2.1.3. *Landsberg-Petela-Press Efficiency*
Let us assume the second medium does not consist of blackbody radiation. Then, we can replace the reversible processes $(0 \to 1)$ and $(2 \to 3)$ with spontaneous (irreversible) processes in which the system does not deliver any work. Consequently:

$$W_{01irrev} = W_{23irrev} = 0 \tag{10}$$

In this case Eqs. (1′), (2′),(4),(6),(7) and (10) yield the Landsberg - Petela - Press (LPP) efficiency (Fig. 2):

$$\eta_{LPP} = 1 - \frac{4}{3}\frac{T_2}{T_1} + \frac{1}{3}\left(\frac{T_2}{T_1}\right)^4 \tag{11}$$

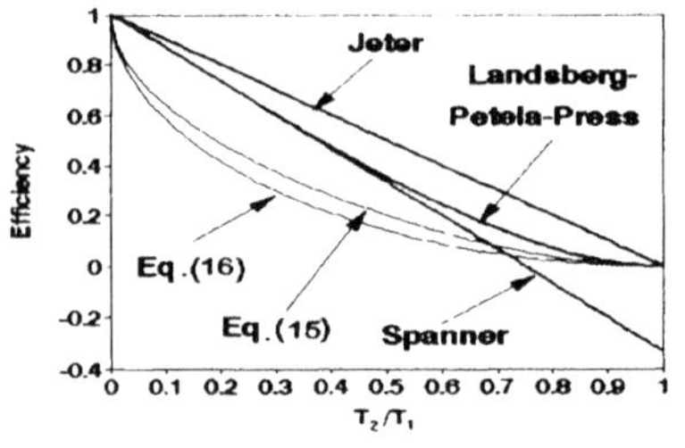

Figure 2. Upper bound efficiencies. T_1 and T_2 - temperatures of hot and cold reservoirs.

The LPP efficiency was first proposed by Petela in 1964 [5] and was re-derived independently twelve years later by Landsberg [6] and Press [7] under different theoretical approaches.

2.2. ENDOREVERSIBLE THERMAL ENGINE

The analysis above can be repeated in case the process $(1 \rightarrow 2)$ represents work generation using an endoreversible thermal engine. For the sake of generality η_{engine} in Eq. (3) is written as

$$\eta_{engine} = 1 - (\frac{T_2}{T})^n \tag{12}$$

where n is a positive underunitary number. If n=1/2 one has the usual case of a Chambadal - Novikov - Curzon - Ahlborn (CNCA) efficiency (for various endoreversible engines see Bejan [8]). Use of Eqs. (3) and (12) yields:

$$W_{12endorev} = aV_1T_1^4[1 - \frac{4}{4-n}(\frac{T_2}{T_1})^n + \frac{n}{4-n}(\frac{T_2}{T_1})^4] \tag{13}$$

This equation may be used to derive three new efficiencies, by repeating the procedure given in section 2.1. A single case is presented here.

2.2.1. *Lower Upper Bound Efficiency*

One replaces the reversible processes $(0 \rightarrow 1)$ and $(2 \rightarrow 3)$ with irreversible processes in which the system does not deliver any work (as in section 2.1.3). Use of Eqs. (2′),(6),(10) and (13) yield:

$$\tilde{\eta} = 1 - \frac{4}{4-n}(\frac{T_2}{T_1})^n + \frac{n}{4-n}(\frac{T_2}{T_1})^4 \tag{14}$$

Of course, when n=1 the Eq. (14) yields the LPP efficiency [Eq. (11)]. In the case n=1/2 (associated with an endoreversible thermal engine) one obtains

$$\tilde{\eta}' = 1 - \frac{8}{7}(\frac{T_2}{T_1})^{\frac{1}{2}} + \frac{1}{7}(\frac{T_2}{T_1})^4 \tag{15}$$

This last upper bound efficiency was derived for the first time in [9]. It is lower than other efficiencies (see Fig. 2) and it is slightly higher than

$$\tilde{\eta}^{\text{sup}} = 1 - \frac{4}{3}(\frac{T_2}{T_1})^{\frac{1}{2}} + \frac{1}{3}(\frac{T_2}{T_1})^2 \tag{16}$$

which is the lowest *simple* upper bound efficiency we know [10].

3. Transferred Radiation

In this section we present a more detailed model based on energy and entropy fluxes [11]. This model allows a simple interpretation of the LPP and Spanner efficiencies. The two heat reservoirs will be called pump (p) and sink (s). The later consists of a fluid transparent for the pump radiation. A solar converter (c) interacts with the pump by interchange of isotropic radiation and with the sink by this and also by thermal convective and/or conductive transfer. In the converter the energy of the blackbody radiation from the pump (temperature T_p) is transformed into heat which is transferred to the thermal engine where it is partially transformed into mechanical work (see Fig. 3 below).

Including the entropy-generation rate $\dot{S}_g^{conv}$, the rates of energy and entropy accumulation per unit converter area ($\dot{E}_{conv}$ and $\dot{S}_{conv}$) are respectively:

$$\begin{aligned} \dot{E}_{conv} &= \varphi_{pc} + \alpha\varphi_{sc} - \varphi_{cs} - \varphi_{cp} - \dot{Q}_{con} - \dot{Q}' && (17) \\ \dot{S}_{conv} &= \psi_{pc} + \alpha\psi_{sc} - \psi_{cs} - \psi_{cp} - (\dot{Q}_{con} + \dot{Q}')/T_c + \dot{S}_g^{conv} && (18) \end{aligned}$$

In Eqs. (17)-(18) T_c is the converter temperature while φ and ψ refer to energy and entropy fluxes, respectively. The suffix pc denotes fluxes from the pump to the converter and cs fluxes from the converter to the sink. Also, we considered the return fluxes from sink towards converter (sc) and from converter towards the pump (cp). The introduction of the parameter α in Eqs. (17)-(18) must be considered only as a mathematical artifice which will be useful during the discussion of the results. Its value will normally be unity. This means that the ambient is a radiative medium. However, in some situations α could be zero. For example, we could imagine the converter as being enveloped by a selective material which is permeable to the pump radiation but is not permeable to the thermal radiation of the ambient. Also, in case the ambient is a transparent non-radiative medium (i.e. a rarefied gas) and we neglect the radiation received by the converter from the more distant bodies we could again put $\alpha = 0$. The converter supplies the heat flux $\dot{Q}'$ to the engine and the heat flux $\dot{Q}_{con}$ to the ambient. These two heat fluxes are accompanied by the entropy fluxes $\dot{Q}'/T_c$ and $\dot{Q}_{con}/T_c$, respectively.

Two balance equations may be written for the accumulation of energy and entropy in the thermal engine (per unit converter area):

$$\dot{E}_{engine} = \dot{Q}' - \dot{Q} - \dot{W} \,, \tag{19}$$

$$\dot{S}_{engine} = \dot{Q}'/T_c - \dot{Q}/T_s + \dot{S}_g^{engine} \tag{20}$$

Here T_s is sink temperature, $\dot{S}_g^{engine}$ is the entropy-generation rate during work production, while the rate of work and the heat flux transferred from the engine to the sink are denoted by $\dot{W}$ and $\dot{Q}$, respectively.

Simple algebra allows to derive the work rate $\dot{W}$. First, we eliminate $\dot{Q}'$ and $\dot{Q}'/T_c$ between Eqs. (17) and (19). and (18) and (20), respectively. Then, we eliminate $\dot{Q}$ between the two equations we just obtained. Finally we derive:

$$\begin{aligned} \dot{W} = {} & \varphi_{pc} - T_s\psi_{pc} - (\varphi_{cs} - T_s\psi_{cs}) + \alpha(\varphi_{sc} - T_s\psi_{sc}) \\ & -(\varphi_{cp} - T_s\psi_{cp}) - (\dot{E}_{conv} - T_s\dot{S}_{conv}) - (\dot{E}_{engine} - T_s\dot{S}_{engine}) \\ & -\dot{Q}_{con}(1 - T_s/T_c) - T_s(\dot{S}_g^{conv} + \dot{S}_g^{engine}) \end{aligned} \tag{21}$$

In case of steady-state ($\dot{E}_{conv} = \dot{E}_{engine} = \dot{S}_{conv} = \dot{S}_{engine} = 0$) the Eq. (21) becomes:

$$\begin{aligned} \dot{W} = {} & \varphi_{pc} - T_s\psi_{pc} - (\varphi_{cs} - T_s\psi_{cs}) + \alpha(\varphi_{sc} - T_s\psi_{sc}) \\ & -(\varphi_{cp} - T_s\psi_{cp}) - \dot{Q}_{con}(1 - T_s/T_c) - T_s(\dot{S}_g^{conv} + \dot{S}_g^{engine}) \end{aligned} \tag{22}$$

The above results apply to any kind of radiation. In the following only steady-state blackbody radiation is considered.

3.1. BLACKBODY RADIATION

When *isotropic* blackbody radiation is considered the energy flux of partially polarized radiation emitted by component i towards component j (i,j=p,c,s) is [11]:

$$\varphi_{ij} = \frac{\sigma}{\pi}\frac{2 - P_i}{2} B_{ij} T_i^4 \ \text{(i,j=p,s,c)} \tag{23}$$

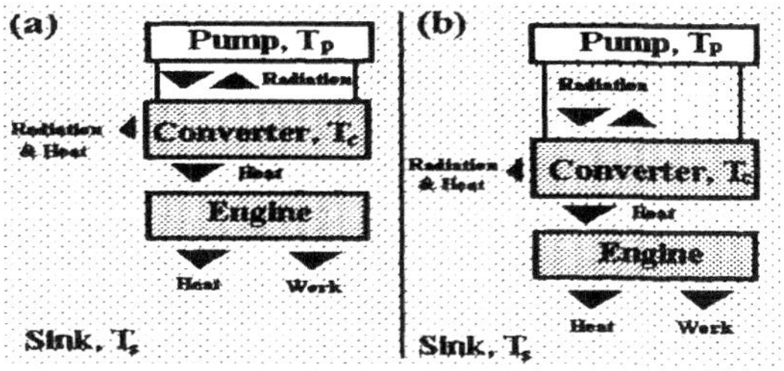

Figure 3. Solar power generation. Pump near (a) and far from (b) the converter.

where B_{ij} are geometric factors while $P_i \in [0,1]$ is the coefficient of polarization. A similar procedure leads to the entropy fluxes:

$$\psi_{ij} = \frac{4}{3}\frac{\sigma}{\pi}\frac{2-P_i}{2}B_{ij}T_i^3 \text{ (i,j=p,s,c)} \tag{24}$$

The reversibility of the individual "rays" has as a consequence the following relations and notation [12]

$$B_{ck} = B_{kc}(k = s, p) \ , \ B_T \equiv B_{cp} + B_{cs} \qquad (25),(25')$$

The energy flux emitted by a fluid layer depends on its thickness. Thus two possible situations may be considered:

(a) the pump is near the converter (i.e. between them there is a *thin* layer of radiative ambient medium). Example: the mirror and the receiver of a concentrating system.

(b) the pump is far from the converter (i.e. between them there is a *thick* layer of radiative ambient medium). Example: the sun and the absorber of a flat-plate collector.

In the case (a) we may neglect the radiation received by the absorber from that (thin) part of the ambient placed in front of the pump (see Fig. 3a). Consequently, we can write:

$$\varphi_{pc} = \frac{2-P_p}{2}\frac{B_{pc}}{\pi}\sigma T_p^4, \ \varphi_{sc} = \frac{2-P_s}{2}\frac{B_T - B_{pc}}{\pi}\sigma T_s^4 \qquad (26),(26')$$

The energy flux emitted by the converter in the total solid angle is:

$$\varphi_{cp} + \varphi_{cs} = \frac{2-P_c}{2}\frac{B_T}{\pi}\sigma T_c^4 \tag{27}$$

In the case (b) the converter receives radiation from the direction pump $\Rightarrow$ converter from both the pump *and* the ambient (Fig. 3b). Consequently:

$$\varphi_{pc} = \frac{2 - P_p}{2}\frac{B_{pc}}{\pi}\sigma T_p^4, \ \varphi_{sc} = \frac{2 - P_s}{2}\frac{B_T}{\pi}\sigma T_s^4 \qquad (28),(28')$$

Again, as in Eq. (27), the energy flux emitted by the converter depends on the total accessible solid angle only.

At this stage notice that the two different physical situations mentioned above can be represented by just one unified formalism if we recast Eqs. (26),(26′),(28) and(28′) in the following form:

$$\varphi_{pc} = \frac{2 - P_p}{2}\frac{B_{pc}}{\pi}\sigma T_p^4, \ \varphi_{sc} = \frac{2 - P_s}{2}\frac{B_T - \delta B_{pc}}{\pi}\sigma T_s^4 \qquad (29)$$

Here $\delta = 1$ or 0, depending on whether the pump is near or far from the converter, respectively.

Similar equations may be written for the entropy fluxes

$$\psi_{pc} = \frac{4}{3}\frac{2 - P_p}{2}\frac{B_{pc}}{\pi}\sigma T_p^3, \ \psi_{sc} = \frac{4}{3}\frac{2 - P_s}{2}\frac{B_T - \delta B_{pc}}{\pi}\sigma T_s^s \qquad (30),(30')$$

$$\psi_{cp} + \psi_{cs} = \frac{4}{3}\frac{2 - P_c}{2}\frac{B_T}{\pi}\sigma T_c^3 \qquad (31)$$

We shall use the following notations

$$r^4 \equiv B_{pc}/B_T \ , \ a \equiv T_s/T_p, \ b \equiv T_c/T_p \qquad (32),(32'),(32")$$

$$(0 \leq r \leq 1 \ ; \ 0 \leq a \leq b \leq 1) \qquad (33)$$

By replacing the Eqs. (27)-(31) in Eq. (22) and using Eqs. (32)-(32") we obtain the steady-state work rate under the form:

$$\dot{W} = \frac{B_T}{\pi}\frac{2 - P_p}{2}\sigma T_p^4[\ (1 - \frac{4}{3}a + \frac{1}{3}\alpha\delta\frac{2 - P_s}{2 - P_p}a^4)r^4 - \frac{2 - P_c}{2 - P_p}(b^4 - \frac{4}{3}ab^3)$$
$$-\frac{1}{3}\alpha\frac{2 - P_s}{2 - P_p}a^4] - [\dot{Q}_{con}(1 - \frac{a}{b}) + T_s(\dot{S}_g^{conv} + \dot{S}_g^{engine})] \qquad (34)$$

3.2. CONVERSION EFFICIENCIES

The efficiency of work production is defined as

$$\eta \equiv \dot{W}/\varphi_{pc} \qquad (35)$$

which is the ratio of two fluxes. Use of Eqs. (29), (32)-(32"), (34) and (35) yields:

$$\eta = 1-\frac{4}{3}a+\frac{1}{3}\alpha\delta\frac{2-P_s}{2-P_p}a^4-\frac{1}{r^4}[\frac{2-P_c}{2-P_p}(b^4-\frac{4}{3}ab^3)+\frac{1}{3}\alpha\frac{2-P_s}{2-P_p}a^4]$$
$$-\frac{2\pi}{(2-P_p)B_T r^4\sigma T_p^4}[\dot{Q}_{con}(1-\frac{a}{b})+T_s(\dot{S}_g^{conv}+\dot{S}_g^{engine})] \quad (36)$$

Equation (36) may be used for a simple interpretation of the LPP and Spanner efficiencies (taking into account that $\dot{S}_g^{conv}+\dot{S}_g^{engine}\geq 0$). Three particular cases are discussed in the following. Unpolarized radiation is considered only ($P_i=0$; i=p,c,s).

3.2.1. *Non-radiative Ambient* ($\alpha = 0$).
Equation (36) reduces to:

$$\eta = 1-\frac{4}{3}a-\frac{1}{r^4}(b^4-\frac{4}{3}ab^3) \quad (37)$$
$$-\frac{\pi}{B_T r^4\sigma T_p^4}[\dot{Q}_{con}(1-\frac{a}{b})+T_s(\dot{S}_g^{conv}+\dot{S}_g^{engine})]$$

We note that

$$\eta < 1-\frac{4}{3}a \equiv \eta_S \text{ subject to } b^3(b-\frac{4}{3}a)>0 \quad (38),(38')$$

3.2.2. *Radiative Ambient* ($\alpha = 1$) *and Pump far from Converter* ($\delta = 0$)
Equation (36) becomes:

$$\eta = 1-\frac{4}{3}a-\frac{b^4}{r^4}[1-\frac{4}{3}\frac{a}{b}+\frac{1}{3}(\frac{a}{b})^4] \quad (39)$$
$$-\frac{\pi}{B_T r^4\sigma T_p^4}[\dot{Q}_{con}(1-\frac{a}{b})+T_s(\dot{S}_g^{conv}+\dot{S}_g^{engine})]$$

The terms in the two square brackets of Eq. (39) are non-negative for $a \leq b$. Consequently, $\eta < 1-\frac{4}{3}a \equiv \eta_S$ without the constraint Eq. (38').

3.2.3. *Radiative Ambient* ($\alpha = 1$) *and Pump near the Converter* ($\delta = 1$)
Equation (36) becomes:

$$\eta = 1-\frac{4}{3}a+\frac{1}{3}a^4-\frac{b^4}{r^4}[1-\frac{4}{3}\frac{a}{b}+\frac{1}{3}(\frac{a}{b})^4] \quad (40)$$
$$-\frac{\pi}{B_T r^4\sigma T_p^4}[\dot{Q}_{con}(1-\frac{a}{b})+T_s(\dot{S}_g^{conv}+\dot{S}_g^{engine})]$$

We see that

$$\eta \leq 1 - \frac{4}{3}a + \frac{1}{3}a^4 \equiv \eta_{LPP} \tag{41}$$

Note that η_{LPP} may be used as an upper bound efficiency in the other two cases (sections 3.2.1 and 3.2.2).

4. Conclusions

Three well-known upper bound efficiencies for solar energy conversion into work are derived in section 2 by using the unifying formalism developed in [1],[2]. The additional assumption of endoreversible power generation yields a new, lower upper bound efficiency [Eq. (15)]. Lower but more *complex* upper bound formulae are treated in other papers [10],[13] - they are more complex as they depend on the heat source temperatures *and* other parameters.

References

1. Bejan, A. (1987) Unification of three different theories concerning the ideal conversion of enclosed radiation, *J. Solar Energy Eng.* **109**, 46-51.
2. Badescu, V. (1988) Discussion on the unification of three different theories concerning the ideal conversion of enclosed radiation, *J. Solar Energy Eng.* **110**, 349-350.
3. Jeter, J. (1981) Maximum conversion efficiency for the utilization of direct solar radiation, *Solar Energy* **26**, 231-236.
4. Spanner, D. C. (1964) *Introduction to thermodynamics*, Academic Press, London, p.218.
5. Petela, R. (1964) Exergy of heat radiation, *J. Heat Transfer* **86**, 187-192.
6. Landsberg, P T. and Mallinson, J R. (1976) Thermodynamic constraints, effective temperatures and solar cells, in Coll. Int. sur l'Electricite Solaire, CNES, Toulouse, pp 27-35.
7. Press, W.H. (1976) Theoretical maximum for energy from direct and diffuse sunlight, *Nature* **264**, 734-735.
8. Bejan, A. (1996) Entropy generation minimization: The new thermodynamics of finite-size devices and finite-time processes, *J. Appl. Phys.* **79**, 1191-1218.
9. Badescu, V. Simple upper bound efficiency for endoreversible conversion of enclosed radiation, to be published.
10. Badescu, V. (1998) Accurate upper bound for the efficiency of converting solar energy into work, *J. Phys. D: Appl. Phys.* **31**, 820-825.
11. Badescu, V. (1992) On the thermodynamics of the conversion of partially polarized black-body radiation, *J. Phys. III France* **2**, 1925-1941.
12. Landsberg, P.T. and Baruch, P. (1989) The thermodynamics of the conversion of radiation energy for photovoltaics, *J. Phys. A: Math. Gen.* **22**, 1911-1926.
13. Badescu, V. (1990) On the thermodynamics of the conversion of diluted radiation, *J. Phys. D : Appl. Phys.* **23**, 289-292.

THERMODYNAMIC APPROACH TO THE OPTIMIZATION OF CENTRAL SOLAR ENERGY SYSTEMS

A. SEGAL
Solar Research Facilities Unit, Weizmann Institute of Science
76100 Rehovot, Israel

1. Introduction

The sun is equivalent to a 5800 K thermal reservoir. If we should be capable to concentrate the sun light on the earth, close to the thermodynamic limit and to use its power almost without losses, we should obtain the most powerful thermal engine with a fantastic efficiency of more than 90%. Unfortunately this seems to be impossible and we must be content with much less. Only a minute process of optimization of all the system components that take the solar energy and transform it in a transportable energy form (i.e. electricity) will increase the actual (rather) low efficiency and will transform it in a competitive efficiency comparative with the non-renewable energy process efficiencies. The central solar energy systems seem to be unique qualified of efficiently producing of high temperatures capable to achieve high thermodynamic efficiencies and therefore the future looks to be more promising for the central receiver (or solar tower) concept. This work is dedicated to an attempt to optimize a central solar energy system from the thermodynamical and optical points of view.

The overall efficiency of a such system is the product of its subsystem efficiencies' components. A central receiver system is composed of a few subsystems: optical collector subsystem, thermal receiver (mounted in a tower or, its challenger, the ground receiver that uses a tower reflector [1, 2]) and final conversion subsystem. Design of this system essentially involves determining the sizes, types and configurations of each of the subsystem's components that would maximize a certain performance criterion (and/or minimize a certain cost criterion) while satisfying the technical specifications and performance requirements. For this central solar receiver system, the computationally feasible approach is to decompose the system and to design each subsystem separately. In that sense the overall system efficiency $\eta_{overall}$ is the product of three subsystem efficiencies:

A. Bejan and E. Mamut (eds.), Thermodynamic Optimization of Complex Energy Systems, 323–334.

$$\eta_{overall} = \eta_{opt}(C) \cdot \eta_{rec}(C, T_{rec}) \cdot \eta_{pb}(T_{rec}, \rho_T) \quad (1)$$

where η_{opt} is the efficiency of the reflecting and concentrating optical components (primary and secondary; η_{rec} is the efficiency of the conversion of sunlight to heat in a receiver and η_{pb} is the efficiency of the conversion of heat to electricity (or mechanical work) in the power block. As is seen in the previous relation, the subsystem efficiencies are interdependent on three main design parameters: the concentration ratio C, the receiver average temperature T_{rec}, and gas turbine pressure ratio ρ_T (when the power block is based on a gas-turbine as will be discussed further). In a first approximation, as the concentration C is increased, η_{opt} decreases and η_{rec} increases; as the receiver average temperature T_{rec} is increased, η_{rec} decreases and η_{pb} increases. The efficiency of the optical subsystem, η_{opt} commonly decreases with concentration, since achieving a high concentration usually requires special devices (secondary concentrators) which will introduce their proper losses due to the imperfections of these devices, inherent absorption, back reflections and so one. Ignoring these losses and considering an ideal machine which converts the heat in the mechanical work, (i.e. $\eta_{opt} = 1$, $\eta_{pb} = \eta_{Carnot}$) Fletcher [3] represented the combined effect of temperature and concentration on the overall system efficiency. He found that this efficiency increases with concentration ratio, with an optimal temperature for each concentration.

Further we will define and will analyze the possibilities (and the limits) for the optimization of the three principal subsystems: the primary concentrator (heliostat field), the receiver together its concentrator (secondary/receiver concentrator) and power block system (from the thermodynamic point of view only).

2. The Optimization of the Collector Field Layout

From all subsystems component a central solar plant, the most expensive item is the collector subsystem. It composed of an individually sun tracking mirrors (heliostats) that continuously tracks the sun during daylight hours so that the specular component of incident direct beam solar radiation is reflected to a common single aim-point or to multiple aim-point array. In the aim-point, either cavity or external receivers can be placed. An advanced configuration, having a hyperboloidal mirror with one of its foci in the aim point of the collector subsystem to reflect the concentrated radiation to the ground, is also a possible optical configuration for a central receiver system [1, 2, 4]. It is therefore economically important to optimize this subsystem and minimize the number of heliostats.

Many attempts have been done to establish a theory that gives a mathematical description of heliostats subsystem [5-7]. However since a field may have thousands of units it is not practical to handle each one separately. It is therefore necessary to simplify the problem and represent the field layout with the least possible number of spacing parameters (preserving however the accuracy of the calculus). Lipps and Vant-Hull [5], adapted this approach. They have divided the field into a number of cells and analyzed the optimum relative position between heliostats in each cell.

Segal and Epstein [8] proposed a new method of heliostat field layout optimization starting with a general accepted algorithm for designing a heliostat field and including actual optimized values for radial and azimuthal separation between the heliostats.

2.1. BACKGROUND

The optimal design of a collector heliostat field involves determining the optimal location of each heliostat in the field and the boundary contour of this field so that the daily, monthly or annually performances of the entire field will optimally meet the requirements.

In general, the figure of merit used for this optimization is the ratio of the total system cost to the total energy delivered by the whole system. In this case, the optimization will consists in finding the minimum for this figure of merit. Referring to the optimization of the collector field subsystem, the previous figure of merit is equivalent to the maximization of the another figure of merit which is the average energy delivered by a single heliostat from the field. (This approach ignores in the first approximation the cost of the land and the wiring). In order to calculate this average energy per heliostat, the coordinates of each heliostat in the field must be determined. At the same time, this optimization cannot be independent of the optimization of the balance of the central solar plant, especially the height of the tower/aim- point. This is important data input for the optimization of the collector subsystem. Therefore, we will express all the lengths in the field in units of tower/aim-point height (in other words the tower height is equal to *one unit* length).

A field of heliostats can be arranged in two fashions. One is cornfield array and the second one is radial stagger array. In the classical cornfield the heliostats are disposed in parallel and equidistant rows. Because of this equidistant positioning of the heliostats, those far of the tower bring about a serious percentage of blocking, so that it is by large agreed that the radial staggered array is a more feasible arrangement. So far, there is only one algorithm for building a precise heliostat field using radial staggered arrangement. In conformity with this algorithm, the heliostats are disposed in concentric circles, the positioning starting with the circle farthest from the tower. Here the heliostats are disposed at equal azimuthal distance starting with the north position. In a inner circle, having a smaller radius from the previous circle with a radial separation, the heliostats are arranged in a staggered array, and so one. When the distance between two neighboring heliostats decreases below a prefixed limit (in general equal to the heliostat diagonal length plus an additional safety distance), one of $n+1$ heliostats is deleted from the row and, in the next row, the azimuthal separation is increased by a uniform repartition of the remaining heliostats. The present code, named RAST, part of the program package WISDOM [9], is capable of calculating the position of each heliostat in the field, based on the prefixed correlations for the radial separation and for the azimuthal separation in the farthest row. The azimuthal distance and radial separations can be determined by various empirical correlations given by Kistler [10] or Winter et al.[11]. These correlations, although mathematically are not likewise, give very closed results. For a given field (i.e. the exterior boundary is fixed at r_{max}) we found that both these correlations leads to remarkable linearity (more than 98%, as can be seen in Fig. 1) between the radial separation and the distance to the tower:

$$\Delta r = (u \cdot r + v) \cdot h_l \tag{2}$$

where u and v are functions of r_{max} and h_l is the typical heliostat dimension. A graphical dependence of these functions vs. r_{max} is shown in Fig. 2.

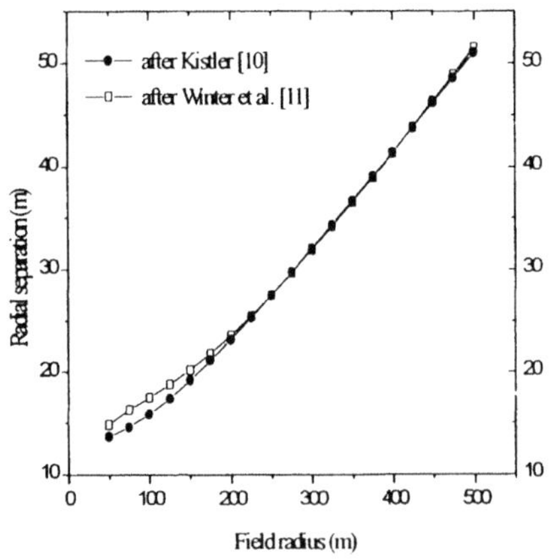

Figure 1. Correlations for the radial separation vs. field maximum radius.

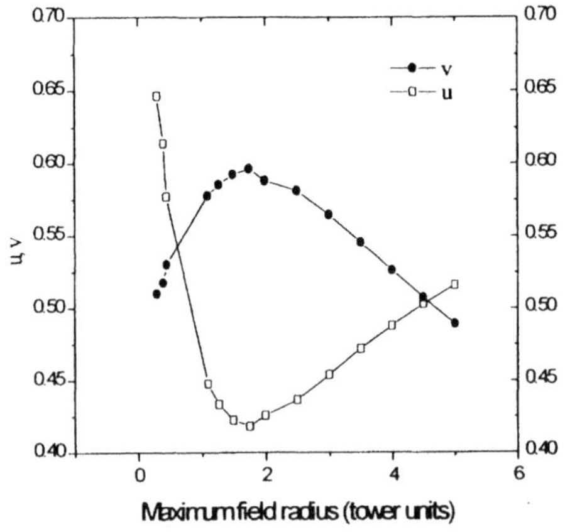

Figure 2. Parameters for linear separation vs. field maximum radius.

2.2. BUILDING OF AN OPTIMUM FIELD OF HELIOSTATS

In view of the above arguments, we believe that an optimal field, arranged radial staggered array, built on basis of the above-described algorithm, can be achieved using the following mathematical formulation:

For a given tower (with a unit height) and a field of N_H heliostats the optimization of the layout of the heliostats in the field (with prefixed boundaries) means the maximization of the ratio:

$$F = E_t / N_H \tag{3}$$

Here E_t is the total *yearly* energy delivered to the target. Initially the field must be sized so that the power delivered at the design point to be larger than the required nominal power P_{nom} which we have to deliver to the target at the design point. The target is considered, for the first iteration, as a horizontal disk situated at the aim-point and having a diameter sufficiently large to intercept all the energy reflected by the field. This initial guess will be improved in the subsequent iterations connected with the following components of the central solar plant.

This simplified model takes into account only the contribution of the heliostat to the cost of energy (the tower height and the target size being prefixed). As a result, it is necessary to couple this model with other computer code (for example DELSOL [10]) which optimize the tower height when the receiver characteristics are given as input.

Therefore, for finding the optimal array of heliostats, we consider a field having the row separation given by Eq.(2) and a fixed initial separation between two heliostats from the same row ($\Delta a_1 = w \cdot h_t$).

This formulation constitutes a problem of optimization in three dimensions (u, v, w) which can be easily solved, assuming that the field was built based on the above described algorithm. After solving this problem, we have an optimal configuration of the field, with maximum average contribution of each heliostat integrated over a year and the power at the design point larger than the required nominal power P_{nom}. We can draw on this field a map of isoenergetic heliostats [8]. At this point, we will select from this layout, only those heliostats ordered according to their annual energetical contribution to the receiver energy, until we achieve the desired power at the design point. The list of heliostats selected in this mode will be the optimal field for the given condition.

This method for building an optimum field of heliostats was exercised in the design of a 30MW$_{th}$ industrial central tower plant for solar cracking of liquid petroleum gas [4].

The optical efficiency can be defined as the ratio of energy delivered at the receiver aperture plane to the total *collectible* solar energy. The general acceptance of the term *collectible* refers to the amount of radiation that primary collectors (heliostats) would have intercepted in some layout of those heliostats. An upper limit of the amount of energy that can be collected from a large field of heliostats is the radiation falling within the boundary of the field. This amount of energy can be collected by a tightly packed collector array arranged so that no ground area is ever exposed to the sun. Of course, such type of field is completely unrealistic from the economic point of view, but this packed field can serve as an idea on the upper limit. Realistic, the collectible field is that field that is beforehand optimized in the sense described in this section. Depend of the field dimension, in that case, the optimum optical efficiency varies in the range 0.65 - 0.72.

3. The Receiver Efficiency

The average primary concentration obtained with a conventional heliostat field and a central tower is not sufficient to reach high temperatures (above 1100K) with maximum receiver efficiency. Therefore, a secondary stage of concentration with a Receiver Concentrator (RC), usually Compound Parabolic Concentrator (CPC), [12], is used to further concentrate the primary level. The receiver and the RC then become a combined unit where both the thermal and the additional optical losses have to be analyzed and considered. The receiver and the RC unit can be placed on a *Tower-Top* (as in the conventional central solar plants) aimed either straight downwards or tilted towards the *center* of a heliostat field section, or it can be placed on the ground, aiming upwards as in the case of a *Tower-Reflector* [1, 2]. In a large plant it is likely that there will be an array or a cluster of receiver/RC units either in the Tower-Top optics as well as in Tower-Reflector optics.

There is a need to redefine the efficiency of the combined receiver/RC unit. The *receiver efficiency,* η_{rec}, is defined as the net power absorbed in receiver reported at the total power arrived at the RC's entrance plane. Therefore, in this definition is included both the optical and thermal efficiencies of the ensemble consisting of the receiver and the RC in front of the receiver.

It is assumed that the RC has a circular entrance with a radius R that intercepts an amount P_{ap} from the total power P_t reflected from the heliostats and reaches the plane of the RC's aperture. The difference between P_{ap} and P_t is the spillage around the RC. The previous definition of the receiver efficiency, η, can be formulated as:

$$\eta_{rec} = \left(\alpha_{eff} \cdot (P_{ap} - P_{rej} - P_{abs}) - P_{rad} - P_{nc}\right) / P_t \tag{4}$$

where α_{eff} is the effective receiver absorptance, P_{ap} is the power intercepted by the RC's aperture, P_{rej} is the amount of power rejected from the RC, P_{abs} is the power absorbed by the RC surface, P_{rad} is the amount of power lost by reradiation from the cavity, P_{nc} is the power lost by natural convection and P_t is the total power arriving to the plane of the RC's aperture. It is assumed that the receiver is well insulated and the losses by conduction can be neglected. An ideally insulated, isothermal receiver, without a transparent window at the receiver entrance, can be considered as a blackbody with the effective emissivity $\varepsilon_{eff} \approx 1$. When a window is added, a spectral discrimination must be performed and ε_{eff} (<1) will be a function of the average temperature of the receiver.

Studying the dependence between the previous components from the Eq. (4) we obtained [1], for a receiver without window, the efficiency given by:

$$\eta_{rec} = \alpha_{eff} f_{ap}(R)\eta_{CPC} - \frac{\sigma T_{rec}^4 \pi R^2 \sin^2\theta}{P_t} - \frac{C_H A_r (T_{rec} - T_a)^{1.25}}{P_t} \quad (5)$$

where: $f_{ap}(R)$ is the fraction of P_t enclosed within a circle having radius R with $f_{ap}(R_{max}) = 1$, R_{max} being the radius in this plane which collects all the power P_t, η_{CPC} is the RC's efficiency (it is assumed that the concentrator is a perfect CPC having the acceptance angle , the entrance radius R and the exit radius $R \sin\theta$ [12]), σ is the Stefan - Boltzmann constant, T_{rec} is the representative (average) receiver working temperature, T_a being the ambient temperature, C_H is a coefficient depending of geometrical receiver parameters and physical air characteristics and A_r is the typical area of the receiver implied in the natural convection process [13]. When a window protects the receiver, the natural convection losses can be reduced substantially although some *window loss* is introduced.

4. Thermodynamical Optimization of the Receiver and Power Block Subsystems

The existing central solar plants (as Luz SEGS plants operating at 300-400°C, or Solar Two from Barstow, CA, operating at 565°C) are primarily based on Rankine-cycle steam turbines. Gas turbines (Brayton-cycle engines) are most commonly used for peak power production only. The efficiency and power of gas-turbine cycle can be noticeably improved by recovering the heat from the exhaust gases to vaporize water and run steam turbines, usually referred to as combine cycles. Combined-cycle plants retain the gas-turbine advantages of quick starting and flexible operation [14,15]. The relatively high efficiency of these plants (over 50%) stems, in part, from the high hot source temperatures (in our case, T_{rec}) that can be exploited, typically in the range 1200-1400 K. The new generation of solar volumetric receivers [16] can achieve easily these temperatures.

We apply basic thermodynamic principles toward evaluating the efficiency of realistic power-block deserving a solar receiver and based on combined cycles. There is also the analytic prediction of optimal operating points (especially the working receiver temperature T_{rec}).

4.1. GAS-TURBINES

The classic Brayton cycle gas-turbine is shown schematically in Fig. 3. The cycle is comprised of: adiabatic compression (1 →2), constant pressure heat input (2 →3), adiabatic expansion (3 →4), and constant pressure heat rejection (4 →1).

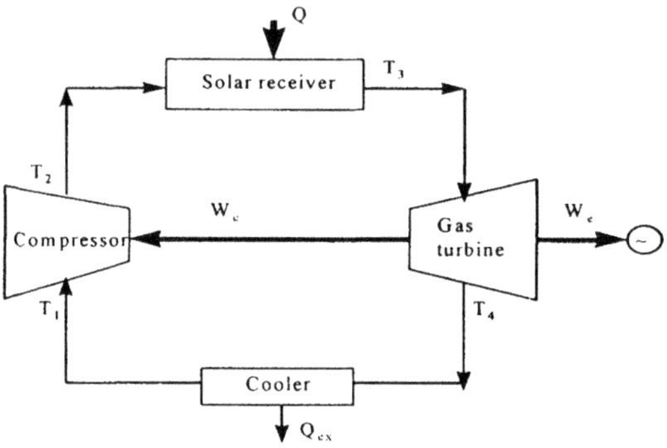

Figure 3. Schematically representation of a Brayton cycle.

The heat received by the gas is:

$$Q = \dot{m} c_p (T_3 - T_2) \tag{6}$$

where $\dot{m}$ is the mass flow rate of the air heated between the temperatures T_3 and T_4. In this case, the Brayton cycle efficiency is [14]:

$$\eta_B = 1 - \left[1 - \frac{\theta \eta_T (\rho_T - 1)}{\rho_T (\theta - 1)}\right] \Bigg/ \left[1 - \frac{\rho_C - 1}{\eta_C (\theta - 1)}\right] \tag{7}$$

where: $\theta = (T_3/T_1) = (T_{max}/T_{min})$, $\rho_T = (p_3/p_4)^{(\gamma-1)/\gamma}$, $\rho_C = (p_2/p_1)^{(\gamma-1)/\gamma}$, $\gamma = c_p/c_v$ is the specific heat ratio, $p_{j\ (j=1,2,3,4)}$ is the air pressure after each stage j, η_T, η_C being the isentropic efficiencies for gas turbine and compressor respectively. If $p_2 = p_3$ and $p_1 = p_4$, $\rho_C = \rho_T = \rho$ and the gas turbine efficiency can be written:

$$\eta_B = \frac{(\rho-1)(\chi-\rho)}{\rho(\xi-\rho)} \tag{8}$$

where $\chi = \eta_T \eta_C \theta$ and $\xi = 1 + \eta_C(\theta - 1)$. The maximum efficiency will be when:

$$\rho = \frac{\chi - [\chi(1-\xi)(\chi-\xi)]^{1/2}}{1+\chi-\xi} \tag{9}$$

The compression work W_c is given by:

$$W_c = \dot{m} c_p T_1 (\rho_C - 1)/\eta_C \tag{10}$$

The expansion work W_e is given by:

$$W_e = \dot{m} c_p T_3 \eta_T (1 - \rho_T^{-1}) \tag{11}$$

Whereas the temperatures T_1 and T_3 are usually known, the intermediate temperatures T_2 and T_4 can be calculated as:

$$T_2 = T_1 + \frac{W_c}{\dot{m} c_p} \qquad T_4 = T_3 - \frac{W_e}{\dot{m} c_p} \tag{12}$$

The overall gas turbine efficiency is:

$$\eta_{gas} = \eta_B \eta_{conv} \tag{13}$$

where η_{conv} is the conversion efficiency from mechanical to electrical energy.

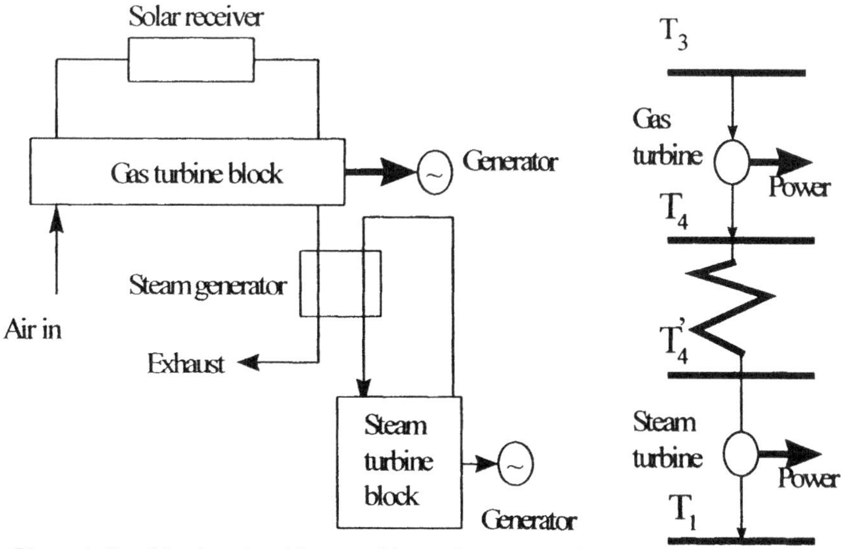

Figure 4. Combined cycle with gas turbine and a steam cycle; temperatures diagram.

4.2. COMBINED CYCLES

A combined cycle is illustrated schematically in Fig. 4. The cycle configuration is one topping cycle (the gas turbine) and one or more (practically, no more than two) bottoming cycles (the steam turbines). A heat exchange couples the gas turbine to the steam cycle(s). Following [15], we will maximize the work output for each cycle individually and then combining the cycles is equivalent to maximizing total work output. From the Fig.4 results that the turbine exhaust is at temperature T_4 and the input for the steam turbine is at temperature T_4' (which is the subject of the optimization). For a single bottoming cycle, the work output is:

$$W^{(1)}_{steam} = \eta_{steam}\eta_{conv}\dot{m}c_p(T_4/T_4' - 1)(T_4' - T_1) \tag{14}$$

where T_4' is the steam temperature at entrance into turbine and T_1 is the temperature of the cold external reservoir. The maximum cycle work output is obtained when [17]:

$$T'_{4,optimal} = (T_4 T_1)^{1/2} \tag{15}$$

so that the maximum cycle work will be:

$$W^{(1)}_{steam,\max} = \eta_{steam}\eta_{conv}\dot{m}c_p(T_4^{1/2} - T_1^{1/2})^2 \tag{16}$$

When two steam bottoming cycles are introduced, the corresponding results are [15, 17]:

$$T_4' = T_4^{2/3}T_1^{1/3}; \qquad T_4'' = T_1^{2/3}T_4^{1/3} \tag{17}$$

with:

$$W^{(2)}_{steam,\max} = \eta_{steam}\eta_{conv}\dot{m}c_p(2T_1 + T_4 - 3T_1^{2/3}T_4^{1/3}) \tag{18}$$

As a result, the work output for the optimized combined cycles W_{max}:

$$W_{\max} = \eta_{conv}(W_e - W_c) + W^{(N)}_{steam,\max} \tag{19}$$

where N=1, rarely 2. The maximum efficiency of the combined cycle (power block) is then:

$$\eta_{pb} = \frac{W_{\max}}{Q} = \frac{W_{\max}}{\dot{m}c_p(T_3 - T_2)} \tag{20}$$

The realistic partial efficiencies presented in previous relations are [15]: η_C = 0.88, $\eta_B = \eta_{steam}$ = 0.85, η_{conv}=0.95. Also other values are: γ = 1.4, $\rho_C = 1.03 \cdot \rho_T$.

5. Observations and Conclusions

We indicated in a previous section that solar collection efficiency decreases with working receiver temperature. But the efficiency of the combined cycles heat engines, however, increases with this operating temperature. Therefore the solar-driven combined-cycle power plants require an additional optimization step to determine the solar working receiver temperature that maximizes the net plant efficiency.

The optimization process for the entire solar system can be broken into three subsystem optimizations: -- optimization of the power block combine cycle for the pressure ratio obtaining $\eta_{pb,optim}(T_{rec})$; -- optimization of the solar receiver for the temperature T_{rec} obtaining $\eta_{rec,optim}(C)$ (we also included here the optimization of the RC); -- finally, the optimization of the solar primary collector in order to obtain the optimum concentration at the receiver entrance at the nominal power required.

The Eq. (5) presents the efficiency of the solar receiver function of the RC efficiency and the receiver temperature T_{rec}. In a first approximation we will put $T_3 = T_{rec}$. Following the strategy for optimization described here we have obtained the results presented in Fig. 5 where we considered $\eta_{opt,max}$=0.7. In [18] is demonstrated that the receiver efficiency can be improved by a partition

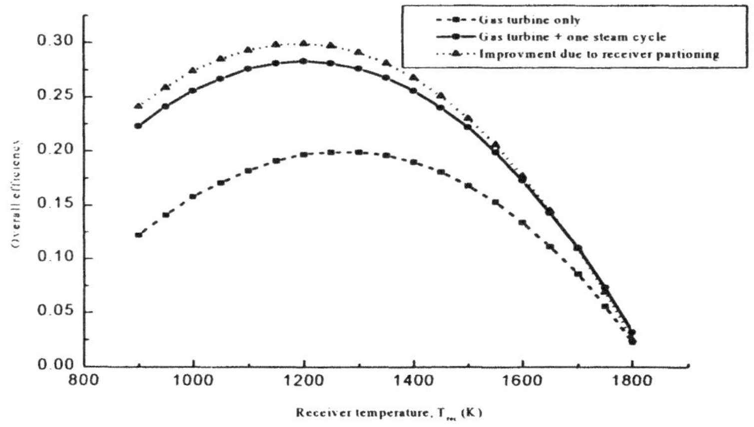

Figure 5. Overall efficiency for a solar-driven combined cycle system.

in the receiver aperture so that the central part which receives the highest flux will have a high temperature T_{rec} while the outer ring receiving lower flux will have a lower temperature T_{per} (in our calculations T_{per} = 0.65 T_{rec}) but this pro portion depending of the initial partition is subject of a further optimization). This improvement contributes with 1-2% at the overall efficiency as can be seen from the upper curve from Fig.5. The optimal working temperature (central receiver) is in the range 1100 - 1300 K and the overall solar plant efficiencies are in 26-28% range.

The central solar energy system is a very complex system. The ultimate foe of the extensive utilization of the solar energy is economic. Therefore, optimization of this system is of paramount importance and the present work represents an attempt in this direction.

6. References

1. Segal, A. and Epstein, M. (1997) Comparative performances of *Tower -Top* and *Tower-Reflector* central solar receivers, submitted to *Solar Energy*.
2 Segal, A., Epstein, M.(1998) The reflective tower as an option for high temperature central receivers, *Solar Thermal Concentrating Technologies*, Proc. of the 9th Intn'l Symposium, Odeillo - Font-Romeu, France.
3. Fletcher, E and Moen, R.L (1977) Hydrogen and oxygen from water- The use of solar energy in one-step effusional process is considered, *Science* **197**, 1050-1056.
4. Segal, A. and Epstein, M. (1997) Modeling of solar receiver for cracking of liquid petroleum gas, *J. of Solar Energy Engineering* **119**, 48-51.
5. Lipps, F.W. and Vant-Hull, L.L. (1978) A cellwise method for the optimization of large central receiver systems, *Solar Energy* **20**, 505-516.
6. Collado, F.J. and Turegano, J.A. (1989) Calculation of the annual thermal energy supplied by a defined heliostat field, *Solar Energy* **42**, 149-165.
7. Elsayed, M. and Fathalah, K.A. (1996) Solar flux-density distribution due to partially shaded/blocked mirrors using the separation of variables superposition technique with polynomial and Gaussian Sunshapes, *J. of Solar Energy Engineering* **118**, 107-114.
8. Segal, A. and Epstein, M. (1996) A model for optimization of a heliostat field layout, in M.Becker and M.Böhmer (eds.), *Solar Thermal Concentrating Technologies*, Proc. of the 8th Int'l Symposium, Köln, **2**, 989-998.
9. Segal, A. (1996) WISDOM - Weizmann Institute Solar Dedicated cOmprehensive Mastercode in Campbell-Howe R. and Wilkins-Crowder B. (eds.) *The Proceedings of Solar'96, The 1996 American Solar Energy Society Annual Conference.*, Asheville, NC, 308-313.
10. Kistler, B.L. (1986) A user's manual for DELSOL3: A computer code for calculating the optical performance and optimal system design for solar thermal central receiver plants, Sandia National Laboratories, SAND86- 8018, Albuquerque, NM.
11. Winter, C. J., Sizmann, R.L. and Vant-Hull, L.L. (Eds.) (1991) *Solar Power Plants: Fundamentals, Technology, Systems, Economics*, Springer- Verlag, Berlin.
12. Welford, W.T. and Winston, R. (1989) *High Collection Nonimaging Optics*, Ch.4, Academic Press, San Diego.
13. Bejan, A., (1995) Convection *Heat Transfer*, Ch.5, 2nd Ed., Wiley, New York.
14. Horlock, J.H. (1992) Combined *Power Plants. Including Combined Cycle Gas Turbine (CCGT) Plants,* Pergamon Press, Oxford.
15. Fraidenraich, N., Gordon, J.M., and Tiba, C. (1992) Optimization of a gas- turbine combined cycles for solar energy and alternative- fuel power generation, *Solar Energy* **48**, 301-307.
16. Karni, J., Kribus, A., Doron, P., Rubin, R., Fiterman, A. and Sagie, D. (1997) The DIAPR - a High-Pressure, High-Temperature Solar Receiver, *J. of Solar Energy Engineering* **119**, 74-78.
17. Bejan, A. (1996) Method of entropy generation minimization, or modeling and optimization based on combined heat transfer and thermodynamics, Rev. Gén. Therm. **35**, 637-646.
18. Ries, H., Kribus, A. and Karni, J. (1995) Nonisothermal Receivers, *J. of Solar Energy Engineering* **117**, 259-261.

THERMODYNAMIC OPTIMIZATION IN OCEAN THERMAL ENERGY CONVERSION

Y. IKEGAMI and H. UEHARA
OTEC Laboratory, Faculty of Science and Eng., Saga University
1 Honjyo-machi, Saga-shi, Saga 840-8502, Japan

1. Introduction

As alternative energy sources to oil and uranium, we can consider well known alternative sources such as solar power, geothermal power and wind power. However when we consider the 21st century energy sources, ocean energy is also an important potential energy source which must not be overlooked.

Among the various forms of ocean energy, the ocean thermal energy associated with the vertical temperature gradient of the ocean is plentiful and very stable. Power generation from ocean thermal energy is referred to as Ocean Thermal Energy Conversion (OTEC). Namely, OTEC is a system of converting heat energy into electricity by using the temperature difference between water at the surface of the ocean and cold water of the depth (600 m～1000 m).

Therefore, due to the small temperature difference (15～26℃) that is available for the OTEC plant, the Rankine cycle efficiency is only 3 - 4 percent [1]. Since the efficiency of an OTEC plant is low, the performance of the whole electric power generating system must be improved in order to reduce the busbar electric cost. This entails accurate estimation of the dimensions of the major components (evaporator, condenser, turbine, pump, valve, etc.) of the electric power plant. Empirical equations for estimating the performance of each component are likewise needed in the optimum design. The authors and other researchers have done numerous theoretical and experimental researches about major components of the OTEC system [1, 2]. Results of these studies show that ammonia and freon 22 are suitable working fluids for a closed cycle OTEC plant. It was also found that the plate-type heat exchangers are best fitted for OTEC. The authors have developed the optimization method for OTEC using Rankine cycle. This method has been used in actual design.

On the other hand, recently the attention has been focused on new cycles invented by Kalina and Uehara, which employs a mixture of ammonia and water as the working fluid [1, 3]. A new stage on the research of OTEC has begun.

In this paper, the principle of thermodynamic design for OTEC is explained and basically the optimum design which is most important for OTEC is presented.

A. Bejan and E. Mamut (eds.), Thermodynamic Optimization of Complex Energy Systems, 335–344.

2. Principles of Thermodynamic Design for OTEC

Figure 1 is a schematic of the closed-cycle OTEC plant that shows the general arrangement of heat exchangers, pumps, pipings, and the turbine generator. The corresponding *T-s* diagram is shown in Fig. 2.

Ocean thermal energy has the following characteristics.

(1) Ocean thermal energy, just like solar energy, is a natural energy and is renewable.

(2) Ocean thermal energy is a clean energy which does not pollute the natural environment.

(3) The size of the energy resource is large but the energy density is low.

As a consequence of these characteristics, power generation from ocean thermal energy requires an approach which is fundamentally different to that employed in the present mainstream resources such as oil, coal and uranium.

Figure 3 shows the flow chart of principles of thermodynamic design for OTEC. As the first step, the optimum design of the OTEC plant is carried out under the given conditions of warm sea water temperature and cold sea water temperature, considering as the design point [4]. The warm sea water temperature fluctuates by the hour, day and month and also on depends on the site. When the sea water temperature is deviated from its design point, the performance of the OTEC plant also differs. As the second step, it is important to predict the optimum operating conditions such as the flow rates of warm sea water, cold sea water and working fluid, when the sea water temperature is deviated from its design point [5, 6]. The annual performance of the OTEC plant which is operated on optimum conditions, depends on the given design point. As the third step, it is valuable to decide the optimum design point for each sea water temperature considering the annual temperature variation of seawater [7]. In this paper the first step of the thermodynamic optimization for the OTEC is presented.

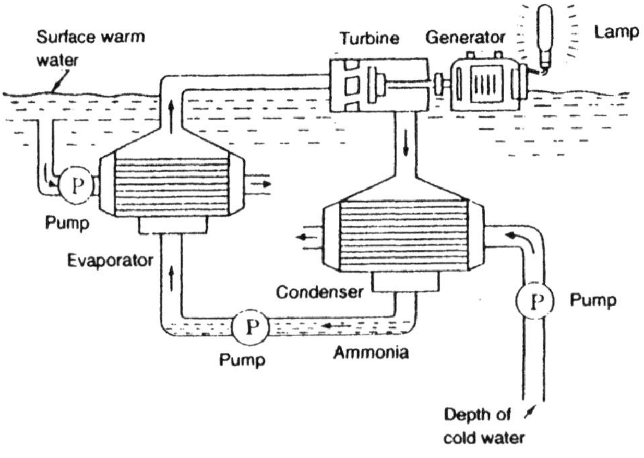

Figure 1. Schematic of the closed-cycle OTEC.

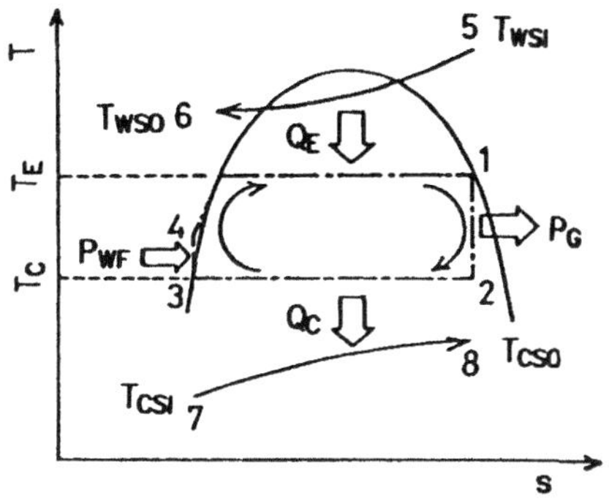

Figure 2. T-s diagram of the closed Rankine cycle.

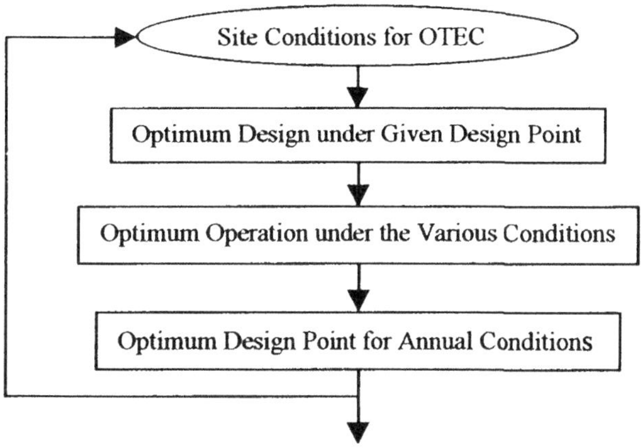

Figure 3. Flow chart of principles of thermodynamic design for the OTEC.

3. Analysis

3.1. OBJECTIVE FUNCTION

The busbar electricity cost is shown in the following equation,

$$y = (\text{total cost})/(\text{net power per unit}) \tag{1}$$

It is desirable to use y as the objective function in the optimum design of the electric power system, but the total cost of the electric power plant will vary depending on the time at which the plant is constructed and on the designer/constructor. This makes it difficult to optimize the OTEC plant using the value of y.

Since the heat exchanger cost (about 25 $\sim$ 50 percent of the plant cost) is one of the major costs of the OTEC plant [1], it is better to use γ as the objective function in the optimum design. The objective function γ is defined by the following equation

$$\gamma = A_T / P_N \tag{2}$$

where A_T is the total heat transfer surface area of the evaporator and condenser and P_N is the net power. The net power is obtained by subtracting the power consumed by the warm sea water pump, cold sea water pump, and working fluid pump from the generated power. The total surface area is calculated using the heat-transfer coefficients of the sea water and working fluid side.

The net power will be slightly lesser than the one in Eq. (2) because auxiliary power is used in the plant for lighting, etc. This power is omitted in the present analysis because it is relatively small and difficult to estimate exactly.

3.2. NET POWER

The net power P_N is defined as

$$P_N = P_G - (P_{WS} + P_{CS} + P_{WF}) \tag{3}$$

where P_G is the turbine generate power, P_{WS} is the warm sea water pumping power, P_{CS} is the cold sea water pumping power, and P_{WF} is the working fluid pumping power.

3.2.1. Turbine Generator Power

The turbine generator power, P_G, is obtained from the product of the working fluid mass flow rate, $\dot{m}_{WF}$, and the adiabatic heat drop between the evaporator and the condenser, and is given as

$$P_G = \dot{m}_{WF} \eta_T \eta_G (h_1 - h_2) \tag{4}$$

where η_T is the turbine efficiency and η_G is the generator efficiency. In this study, the configurations of a single stage reaction axial flow turbine are used [4]. η_T is given as

$$\eta_T = \eta_m \cdot \eta_e \tag{5}$$

where η_m is the mechanical efficiency and η_e is the theoretical efficiency, defined in reference [4] as

$$\eta_e = \frac{H_{ad} - (\Delta h_N + \Delta h_R + \Delta h_{EX} + \Delta h_D + \Delta h_{WET})}{H_{ad}} \tag{6}$$

where H_{ad} is the theoretical adiabatic heat drop, Δh_N is the nozzle loss, Δh_R is the rotor loss, Δh_{EX} is the exhaust loss, Δh_D is the rotary disc loss, and Δh_{WET} is the wetness loss. Δh_N, Δh_R, Δh_{EX}, Δh_D and Δh_{WET} are calculated as reference [4].

3.2.2. Cold Sea Water Pumping Power

The cold sea water pumping power P_{CS} is given as

$$P_{CS} = \dot{m}_{CS} \Delta H_{CS} g / \eta_{CSP} \tag{7}$$

where, η_{CSP} is the cold sea water pump efficiency. ΔH_{CS} is the total pressure difference of the cold sea water piping, defined as

$$\Delta H_{CS} = (\Delta H_{CS})_P + (\Delta H_{CS})_C + (\Delta H_{CS})_d \tag{8}$$

and $(\Delta H_{CS})_P$ is the pressure difference of the cold sea water pipe, defined as

$$(\Delta H_{CS})_P = (\Delta H_{CS})_{SP} + (\Delta H_{CS})_B \tag{9}$$

$(\Delta H_{CS})_{SP}$ (reference [4]) is the friction loss of the straight pipe, defined as

$$(\Delta H_{CS})_{SP} = 6.82 l_{CS} / d_{CS}^{1.17} (V_{CS} / C_{CS})^{1.85} \tag{10}$$

$$C_{CS} = 100 \tag{11}$$

where l_{CS} is the length of the cold sea water pipe, d_{CS} is the diameter of the cold sea water pipe, and V_{CS} is the velocity of the cold sea water. $(\Delta H_{CS})_B$ (reference [4]) is the bending loss of the cold sea water pipe line, defined as

$$(\Delta H_{CS})_B = (\lambda_I + \lambda_V + \lambda_S + \lambda_j + \lambda_O + \lambda_e + \lambda_d + \lambda_r) V_{CS}^2 / 2g \tag{12}$$

where λ_I is the inlet loss coefficient, λ_V is the valve loss coefficient, λ_S is the separating loss coefficient, λ_j is the joint loss coefficient, λ_O is the outlet loss coefficient, λ_e is the elbow loss coefficient, λ_d is the diffuser loss coefficient, and λ_r is the reducer loss coefficient. These loss coefficients are taken from [4].

$(\Delta H_{CS})_d$ is the pressure difference caused by the density difference between the warm and the cold sea water, defined as

$$(\Delta H_{CS})_d = l_{CS} - \frac{1}{\rho_{WS}} \left(\frac{1}{2} (\rho_{WS} + \rho_{CS}) l_{CS} \right) \tag{13}$$

$(\Delta H_{CS})_C$ is the cold sea water pressure difference in the condenser defined as

$$(\Delta H_{CS})_C = \lambda_C \frac{V_{CS}^2}{2g} \frac{l_C}{(D_{eq})_{CS}} \tag{14}$$

where l_C is the length of the condenser plate. The value of λ_C is from [4].

3.2.3. Warm Sea Water and Working Fluid Pumping Power

The warm sea water pumping power P_{WS} and the working fluid pumping power P_{WF} are given as

$$P_{WS} = \dot{m}_{WS} \Delta H_{WS} g / \eta_{WSP} \tag{15}$$

$$P_{WF} = \dot{m}_{WF} \Delta H_{WF} g / \eta_{WFP} \tag{16}$$

where η_{WSP} and η_{WFP} are the warm sea water pump efficiency and the working fluid pump efficiency. ΔH_{WS} and ΔH_{WF} are the total pressure difference of the warm sea water piping and the total pressure difference of the working fluid piping, respectively and defined similar to ΔH_{CS}.

3.3. HEAT TRANSFER SURFACE AREA

The total heat transfer surface area, A_T, is given as

$$A_T = A_E + A_C \tag{17}$$

A_E and A_C are the heat transfer areas of the evaporator and the condenser. In this paper, the shell-and-plate-type heat exchanger is used as the evaporator and the condenser. The heat exchanger configuration and the manifold system are similar to the one reported in reference [1]. The heat transfer enhancement of the evaporator is achieved by using aluminum powder on the plate and providing flutes and drainages. In the condenser only flutes and drainages are provided to enhance the heat transfer. Details are reported in references [8, 9]. Respectively, A_E and A_C are defined as

$$A_E = Q_E / (U_E (\Delta T_m)_E) = \dot{m}_{WS} c_{PWS} (T_{WSI} - T_{WSO}) / (U_E (\Delta T_m)_E) \tag{18}$$

$$A_C = Q_C / (U_C (\Delta T_m)_C) = \dot{m}_{CS} c_{PCS} (T_{CSO} - T_{CSI}) / (U_C (\Delta T_m)_C), \tag{19}$$

where T_{WSI} and T_{WSO} are the inlet and the outlet temperature of warm sea water. T_{CSI} and T_{CSO} are the inlet and the outlet temperature of cold sea water. Q_E and Q_C is the heat transfer rate of the evaporator and the condenser, respectively, defined as

$$Q_E = \dot{m}_{WF} (h_1 - h_4) \tag{20}$$

$$Q_C = \dot{m}_{WF} (h_2 - h_3) \tag{21}$$

h_1, h_2, h_3, and h_4 are the enthalpies indicated by the four points in Fig. 2, respectively. $(\Delta T_m)_E$ and $(\Delta T_m)_C$ are the logarithmic mean temperature differences of the evaporator and the condenser, respectively, defined as

$$(\Delta T_m)_E = [(T_{WSI} - T_E) - (T_{WSO} - T_E)] / \ln[(T_{WSI} - T_E)/(T_{WSO} - T_E)] \tag{22}$$

$$(\Delta T_m)_C = [(T_C - T_{CSI}) - (T_C - T_{CSO})] / \ln[(T_C - T_{CSI})/(T_C - T_{CSO})] \tag{23}$$

where T_E is the evaporation temperature and T_C is the condensate temperature.

$\dot{m}_{WF}$ is the working fluid flow rate, given as

$$\dot{m}_{WF} = P_G / [\eta_T \eta_G (h_1 - h_2)] \tag{24}$$

U_E and U_C are the overall heat-transfer coefficients of the evaporator and the condenser, respectively. U_E can be calculated by using the heat transfer coefficient on the warm sea water side, h_{WS}, the boiling heat-transfer coefficient, h_B, and thermal conductivity of the heat transfer surface, $(k_w)_E$. U_C can be calculated by using the heat transfer coefficient on the cold sea water side h_{CS}, the condensation heat transfer coefficient, h_C, and thermal conductivity of the heat transfer surface, $(k_w)_C$.

The boiling heat-transfer coefficient, h_B, (reference [8]) is calculated from the following equation:

$$Y = 0.716(f_P X)^{0.919} H^{-0.834} \cdot (\rho_L / \rho_V)^{-0.448}, (f_P X \le 14.9) \quad (25)$$

$$Y = 2.218(f_P X)^{0.5} H^{-0.834} \cdot (\rho_L / \rho_V)^{-0.448}, (f_P X > 14.9) \quad (26)$$

The heat transfer coefficient on the warm sea water side, h_{WS} [8] is given as

$$Nu_{WS} = 0.047 \mathrm{Re}_{WS}^{0.8} \mathrm{Pr}_{WS}^{1/3} \quad (27)$$

The condensation heat transfer coefficient, h_C [9] is calculated from

$$Nu_l = 1.77 B_O^{*-0.1} (Gr_l \mathrm{Pr}_l / H)^{1/4} \quad (28)$$

The heat transfer coefficient on the cold sea water side, h_{CS} [9] is given as

$$Nu_{CS} = 0.047 \mathrm{Re}_{CS}^{0.8} \mathrm{Pr}_{CS}^{1/3} \quad (29)$$

4. Optimization Method

4.1. OBJECTIVE FUNCTION AND VARIABLE

Substitution of Eqs. (3) - (29) into (2) gives the objective function γ as a function of the design parameter ***C***, the geometric parameter of the plate-type heat exchanger ***G***, the state parameter of the working fluid ***S***, the design parameter of components of the power station ***D***, the piping parameter of the plant ***P***, and the design parameter of the turbine ***T*** shown as follows:

$$\gamma = f(C, G, S, D, P, T) \quad (30)$$

where ***C, G, S, D, P***, and ***T*** are defined by the following equations:

$$C = f(P_G, T_{WSI}, T_{CSI}, \kappa_W, P_{rop}) \quad (31)$$

$$G = f(l_E, l_C, w_E, w_C, \Delta y_{WS}, \Delta y_{CS}, (\Delta y_{WF})_E, (\Delta y_{WF})_C, \delta_E, \delta_C) \quad (32)$$

$$S = f(T_E, T_C) \quad (33)$$

$$D = f(V_{WS}, V_{CS}, \eta_{WSP}, \eta_{CSP}, \eta_{WFP}, \eta_T, \eta_G) \quad (34)$$

$$P = f(d_{WS}, l_{WS}, d_{CS}, l_{CS}, d_{WF}, l_{WF}) \quad (35)$$

$$T = f(Z_N, Z_R, N_S, D_S, h/D, \alpha_2) \quad (36)$$

where P_{rop} is the property of working fluid. l_E and l_C are the lengths of the evaporator and the condenser plate, w_E and w_C are the widths of the evaporator and the condenser plates, Δy_{WS} and Δy_{CS} are the clearances of warm sea water side in the evaporator and cold sea water side in the condenser, $(\Delta y_{WF})_E$ and $(\Delta y_{WF})_C$ are the clearances of working fluid side in the evaporator and the condenser, δ_E and δ_C are thickness of the evaporator and the condenser plates, respectively. V_{WS} and V_{CS} are the velocities of the warm sea water in the evaporator and the cold sea water in the condenser. Z_N is the number of nozzle blades, Z_R is the number of rotor blades, h/D is the ratio of blade height to diameter, and α_2 is the nozzle outlet angle. The specific speed (N_S) and the specific diameter (D_S) are usually used as the design parameters of a turbine.

4.2. CALCULATION METHOD

If the variables of parameter C are given, a minimum value of the objective function can be calculated by using Eqs. (3) - (29). Furthermore, since l_E, l_C, w_E, w_C, Δy_{WS}, Δy_{CS}, $(\Delta y_{WF})_E$, $(\Delta y_{WF})_C$, δ_E, δ_C, k_W, d_{WS}, l_{WS}, d_{CS}, l_{CS}, d_{WF}, l_{WF}, η_{WSP}, η_{CSP}, η_{WFP}, η_G, Z_N, Z_R and α_2 are given as input data Eq. (30) can be rewritten as follows:

$$\gamma = f(T_E, T_C, V_{WS}, V_{CS}, N_S, D_S, h/D) \tag{37}$$

The objective function becomes a function of only seven independent variables. Calculation is done as follows. First, the objective function (γ) at a certain point is calculated. Again the objective function (γ_1) is calculated for a small change of one variable [for example T_E is changed by ΔT_E] and keeping the other six variables fixed and the value of $(\gamma - \gamma_1)/\Delta T_E$ is obtained. The new value of T_E' at the next step is obtained by multiplying an arbitrary small constant δ with $(\gamma - \gamma_1)/\Delta T_E$ and adding this value to the former T_E ($T_E' = T_E + \delta(\gamma - \gamma_1)/\Delta T_E$). The same procedure is done for the other six variables in Eq. (37).

The objective function is calculated by the combination of these new variables and the minimum value of the objective function can then be obtained. The minimum value of the objective function, γ_{min}, can be calculated by the method of Powell [4].

Figure 4 shows the flow chart of the calculation of optimum design of the OTEC under the given conditions of warm sea water inlet temperature T_{WSI}, cold sea water inlet temperature T_{CSI}, generated turbine power P_G, length and diameter of the cold sea water pipe l_{CS} and d_{CS}.

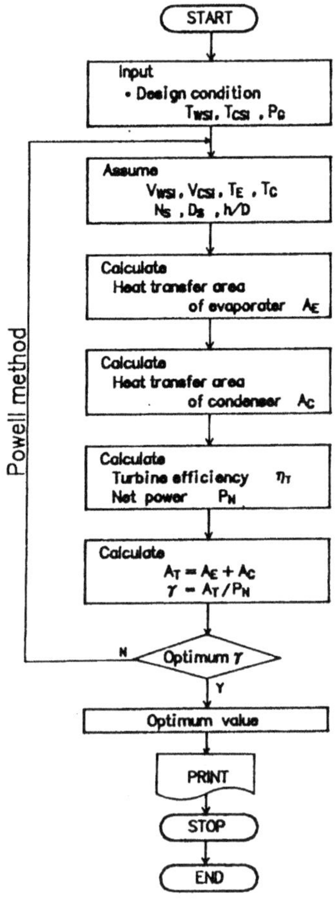

Figure 4. Flow chart of the calculation of optimum design.

5. Results

Optimum design for any OTEC plant can be done using the above method. For example, the optimization results for the OTEC plant at Ho-Ping in Taiwan is shown in Table 1 [4]. The warm sea water temperature varies from 22 to 28.5℃ at Ho-Ping annually. In this paper, the warm sea water inlet temperature of 28℃ is used. Temperature at the 1000-m depth is about 4℃. The length of the cold sea water pipe is 1000 m and the diameter is 11.2 m. Ammonia is used as working fluid. The plates of the heat exchangers are made of titanium. Calculations were done for a 100-MW OTEC plant. The evaporator and the condenser were divided into four units of same size. The piping diagram of designed OTEC is shown in reference [4].

Table 1 Optimization results.

Warm sea water inlet temperature	T_{WSI}	[℃]	28.0
Cold sea water inlet temperature	T_{CSI}	[℃]	4.0
Minimum objective function	γ_{min}	[m^2/kW]	3.75
Rankine cycle efficiency	η_{Ran}	[%]	3.98
Heat transfer area of evaporator	A_E	[×10^{5}m^2]	1.38
Heat transfer area of condenser	A_C	[×10^{5}m^2]	1.30
Net power	P_N	[MW]	71.4
Warm sea water pumping power	P_{WS}	[MW]	11.1
Cold sea water pumping power	P_{CS}	[MW]	15.6
Working fluid pumping power	P_{WF}	[MW]	1.97
Warm sea water flow rate	m_{WS}	[t/s]	206
Cold sea water flow rate	m_{CS}	[t/s]	207
Working fluid flow rate	m_{WF}	[t/s]	2.27
Evaporation temperature	T_E	[℃]	22.3
Condensate temperature	T_C	[℃]	10.2
Warm sea water velocity	V_{WS}	[m/s]	1.009
Cold sea water velocity	V_{CS}	[m/s]	1.068
Specific speed	N_S	[r.p.m., m^3/s, m]	54.6
Specific diameter	D_S	[m, m^3/s, m]	1.67
Ratio of blade to diameter	h/D	[-]	0.244
Nozzle outlet angle	α_2	[deg.]	15.0
Turbine efficiency	η_T	[-]	0.905
Rotational speed	N	[×10^3 r.p.m].	3.135
Nozzle loss	Δh_N	[kJ/kg]	0.317
Rotor loss	Δh_R	[kJ/kg]	1.749
Exhaust loss	Δh_{EX}	[kJ/kg]	0.016
Rotary disc loss	Δh_D	[kJ/kg]	1.534
Wetness loss	Δh_{WET}	[kJ/kg]	0.264
Over all heat transfer coefficient of evaporator	U_E	[kW/m^{2}K]	5.39
Over all heat transfer coefficient of condenser	U_C	[kW/m^{2}K]	4.74

In Table 1 the values of the Rankine cycle efficiency, the net power, the pumping power, the optimum heat transfer area, the optimum sea water velocity in heat exchanger, the optimum flow rates of warm sea water, cold sea water and working fluid, and the pressure loss, etc., which correspond to the objective function are obtained. Although it is not shown in the table, the optimum heat exchanger specifications in more detail and the optimum turbine shapes, etc., are obtained by this optimum design.

6. Conclusion

The principles of thermodynamic design for the OTEC was discussed using the optimum design method under the given conditions and how the optimum design point for each sea water temperature considering the annual temperature variation of seawater. In this paper, as the first step of the thermodynamic optimization for the OTEC, the optimum design method for the OTEC plant under the given conditions of warm sea water temperature and cold sea water temperature is shown. The method and concept shown in this paper can be applied to other conditions and various energy systems using natural energy and lower temperature heat sources (for example, refrigerators, geothermal power plants, power plant using wasted water, etc).

Acknowledgments. The financial support for this study provided by the Japanese Ministry of Education, Science and Culture is gratefully acknowledged.

7. References

1. Uehara, H. (1995) The present status and future of ocean thermal energy conversion, *International Journal of Solar Energy*, **16**, 217-231.
2. Avery, W. and Wu, C. (1994) Renewable Energy from the Ocean, Oxford University Press.
3. Uehara, H., Ikegami, Y., Nishida, T., Kikuchi, S. and Tsuboi, K. (1997) OTEC system using Uehara cycle, *IOA97 International OTEC/DOWA Association Conf. Proceedings*, 29-38.
4. Uehara, H. and Ikegami, Y. (1990) Optimization of a closed-cycle OTEC system, *Trans. of the ASME Journal of Solar Energy Engineering*, **112-4**, 247-256.
5. Ikegami, Y. and Uehara, H. (1992) Performance analysis of OTEC plants at off-design conditions: ammonia as working fluid, *ASME, Solar Engineering 1992*, 633-638.
6. Ikegami, Y. and Uehara, H. (1992) Optimum operating condition for a closed-cycle OTEC system, *Proc. 2nd JSME-KSME Thermal Engineering Conference*, 2·313-2·316.
7. Ikegami, Y. and Uehara, H. (1994) Optimum design point for a closed-cycle OTEC system, *Proc. Fourth International Offshore and Polar Engineering Conf. Osaka*, 383-389.
8. Nakaoka, T. and Uehara, H. (1988) Performance test of a shell-and-plate type evaporator for OTEC, *Experimental Thermal and Fluid Science*, **1-3**, 283-291.
9. Nakaoka, T. and Uehara, H. (1988) Performance test of a shell-and-plate type condenser for OTEC, *Experimental Thermal and Fluid Science*, **1-3**, 275-281.

HOW TO UNIFY SOLAR ENERGY CONVERTERS AND CARNOT ENGINES

A. DE VOS
vakgroep elektronika en informatiesystemen
Universiteit Gent
Sint Pietersnieuwstraat 41
B - 9000 Gent, Belgium

1. Introduction

There exist three ways of converting solar energy into useful work:

- The photothermal way absorbs solar energy as heat and subsequently converts it into work by means of a heat engine.
- The photovoltaic way converts solar energy directly into electrical work.
- The hybrid way combines the photovoltaic and photothermal conversion.

Usually the description of photothermal and photovoltaic conversion is very dissimilar. Photothermal converters are described in terms of absorptive coatings, temperatures, insulation, heat circulation circuits, cyclic engines, etc. Photovoltaic converters are described in terms of semiconductor bandgaps, electron and hole diffusion lengths, electric voltages, etc. Through the wealth of technological details, it is very difficult to see any relationship between the fundamental principles of the two converters.

Nevertheless, one can suspect that both are in some way or another related to the most fundamental energy converter, i.e. the Carnot engine. Indeed, any solar energy converter is a thermodynamic engine working between two heat reservoirs, one at high temperature T_1 and one at low temperature T_2:

$$\begin{aligned} T_1 &= \text{temperature of the Sun} = 5762 \text{ K} \\ T_2 &= \text{temperature of the Earth} = 288 \text{ K} . \end{aligned}$$

But the complicated technological description of solar energy converters stands in strong contrast to the transparency of the Carnot theory. Indeed, let us for a moment assume we could connect a reversible heat engine simultaneously to the Sun's surface and to the Earth's surface: see Fig. 1. Let Q_1 be the heat we extract from the Sun; let Q_2 be the (waste) heat we deposit in the Earth. The produced work is W. From the first law of thermodynamics (i.e. conservation of energy):

$$W = Q_1 - Q_2 ,$$

the second law (i.e. conservation of entropy):

$$\frac{Q_1}{T_1} = \frac{Q_2}{T_2}$$

A. Bejan and E. Mamut (eds.), Thermodynamic Optimization of Complex Energy Systems, 345–362.

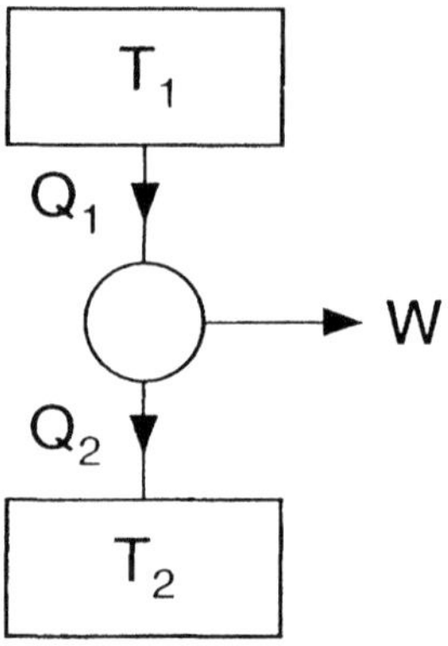

Figure 1: The Carnot engine.

and the definition of efficiency $\eta = W/Q_1$, we immediately find the Carnot formula $\eta = 1 - T_2/T_1$, yielding 95 %. Two questions arise here:

- Why does solar energy conversion, in contrast to Carnot theory, need two complicated descriptions ?
- What is the relationship between solar energy conversion efficiency and the Carnot efficiency ?

2. Endoreversible Thermal Engines

The above puzzles are illuminated by application of endoreversible thermodynamics [1, 2]. Both Novikov [3] and Curzon and Ahlborn [4] proposed a simple model, where an ideal Carnot engine is combined with a linear heat resistor. The latter has a heat-temperature characteristic

$$Q_1 = g\,(T_1 - T_3)\ . \tag{1}$$

See Fig. 2a. Here, g is a constant, called the thermal conductance.

Taking into account the Carnot theorem, we have

$$W = (1 - \frac{T_2}{T_3})\,Q_1\ . \tag{2}$$

Thus:

$$W = g\,(1 - \frac{T_2}{T_3})\,(T_1 - T_3)\ . \tag{3}$$

Now, the crucial step is to consider T_3 in (1) and (3) as a parameter, we can vary freely. Fig. 3a shows the functions $Q_1(T_3)$ and $W(T_3)$. Note that $Q_1(T_3)$ is a straight line, whereas $W(T_3)$ is a hyperbola. We will choose T_3 such that we obtain maximum power output:

$$\frac{dW}{dT_3} = 0 \ .$$

This leads to the maximum-power condition $T_3^2 - T_1T_2 = 0$, a quadratic equation in T_3, yielding immediately

$$(T_3)_{opt} = \sqrt{T_1T_2} \ .$$

Substitution subsequently into (2) and (1) immediately leads to

$$\begin{aligned} W_{max} &= \left(1 - \sqrt{\frac{T_2}{T_1}} \right) Q_1((T_3)_{opt}) \\ &= g \left(\sqrt{T_1} - \sqrt{T_2} \right)^2 . \end{aligned}$$

The factor $1 - \sqrt{T_2/T_1}$ is known as the Curzon–Ahlborn–Novikov efficiency. One can easily demonstrate that (for any T_1 and T_2 obeying $0 < T_2 < T_1$) this number is smaller than the Carnot efficiency $1 - T_2/T_1$. This property is in agreement with the fact that part of the engine is irreversible and with Carnot's law which states that any irreversible engine is less efficient than the reversible engine working between the same two temperatures.

We now relax the condition (1). This means we do not restrict ourselves anymore to linear heat transport laws. We assume that the heat flux is given by

$$Q_1 = f(T_1) - f(T_3) \ , \tag{4}$$

with $f(x)$ some mathematical function, not yet specified. This expression automatically guarantees that no heat is exchanged if no temperature difference exists (i.e. $Q_1 = 0$ whenever $T_1 = T_3$). We assume $f(x)$ monotonously increasing (in order to guarantee that heat always flows from high to low temperature). The special case $f(x) = gx$, of course, recovers the Curzon–Ahlborn–Novikov case.

If we assume that heat is exchanged not by conduction but instead by radiation, we have to apply the appropriate law. The simplest model is the case of black bodies. The heat radiated by a black body is proportional to the fourth power of its temperature, according to the Stefan–Boltzmann law. We therefore apply $f(x) = gx^4$, such that the net exchanged heat is

$$Q_1 = g \left(T_1^4 - T_3^4 \right) .$$

Thus the characteristic $Q_1(T_3)$ is a fourth-power curve, replacing the straight line of Fig. 3a. The corresponding $W(T_3)$ characteristic is a fifth-power curve:

$$W = g \left(1 - \frac{T_2}{T_3}\right)(T_1^4 - T_3^4) \ .$$

The corresponding endoreversible model is the Müser model [5]. Maximum power is extracted, if we satisfy

$$\frac{d}{dT_3} \left[\left(1 - \frac{T_2}{T_3}\right)(T_1^4 - T_3^4) \right] = 0 \ ,$$

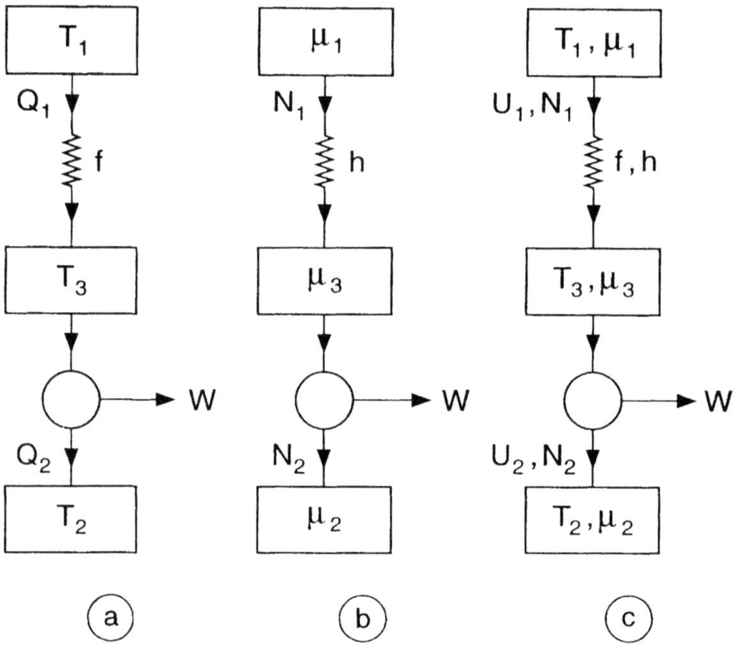

Figure 2: Endoreversible engines: (a) thermal engine, (b) chemical engine, and (c) thermochemical engine.

leading to the fifth-degree equation

$$4T_3^5 - 3T_2T_3^4 - T_1^4T_2 = 0 \ .$$

Such a polynomial equation cannot be solved analytically. We thus have to recur to numerical calculations. For $T_1 = 5762$ K and $T_2 = 288$ K, we find $T_3 = 2443$ K. If we calculate the efficiency $\eta = \frac{W}{Q_1} = 1 - \frac{T_2}{T_3}$, we find 88.2 %. However, in solar energy conversion, the tradition exists to define the efficiency as the ratio of the produced power to the incoming solar radiation alone:

$$w = \frac{W}{Q_{in}} \ .$$

Whereas the denominator Q_1 of η is $g(T_1^4 - T_3^4)$, here the denominator Q_{in} of w is merely gT_1^4, i.e. a constant. Therefore w is like a dimensionless W. This solar conversion efficiency w amounts to

$$\frac{(1 - \frac{T_2}{T_3})\ g(T_1^4 - T_3^4)}{gT_1^4} = (1 - \frac{T_2}{T_3})(1 - \frac{T_3^4}{T_1^4}) \ .$$

Substitution of the T_1, T_2, and $(T_3)_{opt}$ values leads to an efficiency w of 85.4 %.

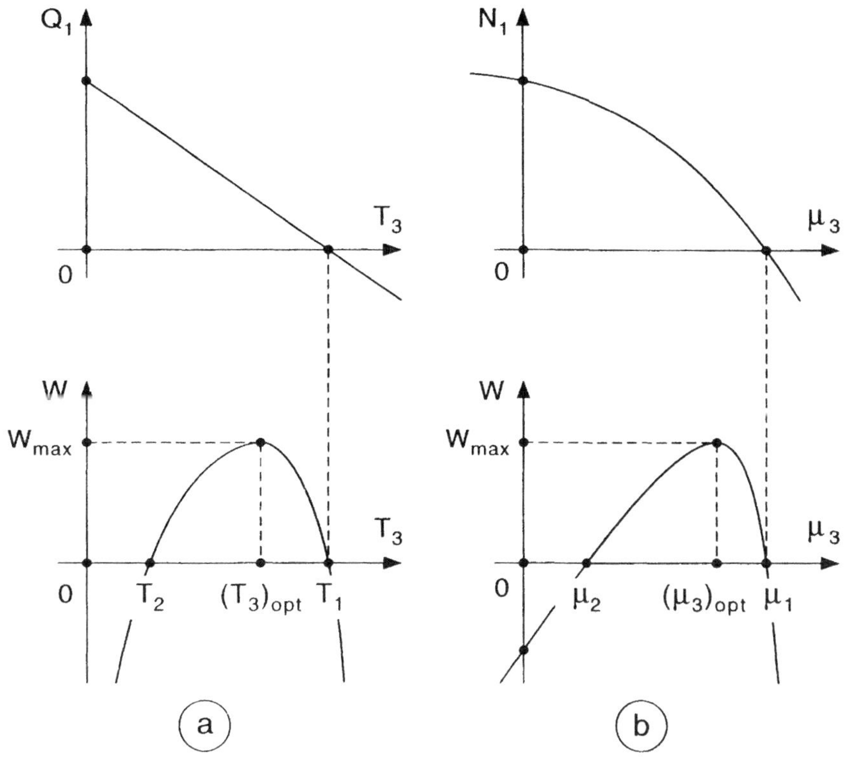

Figure 3: Characteristics of endoreversible engines: (a) heat-temperature and work-temperature characteristics of a thermal engine, and (b) particle-potential and work-potential characteristics of a chemical engine.

We now go one step further: instead of black-body radiation, we assume selective black-body radiation:

$$f(T) = \int_{E_g}^{\infty} E\, N(E,T)\, dE \ ,$$

where E is the photon energy and where the spectral density $N(E,T)$ is assumed to obey Planck's law:

$$N(E,T) = \frac{2\pi}{c^2 h^3} \frac{E^2}{\exp(\frac{E}{kT}) - 1} \ , \tag{5}$$

with c the speed of light, h Planck's constant, and k Boltzmann's constant. This constitutes a generalization of the black-body case, as the Stefan–Boltzmann law corresponds to the special case $E_g = 0$. Indeed, because of

$$\int_0^\infty \frac{x^3\,dx}{\exp(x)-1} = \frac{\pi^4}{15} ,$$

we have

$$\frac{2\pi}{c^2h^3} \int_0^\infty \frac{E^3}{\exp(\frac{E}{kT})-1}\,dE = \sigma\,T^4 ,$$

where $\sigma = \frac{2\pi^5k^4}{15c^2h^3}$ is called the Stefan–Boltzmann constant. For $E_g > 0$, we have

$$Q_1 = g\,[\int_{E_g}^\infty \frac{E^3}{\exp(\frac{E}{kT_1})-1}\,dE - \int_{E_g}^\infty \frac{E^3}{\exp(\frac{E}{kT_3})-1}\,dE\,]$$

and thus

$$W = g\,(1-\frac{T_2}{T_3})\,[\int_{E_g}^\infty \frac{E^3}{\exp(\frac{E}{kT_1})-1}\,dE - \int_{E_g}^\infty \frac{E^3}{\exp(\frac{E}{kT_3})-1}\,dE\,] .$$

Again we consider T_3 (which appears twice in the right-hand side of the latter equation) as a free parameter. The maximum power condition $dW/dT_3 = 0$ now leads to a transcendental equation in T_3. Numerical calculations (for the above values of T_1 and T_2) lead to Fig. 4a, showing the efficiency w as a function of the material's bandgap E_g. We see that the optimum bandgap is zero. The maximum efficiency is thus the black-body efficiency found earlier, i.e. 85.4 %. A finite bandgap E_g is not advantageous for a photothermal solar converter. Nevertheless, the introduction of a bandgap E_g will turn out to be fruitful, in later sections.

To model a photovoltaic solar converter is somewhat more complicated. Indeed, a solar cell is not a thermal engine but a thermochemical engine, i.e. a combination of a thermal engine and a chemical engine. Thus we first have to introduce the endoreversible chemical engine.

3. Endoreversible Chemical Engines

In a chemical engine [6] the intensive variable is the chemical potential μ, playing a role similar to temperature T in the thermal engine. The extensive variable exchanged between reservoirs is the particle flow N, playing a role similar to heat flow Q in the thermal engine. See Fig. 2b. We assume the following type of transport law holds:

$$N_1 = h(\mu_1) - h(\mu_3) ,$$

where $h(x)$ is some (monotonously increasing) mathematical function, not yet specified. This equation replaces Eq. (4), the law of conservation of energy being replaced by the law of conservation of particle number:

$$N_1 - N_2 = 0 ,$$

and the Carnot formula (2) being replaced by Gibbs's law

$$W = (\mu_3 - \mu_2)\,N_1 .$$

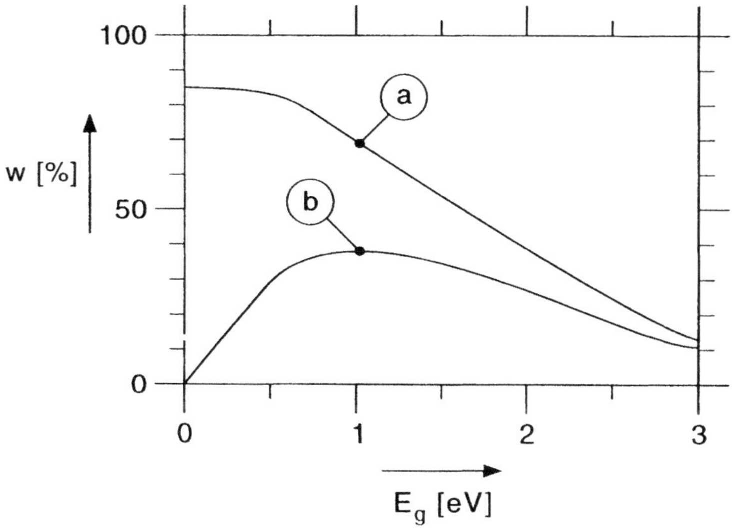

Figure 4: Efficiency of solar energy converters: (a) photothermal converter, (b) photovoltaic converter.

Thus:

$$W = (\mu_3 - \mu_2)\ [\ h(\mu_1) - h(\mu_3)\ ]\ .$$

In the same way as T_3 was the free parameter in the endoreversible thermal engine, now μ_3 is the parameter of the endoreversible chemical engine. See Fig. 3b. Maximum power output is found for

$$\frac{dW}{d\mu_3} = 0\ .$$

Applications are e.g. the linear law $h(x) = gx$ and the exponential law $h(x) = g\exp(\frac{x}{kT})$. But we will not go into the details of such models. Instead, we will, in the next section, immediately combine the chemical engine with a thermal engine.

4. Endoreversible Thermochemical Engines

In a thermochemical system, both energy and particles are exchanged between the two reservoirs # 1 and # 3. The energy current U_1 and the particle current N_1 are converted into a work flow W, with conservation of energy, i.e.

$$W = U_1 - U_2\ ,$$

conservation of particles, i.e.

$$N_1 - N_2 = 0\ ,$$

and conservation of entropy (in the reversible part):

$$\frac{U_1 - \mu_3 N_1}{T_3} = \frac{U_2 - \mu_2 N_2}{T_2} .$$

We easily deduce the unified Carnot–Gibbs law, i.e.

$$W = (1 - \frac{T_2}{T_3})\ U_1 + (\frac{T_2}{T_3}\mu_3 - \mu_2)\ N_1 .$$

Here the input currents U_1 and N_1 still have to be replaced by their appropriate transport expressions

$$\begin{aligned} U_1 &= f(T_1, \mu_1) - f(T_3, \mu_3) \\ N_1 &= h(T_1, \mu_1) - h(T_3, \mu_3) , \end{aligned}$$

such that W is a function of two variables, i.e. T_3 and μ_3. See Fig. 2c.

Maximum power is guaranteed if

$$\begin{aligned} \frac{\partial W}{\partial T_3} &= 0 \\ \frac{\partial W}{\partial \mu_3} &= 0 . \end{aligned}$$

We now investigate the case where the exchanged particles are photons. For this purpose, we apply a generalization [7, 8, 9] of Planck's spectrum (5):

$$N(E, T, \mu) = \frac{2\pi}{c^2 h^3} \frac{E^2}{\exp(\frac{E-\mu}{kT}) - 1} ,$$

such that

$$h(T, \mu) = g \int_{E_g}^{\infty} \frac{E^2}{\exp(\frac{E-\mu}{kT}) - 1}\, dE .$$

For the exchanged energy, we need to multiply the number of photons by the photon energy E, such that the E^2 factor becomes an E^3 one:

$$f(T, \mu) = g \int_{E_g}^{\infty} \frac{E^3}{\exp(\frac{E-\mu}{kT}) - 1}\, dE .$$

Thus

$$\begin{aligned} W = \quad & g\ (1 - \frac{T_2}{T_3})\ [\int_{E_g}^{\infty} \frac{E^3\, dE}{\exp(\frac{E-\mu_1}{kT_1}) - 1} - \int_{E_g}^{\infty} \frac{E^3\, dE}{\exp(\frac{E-\mu_3}{kT_3}) - 1}] \\ + \ & g\ (\frac{T_2}{T_3}\mu_3 - \mu_2)\ [\int_{E_g}^{\infty} \frac{E^2\, dE}{\exp(\frac{E-\mu_1}{kT_1}) - 1} - \int_{E_g}^{\infty} \frac{E^2\, dE}{\exp(\frac{E-\mu_3}{kT_3}) - 1}] . \end{aligned}$$

The function $W(T_3, \mu_3)$ is a surface, displayed in Fig. 5.

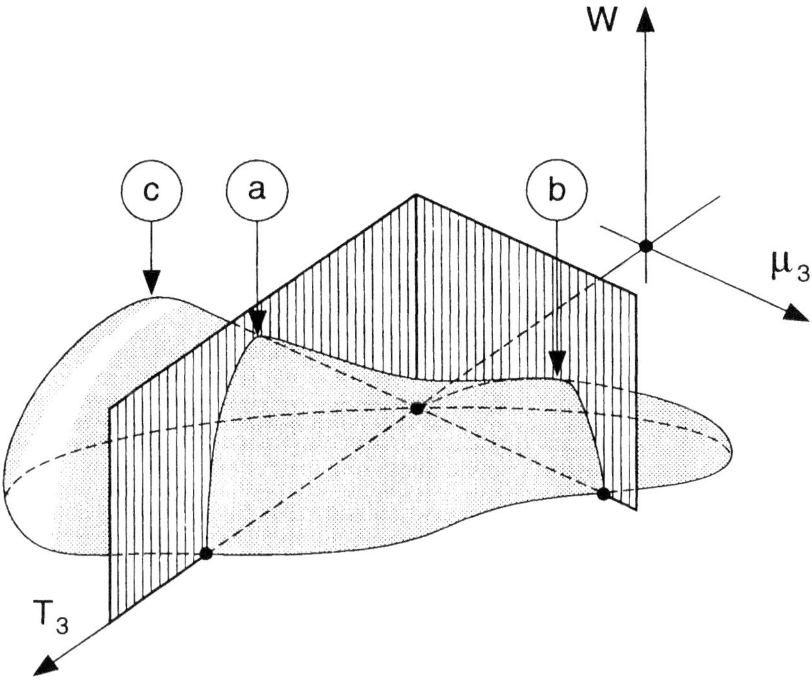

Figure 5: Solar energy conversion: (a) photothermal maximum-power point, (b) photovoltaic maximum-power point, (c) hybrid photothermal-photovoltaic maximum-power point.

For $(T_1, \mu_1) = (5762\text{ K}, 0)$ and $(T_2, \mu_2) = (288\text{ K}, 0)$, this yields a maximum w of 86.6 per cent at $(T_3, \mu_3) = (3890\text{ K}, -0.75\text{ eV})$. Because we have here two degrees of freedom (T_3 and μ_3), it is no surprise we attain higher efficiencies than the 85.4 % we can obtain with a single degree of freedom (T_3), as in Section 2. The hybrid cell functions at a high temperature T_3, but thanks to the negative potential μ_3 the radiation of the semiconductor is limited [10, 11] and thus also the (negative) terms of W containing the factor $\frac{1}{\exp(\frac{E-\mu_3}{kT_3})-1}$.

5. Solar Cells

In the previous section, we have investigated the optimum performance of the thermochemical solar energy converter, by imposing

$$\begin{aligned} \frac{\partial W}{\partial T_3} &= 0 \\ \frac{\partial W}{\partial \mu_3} &= 0\,. \end{aligned} \tag{6}$$

This results in a maximum efficiency w of 86.6 % for this so-called hybrid device. We will now investigate the performance of purely thermal as well as purely photovoltaic solar energy converters.

In a purely photothermal converter, we do not use the μ degree of freedom. We choose μ_3 equal to the terrestrial value $\mu_2 = 0$. Thus the set of conditions (6) is replaced by the 'less optimal' conditions

$$\begin{aligned} \frac{\partial W}{\partial T_3} &= 0 \\ \mu_3 &= \mu_2 \, . \end{aligned}$$

We thus intersect the $W(T_3, \mu_3)$-hill of Fig. 5 with a plane $\mu_3 = \mu_2$, i.e. $\mu_3 = 0$, and search for the maximum on the resulting $W(T_3)$-curve. It turns out to amount to 85.4 %, in accordance with Section 2.

In a purely photovoltaic converter, we do not use the T degree of freedom. We choose T_3 equal to the terrestrial value $T_2 = 288$ K. Thus the set of conditions (6) is replaced by the 'less optimal' conditions

$$\begin{aligned} T_3 &= T_2 \\ \frac{\partial W}{\partial \mu_3} &= 0 \, . \end{aligned}$$

We thus intersect the $W(T_3, \mu_3)$-hill of Fig. 5 with a plane $T_3 = T_2$ and search for the maximum on the resulting $W(\mu_3)$-curve:

$$W = g\ \mu_3 \left[\int_{E_g}^{\infty} \frac{E^2}{\exp(\frac{E}{kT_1}) - 1}\, dE - \int_{E_g}^{\infty} \frac{E^2}{\exp(\frac{E-\mu_3}{kT_2}) - 1}\, dE \right] \tag{7}$$

with $dW/d\mu_3 = 0$. The results of the calculations are displayed in Fig. 4b. We see that, in contrast to photothermal conversion, the photovoltaic conversion really needs a non-zero bandgap value in order to yield a non-zero power W. The optimal bandgap is 1.15 eV and the corresponding maximum efficiency is 40.8 %, i.e. far less than the photothermal converter.

We now remark that the chemical potential μ_3 of the emitted photons is changed by varying the bias voltage V of the semiconductor, according to the simple relation $\mu_3 = qV$. This means that by an electric voltage V we influence μ_3 and thus both intensity and spectral distribution of the emitted radiation. This is what we call electroluminescence. Applying $\mu_3 = qV$ allows us to rewrite Eq. (7) as $W(V) = V \times I(V)$ with

$$I(V) = qg \left[\int_{E_g}^{\infty} \frac{E^2}{\exp(\frac{E}{kT_1}) - 1}\, dE - \int_{E_g}^{\infty} \frac{E^2}{\exp(\frac{E-qV}{kT_2}) - 1}\, dE \right] . \tag{8}$$

This current-voltage characteristic leads to the classical $I(V)$ characteristic of a solar cell by replacing 'Bose statistics with chemical potential' by 'Bose statistics without chemical potential multiplied by Boltzmann statistics':

$$\frac{1}{\exp(\frac{E-qV}{kT}) - 1} \approx \exp(\frac{qV}{kT}) \times \frac{1}{\exp(\frac{E}{kT}) - 1} \, .$$

Indeed, this approximation converts (8) into

$$I(V) \approx qg \left[\int_{E_g}^{\infty} \frac{E^2}{\exp(\frac{E}{kT_1}) - 1} \, dE - \exp(\frac{qV}{kT_2}) \int_{E_g}^{\infty} \frac{E^2}{\exp(\frac{E}{kT_2}) - 1} \, dE \right] ,$$

or

$$I(V) \approx I_l - I_s \left[\exp(\frac{qV}{kT_2}) - 1 \right] ,$$

where I_l is called the light current and I_s the diode saturation current.

6. Tandem Solar Cells

We recall that the above calculations demonstrate that photovoltaic conversion is far less efficient than photothermal conversion. There is however a way of boosting the photovoltaic result $w = 40.8$ % of Section 5 towards higher efficiencies. This happens by introducing more than one bandgap. This is realized by combining different semiconductors, each having its own bandgap. This leads to multi-gap solar cells or 'tandem solar cells' [12, 13].

We assume that the n semiconductors are arranged in order of bandgap, the largest bandgap at the sun-side:

$$E_{g1} > E_{g2} > \ldots > E_{g(i-1)} > E_{gi} > E_{g(i+1)} > \ldots > E_{gn} .$$

See Fig. 6. The most energetic photons ($E > E_{g1}$) are absorbed in the uppermost semiconductor. Less energetic photons are absorbed in subsequent materials underneath. The last semiconductor absorbs photons with little energy ($E_{gn} < E < E_{g(n-1)}$).

If each subcell of the tandem has its own voltage V_i, then the i th semiconductor emits a spectrum

$$g \, \frac{E^2}{\exp(\frac{E - qV_i}{kT_2}) - 1} \qquad \text{for } E > E_{gi} .$$

We remark that the subcell absorbs two spectra:

- a part of the solar spectrum that transverses all semiconductors 1 to $i - 1$:

$$g \, \frac{E^2}{\exp(\frac{E}{kT_1}) - 1} \qquad \text{for } E_{gi} < E < E_{g(i-1)}$$

- a part of the spectrum emitted by the layer beneath:

$$g \, \frac{E^2}{\exp(\frac{E - qV_{i+1}}{kT_2}) - 1} \qquad \text{for } E > E_{gi} .$$

Each subcell not only has its own electric voltage V_i but also its own electric current I_i and thus produces its own electric power $W_i = V_i I_i$:

$$W_i = g \, qV_i \left[\int_{E_{gi}}^{E_{g(i-1)}} \frac{E^2 dE}{\exp(\frac{E}{kT_1}) - 1} + \int_{E_{gi}}^{\infty} \frac{E^2 dE}{\exp(\frac{E - qV_{i+1}}{kT_2}) - 1} - \int_{E_{gi}}^{\infty} \frac{E^2 dE}{\exp(\frac{E - qV_i}{kT_2}) - 1} \right] .$$

Note that this expression is similar to Eq. (7) with two modifications:

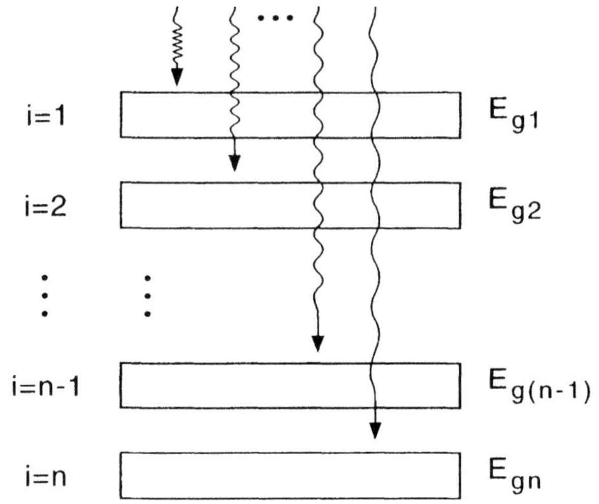

Figure 6: Tandem solar cell.

- a different upper bound of the first integral, and
- an additional term (i.e. the spectrum incident from the layer underneath).

The total work delivered per unit time is $W = \Sigma_i W_i$. We will not go into the calculation details. Suffice it to give the results [13]. Fig. 7 shows the results for $n = 2$: the efficiency as a function of the two bandgaps E_{g1} and E_{g2}. We find a maximum of 55.7 % for $E_{g1} = 1.7$ eV and $E_{g2} = 0.8$ eV. The maximum deliverable work is a strongly increasing function of n, the number of layers in the system:

- $w = 40.8$ per cent for $n = 1$,
- $w = 55.7$ per cent for $n = 2$,
- $w = 63.9$ per cent for $n = 3$,
- $w = 68.9$ per cent for $n = 4$, etc.

In the limit $n \to +\infty$, all bandgap increments $E_{gi} - E_{g(i-1)}$ become infinitesimally small. And so do the work contributions W_i. After some manipulations, we find that each subsystem acts as a monochromatic solar cell:

$$dW = V\,dI = gqV \left[\frac{E_g^2}{\exp(\frac{E_g}{kT_1}) - 1} - \frac{E_g^2}{\exp(\frac{E_g - qV}{kT_2}) - 1} \right] dE_g \,.$$

Eventually, the sum $\Sigma_i W_i$ becomes an integral:

$$W = \int_{E_g=0}^{\infty} dW = qg \int_0^{\infty} V \left[\frac{E_g^2}{\exp(\frac{E_g}{kT_1}) - 1} - \frac{E_g^2}{\exp(\frac{E_g - qV}{kT_2}) - 1} \right] dE_g \,. \qquad (9)$$

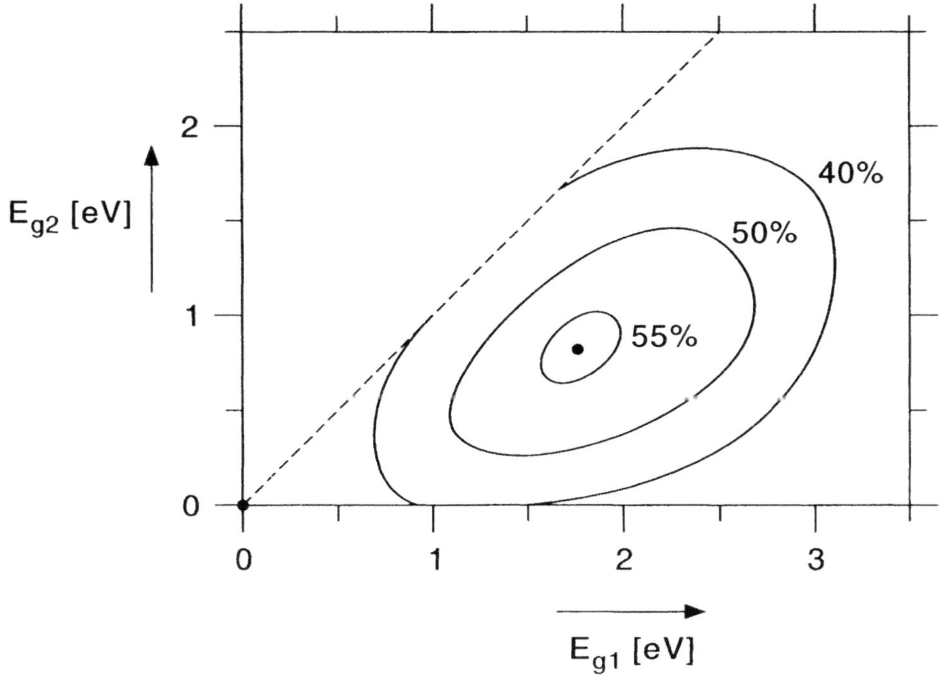

Figure 7: Efficiency of a tandem photovoltaic converter.

The maximum of the functional W is obtained for the optimal choice of the function $V(E_g)$. One finds 86.8 %.

We conclude this section with the formula for the upper bound of photovoltaic energy conversion for an arbitrary temperature ratio $t = T_2/T_1$. By merely introducing the dimensionless bandgap $u = E_g/kT_2$ and the dimensionless voltage $x = qV/kT_2$, formula (9) becomes

$$w = \frac{15t^4}{\pi^4} \int_0^\infty x \left[\frac{1}{\exp(tu) - 1} - \frac{1}{\exp(u - x) - 1} \right] u^2 du$$

where $x(u)$ satisfies

$$\frac{d}{dx} \left\{ x \left[\frac{1}{\exp(tu) - 1} - \frac{1}{\exp(u - x) - 1} \right] \right\} = 0 ,$$

i.e. where $x(u)$ is the solution of

$$\frac{(1 + x)\exp(u - x) - 1}{[\exp(u - x) - 1]^2} = \frac{1}{\exp(tu) - 1} .$$

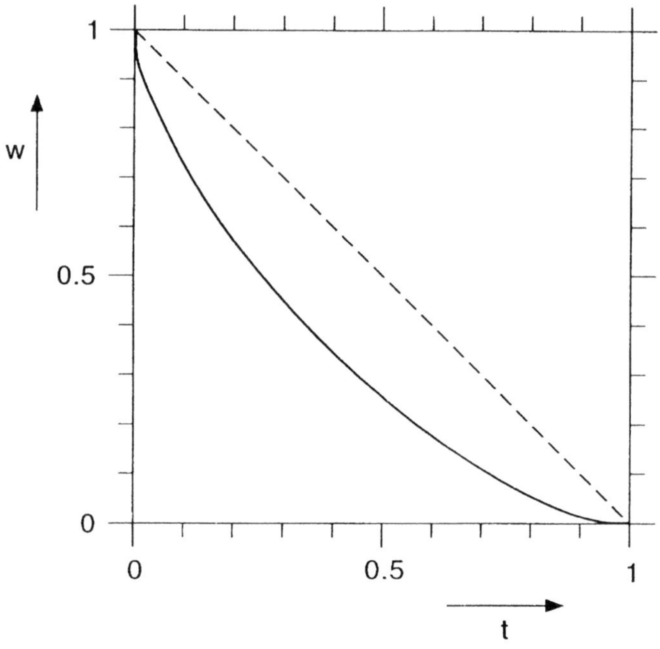

Figure 8: Efficiency of solar energy conversion.

Fig. 8 displays this function $w(t)$ and compares it with the Carnot function $1 - t$. The curve $w(t)$ is tangent to the ordinate axis for $t = 0$ and tangent to the abscissa for $t = 1$. A good approximation [14, 15] is given by:

$$w(t) \approx (1 - t)^2 \, [\, 1 + 0.3\, t \log(t) \,] \, .$$

E.g. for t = 288 K/5762 K, we find $w \approx 0.862$, close to the exact value 0.868.

7. Tandem Photothermal Converters

Not only photovoltaic converters can gain efficiency by introduction of more than one bandgap. Also photothermal converters gain from additional bandgaps. Fig. 9 shows the computational results [17] for $n = 2$: the efficiency as a function of the two bandgaps E_{g1} and E_{g2}. We find a maximum of 86.1 % for E_{g1} = 1.7 eV and E_{g2} = 0.0 eV. The maximum deliverable work is a weakly increasing function of n, the number of layers in the system:

- w = 85.4 per cent for $n = 1$,
- w = 86.1 per cent for $n = 2$,
- w = 86.3 per cent for $n = 3$,
- w = 86.4 per cent for $n = 4$, etc.

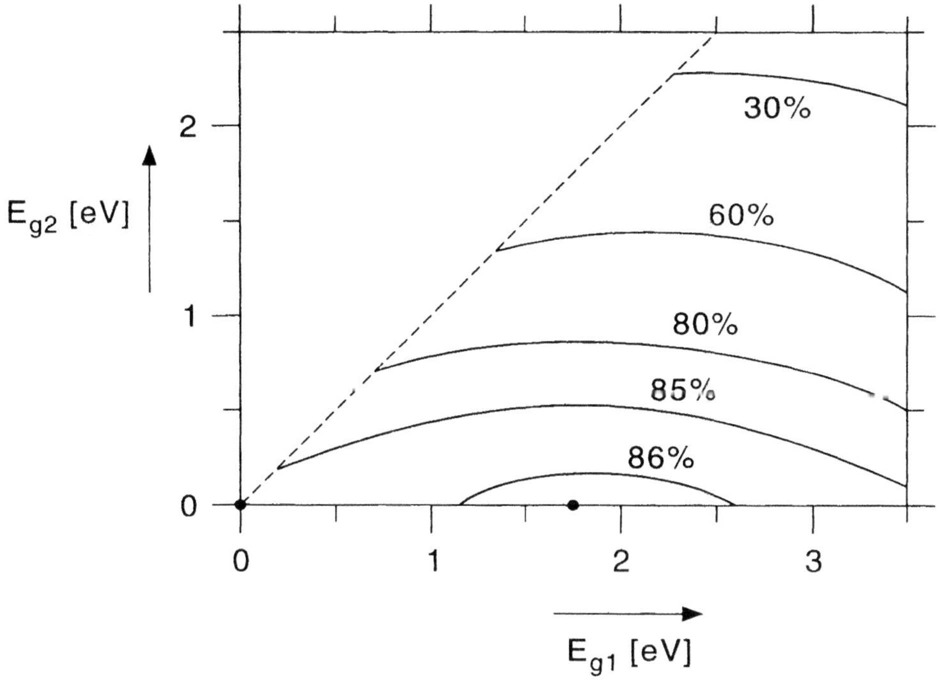

Figure 9: Efficiency of a tandem photothermal converter.

Thus the benefits from large n are not as important as in the photovoltaic case.

The case $n \to \infty$ results in an efficiency of 86.8 %, i.e. the same number as for solar cells. This is no surprise. Indeed, making all bandgap increments infinitely small results in a total power of

$$W = \int_{E_g=0}^{\infty} dW = g \int_0^{\infty} (1 - \frac{T_2}{T_3}) \left[\frac{E_g^3}{\exp(\frac{E_g}{kT_1}) - 1} - \frac{E_g^3}{\exp(\frac{E_g}{kT_3}) - 1} \right] dE_g \, .$$

In dimensionless form (with $y = T_2/T_3$) this becomes

$$w = \frac{15t^4}{\pi^4} \int_0^{\infty} (1 - y) \left[\frac{1}{\exp(tu) - 1} - \frac{1}{\exp(yu) - 1} \right] u^3 du$$

where $y(u)$ satisfies

$$\frac{d}{dy} \left\{ (1 - y) \left[\frac{1}{\exp(tu) - 1} - \frac{1}{\exp(yu) - 1} \right] \right\} = 0 \, ,$$

i.e. where $y(u)$ is the solution of

$$\frac{(1 + u - yu)\exp(yu) - 1}{[\, \exp(yu) - 1 \,]^2} = \frac{1}{\exp(tu) - 1} \, .$$

It is clear that this expression is mathematically equivalent to its photovoltaic counterpart in the previous section. This is verified by applying the substitution $x = (1 - y)u$ or $y = 1 - \frac{x}{u}$. Physically this means the following: each partial converter converts a monochromatic slice of incident solar energy, where the argument of the Bose factor $\frac{1}{\exp(\ldots)-1}$ is

$$\frac{E}{kT_1} .$$

For the monochromatic light radiated back this factor is

$$\text{either} \quad \frac{E - \mu_3}{kT_2} \quad \text{or} \quad \frac{E}{kT_3} ,$$

either by applying a chemical potential μ_3 or by changing the temperature from T_2 to T_3, respectively. The effect can be made identical. It suffices to choose μ_3 such that $\mu_3 = (1 - \frac{T_2}{T_3})E$ or to choose T_3 such that $T_3 = \frac{E}{E-\mu_3} T_2$. Such a choice is possible, as E is a number (not an interval) in the monochromatic case.

As there are only two parameters (μ and T) for influencing the Bose statistics, the above proofs that the 86.8 % result is not only the common photovoltaic and photothermal upper bound, but actually constitutes the upper bound for any conceivable solar energy converter (working between the temperatures of 5762 K and 288 K). The fact we cannot get the Carnot efficiency $1 - \frac{288}{5762} = 95.0$ % itself, is caused by the fact that absorption of radiation without simultaneous emission of radiation with the same spectrum, is inevitably an irreversible process [16].

8. Conclusion

In the present chapter, we have revealed the relationship between a solar energy converter and a Carnot engine. We have demonstrated that a solar energy converter can be modelled as an endoreversible thermochemical engine. The latter consists of two separate parts:

- an irreversible part taking care of the transport of particles (i.e. photons) and energy, by means of black body radiation and
- a reversible part taking care of the proper energy conversion.

We have found fundamental upper bounds for solar energy conversion efficiency:

- if we use a single semiconductor material:
  - 40.8 % in case of photovoltaic conversion
  - 85.4 % in case of photothermal conversion
  - 86.6 % in case of hybrid conversion;
- if we use many semiconductor materials: 86.8 % for any of the three ways of conversion.

It is necessary to stress that all theory in the present chapter was developed for sunlight combined with an ideal light concentrator, consisting either of ideal mirrors, ideal lenzes or a combination of both. For natural sunlight and for moderately concentrated sunlight, the mathematics are somewhat less transparent. For details the reader is refered to Reference [1]. Suffice it here to give the results for natural, i.e. unconcentrated, sunlight. The theoretical upper bounds are:

- if we use a single semiconductor material:
  - 31.0 % in case of photovoltaic conversion
  - 53.6 % in case of photothermal conversion
  - 67.5 % in case of hybrid conversion;
- if we use many semiconductor materials: 68.2 % for any of the three ways of conversion.

We finally like to stress that endoreversible thermodynamics not only is useful for describing solar energy conversion, but also has been successfully applied in numerous other fields, like chemistry [6, 18], climatology [19, 20], economics [21, 22, 23], computing [24, 25, 26], etc.

References

[1] De Vos, A. (1992) *Endoreversible Thermodynamics of Solar Energy Conversion*, Oxford University Press, Oxford.

[2] De Vos, A. (1993) The endoreversible theory of solar energy conversion : a tutorial, *Solar Energy Materials and Solar Cells* **31**, 75-93.

[3] Novikov, I. (1957) Effektivyj koefficient poleznovo deystvia atomnoy energeticeskoj ustanovki, *Atomnaya Energiya* **3**, 409-412; in English translation (1958) The efficiency of atomic power stations (a review), *Journal of Nuclear Energy II* 7, 125-128.

[4] Curzon, F. and Ahlborn, B. (1975) Efficiency of a Carnot engine at maximum power output, *American Journal of Physics* **43**, 22-24.

[5] Müser, H. (1957) Behandlung von Elektronenprozessen in Halbleiter-Randschichten, *Zeitschrift für Physik* **148**, 380-390.

[6] De Vos, A. (1991) Endoreversible thermodynamics and chemical reactions, *Journal of Physical Chemistry* **95**, 4534-4540.

[7] De Vos, A. and Pauwels, H. (1981) On the thermodynamic limit of photovoltaic energy conversion, *Applied Physics* **25**, 119-125.

[8] Landsberg, P. (1981) Photons at non-zero chemical potential, *Journal of Physics C: Solid State Physics* **14**, L 1025-1027.

[9] Würfel, P. (1982) The chemical potential of radiation, *Journal of Physics C: Solid State Physics* **15**, 3967-3985.

[10] De Vos, A. and Landries, J. (1992) Endoreversible thermodynamics of the hybrid photothermal-photovoltaic converter, *11 th European Photovoltaic Solar Energy Conference*, Montreux, 363-366.

[11] Spirkl, W. and Ries, H. (1995) Luminescence and efficiency of an ideal photovoltaic cell with charge carrier multiplication, *Physical Review B* **52**, 11,319-11,325.

[12] Jackson, E. (1955) Areas for improvement of the semiconductor solar energy converter, *Conference on the Use of Solar Energy*, Tucson, 122-126.

[13] De Vos, A. (1980) Detailed balance limit of the efficiency of tandem solar cells " *Journal of Physics D: Applied Physics* **13**, 839-846.

[14] Grosjean, C. and De Vos, A. (1981) On the upper limit of the energy conversion efficiency in tandem solar cells, *Journal of Physics D: Applied Physics* **14**, 883-894.

[15] De Vos, A., Grosjean, C., and Pauwels, H. (1982) On the formula for the upper limit of photovoltaic solar energy conversion efficiency, *Journal of Physics D: Applied Physics* **15**, 2003-2015.

[16] De Vos, A. and Pauwels, H. (1983) Comment on a thermodynamical paradox presented by Würfel, *Journal of Physics C: Solid State Physics* **16**, 6897-6909.

[17] De Vos, A. and Vyncke, D. (1983) Solar energy conversion : photovoltaic versus photothermal conversion, *5 th E.C. Photovoltaic Solar Energy Conference*, *Αϑιναι*, 186-190.

[18] De Vos, A. (1995) Thermodynamics of photochemical solar energy conversion, *Solar Energy Materials and Solar Cells* **38**, 11-22.

[19] De Vos, A. and Flater, G. (1991) The maximum efficiency of the conversion of solar energy into wind energy, *American Journal of Physics* **59**, 751-754.

[20] De Vos, A. and van der Wel, P. (1993) The efficiency of the conversion of solar energy into wind energy by means of Hadley cells, *Theoretical and Applied Climatology* **46**, 193-202.

[21] De Vos, A. (1995) Endoreversible thermoeconomics, *Energy Conversion and Management* **36**, 1-5.

[22] De Vos, A. (1997) Endoreversible economics, *Energy Conversion and Management* **38**, 311-317.

[23] De Vos, A. (1998) Endoreversible thermodynamics versus economics, *Energy Conversion and Management*, to be published.

[24] De Vos, A. (1995) Reversible and endoreversible computing, *International Journal of Theoretical Physics* **34**, 2251-2266.

[25] De Vos, A. (1996) Introduction to r-MOS systems, *4 th Workshop on Physics and Computation*, Boston, 92-96.

[26] De Vos, A. (1997) Towards reversible digital computers, *European Conference on Circuit Theory and Design*, Budapest, 923-931.

OPTIMAL CONTROL FOR MULTISTAGE ENDOREVERSIBLE ENGINES WITH HEAT AND MASS TRANSFER

S. SIENIUTYCZ
Faculty of Chemical Engineering, Warsaw University of Technology
00-645 Warsaw, 1 Warynskiego Street, Poland

1. Multistage Endoreversible Process with Coupled Heat and Mass Transfer

This work makes an effort to synthesize a number of recent results obtained within the theory of active (work-producing) heat processes and to generalize these results to the so-far unsolved case of active simultaneous heat and mass transfer. Reviews on the theory of engines with pure heat transfer are available [1-3] which display various contributions to the classical theory of the Novikov-Curzon-Ahlborn process [4, 5]. A description of an engine process with active simultaneous transfer of heat and mass is, however, more complicated than that for an engine with pure heat transfer, mainly due to the complexity of the relation describing the continuity of the entropy flux through the internal reversible part of the engine. This complexity makes derivation of the power generation function difficult, and, even if the function is finally found, is so complex that an analytical solution to a work maximization problem cannot be obtained. Thus certain effective numerical approaches need to be sought. Consequently, the main task of this work is twofold: i) to find complex analytical characteristics of a single-stage system and ii) to develop effective numerical algorithms for optimization of multistage cascades, which allow to exploit available energy contained in finite resources more efficiently than in the single-stage objects. In this paper we accomplish these goals under the assumption of the Lewis analogy between the heat transfer and the mass transfer [6, 7]. Since, to our knowledge, most of the currently-published results on FTT of chemical engines are restricted to isothermal transfer [3, 8, 9], we believe that a successful approach to nonisothermal transfer of energy and matter should be an important step towards the development of a realistic theory of nonisothermal chemical engines.

As previously [1] we are dealing with a finite-time extension of the classical problem of maximizing the work produced by a nonequilibrium engine system operating with two flowing fluids or a fluid and a bath, i.e. an infinite reservoir. The fluids have finite thermal conductivity and diffusivity, and hence nonvanishing resistances in their boundary layers and the irreversible properties of any such heat-transfer system. The process operates at a steady state, as represented in the multistage system shown schematically in Fig. 1 for the case of a fluid and a bath. For the purpose of optimization, we work with a standard representation of this multistage process as a discrete optimal control problem in terms of a set of difference equations. The multistage process is, in fact, a sequence of *generalized* Novikov-Curzon-Ahlborn processes (NCA processes; [3-5]). These repeat in all the stages of the cascade; the only differences among them are the temperatures of each

A. Bejan and E. Mamut (eds.), Thermodynamic Optimization of Complex Energy Systems, 363–384.

stage, which must be consistent with energy balance. The multistage syster contains: the driving fluid, which flows with temperatures $T^1..T^N$ and concentration of a single active component $X^1..X^N$ through the stages; the environment at th constant temperature T^e and concentration X^e; the boundary layers which act as th thermal conductances; and the set of the reversible engines, $R^1..R^N$, generalize Carnot machines which generate the mechanical work at each stage n.

At each stage the bulk of the fluid is well mixed, and a mixing scale is assigned t the fluid flow. The fluxes of the work produced in stages are summed up, so that cumulative average power output W is obtained at the last stage. The power output pe unit flux of the driving fluid, W = W/G, has units of work per unit mole of the iner fluid, and describes the molar work associated with the steady-flow process of powe production. In classical thermodynamics such processes involve fluids and Carnc engines only; no boundary layers are considered. Their continuous quasistatic limi leads to the classical exergy for a reversible process; it has no rate term. It is th irreversible multistage process with sequence of endoreversible machines of NC/ type which leads to a finite-time counterpart of the available energy (exergy) of th driving fluid.

By definition, the work released in the engine mode is positive. An optimizatio problem for the cascade can be stated for the extremum work W= W/G, consisten with extremum of power W. For the cascade of engines, where the work is releasec the work W has to be maximized. For the cascade of heat pumps, where the work i supplied to the system, the minimization of the negative of the work, (-W), i required. When one of the boundary states of the resource fluid is the state o equilibrium with the environment (e.g. $T^N = T^e$, $X^N = X^e$), the maximal specifi work represents an extended or finite-time exergy of the flowing fluid. We shall se that this extended exergy simplifies to the classical thermal exergy in the reversibl limit when the overall-stage conductances, g^n, tend to infinity.

An active, work-producing, transfer process between two fluids causes destructio of initial non-equilibrium structure and approach of the resource fluid to equilibriun with the reservoir fluid. For this process the problem of the maximum work delivere in a finite time is a relevant optimization problem. This (first) version of the proces is called the engine mode. In the second version, called the heat-pump mode, th system leaves thermodynamic equilibrium and work is added. The system works the as a heat pump, and the associated optimization problem is that of minimizing th work supplied in a finite time.

In classical, reversible thermodynamics, the associated magnitudes of th mechanical work are the same for the two operations. Finite-time thermodynamics, theory for irreversible processes, is a schema showing that (and why) the magnitud of the heat-pump-mode work is always larger than that of the engine-mode work.

We assume a steady-state process in the cascade. Each stage n, which is describe by its difference equations (Sec.6) as a distinct "black-box" unit, has nonetheless it own internal structure shown in Fig. 1 characteristic of that stage. This structure i composed of the following elements: a well-mixed cell of bulk fluid (the shadowe upper part of the stage n, with the inlet temperature T^{n-1} and outlet temperature T^n the thermal resistances of the two boundary layers, a reversible engine R^n, and contact zone with an isothermal bath (the shadowed lower part of the stage). Th bath is assumed to be thermally homogeneous at the outset and retains it homogeneity for all time because it is infinite. The mixing is assumed ideal withi each nth fluid cell, consistent with the thermal homogeneity for pipeline fluid within each stage and at the outlet from the stage

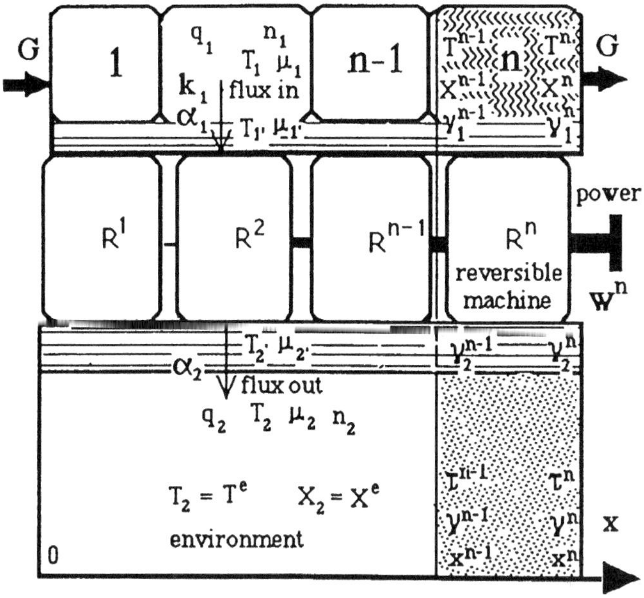

Figure 1. Scheme of work production when a moist gas exchanges the heat and matter with an infinite environment in a sequence of endoreversible thermal machines.

2. An Analysis of a Single-Stage Process to Obtain the Power Formula

We consider balances of energy, entropy and mass assuming that a single active component (e.g. moisture) and the sensible heat q can be transferred in an inert gas whose molar flow is G. The energy balance,

$$\varepsilon_1 = \varepsilon_2 + w , \tag{1}$$

and the mass balance in terms of molar fluxes,

$$n_1 = n_2 , \tag{2}$$

are combined with an equation describing the continuity of the entropy flux in the reversible part of the system

$$(\varepsilon_1 - \mu_{1'} n_1)/T_{1'} = (\varepsilon_2 - \mu_{2'} n_2)/T_{2'} . \tag{3}$$

This yields the power expression

$$w = \varepsilon_1 - \varepsilon_2 = q_{1'}(1 - T_{2'}/T_{1'}) + T_{2'}(\mu_{1'}/T_{1'} - \mu_{2'}/T_{2'})n_1 . \tag{4}$$

In terms of the heat flux q, the energy flux in the resistive parts of the system has the continuity property, for example,

$$\varepsilon_1 = q_1 + h_1 n_{1'} = q_{1'} + h_{1'} n_{1'} = \varepsilon_{1'} . \tag{5}$$

Using the primed part of this equation in the power production formula one finds

$$w = (1 - T_{2'}/T_{1'})\, q_{1'} + [(h_{1'} - h_{2'}) - T_{2'}(s_{1'} - s_{2'})]n_{'} . \tag{6}$$

We conclude that the combination of the second law and the reversible balance of the entropy yields the power expression (6) in which the thermal efficiency of an endoreversible process

$$\eta \equiv 1 - T_{2'}/T_{1'} \tag{7}$$

is the component of a two-dimensional vector of efficiency. The second component of this vector is the exergy-like function of the active component evaluated for the primed state 2 as the reference state

$$\beta' \equiv [h_{1'} - h_{2'} - T_{2'}(s_{1'} - s_{2'})]n_{1'} . \tag{8}$$

The primed quantities and equations will further be used in a transformed form expressinfg all physical quantities in terms of the bulk state variables of both fluids and certain control variables. The latter will be related to fluxes of heat and matter in a simple way as shown in the following section.

Using in Eq. (6) the energy flux continuity (5) we tend to obtain the power formula in a 'flux representation'

$$w = \eta(q_1, n_1)q_1 + \beta(q_1, n_1)n_1 \tag{9}$$

rather than in an 'efficiency representation'

$$w = \eta q_1(\eta, \beta) + \beta n_1(\eta, \beta) . \tag{10}$$

The energy flux continuity (5) is applied in Eq. (6) to eliminate the flux $q_{1'}$ on account q_1. This yields

$$w = q_1(1 - T_{2'}/T_{1'}) + \beta n_1 , \tag{11}$$

where β is defined as

$$\beta = [c_p(T_1 - T_{1'})(1 - T_{2'}/T_{1'}) + (h_{1'} - h_{2'}) - T_{2'}(s_{1'} - s_{2'})]n_1 \tag{12}$$

and, as η, it should be expressed in terms of q_1 and n. By this the power w can finally be expressed in terms of q_1 and n. These variables are good indicators of the process intensity; their simple rate counterparts will be accepted as control variables in all power optimization problems formulated in this work.

3. Description of Transfer Processes with Rates as Control Variables

When considering the simultaneous heat and mass transfer we use mainly the transfer conductances in units of transferred mass, g rather than the heat transfer conductances, g', used frequently in descriptions of pure heat transfer. They both are connected by the relationship g'= cg. In the formulae below we exploit the analogy between the heat and mass transfer which yields for the conductances

$$d\gamma = \frac{\alpha'}{c} dA = k dA = \frac{\alpha'}{c} a_v F dx = k a_v F dx = \frac{\alpha'}{c} a_v F v dt \tag{13}$$

We also define $d\gamma' = c d\gamma$, where $d\gamma'$ is the cumulative conductance in units of heat. Now we introduce the so-called 'height of the transfer unit'

$$\frac{Gdx}{d\gamma} = \frac{GdV}{Fd\gamma} = \frac{GcdV}{F\alpha' dA} = \frac{GdV}{FkdA} = \frac{Gc}{\alpha' a_v F} = \frac{G}{k a_v F} = H_{TU} \tag{14}$$

and the differential of a nondimensional length called the 'number of transfer units'

$$d\tau \equiv \frac{dx}{H_{TU}} = \frac{d\gamma}{G} = \frac{\alpha' a_v F}{Gc} dx = \frac{k a_v F}{G} dx = \frac{k a_v F v}{G} dt \ . \tag{15}$$

Note that a finite increment of τ, designated by θ, equals g/G. From the mass balance, the driving mass flux, N, satisfies dN = - GdX, where dX is the differential change of the moisture concentration in the first fluid along the length dx. Hence the following equations describe the change of the cumulative heat and mass fluxes, dQ_1 and dN_1 along the coordinates of cumulative conductances, γ' and γ

$$\frac{dQ}{d\gamma'} = -\frac{GcdT}{\alpha' dA} = -\frac{GcdT}{\alpha' a_v F dx} = -\frac{dT}{d\tau} \equiv -u \tag{16}$$

and

$$\frac{dN}{d\gamma} = -\frac{GdX}{kdA} = -\frac{GdX}{k a_v F dx} = -\frac{dX}{d\tau} \equiv -v \ . \tag{17}$$

Subscript 1 refers to the first fluid (a finite resource fluid). In Section 6 reference is made to such formulae for cascades composed of finite number of finite-size stages. The above equations prove that the rate of change of T_1 and X_1 in *nondimensional* time τ may be chosen as the process control variables. The control u has the temperature units, and the control v has nondimensional units of mole of the active component per mole of inert gas, as the concentration X. They both are components of the control vector **u** = (u, v).

In terms of the mass conductance g_1 an equation for the first intermediate temperature $T_{1'}$ can be derived. Each of the primed parameters should be now expressed in terms of the state variables of both fluids and the accepted control variables. The operations use the continuity of the energy flux and yield successively

$$T_{1'} = T_1 - q_{1'}/(c_1 g_1) \tag{18}$$

$$T_{1'} = T_1 - q_1 + c_p n(T_1 - T_{1'})/(c_1 g_1) \tag{19}$$

whence the heat flux q_1 follows as

$$q_1 = (c_1 g_1 - c_1 n)(T_1 - T_{1'}) \tag{20}$$

and

$$T_{1'} = T_1 - q_1/(c_1 g_1 - c_p n) \ . \tag{21}$$

For a process of pure heat transfer (n=0) one has

$$\begin{aligned} T_{1'} &= T_1 - \frac{q_1}{cg_1} = T_1 - \frac{dQ_1}{cd\gamma_1} = T_1 - \left(\frac{dQ_1}{cd\gamma}\right)\left(\frac{d\gamma}{d\gamma_1}\right) \\ &= T_1 + \left(\frac{GcdT_1}{\alpha a_v F dx}\right)\left(\frac{g}{g_1}\right) = T_1 + \frac{g}{g_1}\left(\frac{dT_1}{d\tau}\right) \end{aligned} \tag{22}$$

which means that

$$q_1 = - cgdT_1/d\tau = - cgu \equiv - g'u \ . \tag{23}$$

A derivation of the related formula for $T_{2'}$ is more difficult. Here only an outline of the derivation is given and a detailed derivation will be presented elsewhere. This temperature follows from the reversible balance of the entropy through the internal engine

If Newton's law of the heat exchange holds, the temperature $T_{1'}$ in terms of the heat flux q_1 is given above by Eq. (21). For the primed concentrations analogous relations follow from the Fick law of mass transfer. They are

$$X_{1'} = X_1 - n/g_1 \tag{24}$$

and

$$X_{2'} = X_2 + n/g_2 \ . \tag{25}$$

To obtain the second intermediate temperature $T_{2'}$ and concentration $X_{2'}$ one substitutes Eqs. (21), (24) and (25) into the reversible entropy balance

$$q_{1'}/T_{1'} + s_{1'}(T_{1'}, X_{1'})n = q_{2'}/T_{2'} + s_{2'}(T_{2'}, X_{2'})n \tag{26}$$

in which the mass balance (2) was taken into account. For the accepted equations of heat and mass transfer, the equation of the continuity for the entropy flux through the engine becomes

$$g_1 c(T_1 - T_{1'})/T_{1'} + s_{1'} g_1 (X_1 - X_{1'}) = g_2 c(T_{2'} - T_2)/T_{2'} + s_{2'} g_2 (X_{2'} - X_2) \tag{27}$$

Using the primed flux $q_{1'}$ as the process variable this equation can be written as

$$\frac{q_{1'}}{T_1 - q_{1'}/(c_1 g_1)} + s_1'(T_1 - \frac{q_{1'}}{c_1 g_1}, X_1 - \frac{n}{g_1})n$$
$$= g_2 c(T_{2'} - T_2)/T_{2'} + s_2'(T_{2'}, X_2 + \frac{n}{g_2})n . \tag{28}$$

Yet, this formula should be expressed in terms of the heat q_1 which is a direct indicator of temperature changes in the fluid bulk. The heat q_1 is related to the heat $q_{1'}$ driving the engine by the formula

$$q_{1'} = q_1 + c_1(T_1 - T_{1'})n \tag{29}$$

the formula describing the continuity of entropy flux is

$$\frac{q_1 + c_p(T_1 - T_{1'})n}{T_1 - (c_1 g_1 - c_p n)^{-1} q_1} + s_1'(T_1 - \frac{q_1}{c_1 g_1 - c_p n}, X_1 - n/g_1)n$$
$$= g_2 c(T_{2'} - T_2)/T_{2'} + s_2'(T_{2'}, X_2 + n/g_2)n . \tag{30}$$

This equation along with Eqs. (18) and (29) defines the temperature $T_{2'}$ as an implicit function of the fluxes q_1 and n. To work with this equation we assume an isobaric entropy function for a humid gas in the form [7]

$$S(T, X) = c\ln(T/T^*) + R\Phi(X) , \tag{31}$$

where

$$\Phi(X) \equiv (1+ X)\ln(1+X) - X\ln X . \tag{32}$$

For this structure of the entropy we derive an equation for the temperature $T_{2'}(q_1, n)$ in the form

$$T_{2'} = T_2\left(\frac{T_1 - q_1(c_1 g_1 - c_p n)^{-1}}{T_1 - (g_1/g)(c_1 g_1 - c_p n)^{-1} q_1}\right)\left(\frac{(1+X_1)^{(1+X_1)}(X_1 - \frac{n}{g_1})^{(X_1 - \frac{n}{g_1})}}{X_1^{X_1}(1+X_1 - \frac{n}{g_1})^{(1+X_1 - \frac{n}{g_1})}}\right)^{(\frac{g_1 R}{g_2 c_2})}$$
$$\left(\frac{(1+X_2)^{(1+X_2)}(X_2 + \frac{n}{g_2})^{(X_2 + \frac{n}{g_2})}}{X_2^{X_2}(1+X_2 + \frac{n}{g_2})^{(1+X_2+\frac{n}{g_2})}}\right)^{(-\frac{g_2 R}{g_2 c_2})} . \tag{33}$$

The associated thermal efficiency follows in the form

$$\eta = 1 - \frac{T_2}{T_1 - (\frac{c_1 g_1}{c_1 g_1 - c_1 n})\frac{q_1}{g'}} \left(\frac{(1+X_1)^{(1+X_1)}(X_1 - \frac{n}{g_1})^{(X_1 - \frac{n}{g_1})}}{X_1^{X_1}(1+X_1 - \frac{n}{g_1})^{(1+X_1 - \frac{n}{g_1})}} \right)^{(\frac{g_1 R}{g_2 c_2})} \left(\frac{(1+X_2)^{(1+X_2)}(X_2 + \frac{n}{g_2})^{(X_2 + \frac{n}{g_2})}}{X_2^{X_2}(1+X_2 + \frac{n}{g_2})^{(1+X_2 + \frac{n}{g_2})}} \right)^{(-\frac{g_2 R}{g_2 c_2})} . \tag{34}$$

Our approach takes also into account the primed quantities which characterize the mass transfer. In view of the equality $n_1 = n_2$, we have

$$X_{1'} = X_1 - \frac{n_1}{g_1} = X_1 - \frac{dN_1}{d\gamma_1} = X_1 - (\frac{dN_1}{d\gamma})(\frac{d\gamma}{d\gamma_1}) = X_1 + (\frac{GdX_1}{ka_vFdx})\frac{g}{g_1} = X_1 + (\frac{GdX_1}{ka_vFdx})\frac{g}{g_1} = X_1 + \frac{g}{g_1}(\frac{dX_1}{d\tau}) \tag{35}$$

and

$$X_{2'} = X_2 + \frac{n_1}{g_2} = X_2 + \frac{dN_1}{d\gamma_2} = X_2 + (\frac{dN_1}{d\gamma})(\frac{d\gamma}{d\gamma_2}) = X_2 - (\frac{GdX_1}{ka_vFdx})\frac{g}{g_2} = X_2 - (\frac{GdX_1}{ka_vFdx})\frac{g}{g_2} = X_2 - \frac{g}{g_2}(\frac{dX_1}{d\tau}) . \tag{36}$$

Therefore the continuity equation for the transferred active component

$$n_1 = g_1(X_1 - X_{1'}) = g_2(X_{2'} - X_2) = n_2 \tag{37}$$

is identically satisfied by the formulae for $X_{1'}$ and $X_{2'}$ as shown by equations

$$n_1 = g_1(X_1 - X_{1'}) = - g(\frac{dX_1}{d\tau}) = - gv \tag{38}$$

$$n_2 = g_2(X_{2'} - X_2) = - g(\frac{dX_1}{d\tau}) = - gv . \tag{39}$$

Because of coupling between the heat transfer and the transfer of mass, an equation for the temperature $T_{1'}$ takes into account the effect of transfer of moisture vapor

$$T_{1'} = T_1 - \frac{q_1}{cg_1 - c_p n} = T_1 - \frac{dQ_1}{cd\gamma_1 - c_p dN} = T_1 - \frac{(d\gamma/d\gamma_1)(dQ_1/cd\gamma)}{1 - (c_p/c)(dN/d\gamma)(d\gamma/d\gamma_1)} = T_1 + \frac{(g/g_1)dT_1/d\tau}{1 + (c_p/c)(g/g_1)(dX/d\tau)} . \tag{40}$$

This equation holds along with the mass transfer equation above. Hence, in terms of the accepted controls, u and v, we obtain the following set of equations for the primed quantities

$$T_{1'}(u, v) = T_1 + \frac{ug/g_1}{1 + (c_p/c)(g/g_1)v} = T_1 + \frac{u}{g_1/g + (c_p/c)v} , \tag{41}$$

$$T_{2'}(u, v) = T_2\left(\frac{T_1 + (c_1 g_1/g + c_p v)^{-1} c_1 u}{T_1 + (g_1/g)(c_1 g_1/g + c_p v)^{-1} c_1 u}\right)$$

$$\left(\frac{(1+X_1)^{(1+X_1)}(X_1 + \frac{gv}{g_1})^{(X_1 + \frac{gv}{g_1})}}{X_1^{X_1}(1+X_1 + \frac{gv}{g_1})^{(1+X_1+\frac{gv}{g_1})}}\right)^{(\frac{g_1 R}{g_2 c_2})}\left(\frac{(1+X_2)^{(1+X_2)}(X_2 - \frac{gv}{g_2})^{(X_2 - \frac{gv}{g_2})}}{X_2^{X_2}(1+X_2 - \frac{gv}{g_2})^{(1+X_2-\frac{gv}{g_2})}}\right)^{(-\frac{g_2 R}{g_2 c_2})}. \tag{42}$$

Moreover, for the primed concentrations

$$X_{1'} = X_1 + gv/g_1 , \quad X_{2'} = X_2 - gv/g_2 . \tag{43, 44}$$

From the last two equations we conclude that whenever process satisfies the equality $X_{1'} = X_{2'}$ then the concentration change $v = dX_1/d\tau$ satisfies

$$- v = X_1 - X_2 . \tag{45}$$

Along with the equality $T_{1'} = T_{2'}$ this defines the so-called 'short-circuit point', the point at which the production of the power in the system equals zero. This production vanishes also at the point of vanishing fluxes. Thus, for some moderate values of the fluxes, there exists a point in the system at which the power is maximum.

4. Flux and Rate Representations of Power and Related Quantities

The formulae for the temperature $T_{2'}$, given above, correspond to the second heat flux at the interface contacting the engine

$$q_{2'}(q_1, n) = g_2 c_2 (T_{2'}(q_1, n) - T_2) . \tag{46}$$

Absence of phase changes is assumed here. We should work, however, with the flux q_2 to describe an effect of the partial enthalpy of vapor of the active component

$$q_2 = q_{2'} + c_p(T_{2'} - T_2)n . \tag{47}$$

Hence, using the expression for $T_{2'}$, one obtains the second heat flux in the form

$$q_2(q_1, n) = (g_2 c_2 + c_p n)(T_{2'}(q_1, n) - T_2) = (g_2 c_2 + c_p n)T_2\Bigg\{ \left(\frac{T_1 - q_1(c_1 g_1 - c_p n)^{-1}}{T_1 - \frac{g_1}{g}(c_1 g_1 - c_p n)^{-1} q_1}\right)$$

$$\left(\frac{(1+X_1)^{(1+X_1)}(X_1 - \frac{n}{g_1})^{(X_1 - \frac{n}{g_1})}}{X_1^{X_1}(1+X_1 - \frac{n}{g_1})^{(1+X_1-\frac{n}{g_1})}}\right)^{(\frac{g_1 R}{g_2 c_2})}\left(\frac{(1+X_2)^{(1+X_2)}(X_2 + \frac{n}{g_2})^{(X_2 + \frac{n}{g_2})}}{X_2^{X_2}(1+X_2 + \frac{n}{g_2})^{(1+X_2+\frac{n}{g_2})}}\right)^{(-\frac{g_2 R}{g_2 c_2})} - 1\Bigg\}. \tag{48}$$

Consequently the second energy flux

$$\varepsilon_2 = q_2 + h_2 n = q_2 + c_p T_2 n \tag{49}$$

follows in the form

$$\varepsilon_2(q_1, n) = (g_2 c_2 + c_p n)(T_{2'}(q_1, n) - T_2) + h_2 n\,, \tag{50}$$

where $T_{2'}$ is defined by Eq. (33). This serves to find the power produced. Since

$$\varepsilon_1 = q_1 + h_1(T_1)n = q_1 + c_p T_1 n \tag{51}$$

and

$$\varepsilon_2(q_1, n) = (g_2 c_2 + c_p n)(T_{2'}(q_1, n) - T_2) + h_2(T_2)n \tag{52}$$

the power output can be obtained as

$$w_1 = \varepsilon_1 - \varepsilon_2 = (q_1 + h_1 n) - (q_2 + h_2 n) = q_1 + h_1(T_1)n - \varepsilon_2(q_1, n)\,. \tag{53}$$

From this equation and Eq. (50) the power expression in terms of fluxes follows for $T_{2'}$ defined by Eq. (33),

$$w(q_1, n) = q_1 + c_1 T_1 n + g_2 c_2 T_2 - (g_2 c_2 + c_p n) T_2 \Big\{ \Big(\frac{T_1 - q_1(c_1 g_1 - c_p n)^{-1}}{T_1 - (g_1/g)(c_1 g_1 - c_p n)^{-1} q_1} \Big)$$
$$\Big(\frac{(1+X_1)^{(1+X_1)}(X_1 - \frac{n}{g_1})^{(X_1 - \frac{n}{g_1})}}{X_1^{X_1}(1+X_1 - \frac{n}{g_1})^{(1+X_1 - \frac{n}{g_1})}} \Big)^{(\frac{g_1 R}{g_2 c_2})} \Big(\frac{(1+X_2)^{(1+X_2)}(X_2 + \frac{n}{g_2})^{(X_2 + \frac{n}{g_2})}}{X_2^{X_2}(1+X_2 + \frac{n}{g_2})^{(1+X_2 + \frac{n}{g_2})}} \Big)^{(-\frac{g_2 R}{g_2 c_2})} \Big\}\,. \tag{54}$$

The negative ratio of the power w and the conductance g defines the Lagrangian L of the problem in terms of the controls u and v. It has units of the energy per mole. We also use the quantity $f_0 = -L$ i.e. the power output per unit molar conductance $g = G\theta$

$$dW/d\tau = w/g = w/(\theta G) \equiv f_0 = -L\,. \tag{55}$$

To obtain the work produced per unit mole of the fluid, we integrate over τ the function f_0 expressed in terms of controls and state. Substituting $q = -gcu$ and $n = -gv$ to the power formula and using the nondimensional conductance $\theta = g/G$ yields

$$f_0(u, v) = \frac{w}{g} = c_1 u - c_1 T_1 v + \frac{g_2}{g} c_2 T_2 - (\frac{g_2}{g} c_2 - c_p v) T_2 \Big\{ \Big(\frac{T_1 + (c_1 \frac{g_1}{g} + c_p v)^{-1} c_1 u}{T_1 + \frac{g_1}{g}(c_1 \frac{g_1}{g} + c_p v)^{-1} c_1 u} \Big)$$
$$\Big(\frac{(1+X_1)^{(1+X_1)}(X_1 + \frac{gv}{g_1})^{(X_1 + \frac{gv}{g_1})}}{X_1^{X_1}(1+X_1 + \frac{gv}{g_1})^{(1+X_1 + \frac{gv}{g_1})}} \Big)^{(\frac{g_1 R}{g_2 c_2})} \Big(\frac{1+X_2)^{(1+X_2)}(X_2 - \frac{gv}{g_2})^{(X_2 - \frac{gv}{g_2})}}{X_2^{X_2}(1+X_2 - \frac{gv}{g_2})^{(1+X_2 - \frac{gv}{g_2})}} \Big)^{(-\frac{g_2 R}{g_2 c_2})} \Big\}\,. \tag{56}$$

This equation allows us to find a maximum for the cumulative mechanical work W when a finite-resource fluid changes its thermodynamic parameters in a finite time between two assumed states. However, we will first show that the first power formula reduces *exactly* to that of pure heat transfer when n = 0.

5. Correspondence with Pure Heat Transfer

For the vanishing flows q_1 and n the power w equals zero. In absence of mass transfer (n = 0) one obtains from the above equation

$$w(q_1, 0) = q_1 - g_2c_2 T_2\{(\frac{T_1 - q_1(c_1g_1)^{-1}}{T_1 - (g_1/g)(c_1g_1)^{-1}q_1}) - 1\} \qquad (57)$$

or, equivalently

$$w(q_1, 0) = q_1 - g_2c_2 T_2\{\frac{q_1}{\{T_1 - (g_1/g)(c_1g_1)^{-1}q_1\}/\{(g_1/g - 1)(c_1g_1)^{-1}\}}\} . \qquad (58)$$

By transforming the denominator a simpler form of the above result can be obtained

$$w(q_1, 0) = q_1 - g_2c_2 T_2\{\frac{q_1}{g_2c_2(T_1 - q_1/(g_1c_1)) - q_1}\} . \qquad (59)$$

Consistently, an equation describing the pure heat flux $q_2 = q_1 - w$ takes the form

$$q_2(q_1, 0) = g_2c_2(T_{2'} - T_2) = g_2c_2T_2\frac{q_1}{g_2c_2(T_1 - q_1/(g_1c_1)) - q_1} . \qquad (60)$$

This was obtained previously in the context of thermal multistage NCA processes [1]. It can also be shown that, in absence of the mass transfer flux, n = 0, the analytical solution with respect to the temperature $T_{2'}$ is obtained in the form

$$T_{2'} = \frac{T_2}{1 - q_1[g_2c_2(T_1 - q_1/(c_1g_1))]^{-1}} \qquad (61)$$

which is consistent with the above equation for the heat flux q_2.

The above power formula can also be transformed into the form

$$w = q_1(1 - \frac{g_2c_2T_2}{g_2c_2(T_1 - q_1/(g_1c_1)) - q_1}) = q_1(1 - \frac{T_2}{T_1 - q_1((g_1c_1)^{-1} + (g_2c_2)^{-1})}) , \qquad (62)$$

where the last expression contains the overall conductance of standard inactive heat transfer. This conductance is defined in the standard way as

$$g' = \left(\frac{1}{g_1c_1} + \frac{1}{g_2c_2}\right)^{-1} = \frac{g_1c_1g_2c_2}{g_1c_1 + g_2c_2} = gc \ . \tag{63}$$

Consequently, in absence of mass transfer, one arrives at a simple equation linking the power w with the driving heat flux $q_1 = -gcu$

$$w = q_1\left(1 - \frac{T_2}{T_1 - gcq_1^{-1}}\right) = -gc\left(1 - \frac{T_2}{T_1 + u}\right)u \tag{64}$$

and the corresponding first-law efficiency of the stage ($g' = gc$)

$$\eta = \frac{w}{q_1} = 1 - \frac{T_2}{T_1 - g'q_1^{-1}} = 1 - \frac{T_2}{T_1 + u} \ . \tag{65}$$

and the heat q_1 follows from this equation in the form

$$q_1 = g(T_1 - T_2/(1-\eta)) \ . \tag{66}$$

In the efficiency representation, the power produced, $w = \eta q_1(\eta)$, becomes

$$w = g\eta(T_1 - T_2/(1 - \eta)) \ . \tag{67}$$

in agreement with ref. [3]. The Lagrangian of the thermal process in an infinite cascade of NCA stages follows as the negative power w per unit conductance g, or the work obtained from the one mole of the resource fluid at flow

$$L \equiv -f_0 = -\frac{w}{g} = -\frac{q_1}{g}\left(1 - \frac{T_2}{T_1 - gcq_1^{-1}}\right) = c\left(1 - \frac{T_2}{T_1 + u}\right)u \ . \tag{68}$$

The efficiency of an engine stage deviates from the Carnot efficiency with the finite heat flux q_1. For a quasistatic transfer, the efficiency becomes the Carnot efficiency. Otherwise, for a sufficiently large q_1, equal to $g'(T_1 - T_2)$, which corresponds with the pure heat conduction (no power) the efficiency vanishes. The efficiency formula shows that, in a finite rate process, the efficiency is Carnot-like, with an effective temperature of the upper source $T' = T_1 + u$. In the engine mode , where q_1 is positive (the resource fluid is cooled), the quantity T' is reduced due to the finite flux q_1, and the efficiency decreases with q_1. The power expression, Eq. (68), shows the deviation from the Carnot theory caused by the nonvanishing heat current. The Carnot efficiency is achieved when the effect of the overall resistance $(g')^{-1}$ is negligible or the flux q_1 is very low. The maximum power at the stage corresponds to q_1 satisfying

$$\frac{\partial w}{\partial q_1} = \frac{(T_1 - q_1/g')^2 - T_1T_2}{(T_1 - q_1/g')^2} = 0 \ . \tag{69}$$

Whence, the driving heat flux at the maximum power conditions (subscript 0)

$$(q_1)_0 = g'[T_1 - (T_1T_2)^{1/2}] . \tag{70}$$

It may be verified that the second derivative of w is negative at the extremum, so that the extremum is a maximum. Note that the extremum q_1 is lower than the purely conductive heat flux $q_1 = g'(T_1 - T_2)$ which takes place at the zero efficiency η. When the result described by Eq. (70) is substituted into the efficiency formula, Eq. (65), the well-known efficiency of the Novikov engine at maximum power,

$$\eta_0 = 1 - (T_2/T_1)^{1/2} \tag{71}$$

is obtained. Equation (69) leads to the following equation for the maximum power

$$w_0 = g'[(T_1)^{1/2} - T_2)^{1/2}]^2 \tag{72}$$

in agreement with the standard theory of thermal machines which uses the heat transfer conductance $g' = gc$. When mass transfer plays a role, the results for optimal values for the fluxes n, q_1, q_2 and w should be found numerically.

Thus we have proven the correspondence of our theory of active simultaneous heat and mass transfer with the classical theory of the Novikov-Curzon-Ahlborn processes. Change of control variable yields an equivalent description. Thus it is permissible to choose a convenient control variable, typically one preferred for technical reasons. Use of the driving heat as a control has some virtues for making physical interpretation simple: with this choice, the reversible point is the point at which the control variable q_1 is zero, and the irreversibility description in terms of q_1 leads to formulae similar to those used frequently for describing 'inactive' transfer. On the other hand, transition from one control to another is possible.

6. Applying Single-Stage Formulae to a Multistage Process

Now we return to the process of simultaneous heat and mass transfer to consider cascade processes with finite number of stages. The state changes between the boundary states of a stage are no longer infinitesimal. We call such processes genuine discrete processes. We add the stage number superscript, n, to all symbols in the above formulae when we apply them to a cascade. Yet, since in the infinite reservoir case T_2 and X_2 are only parameters and not state variables, we can also simplify our designation for the state variables T_1^n, X_1^n and the heat q_1^n by rejecting the subscript 1, unnecessary at the present time. Thus we will use the symbols T^n and q^n for the resource fluid temperature and driving heat flux at the stage n. Also we will use the symbols X^n and n^n for the resource fluid concentration and mass flux at the stage n. We shall also designate the constant thermal reservoir temperature as T^e rather than T_2^n and its concentration as X^e rather than X_2^n.

In genuine discrete processes we use the cumulative power per unit mass flow, $W^n \equiv \Sigma w^n/G$, a quantity whose units are those of work per mole. When an equation describing the power is applied at the stage n with the identifications $T_2 = T^e$, $T_1 = T^n$, $X_2 = X^e$ and $X_1 = X^n$ and we introduce a nondimensional conductance θ^n at the stage n, such that

$$\theta^n \equiv g^n/G\ , \tag{73}$$

then $u^n = -(g'^n)^{-1}q^n = -(Gc\theta^n)^{-1}q^n$ satisfies

$$q^n = -cu^n g^n = -Gcu^n\theta^n\ . \tag{74}$$

Similarly as for continuous processes, we also define a quantity v^n as the mass flux per unit mass flow G and unit nondimensional time θ^n by the equation

$$n^n = -v^n g^n = -Gv^n\theta^n\ . \tag{75}$$

We also obtain a state equation for the molar work in terms of the controls u^n and v^n

$$W^n - W^{n-1} = f_0^n\theta^n\ , \tag{76}$$

where f_0 is the function appearing in Eq. (56).

Modeling of cascades with finite number of stages is precisely a discrete counterpart of that for continuous processes, described earlier. The total work is the sum of the expressions $f_0^n\theta^n$ over the stages; this sum replaces the work integral of a continuous process, which is the integral over $f_0 d\tau$. Again, taking into account that the conductance g^n may be expressed as the product of an overall coefficient of mass transfer and a corresponding overall transfer area, we identify the nondimensional time interval θ^n as the number of transfer units at the stage n, $\theta^n = \Delta\tau^n$. The ratio of the stage length ℓ^n to θ^n is a quantity with units of length; it is identified with the height of the transfer unit at the stage n, H_{TU}^n. From its definition and Eq. (73), H_{TU}^n equals $G\ell^n/g^n$. The smallness of this quantity is a good indicator of the quality of transfer at the stage; the quantities H_{TU}^n exhibit very small values for large molar conductances g^n. However it will be sufficient to use the dimensionless variable θ^n only. One may note that the variable, θ^n, appears linearly in our discrete model of optimization, a fact we will use in optimizing the model. To do this, we first need the discrete equation of state, which describes the driving heat.

To find the equation of state for the temperature T^n, we define the cumulative driving heat Q^n for the first n stages of the cascade: $Q^n = \Sigma q^k$, where k = 1, 2..n. This quantity satisfies the equality $Q^n - Q^{n-1} = -Gc(T^n - T^{n-1}) = q^n$, where, because of Eq. (74), the heat q^n is the product of $-g'^n$ ($= -Gc\theta^n$) $= \alpha' a^n$ and u^n. The comparison of these two expressions yields a state equation for the temperature of the resource fluid

$$T^n - T^{n-1} = u^n\theta^n\ . \tag{77}$$

This equation tells us that $u^n = -q^n/g'^n$, i.e. the negative of the heat received by the $n^{\underline{th}}$ reversible engine (in units of temperature), at the same time plays the role of the of the discrete rate of the temperature change, $\Delta T^n/\Delta\tau^n$.

To find the equation of state for the concentration X^n an analogous reasoning holds. The cumulative molar flux N^n for the first n stages of the cascade: $N^n = \Sigma n^k$, where k = 1, 2..n. This quantity satisfies the equality $N^n - N^{n-1} = -G(X^n - X^{n-1}) = n^n$, where, because of Eq. (75), the mass flux n^n is the product of $g^n = G\theta^n$ and v^n. The comparison of these two expressions yields the equation of state. Thus an equation of state for the concentration changes follows in the form

$$X^n - X^{n-1} = v^n\theta^n, \tag{78}$$

where v^n is the rate change of the concentration per unit of the nondimensional time τ. To close the model, we recall that θ^n is the increment of the independent variable τ^n. This variable is nondimensional, of definite (positive) sign, and measures the cumulative number of heat transfer units. The variable τ^n satisfies the definition $\tau^n = \Sigma\theta^k$ for k = 1, 2..n. Thus last discrete equation of state is simply

$$\tau^n - \tau^{n-1} = \theta^n . \tag{79}$$

Equation (79) is essential in problems with the constrained sum of θ^n. This corresponds to the constrained total transfer area, because the single-stage transfer areas are contained in g^n and hence in θ^n. This also corresponds to a constraint on the total length. On comparing these results with those for continuous processes, we conclude that, in finite-n cascades, the discrete state equations follow from a simple discretizing operation applied for the derivatives $dT/d\tau = u$, $dX/d\tau = v$ and $d\tau/d\tau \equiv 1$.

In accordance with the terminology used in optimization, Eqs. (76) - (79) are the discrete equations of state for the cascade. They contain on their right-hand sides the state variables (in f_0^n, Eq. (56)) and controls. An optimizing procedure for the cascade requires specification of an optimization criterion, a complete set of the state equations (as above), and possibly some local constraints at the stage.

7. General Optimal Control for Discrete Systems Linear in Holdup Times

For the endoreversible process considered here, we now show the application of two efficient algorithms, of which the first is based on the dynamic programming and the second on the so-called discrete maximum principle with a constant Hamiltonian [10-13]. Each can be applied to endoreversible cascades with complex function f_0. The existence of the maximum principle has been shown for discrete models which are linear and homogeneous with respect to a particular unconstrained control variable. This control may be an unconstrained interval of one of the state variables, such as our interval of the dimensionless time, θ^n, or an unconstrained interval of an extensive parameter. The model's linearity with respect to θ^n is crucial for the formal similarity of the necessary conditions for optimality in discrete and continuous processes because, broadly speaking, this linearity causes vanishing the second order and higher terms in Taylor expansions of characteristic functions for discrete processes. Consequently discrete formulae acquire the structures already known for continuous processes.

Our aim here is a brief analysis of these properties for *discrete* (multistage) processes with finite numbers of stages. For discrete processes, the standard discrete theory of optimal control [14, 15] does not predict any special similarity of discrete and continuous descriptions, e.g. such characteristic features of the continuous theory as constancy of an autonomous Hamiltonian or the Hamilton-Jacobi equation [16, 17].

This is because discrete descriptions are generally not reducible to continuous ones in the limit of an infinite number of discrete units. To assure that a discrete model converges to the continuous limit and symplectic properties, one must restrict oneself to a special class of discrete processes in which finite intervals of state variables are not constrained explicitly, and the allowed constraints can affect only ratios of differences of the state variables at any stage. To satisfy these requirements, a discrete set of the state equations and an optimization algorithm must be linear with respect to a particular decision, θ^n. Whenever a discrete model has a structure linear in θ^n, a remarkable similarity emerges between necessary optimality conditions in both continuous and discrete cases. In particular, a discrete maximum principle emerges in in a form analogous to that known for continuous systems. In the case of endoreversible cascades, it is our optimization model, Eqs. (76)-(79) which shows the similarity discussed. The outline of an abstract analytical description with occasional references to the endoreversible cascades is given below.

Applying the usual description based on ordinary difference equations, the optimization theory for Eqs. (76) - (79) of the endoreversible cascade can be represented by a general set set of equations

$$x_i^n - x_i^{n-1} = f_i^n(\mathbf{x}^n, \mathbf{u}^n)\theta^n \tag{80}$$

Here i = 0, 1...s, s+1 and $f_{s+1}^n = 1$ for the (s+1)-th coordinate of the state vector $\mathbf{x}^n$, consistent with Eqs. (76)- (79) for θ^n which is the unconstrained interval of the variable $\tau \equiv x_{s+1}^n$. In the case of our endoreversible cascade, for which s=2, the state vector $\mathbf{x}^n$ is four-dimensional: $x_0^n \equiv W^n$, $x_1^n \equiv T^n$, $x_2^n = X_2^n$ and $x_3^n \equiv \tau^n$, for each stage n. The set (80) is now considered as a proper vector representation of the original model of our endoreversible cascade. For this set the optimization problem is stated as maximizing the zero-th state coordinate for n = N when the initial states of $\mathbf{x}^n$ are fixed. To derive necessary optimality conditions for for Eqs. (76)-(74) or general Eq. (50), Bellman's method of dynamic programming (DP) can be applied [17-19]. This allows one to pass from the DP algorithm to the algorithm of discrete maximum principle, both being effective computational algorithms.

Let us define the optimal performance function for a general n-stage process

$$I^n(\mathbf{x}^n) \equiv \max \sum_1^n f_0^n(\mathbf{x}^n, \mathbf{u}^n)\theta^n = -\min \sum_1^n -f_0^n(\mathbf{x}^n, \mathbf{u}^n)\theta^n, \tag{81}$$

where f_0^n is the work generation function described by Eq. (56) with added stage index n. In Eq. (81) one assumes that $\mathbf{x} = (x_1, x_2, ...x_{s+1} = \tau)$ i.e. that the performance coordinate x_0^n has been excluded from the coordinates of the state vector. It may be shown that the function I^N incorporates the minimum entropy production with a negative sign, and , for fixed end states, the work maximization is equivalent with minimization of the entropy production. This is in agreement with general thermodynamic principles [2, 21, 22].

For a general n-stage process, in the stages 1, 2,...,n described by Eqs. (80) and (81), Bellman's recurrence equation is

$$I^n(\mathbf{x}^n) = \max_{u^n, \theta^n}\{f_0^n(\mathbf{x}^n, \mathbf{u}^n)\theta^n + I^{n-1}(\mathbf{x}^n - \mathbf{f}^n(\mathbf{x}^n, \mathbf{u}^n)\theta^n)\} . \tag{82}$$

This corresponds with the so-called forward algorithm of dynamic programming where the characteristic function I is generated in terms of the final states. This algorithm is illustrated in Figure 2 which explains the principle of generation of optimal profits.

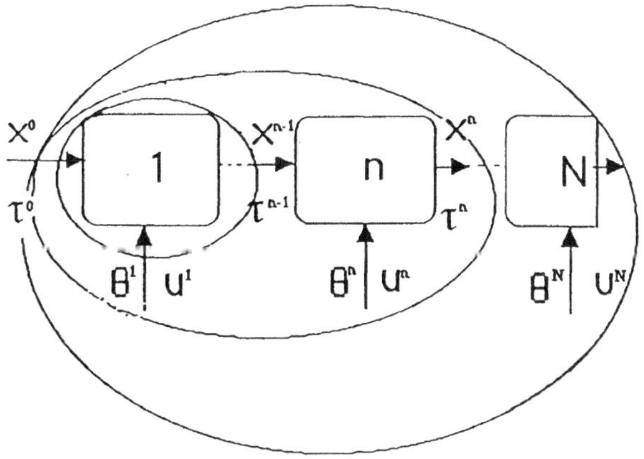

Figure 2. Application of Bellman's principle of optimality for the forward algorithm of the dynamic programming method. Elipse-shaped balance areas refer to sequential computational subprocesses, which evolve by inclusion of remaining stages.

Starting with $I^0=0$, the sequence of the optimal functions $I^1, I^2, \ldots I^n \ldots I^N$ can be obtained with a computer by a well-established recurrence procedure [18, 19] in which extremizations on the right sides are carried out with respect to controls at constant coordinate values of the final state $\mathbf{x}^n$. The tables describing the computer output contain the numerical values of I^n, u^n, v^n and θ^n, in terms of T^n, X^n and τ^n. Dimensionality reduction is possible [1] for τ^n, which improves the data accuracy.

Now we outline the passage to the second solving method, a maximum principle. To pass to a maximum principle one writes Eq. (82) in the form

$$\max_{u^n, \theta^n}\{f_0^n(\mathbf{x}^n, \mathbf{u}^n)\theta^n - (I^n(\mathbf{x}^n) - I^{n-1}(\mathbf{x}^n - \mathbf{f}^n(\mathbf{x}^n, \mathbf{u}^n)\theta^n)\} = 0 \tag{83}$$

which enables us to include variations of final coordinates of state [11, 20]. For an unconstrained θ^n and possibly constrained $\mathbf{u}^n$, the set equivalent to Eq. (83) has the form of the three equations

$$f_0^n(\mathbf{x}^n, \mathbf{u}^n)\theta^n - (I^n(\mathbf{x}^n) - I^{n-1}(\mathbf{x}^n - \mathbf{f}^n(\mathbf{x}^n, \mathbf{u}^n)\theta^n) = 0 , \tag{84}$$

$$f_0^n(\mathbf{x}^n, \mathbf{u}^n) - (\partial I^{n-1}/\partial \mathbf{x}^{n-1}).\mathbf{f}^n(\mathbf{x}^n, \mathbf{u}^n) = 0 , \tag{85}$$

$$\max_{u^n}\{ f_0^n(\mathbf{x}^n, \mathbf{u}^n)\theta^n - (I^n(\mathbf{x}^n) - I^{n-1}(\mathbf{x}^n - \mathbf{f}^n(\mathbf{x}^n, \mathbf{u}^n)\theta^n)) \} = 0 . \tag{86}$$

In Eq. (85) the stationarity condition for the extremal intervals θ^n was applied. Whenever θ^n is finite and positive, then, after using Eqs. (84) and (85) in (86), the last equation can be transformed to

$$\max_{u^n}\{ f_0^n(\mathbf{x}^n, \mathbf{u}^n) - (\partial I^{n-1}/\partial \mathbf{x}^{n-1}).\mathbf{f}^n(\mathbf{x}^n, \mathbf{u}^n)\} = 0 \tag{87}$$

which is the strong maximum principle for the so-called enlarged Hamiltonian

$$H^{n-1} \equiv f_0^n(\mathbf{x}^n, \mathbf{u}^n) - \frac{\partial I^{n-1}}{\partial \mathbf{x}^{n-1}}.\mathbf{f}^n(\mathbf{x}^n, \mathbf{u}^n) = f_0^n - \sum_{i=1}^{s} \frac{\partial I^{n-1}}{\partial x_i^{n-1}} f_i^n - \frac{\partial I^{n-1}}{\partial \tau^{n-1}} \tag{88}$$

with respect to the controls $\mathbf{u}^n$. As f_{s+1} (=1) is $\mathbf{u}$-independent, this optimality condition can also be expressed in terms of the energylike Hamiltonian which does not contain the partial derivative of I^{n-1} with respect to the time,

$$H^{n-1} \equiv f_0^n - \sum_{i=1}^{s} (\partial I^{n-1}/\partial x_i^{n-1}) f_i^n . \tag{89}$$

Equation (89) represents the Hamiltonian (88) without the $\mathbf{u}$-independent term $\partial I^{n-1}/\partial \tau^{n-1}$. Next the so-called adjoint variables are defined which are the partial derivatives of I^n with respect to the state coordinates x_i^n,

$$z_i^{n-1} \equiv - \partial I^{n-1}(\mathbf{x}^{n-1})/\partial x_i^{n-1} . \tag{90}$$

(i = 1...s, s+1.) The Hamiltonian (89) must be a maximum with respect to the controls $\mathbf{u}^n$ which maximize the work production, $W^N = \Sigma f_0^n \theta^n$, whose optimal function I^N is defined by Eq. (81) in which the work intensity f_0^n is described by Eq. (56).

Differentiating the bracketed expression in Eq. (83) to determine its stationarity conditions with respect to the final space and time coordinates, we obtain an optimal difference set which is canonical with respect to two sort of equations, one defining the changes of state and one the changes of the adjoints. Using the energylike Hamiltonian, Eq. (89), expressed in terms of the adjoint variables,

$$H^{n-1}(\mathbf{x}^n, \mathbf{z}^{n-1}, \mathbf{u}^n, \tau^n) \equiv f_0^n(\mathbf{x}^n, \mathbf{u}^n, \tau^n) + \sum_{i=1}^{s} z_i^{n-1} f_i^n(\mathbf{x}^n, \mathbf{u}^n, \tau^n) \tag{91}$$

the computational algorithm of maximum principle is obtained in the canonical form

$$\frac{x_i^n - x_i^{n-1}}{\theta^n} = \frac{\partial H^{n-1}}{\partial z_i^{n-1}} , \qquad \frac{z_i^n - z_i^{n-1}}{\theta^n} = - \frac{\partial H^{n-1}}{\partial x_i^n} , \tag{92, 93}$$

$$\frac{H^n - H^{n-1}}{\theta^n} = \frac{\partial H^{n-1}}{\partial \tau^n}, \tag{94}$$

$$z_\tau^{n-1} + \max_{u^n} H^{n-1}(x^n, z^{n-1}, u^n, \tau^n) = 0 \ . \tag{95}$$

(n=1,...N ; i=1,...s and l =1,...r.) Equation (92) constitutes the Hamiltonian form of the state equations, and Eq. (93) is its adjoint equation. Equation (94) sets the Hamiltonian interval at the stage *n*, whereas Eq. (95) states that the enlarged Hamiltonian $H^{n-1} = H^{n-1} + z_\tau^{n-1}$ of the extremal process is always constant and equal zero. Equation (95) describes the discrete maximum principle in the allowable range U. This is a very powerful computational algorithm which can effectively be used even in case of very complex functions f_0, such as that given by Eq. (56).

8. Work Maximizing in Endoreversible Cascades and Finite-Time Exergy

Since we possess an analytical expression for the power generation, f_0, either Bellman's Eq. (82) or discrete maximum principle, Eqs. (92) - (95), can be used to numerically generate optimal controls and optimal profit functions. When one of the end states is that of ambient a finite-time exergy, A, is generated

$$\begin{aligned} A(T, X) = {} & (c_g + Xc_p)[(T - T^e) - T^e \ln\frac{T}{T^e}] + RT^e\{X\ln[\frac{X(1 + X^e)}{(1 + X)X^e}] \\ & + \ln[\frac{1 + X^e}{1 + X}]\} \pm T^e S_\sigma(T, X, T^e, X^e, \tau^f) \ . \end{aligned} \tag{96}$$

Of the two parts of the finite-time exergy the classical one is known from reversible thermodynamics, thus one may use the tables of A to evaluate also the associated mimimal entropy production for the process. Equation (96) refers to a sufficiently large N, approximating well the continuous limit. Its last term contains the minimum entropy production S_σ as a function of the end states and the non-dimensional duration, τ^f, which is the number of mass transfer units. This last term is the nonclassical or path-dependent term which vanishes for infinite durations and which should be distinguished from the first or classical term that has potential properties. The upper (plus) sign at the last term refers to processes departing from the equilibrium and the lower (minus) sign to the proceses approaching to the equilibrium. With the knowledge of the classical exergy, shown in the above equation, our numerical procedure can generate both data for A and S_σ. The enhanced bounds resulting from the finite-time exergy are illustrated in Fig. 3.

9. Concluding Remarks

Enhanced thermodynamic bounds limiting the work delivered or consumed in a finite time constitute the main benefit resulting from generalized exergies. Because they include the role of dissipative factors and the dynamic nature of bounds, these are stronger, more useful and informative than those stemming from the classical exergy.

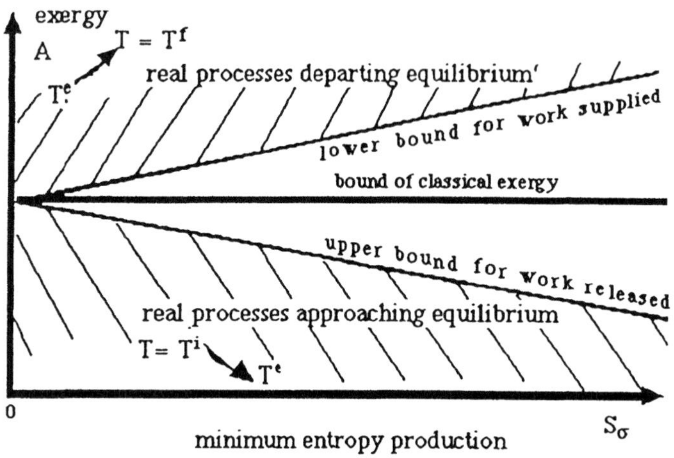

Figure 3. Finite-time exergy A of a limiting continuous processes ($N \rightarrow \infty$) prohibits processes from operating below the heat-pump mode line which is the lower bound for work supplied (lbws) and or above the engine mode line which is the upper bound for work produced (ubwp). These bounds, more realistic than those based on the classical, reversible definition of exergy, are the result of finite rates consistent with finite duration of the process.

To understand the problem of bounds and their distinction for the work production and consumption, recall that the work-producing process is the inverse of the work-consuming process (the final state of the second process is the initial state of the first, and conversely), when we have fixed the durations of the two processes to be the same. In thermostatics the two bounds on the work, the bound on the work produced and that on the work consumed, coincide. However the static limits are often too far from reality to be really useful. The generalized exergy provides bounds stronger than those predicted by the classical exergy. They do not coincide for processes of work production and work consumption, and they are 'thermokinetic' rather than 'thermostatic' bounds. Only for infinitely long durations or for processes with excellent transfer (an infinite number of transfer units) do the thermokinetic bounds reduce to their classical thermostatic limits. The hysteresis, or divergence of the bounds (Fig. 3), proves that many idealized processes allowed by classical thermostatics are prohibited by the more severe and realistic constraints of thermokinetics. The hysteretic effect, caused by the dissipation and the associated increase of the exergy supplied in the pump mode (and decrease of exergy released in the engine mode), reveals the extent to which thermokinetic bounds are stricter than thermostatic bounds. The greater the mean rate we demand, the greater is the range of performance excluded by limits predicted by the generalized exergy.

A real process which does not apply the optimal protocol but has the same boundary states and duration as the optimal sequence, requires a real work supply that can only be larger than the finite-rate limit obtained by optimization. Similarly, the real work delivered from a nonequilibrium system (with the same boundary states and duration but with a suboptimal control) can only be lower that the corresponding finite-rate limit. Thus the two bounds for a process and its inverse, which coincide in thermostatics, diverge in thermodynamics, at a rate that grows with any of the indices of deviation from static behavior, f_0, S_σ, L or H. For sufficiently high values of rate indices, the work consumed may far exceed the classical work; the work produced can even vanish. Thus we can confront and surmount the limitations of applying classical thermodynamic bounds to real processes. This is a direction with open opportunities, especially for chemical or electrochemical systems.

Acknowledgments. The author acknowledges an invitation and involvement of Profs. A. Bejan and R.S. Berry, and a support from NATO ASI and the EC program 'Carnet'.

Glossary of Symbols

A - exergy; a_v - specific area; c- molar heat capacity; f_0- work generation function; g_1, g_2 - partial conductances; g- overall conductance; G- molar flux of resource fluid; H- hamiltonian; H_{TU}- height of transfer unit; h- partial molar enthalpy of component; I- optimal performance function; k-molar mass transfer coefficient; n- molar flux of active component; q_1-driving heat; Q-cumulative heat; t-usual time; T_1, T_2 -temperatures of fluid and reservoir (usually $T_2 = T^e$); T^e -constant temperature of infinite reservoir; $T_{1'}$, $T_{2'}$ -upper and lower temperatures of engine fluid; T^{n-1} - temperature of fluid entering the stage n; $\mathbf{u}^n$- control vector; $u^n = \Delta T^n/\Delta\tau^n$ - rate of temperature change; $v^n = \Delta X^n/\Delta\tau^n$ - rate of concentration change; w^n - power delivered at stage n; W-total molar work or power per unit molar flux; X-concentration (inert basis); X^e constant reference concentration; $\mathbf{x}^n$ -state vector of a general process; x- length coordinate; $\mathbf{z}^n$ - adjoint vector; z- temperature adjoint; α' - heat transfer coefficient; γ- cumulative conductance; η = thermal efficiency; τ -dimensionless time (x/H_{TU}); θ^n- nondimensional conductance.

References

1. Sieniutycz, S. and Berry, R. S. (1999) Discrete Hamiltonian analysis of endoreversible thermal cascades, in S. Sieniutycz and A. de Vos (eds.), *Thermodynamics of Energy Conversion and Transport*, Springer, New York, Chap. 6.
2. Bejan, A. (1995) *Entropy Generation Minimization: The Method of Thermodynamic Optimnization of Finite-Size Systems and Finite-time Processes*, CRC Press, Boca Raton.
3. De Vos, A. (1992) *Endoreversible Thermodynamics of Solar Energy Conversion*, Clarendon Press, Oxford, pp. 29-51.
4. Novikov, I. I. (1958) The efficiency of atomic power stations (a review), *Journal of Nuclear Energy II* **7**, 125-128. English translation from *At. Energ.* (1957) **3**, 409-412.
5. Curzon, F. L. and Ahlborn, B. (1975) Efficiency of Carnot engine at maximum power output, *Amer. J. Phys.* **43**, 22-24.
6. Spalding, D. B. (1963) *Convective Mass Transfer*, Edward Arnold, London.
7. Ciborowski, J. (1973) *Process Engineering*, Wydawnictwa Naukowo Techniczne, Warszawa.
8. Gordon J. M. and Orlov, V. (1993) Performance characteristics of endoreversible chemical engines, *J. Appl. Phys.* **74**, 5303-5309.
9. Chen, L., Sun, F., Wu Ch., and Wu, J. (1997) Performance characteristic of endoreversible chemical engines, *Energy Convers. Mgmt* **38**, 1841-1846.
10. Sieniutycz, S. (1973) The constancy of hamiltonian in a discrete optimal process, *Reports of Inst. of Chem. Engng. Warsaw Tech. Univ.* **2**, 399-429.
11. Sieniutycz, S. (1978) *Optimization in Process Engineering*, 1-st edn., Wydawnictwa Naukowo Techniczne, Warszawa.
12. Szwast, Z. (1979) *Discrete Algorithms of Maximum Principle with Constant Hamiltonian and their Selected Applications in Chemical Engineering*, PhD Thesis, Institute of Chemical Engineering at the Warsaw University of Technology, Warsaw.

13. Sieniutycz, S. and Szwast, Z. (1983) A discrete algorithm for optimization with a constan hamiltonian and its application to chemical engineering", *Intern. J. Chem. Engng.* **23**, 155-166
14. Fan, L. T. and Wang, C. S. (1964) *The Discrete Maximum Principle, A Study of Multistag System Optimization*, Wiley, New York.
15. Boltyanski, V. G. (1973) *Optimal Control of Discrete Systems*, Nauka, Moscow.
16. Fan, L. T. (1966) *The Continuous Maximum Principle, A Study of Complex Systen Optimization*, Wiley, New York.
17. Findeisen, W., Szymanowski, J., and Wierzbicki, A. (1980) *Theory and Computationc Methods of Optimization*, Panstwowe Wydawnictwa Naukowe, Warsaw.
18. Bellman, R. E. (1961) *Adaptive Control Processes: a Guided Tour*, Princeton University Pres: Princeton.
19. Aris, R. (1964) *Discrete Dynamic Programming*, Blaisdell, New York.
20. Boltyanski, V. G. (1971) *Maximum Principle and Optimal Control of Continuous Systems* Nauka, Moscow.
21. Bejan, A. (1996) Models of power plants that generate minimum entropy while operating a maximum power, *Am. J. Phys.* **64**, 1054-1059.
22. Bejan, A. (1996) Entropy generation minimization: The new thermodynamics of finite-siz devices and finite-time processes, J. *Appl. Phys.* **79**, 1191-1218.
23. Sieniutycz, S. (1997) Hamilton-Jacobi-Bellman theory of dissipative thermal availability *Phys. Rev. E* **56**, 5051-5064.

THERMODYNAMICS AND OPTIMIZATION OF REVERSE CYCLE MACHINES

M.L. FEIDT
G.E.S.P.E., L.E.M.T.A., University Henri Poincaré of Nancy 1
2, avenue de la Forêt de Haye, 54516 Vandoeuvre les Nancy, France

1. Introduction

The interest in thermodynamic modeling and optimization of the reverse cycle machines has been recently renewed, due to the C.F.C. replacement. This new consideration had probably the start in Blanchard's paper [1]. Since that time, the endoreversible models have been extensively considered, particularly for the two heat-source machines [2-7]. Attempts have been made in order to generalize, more or less, the particular results [8-11]. Also internal irreversibilities have been considered, as well as heat losses, according to Bejan's approach on entropy analysis [12-14]. Numerous papers have been published, namely on Carnot machines (refrigerator or heat pump) [15-28]. Most of these studies suppose that the source and the sink are isothermal. Some of them consider the effect of the heat transfer law on the machine performance [24, 26, 27] when the same law form is supposed at the source and the sink. Performance of more sophisticated configurations has also been investigated recently: combined refrigeration cycle [29, 30], cascade heat pump [31-32], Brayton machine [33, 34], solar air conditioning system [35, 36].

We propose here a tentative of model synthesis. In the first paragraph the methodology is applied to two heat-reservoir machines with perfect regeneration (Carnot, Ericsson, Stirling). Various optimization cases of these machines are listed, exemplified and related to existing results. The effect of internal irreversibilities, heat losses and form of the heat transfer laws is considered. Particular attention is given to the optimal allocation of the heat transfer areas. Extension to the machines with real internal fluids is presented. Finally, the case of three heat-source machines is considered showing its relation with the preceding considerations, and how it can restore the preceding results as particular limits.

2. Machines Following Carnot, Ericsson, Stirling Cycles

2.1. GENERAL FORMULATION OF THE OPTIMIZATION PROBLEM

The classical cycles mentioned above are shown in Fig. 1. Perfect mechanical or thermal regeneration is supposed. It results the same basic equations for the three machines. The First Law of Thermodynamics is written

A. Bejan and E. Mamut (eds.), Thermodynamic Optimization of Complex Energy Systems, 385–401.

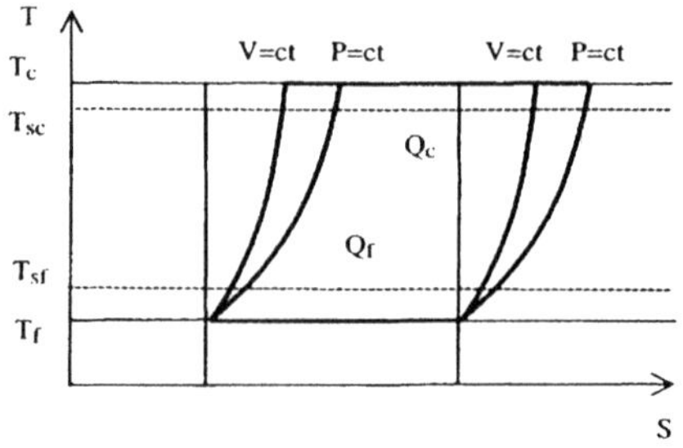

Figure 1. Endoreversible Carnot, Ericsson, Stirling reverse cycles.

$$q_C + q_F + W = 0 \tag{1}$$

where q_C is the energy delivered at the hot side of the machine (thermostat of temperature T_{SC} ; $q_C < 0$) ; q_F is the energy received at the cold end of the machine (thermostat of temperature T_{SF} ; $q_F > 0$) ; W is the energy consumption of the machine ($W < 0$). The Second Law of Thermodynamics is expressed by

$$\frac{q_C}{T_C} + \frac{q_F}{T_F} + S = 0 \tag{2}$$

In this algebraic form we notice : T_C, temperature of the cycled fluid at the hot end of the machine ; T_F, temperature of the cycled fluid at the cold end of the machine ; S, entropy generation due to the internal irreversibility of the machine. As a first step, this last term will be supposed constant. It can be detailed if necessary by a careful analysis, as suggested Bejan [12]. The heat transfer description must be added to the preceding relations, for both ends of the machine. We state that the general form of the heat transfer laws at the source and the sink is

$$q_i = F_i(T_i, T_{Si}) \tag{3}$$

where T_i is the cycled medium temperature at the source (or sink) i, T_{Si} is the temperature of the source i, and q_i is the energy transferred to the medium at the source (or sink) i.

For generality, we illustrate the method by the formal description of two heat-source machine having direct heat leak between the source and the sink. A general form of the heat loss energy, q_l, transferred directly from the source to the sink is

$$q_l = G(T_{SC}, T_{SF}) \tag{4}$$

The general approach taking account of all the preceding aspects uses the same method as previously proposed [8, 9].

We precise three major points of interest of the proposed method. First, this approach is valid for reverse cycle machine, as well as for direct ones. Second, this approach is valid for discontinuous cycle machine (time duration for each thermodynamic process), but also for continuous ones (i.e., in two heat-source machines, the contact durations of the cycled medium with the heat sources are replaced by the heat exchange areas). Third, the model is valid for any form of the heat transfer law used in

Eq. (3). Unlike the literature, where the heat transfer law is supposed to be the same at the source and the sink, we note that Eq. (3) permit to use different heat transfer laws in the models.

2.2. PARTICULAR CASE OF REFRIGERATOR AND HEAT PUMP

The objective function for a refrigerating machine can be (1) the minimum of the consumed energy (case R1) or (2) the maximum of the useful heat coming out of the cold source ($\phi > 0$, case R2). For this refrigerating machine a new constraint has to be added at the cold source

$$q_F = \phi + q_l \tag{5}$$

The objective function for a heat pump can be (1) the minimum of the consumed energy (case H1) or (2) the maximum of the useful heat delivered to the hot sink ($\phi < 0$, case H2). For the heat pump a new constraint appears also, expressed at the hot sink as

$$q_C = \phi - q_l \tag{6}$$

In addition to the above mentioned constraints, a new one must be considered. It is given by the imposed ϕ (case R1 and H1) or imposed W (case R2 and H2). Generally, for a refrigerating machine (respectively for a heat pump) T_{SF} (respectively T_{SC}) is imposed. For the two machines, we can also consider the cases with both T_{SF} and T_{SC} parameters. In these cases remain only two variables related to the two constraints (2) and (4), and there is only a thermodynamic state that can be attained, not an optimum.

TABLE 1. The equations for the optimum of the two-source machines. Variables are the two cycled medium temperatures.

Case	Equation
R1	$\frac{1}{T_F^2 F_{F.F}}\left(T_F F_{F.F} - F_F\right) = \frac{1}{T_C^2 F_{C.C}}\left[T_C F_{C.C} - F_C\left(1+\frac{F_{C.S}}{G_{SC}}\right)\right]$
R2	$\frac{F_F}{T_F^2 F_{F.F}} - \frac{1}{T_F} = \frac{F_C}{T_C^2 F_{C.C}}\left(1+\frac{F_{C.S}}{G_{SC}}\right) - \frac{1}{T_C}$
H1	$\frac{1}{T_F^2 F_{F.F}}\left[T_F F_{F.F} - F_F\left(1-\frac{F_{F.S}}{G_{SC}}\right)\right] = \frac{1}{T_C^2 F_{C.C}}\left(T_C F_{C.C} - F_C\right)$
H2	$\frac{F_F}{T_F^2 F_{F.F}}\left(1-\frac{F_{F.S}}{G_{SF}}\right) - \frac{1}{T_F} = \frac{F_C}{T_C^2 F_{C.C}} - \frac{1}{T_C}$

In the general case, applying the Lagrange's multipliers method leads to a system with three equations having the three cited unknowns (Table 1)

$$F_{i.i} = \frac{\partial F_i}{\partial T_i} \quad ; \quad F_{i.S} = \frac{\partial F_i}{\partial T_{Si}} \quad ; \quad G_{Si} = \frac{\partial G}{\partial T_{Si}}$$

When the heat losses are not considered, all the equations of Table 1 become identical.

Owing to the presence of Eq. (2) in any cases, it results a unique system of two equations that yields the optimum values of T_C and T_F as functions of the other parameters. Particularly, one can see that the optimum T_C and T_F depend on the internal irreversibility, Eq. (2). If S vanishes (endoreversible case), the system restores the endoreversible results currently exposed in the literature. There must be noticed that the final solution differs from case to case, due to the specificity of the last equations used ((1), (5), or (6)). These equations can be solved analytically in some cases, and only numerically in the remaining ones.

2.3. OPTIMAL ALLOCATION OF HEAT TRANSFER CONDUCTANCES

Following the approach presented in a previous paper [8], the same Lagrange's method can be applied to solve the cases where the conductances of the heat exchangers (or contact duration times) are variables. In these cases, one more relation must be added to the two preceding ones. It will relate the two new variables which are the overall heat transfer conductances at the source and the sink

$$K_C + K_F = K \qquad \text{or} \qquad t_C + t_F = t \tag{7, 8}$$

where K_C is the thermal conductance at the hot end, K_F is the thermal conductance at the cold end, t_C is the thermal contact duration at the hot end, and t_F is the thermal contact duration at the cold end of the machine.

For the continuous cycles, Eq. (7) expresses the total heat transfer conductance to be distributed between the hot and the cold end of the machine that is considered in a stationary state. Eq. (8) is useful for discontinuous cycles and supposes that the duration of the transfer between the source and the sink is negligible.

TABLE 2. Extended optimization of two heat-source machines. Variables are the two cycled medium temperatures, and the overall heat transfer conductances at the source and the sink.

Problem	Objective function	Constraints
R1E	min W	(2), (5), (7) or (8)
R2E	MAX ϕ	(2), (1), (7) or (8)
R3E	min K or t	(2), (5)
R4E	min K or t	(2), (1)
H1E	min W	(2), (6), (7) or (8)
H2E	min ϕ	(2), (1), (7) or (8)
H3E	min K or t	(2), (6)
H4E	min K or t	(2), (1)

With the two supplementary variables related by Eq. (7) or (8), the optimization problem will have two degrees of freedom. The corresponding form of the heat transfer laws at the source and the sink becomes

$$q_i = K_i f_i\left(T_i, T_{Si}\right) = K_i f_i \tag{9}$$

Table 2 shows that the added Eqs. (7) or (8) make possible two new optimization processes (R2E, R4E for a refrigerating machine, and H3E, H4E for a heat pump) for each machine, as originally was proposed by Bejan [19]. We notice that the cases R4E and H4E are formally identical. They differ only by the third variable, (T_{SC}, T_{SF}), if it is considered. By derivation and direct elimination, one can derive a general relation between the two variables T_C and T_F for minimum of K

$$\frac{T_F S}{f_F}+(W-ST_F)\frac{f_C-T_C f_{C.C}}{f_C^2}=\frac{T_C S}{f_C}+(W-ST_C)\frac{f_F-T_F f_{F.F}}{f_F^2} \tag{10}$$

For these cases, it appears that the minimum of K depends only on the internal irreversibility, and not on the heat losses. This general expression restores easily the known particular results.

For the case R3E (H3E), the same approach shows that K is a monotone function of T_C (T_F), and these two variables are related by

$$\left(\frac{T_F}{f_F}\right)^2 f_{F.F}=\frac{T_C}{f_C} \tag{11}$$

or

$$\left(\frac{T_C}{f_C}\right)^2 f_{C.C}=\frac{T_F}{f_F} \tag{12}$$

2.4. OPTIMAL ALLOCATION OF AREAS

This extension is valid essentially for continuous cycle machines (stationary operating). It uses a more realistic relation than Eq. (7) given by

$$A=A_C+A_F \tag{13}$$

where A is the total heat transfer area to be allocated, A_C is the heat transfer area at the hot end, and A_F is the heat transfer area at the cold end of the machine.

Using these new variables, the generalized overall heat transfer coefficient at the hot end k_C, respectively at the cold end k_F, must be known. As a first step, we suppose that these coefficients are constants. Consequently, the general form of the heat transfer laws at the source and the sink becomes

$$q_i=k_i A_i f_i\left(T_i, T_{Si}\right) \tag{14}$$

Further, the optimization procedure can be extended taking account of the investment cost of the heat exchangers. As a first approximation, the investment cost is expressed with reference to the heat exchanger areas

$$v=v_C A_C+v_F A_F \tag{15}$$

where v_C is the cost by unit area at the hot end, v_F is the cost by unit area at the cold end of the machine. Even if detailed cost correlations are available [46], this approach offers a good overview of the corresponding tendencies.

Eq. (15) will replace Eq.(7) and it can be used also to find the optimal allocation of the conductances, when the heat transfer coefficients, k_i ,are equals to unity. Similarly, if the unity costs v_C and v_F are considered equals to unity, Eq. (15) restore Eq. (13). The problem considering v_C and v_F as parameters has been already studied [2].

Combining Eqs. (15) and (2) results the general expressions of the heat transfer areas at the optimum

$$A_C = \frac{(v k_F f_F + S v_F T_F) T_C}{v_C T_C k_F f_F - v_F T_F k_C f_C} \tag{16}$$

$$A_F = \frac{-(v k_C f_C + S v_C T_C) T_F}{v_C T_C k_F f_F - v_F T_F k_C f_C} \tag{17}$$

where the values of T_C, T_F correspond to those obtained from the system having Eqs. (1), (5) or (6) as constraints and the objective function owing to the current problem (case 1 or 2). If the objective function is the investment cost v (case 3 or 4), the corresponding expressions are given in Table 3. It results the effect of Φ, G, S variation and of the cycled medium temperature rise (T_C-T_F) on the investment cost in various cases. We consider this approach particularly important because of its potential to differentiate wide-spread thermal machines such as air-air or water-water systems.

TABLE 3. Thermoeconomic optimization of two heat-source machines : minimum of investment cost for heat exchangers.

Problem	Expression o the objective function
R3G	$v = (\Phi + G)\left(\frac{v_F}{k_F f_F} - \frac{v_C}{k_C f_C} \cdot \frac{T_C}{T_F}\right) - S\frac{v_C T_C}{k_C f_C}$
R4G	$v = \frac{1}{T_C - T_F}\left[\frac{v_C T_C}{k_C f_C}(-w + S T_F) + \frac{v_F T_F}{k_F f_F}(w - S T_C)\right]$
H3G	$v = (\Phi - G)\left(\frac{v_C}{k_C f_C} - \frac{v_F}{k_F f_F} \cdot \frac{T_F}{T_C}\right) - S\frac{v_F T_F}{k_F f_F}$
H4G	the same eq. as for R4G

These systems are supposed to use similar heat exchangers technologies at both ends of the machine, while the air-water or water-air machines will have different specific costs for the air system, v_a, respectively, water system, v_w. Studies on this subject are now under development in our research group [47].

2.5. SUMMARY OF RESULTS

In order to illustrate the general methodology proposed in the preceding paragraphs, we propose hereafter a tentative synthesis of the available results. Table 4 presents the optimum values of the four variables for a refrigerating machine, when the heat losses and internal irreversibilities are considered. The studied cases correspond to a fixed useful effect and to K_P, T_{SC}, the model parameters.

where $k = \frac{k_C}{k_F}$; $s = \frac{S}{Ak_F}$; $\varphi = \frac{\Phi + K_P(T_{SC} - T_{SF})}{Ak_F T_{SF}}$;

$\theta_C = \frac{T_{SC}}{T_{SF}}$ is a hidden parameter in φ.

TABLE 4. Optimum operating point of a refrigerating machine with heat losses and internal irreversibilities.

Conductances	Surfaces
$\frac{T_F}{T_{SF}}=\frac{-2\varphi+1}{1+s}$ $\frac{T_C}{T_{SC}}=\frac{2\varphi-1}{4\varphi+s-1}$	$\frac{T_F}{T_{SF}}=\frac{k^{1/2}-\varphi\left(1+k^{1/2}\right)}{k^{1/2}+s}$ $\frac{T_C}{T_{SC}}=\frac{k^{1/2}\left[\varphi\left(1+k^{1/2}\right)-k^{1/2}\right]}{s-k^{1/2}+\varphi\left(1+k^{1/2}\right)^2}$
$\frac{K_F}{K}=\frac{\varphi(1+s)}{2\varphi+s}$ $\frac{K_C}{K}=\frac{\varphi(1-s)+s}{2\varphi+s}$	$\frac{A_F}{A}=\frac{\varphi\left(k^{1/2}+s\right)}{\varphi\left(1+k^{1/2}\right)+s}$ $\frac{A_C}{A}=\frac{\varphi(1-s)+s}{\varphi\left(1+k^{1/2}\right)+s}$

These analytical results show clearly the influence of φ and s on the optimal conductance distribution. We notice that generally conductance equipartition is not valid. It is true only for the endoreversible case.

If the optimization is extended to the area distribution, the influence of the global heat transfer coefficients appears through a square root (Table 2, 2nd row). In the same manner, it is easy to deduce the optimum operating point for a refrigerating machine when W is imposed, the corresponding results for a heat pump, and the associated values of the COP at optimum. The results are available, but they will not be presented here for brevity.

Table 5 illustrate the optimization of a reverse cycle machine for two generalized heat transfer laws. Law 1 is relative to a generalized radiative law

$$q_i = k_i A_i \left(T_{Si}^n - T_i^n\right) \tag{18}$$

Law 2 is relative to a generalized convective law

$$q_i = \varepsilon_i k_i A_i \left(T_{Si} - T_i\right)^m \tag{19}$$

where ε_i is chosen according to the convention stating that q_C remains always negative, and q_F positive, whatever the exponent m values would be, and $X_i = T_{Si} - T_i$.

The most common value of the exponent n is 4, while m could indicate natural convection (m = 0.24), or pool boiling (m = 3). Also, the value n= - 1 can be find in the literature. It corresponds to the conductive phenomenological law of the linear irreversible thermodynamics. The two forms given by Eqs. (18) and (19) admit the same limit, for n = m = 1 which corresponds to Newton's convection law, well developed and known case in the literature.

From Table 5 results that the optimization problem can be generalized. The system of four equations to be solved with respect to the variables A_C, A_F, T_C or X, T_F or Y is constituted for any case of the general equation

TABLE 5. The system of four equations to be solved for the optimum (W_0, q_0, A).

Law 1	Law 2
Refrigerating machine	
$\frac{k_F\left(T_{SF}^m - T_F^m\right)^2}{mT_F^{m+1}} - \frac{k_C\left(T_{SC}^n - T_C^n\right)^2}{nT_C^{n+1}} = 0$ $\frac{k_FA_F\left(T_{SF}^m - T_F^m\right)}{T_F} + \frac{k_CA_C\left(T_{SC}^n - T_C^n\right)}{T_C} + S = 0$ $A_C + A_F - A = 0$ $W_0 + k_CA_C\left(T_{SC}^n - T_C^n\right) + k_FA_F\left(T_{SF}^m - T_F^m\right) = 0$ or $q_0 - k_FA_F\left(T_{SF}^m - T_F^m\right) = 0$	$\frac{(-1)^n k_C X^{n+1}}{(T_{SC} - X)^2} + \frac{k_F Y^{m+1}}{(T_{SF} - Y)^2} = 0$ $-\frac{(-1)^n X^n k_C A_C}{T_{SC} - X} + \frac{Y^m k_F A_F}{T_{SF} - Y} + S = 0$ $A_C + A_F - A = 0$ $W_0 - (-1)^n k_C A_C X^n + k_F A_F Y^m = 0$ or $q_0 - k_F A_F Y^m = 0$
Heat pump	
$\frac{k_F\left(T_{SF}^m - T_F^m\right)^2}{mT_F^{m+1}} - \frac{k_C\left(T_{SC}^n - T_C^n\right)^2}{nT_C^{n+1}} = 0$ $\frac{k_FA_F\left(T_{SF}^m - T_F^m\right)}{T_F} + \frac{k_CA_C\left(T_{SC}^n - T_C^n\right)}{T_C} + S = 0$ $A_C + A_F - A = 0$ $W_0 + k_CA_C\left(T_{SC}^n - T_C^n\right) + k_FA_F\left(T_{SF}^m - T_F^m\right) = 0$ or $q_0 - k_CA_C\left(T_{SC}^n - T_C^n\right) = 0$	$\frac{k_F Y^{m+1}}{(T_{SF} - Y)^2} - \frac{(-1)^n k_C X^{n+1}}{(T_{SC} - X)^2} = 0$ $\frac{k_F A_F Y^m}{T_{SF} - Y} - \frac{(-1)^n k_C A_C X^n}{T_{SC} - X} + + S = 0$ $A_C + A_F - A = 0$ $W_0 - (-1)^n k_C A_C X^n + k_F A_F Y^m = 0$ or $q_0 - (-1)^n k_C A_C X^n = 0$

$$\left(\frac{f_F T_C}{f_C T_F}\right)^2 = \frac{k_C}{k_F}\frac{f_{F,F}}{f_{C,C}} \tag{20}$$

The second equation is the expression of the second law of thermodynamics. The two other equations are relative to the total area to be allocated when W or q_i is fixed. When the minimum of A is sought, there is only one constraint (W or q_i fixed), as it was told previously.

Important and interesting results can be deduced from Table 5. When W or q_i is fixed and S = 0, the general distribution of the heat transfer area is given by

$$A_C = A\frac{1}{1+B} \qquad A_F = A\frac{B}{1+B} \tag{21, 22}$$

where B is a given function of $(k_C / k_F)^{1/2}$, T_C, T_F and of the heat transfer law form.

It is obvious that the equipartition of the heat transfer conductances is not a optimum distribution. The expression of the ratio K_C/K_F relative to the non symmetrical generalized radiative laws illustrate this fact

$$\frac{K_F}{K_C} = \left(\frac{n}{m}\cdot\frac{T_C^{(n-1)/2}}{T_F^{(m-1)/2}}\right)^{1/2} \tag{23}$$

It appears that, even for endoreversible machine, the equipartition of the heat transfer conductances is obtained only for the symmetrical case of the linear heat transfer law (n = m = 1). If the heat transfer law is symmetrical, but not linear, the conductance ratio depends generally on the temperature rise of the cycled medium.

3. Machines with Real Internal Fluid

3.1. PHYSICAL PROPERTIES OF THE REFRIGERANT FLUID

A new step in the model development consists of considering the real properties of the refrigeration fluids. This corresponds to a recent work done in collaboration with V. Radcenco [46], now in press.

Correlations for three fluids (R12, R22, R134a) are proposed. R12 is a C.F.C. not allowed by now, R22 is a H.C.F.C. for transition, and R134a is the most common H.F.C. proposed to replace the preceding ones.

The model requires the form of the latent heat of boiling or condensing, proposed as

$$L(T) = \alpha + \beta T + \chi T^2 \tag{24}$$

Also, the saturation pressure correlation, with reference to the normal boiling temperature T_N (at $P_N = 1.03105$ Pa) is needed

$$\ln(P) = A\left(\frac{1}{T} - \frac{1}{T_N}\right) + B\ln\left(\frac{T}{T_N}\right) + C(T - T_N) \tag{25}$$

where α, β, γ, A, B, C are the coefficients given in [48].

3.2. REAL MACHINE MODEL WITH VAPOR COMPRESSION

The vapor compression machine is one with regeneration, so that the subcooling of the liquid refrigerant (3-3") is done by the vapor superheating at the exit of the boiler (1-1'). Fig.2 illustrates the complete evolution of the refrigerant during the cycle, where T_a is the ambient temperature, T_s is the temperature of the superheated vapor entering the compressor. The heat exchanged in the boiler and the condenser is expressed according to Newton's heat transfer law, in term of conductances or areas. The compressor is supposed adiabatic.

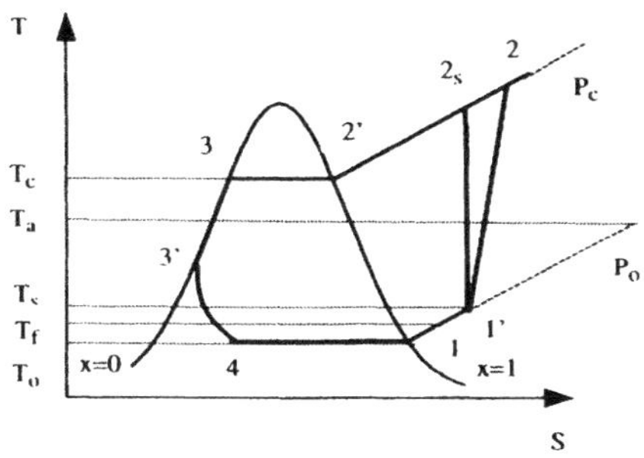

Figure 2. Temperature-entropy diagram for a real refrigerating machine.

Introducing the mean entropic temperature

$$T_{m_i} = \frac{q_i}{\Delta_{Si}} \tag{26}$$

the entropy balance for a cycle results

$$\frac{\dot{q}_C}{T_m} - \frac{q_0}{T_0} = \frac{1}{C_l T_0} \cdot$$

$$\frac{\dot{q}_0}{\frac{L_v(T_0)}{C_l T_0} - \frac{T_r}{T_0} + 1}\left[C_l\left[\frac{T_r}{T_0} - 1 - \ln\left(\frac{T_r}{T_0}\right)\right] + C_p \ln\left(\left(1 + \frac{H^{\frac{k-1}{k}} - 1}{\Im_C}\right) H^{\frac{1-k}{k}}\right) + C_p \ln\left(\frac{T_S}{T_0}\right) + C_l \ln\left(\frac{T_r}{T_0}\right)\right] \tag{27}$$

where T_r is the temperature of the liquid refrigerant at the exit of the recuperator, H is the pressure ratio ($H = P_c/P_o$), and $\Im_C$ is the isentropic efficiency of the compressor.

Based on the entropy analysis methodology, an expression of the entropy flux generation for stationary operating of the machine is deduced in this case. For brevity the details regarding the nondimensional parameters and variables will be omitted, as well as the constraints and the intermediate function f_i (i = 1 to 7). The objective function to be minimized is

$$P = \left(1 - \frac{q_0}{\theta_0}\right) \cdot \theta_m \tau - q_0 \tag{28}$$

It depends on the three reduced variables

$$\theta_0 = \frac{T_f - T_0}{T_f}, \qquad \theta_c = \frac{T_c - T_a}{T_a}, \qquad \theta_m = \frac{T_m - T_a}{T_a}$$

with the following explicit parameters

$$\tau = \frac{T_a}{T_r}, \qquad q_0 = \frac{\dot{q}_0}{KT_f}$$

Actually the optimization problem has five variables if we also add K_0, K_C and twelve parameters. Two of the parameters are related to the application (τ, q_0), other two to the machine ($\tau_S = \Delta T_S / T_f, \Im_c$), and eight are related to the refrigerant fluid. The variables K_i can be eliminated as it was indicated previously, and there remain only two equations relating the three temperatures

$$\frac{1+\theta_m}{1+\theta_c} = \frac{c(f_7 - 1) + f_6}{c\ln(f_7) + f_6} \tag{29}$$

$$\left(1 - \frac{q_0}{\theta_0}\right)\frac{\theta_m}{1+\theta_m} - \frac{q_0}{1-q_0}\left(1 + \frac{f_1 - 1 - \ln(f_2) + c\ln(f_3)}{f_5 - f_1 + 1}\right) = 0 \tag{30}$$

From Eq. (29), θ_m can be expressed as a function of θ_c, θ_0, with $c = C_p/C_l$. It remains only one objective function, Eq. (28), and one constraint, Eq. (30). Due to the non linearity of the entropy balance, only numerical solution of the two equations system can be obtained.

3.3. SENSITIVITY ANALYSIS OF THE OPTIMUM

There is well known the temperature T_f range for air conditioning, 283 K-293 K, for refrigeration, 268 K-278 K, and for freezing, 243 K-263 K. The vapor superheating varies from 1°C to 10 °C. If the ambient temperature reference is supposed to be 298 K, we found that $3.6\ 10^{-3} < \tau_s < 4\ 10^{-2}$, and $1.07 < \tau < 1.35$. The isentropic efficiency of the compressor varies from 0.6 to 1 (the limit case). A realistic range of variation for q_0 is $10^{-3} < q_0 < 10^{-1}$.

Here after the optimum values variation of the K_0/K, θ_0, θ_c versus q_0 are presented when $\tau = 298/273$, $\tau_s = 5/273$, and $\Im_c = 0.8$, for the three proposed refrigerant fluids.

Fig. 3 shows that when q_0 increases, the departure from the equipartition of the heat transfer conductance increases. The results for R134a and R22 are very closed, but for R12 the curve shows that less conductance is needed at the cold end. The pinch is practically the same for the three refrigerants at the cold end of the machine. It is not the same at the hot end, where for R22 the pinch values are smaller. In each case, θ_0, and θ_c are increasing function of q_0.

The trend of reestablishing the equipartition of the conductances, when $\Im_c$ increases (lowering the internal irreversibility of the machine) has been confirmed.

4. Presentation of Other Machine Configurations

4.1. INVENTORY OF STUDIED MACHINE CONFIGURATIONS

Most of the recent published studies are focused on Carnot refrigerating machine or heat pump with different driving systems [36-38]. An attempt was also made for a cascade machine [31], but generally these studies are limited to endoreversible system. An interesting confrontation between experimental and industrial available results and a thermodynamic model was developed by Gordon and co workers since 1993 [40-43]. This comparison is relative to the reciprocating chillers, being also extended to absorption chillers.

Before developing this last machine optimization, just a comment on the other studied configurations appearing recently in the literature. Chen and al. [34] considered an endoreversible Brayton refrigerating machine and the relation between the COP or specific load, and the pressure ratio. The same approach was developed by Erbay and Yavuz [47, 48], but relative to the Ericsson and Stirling machines. Similar studies were performed in our research group [49] in order to obtain the minimum W for given q_F for a refrigerating machine or q_C for a heat pump. Brayton, Ericsson and Stirling cycles were considered with perfect heat regeneration, but with global internal irreversibility (S), and heat leak between the hot end and the cold end of the machine. Due to its growing practical interest, the case of three heat-source machines was also envisaged, and it is shortly reported hereafter.

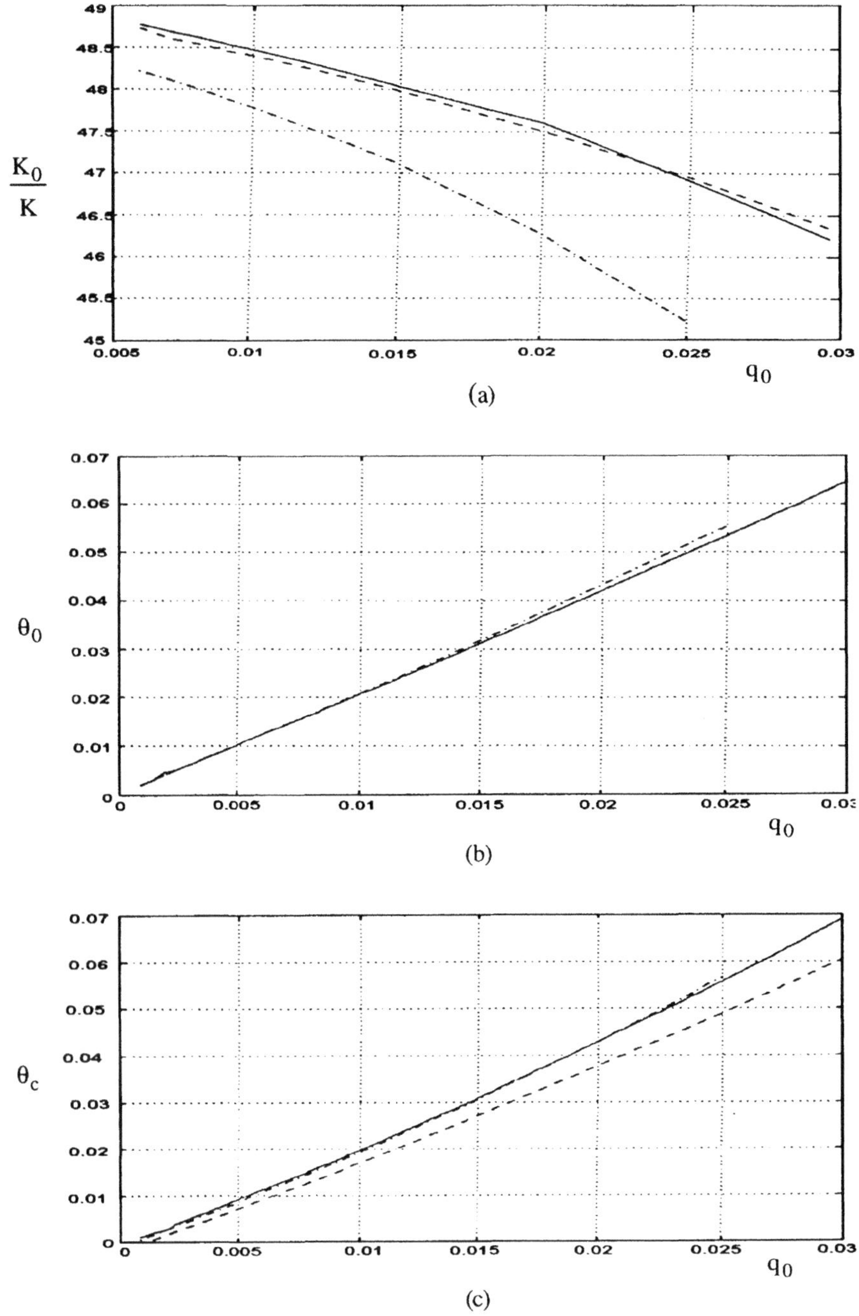

Figure 3. Evolution of K_0/K (a), θ_0 (b), θ_c (c) versus q_0, for R134a, - - R22, -.- R12.

4.2. REFRIGERATORS AND HEAT PUMPS IN CONTACT WITH 3 OR 4 HEAT RESERVOIRS

4.2.1. COEFFICIENT OF PERFORMANCE VERSUS USEFUL EFFECT

Several papers were published on the three heat-reservoir machines [52-54]. Generally the intermediate contacts are assumed with the same reservoir (reduction from 4 to 3 heat-reservoir machines). This assumption constitutes a good approximation of the technological configurations. We notice only a paper of Chen and Yan proposing a unified description of the endoreversible cycles [51]. A recent review paper [54] focuses on a more general model of the three heat-source machines that can be conceived owing to the published materials.

The general relation between the coefficient of performance and the useful effect was derived using the algebraic forms of the first and second law of thermodynamics

$$W + q_C + q_I + q_F = 0 \tag{31}$$

$$\frac{q_C}{T_C} + \frac{q_I}{T_I} + \frac{q_F}{T_F} + S = 0 \tag{32}$$

The coefficient of performance, COP, is defined as the ratio of the useful effect and the energetic cost. It is expressed respectively

for a refrigerating machine $$\Sigma = \frac{q_F}{q_C + W} \tag{33}$$

for a heat pump $$\Sigma' = \frac{-q_I}{q_C + W} \tag{34}$$

for a heat transformer $$\Sigma'' = \frac{q_C}{q_I} \tag{35}$$

Combining Eqs. (31), (32) with the definitions given by Eqs. (33)-(35) we derive for real irreversible machines

$$\Sigma_I = \frac{T_I^{-1} - T_C^{-1}}{T_F^{-1} - T_I^{-1}} \left[1 + \frac{W - T_C S}{W + q_C} \cdot \frac{T_I}{T_C - T_I} \right] \tag{36}$$

$$\Sigma'_I = \frac{T_F^{-1} - T_C^{-1}}{T_F^{-1} - T_I^{-1}} \left[1 + \frac{W - T_C S}{W + q_C} \cdot \frac{T_F}{T_C - T_F} \right] \tag{37}$$

$$\Sigma''_I = \frac{T_F^{-1} - T_I^{-1}}{T_F^{-1} - T_C^{-1}} \left[1 + \frac{W - T_I S}{W + q_I} \cdot \frac{T_F}{T_C - T_I} \right] \tag{38}$$

Complementary to the proposition of Chen et al and Gordon et al, the COP's appears to be naturally expressed in term of the heat input. It is an interesting result, especially when the heat input is fixed (constraint). Chen et al and Gordon et al consider the useful effect as constraint. We propose here the corresponding relation between the COP and the useful effect

$$\frac{1}{\Sigma_I} = \frac{1}{\psi} - \frac{W - T_C S}{q_F} \cdot \frac{T_I}{T_C - T_I} \quad (39)$$

$$\frac{1}{\Sigma'_I} = \frac{1}{\psi'} - \frac{W - T_C S}{q_I} \cdot \frac{T_F}{T_C - T_F} \quad (40)$$

$$\frac{1}{\Sigma''_I} = \frac{1}{\psi''} - \frac{W - T_I S}{q_C} \cdot \frac{T_F}{T_I - T_C} \quad (41)$$

where ψ, ψ', ψ'' correspond to the three thermal machines COP, generally indicated in the literature (endoreversible machines with W contribution neglected).

4.2.2. OPTIMAL AREA DISTRIBUTION OF THE MACHINE

The three heat transfer quantities are given in a general form by Eq. (3)

$$q_C = k_C A_C f_C(T_{SC}, T_C) \quad q_I = k_I A_I f_I(T_{SI}, T_I) \quad q_F = k_F A_F f_F(T_{FC}, T_F) \quad (42, 43, 44)$$

It is obvious that in this case the size of the system is finite (or the duration of the machine cycles is finite, the formalism being strictly identical, so A_i becomes t_i)

$$A = A_C + A_I + A_F W \quad (45)$$

Also we notice that Eqs. (31), (32), and (45) constitute a system of equations, with three unknowns, A_C, A_I, A_F. These variables are expressed in function of the temperatures and the case parameters

$$A_C = \frac{1}{\Delta}\left[A k_I f_I k_F f_F \left(\frac{1}{T_F} - \frac{1}{T_I}\right) + W\left(\frac{k_F f_F}{T_F} - \frac{k_I f_I}{T_I}\right) - S(k_F f_F - k_I f_I)\right] \quad (46)$$

$$A_I = \frac{1}{\Delta}\left[A k_F f_F k_C f_C \left(\frac{1}{T_C} - \frac{1}{T_F}\right) + W\left(\frac{k_C f_C}{T_C} - \frac{k_F f_F}{T_F}\right) - S(k_C f_C - k_F f_F)\right] \quad (47)$$

$$A_F = \frac{1}{\Delta}\left[A k_C f_C k_I f_I \left(\frac{1}{T_I} - \frac{1}{T_C}\right) + W\left(\frac{k_I f_I}{T_I} - \frac{k_C f_C}{T_C}\right) - S(k_I f_I - k_C f_C)\right] \quad (48)$$

with $$\Delta = k_I f_I k_F f_F \left(\frac{1}{T_F} - \frac{1}{T_I}\right) + k_F f_F k_C f_C \left(\frac{1}{T_C} - \frac{1}{T_F}\right) + k_C f_C k_I f_I \left(\frac{1}{T_I} - \frac{1}{T_C}\right)$$

The above equations show that the area distribution depends on the three variables (T_C, T_I, T_F) and on the model parameters (T_{SC}, T_{SI}, T_{SF}, k_C, k_I, k_F, W, S).We renew that the heat transfer laws are not necessary the same at the three thermal contacts, even if the assumption of similarity is always used in the literature. The influence of the heat transfer law form appears through the functions f_C, f_I, f_F in Eqs (46)-(48). It remains to precise the objective function and the constraint(s), or eventually the COP, in order to optimize the machine.

Eqs. (46)-(48) represents the generalized formulas of the area distribution in the machine, indicating the influence of W and S on that. If W and S are equal to zero, the particular results published previously [54] are obtained.

4.2.3. PARTICULAR CONFIGURATIONS

As it was demonstrated in a recent paper, it is easy to restore the form of Eq. (16), (17) corresponding to the limit of two heat-reservoir refrigerating machine or heat pump. In these relations W vanishes, T_{SC} and T_C tends to infinity, q_H becomes pure mechanical energy. The corresponding positive value of W results from the finite value of the undetermined form A_C (or t_C) $\rightarrow 0$, multiplied by $K_C f_C$ that both tends to infinity.

In the same manner, it is possible to show that the two heat-reservoir heat engine corresponds to the limit of a heat pump, when T_{SI} tends to infinity [54].

4.2.4. ILLUSTRATION OF SOME OBTAINED RESULTS

Based on similar, but particular approach, there have been studied more sophisticated machines, as heat pumps driven by thermal machines or solar three heat-reservoir machines. We focus here on the finite thermal capacities machines. The preceding analysis of a solar system indicated that generally, the heated fluid in the solar collector is not subject to phase change. It means that it cannot be considered as isothermal ($T_C \neq ct$). This fact has been considered [51] taking account of the finite thermal capacity of a system under transient conditions.

The same approach was adopted for a stationary system [4], where it was taking account of the finite mass fluxes of the external fluids. The corresponding results are available [53].

5. Conclusion

The proposed paper intended to achieve three goals. First, it proposes a review of the major fields of interest in recent thermodynamics approach and optimization of the reverse cycle machine. This interest is probably due to the replacement of C.F.C. that requires the reconsideration of the conception of these machines. It is noticed that thermodynamics was often limited to endoreversible consideration until now.

Second, this paper aims to offer a general methodology. The exposed general model is able to take account of (1) the internal dissipation of the energy, by a global or refined form (paragraph 4.2), (2) the heat losses between the source and the sink, by a general linear form, (3) the general heat transfer laws that can be different between the hot (intermediate) and the cold end, when in the literature there is explored only the symmetrical case.

Third, there have been developed some examples in order to enlighten the proposed method. These examples have been particularly focused on the optimal allocation of the heat transfer areas, subject to various objectives and constraints. It has been shown that this approach can be easily transposed to the contact duration or heat transfer conductances, and that the methodology is also valid for thermoeconomic analysis of these machines.

6. References

1. Blanchard, C.H. (1980) Coefficient of performance for finite speed heat pumps, *J. Appl. Phys.*, **51**, 2471-2472.
2. Goth, Y. and Feidt, M. (1986) Conditions optimales de fonctionnement des pompes à chaleur ou machines à froid associés à un cycle de Carnot endoréversible, *C.R.A.S.*, T. 303, série II, 19-24.
3. Feidt, M. (1986) Un outil d'aide à la conception des machines à froid et pompes de chaleur à partir de la modélisation des cycles de machines à deux et trois sources de chaleur. Optimisation résultante, compte tenu des contraintes en temps de cycle ou en surface d'échange, *Rencontre S.F.T.*, Marseille, France.
4. Feidt, M. (1988) Finite time thermodynamics applied to optimization of heat pumps and refrigerating machine cycles, *12th IMACS World Congress on Scientific Computation*, Paris, France.
5. Philippi, I. and Feidt, M. (1991) Optimisation des machines thermiques à deux sources, cas des faibles pincements, *Entropie*, **164**, 113-114.
6. Bejan, A. (1989) Theory of heat transfer irreversible refrigeration plants, *Int. J. Heat and Mass Transfer*, **32**, 1631-1638.
7. Gordon, J.M. (1990) The impact of heat transfer mode on the efficiency of heat engine operating at maximum power, *9th.Int. Heat Transfer Conference*.
8. Feidt, M. and Philippi, I. (1992) Basis of a general approach for finite time thermodynamics applied to two heat-source machines, *Proc. of ECOS'92*, Zaragoza, Spain.
9. Feidt, M. (1993) Thermodynamique en temps fini des cycles moteurs ou générateurs de froid et de chaleur, *Proc. of Thermodynamics National Conference*, Bucharest, Romania.
10. Philippi, I. and Feidt, M. (1991) Finite time thermodynamics applied to inverse cycle machine, *XVIII-ème Congrés Inernational du Froid*, Montréal, Canada.
11. Feidt, M. and al. (1994) Synthesis unification and generalization of preceding finite time thermodynamics studied devoted to Carnot cycles, *Proc. of FLOWERS'94*, 429-441.
12. Bejan, A. (1994) *Entropy Generation Through Heat and Fluid Flow*, Wiley & Sons, New York.
13. Feidt, M. (1996) *Thermodynamique et Optimisation Énergétique des Systèmes et Procédés*, Tech & Doc, Paris.
14. Feidt, M. (1997) Sur une systématique des cycles imparfaits, *Entropie*, **205**, 53-61.
15. Petrescu, S. and al. (1993) The study for optimization of the Carnot cycle which develop with finite speed, *Proc. of the Inter. Sympos. ECOS'93*, Kracow, Poland.
16. Crazzini, G. (1993) Irreversible refrigerators with isothermal heat exchangers, *Rev. Int. Froid*, **15**, 101-106.
17. Yan, Z. and Chen, J. (1990) Coefficient of performance for an irreversible Carnot heat pump, *Proc. of FLOWERS'90*, 193-200.
18. Klein, S.A. (1992) Design considerations for refrigeration cycles, *Rev. Int. Froid*, **15**, 181-185.
19. Bejan, A. (1993) Power and refrigeration plants for minimum heat exchangers inventory, *J. of Energy Resources Technology*, **115**, 148-150.
20. Yan, Z. and Chen, J. (1990) A class of irreversible Carnot refrigeration cycles with general heat transfer law, *J.of Appl. Phys.*, **23**, 136-141.
21. Agrawal, D.C. and Meson, V.J. (1990) Performance of a Carnot refrigerator at maximum cooling power, *J. of Appl. Phys.*, **23**, 136-141.
22. Agrawal, D.C. and Meson, V.J. (1993) Finite Carnot refrigeration with wall gain and products loads, *J. of Appl. Phys.*, **74**, 2153-2158.
23. Chen, L. and al. (1997) A generalized model of a real refrigerator and its performance, *Appl. Therm. Eng.*, **17**, 401-412.
24. Chen, L. and al. (1996) The influence of a heat transfer law on the endoreversible Carnot refrigerator, *J. Int. Energy*, **69**, 96-100.
25. Wu, C. and al. (1996) General performance characteristics of a finite speed Carnot refrigerator, *Appl. Therm. Eng.*, **16**, 299-303.
26. Chen, L. and al. (1996) Heat transfer effect on the specific cooling load and refrigerators, *Appl.Therm. Eng.*, **16**, 989-997.
27. Chen, L. and al. (1996) Heat transfer effect on the specific cooling load and refrigerators, *Appl.Therm. Eng.*, **17**, 103-110.
28. Sun, F. and al. (1997) Optimal performance of an endoreversible Carnot heat pump, *Energy Convers. Mgmt.*, **38**, 1439-1443.

29 Chen, L. and al. (1997) Steady flow combined refrigeration cycle performance with heat leak, *Appl.Therm. Eng.*, **17**, 639-645.

30 Chen, L. and al. (1995) Optimization of the specific rate of refrigeration in combined refrigeration cycle, *Energy.*, **20**, 1049-1053.

31 Chen, L. and Wu, C. (1995) Performance of a cascade endoreversible heat pump system, *Int. J. of Ambient Energy.*, **68**, 137-141.

32 Wu, C. and al. (1997) Finite time thermodynamic performance for a class of irreversible heat pumps, *ASME, ASIA'97 Conference*, Singapore.

33 Wu, C. and al. (1996) Optimization of a steady flow refrigeration cycles, *Int. J. of Ambient Energy.*, **17**, 199-206.

34 Chen, L. and al. (1997) Finite time thermodynamic performance of an isentropic closed regenerated Brayton refrigeration cycle, *Int. J. of Energy Environment Economics*, **4**, 261-274.

35 Wu, C. (1997) Specific heating load of an endoreversible Carnot heat pump, *Int. J. of Ambient Energy.*, **14**, 25-28.

36 Wu, C. (1993) Performance of a solar engine driven air conditioning system, *Int. J. of Ambient Energy.*, **14**, 77-82.

37 Davis, G.W. (1995) Finite time thermodynamic analysis of an internally irreversible heat engine driven heat pump system utilizing low grade energy, *Proc. of Thermodynamics and Design, Analysis and Improvement of Energy Systems, AES-35*, ASME, 93-98.

38 D'Aponte, F. and al. (1995) Influence of irreversible heat transfer on heat engine driven heat pump performance, *Proc. of Thermodynamics and Design, Analysis and Improvement of Energy Systems, AES-35*, ASME, 29-34.

39 Lu, P.C. (1995) Thermodynamics with finite heat transfer area of finite surface, *Proc. of Thermodynamics and Design, Analysis and Improvement of Energy Systems, AES-35*, ASME, 51-60.

40 Gordon, J.M. and al. (1997) Optimizing chiller operation based on finite time thermodynamics : universal modeling and experimental confirmation, *Int. J. Refrig.*, **20**, 191-200.

41 Chua, H.T. and al. (1996) Experimental study of the fundamental properties of reciprocating chillers and their relation to thermodynamic modeling and chiller design, *Int. J. of Heat and Mass Transfer*, **39**, 2195-2204.

42 Gordon, J.M., NG, K.C. (1995) Predictive and diagnostic aspects of a universal thermodynamic model for chillers, *Int. J. of Heat and Mass Transfer*, **38**, 807-818.

43 Gordon, J.M., NG, K.C. (1993) Thermodynamic modeling of reciprocating chillers, *J. of Appl. Phys.*, **75**, 2769-2774.

44 Grosu, L. and al. (1997) Contribution à l'étude exergoéconomique des systèmes de réfrigération, *Proc. of BIRAC'97 Conf.*, Bucharest, Romania.

45 Grosu, L. (1998) Contribution to exergo economy of refrigerating machine, *Proc. of ECOS'98*, Nancy, France.

46 Metrot, A. (1996) Étude comparative de l'optimisation des machines à cycle inverses, *Engineering student report*, ENSEM, Nancy. France.

47 Erbay, L. and Yavuz, H. (1998) Maximum heating density of a Stirling heat pump, *Proc. of ECOS'98*, Nancy, France.

48 Yavuz, H. and Erbay, L. (1998) General performance characteristics of an Ericsson refrigerator, *Proc. of ECOS'98*, Nancy, France.

49 Feidt, M. and al. (1995) Optimisation des machines à froid et pompe à chaleur fonctionnant selon des cycles variés (Carnot, Stirling, Ericsson, Brayton ; machines à 3 sources), *Internal report*, ENSEM, Nancy, France

50 Diny, M. and al. (1995) Optimisation thermodynamique des machines à froid trithermes, *Proc. of the 19th I.I.R.Congress*, Hague, Netherlands, 147-154.

51 Chen, J. and Yan, Z. (1989) Unified description of endoreversible cycles, *Phys. Rev.*, **39**, 4140-4147.

52 Feidt, M. and Benelmir, R. (1992) Première approche de l'optimisation des échangeurs de machines trithermique, *Journées S.F.T.*, Paris, France

53 Diny, M. and al. (1998) Optimal performance of three heat-source machines, *Cryogenics*, in press.

54 Feidt, M. (1998) Tentative unified description of engines, refrigerators and heat pumps in contact with two or three heat reservoirs, *J. of Energy, Environment, Economics*, in press.

SYNTHESIS ON STIRLING ENGINE OPTIMIZATION

M. COSTEA[1], M. FEIDT[2] and S. PETRESCU[3]

[1] *Department of Applied Thermodynamics*
Polytechnic University of Bucharest
Splaiul Independentei 313, Bucharest, Romania

[2] *L.E.M.T.A., University Henri Poincaré of Nancy 1*
2, avenue de la Forêt de Haye, 54516 Vandoeuvre, France

[3] *Department of Mechanical Engineering*
Bucknell University, Lewisburg, PA 17837, USA

1. Introduction

The Stirling cycle analysis is a topic of interest to those concerned with the performance of thermal machines. The current trend in this analysis is given by the evaluation of the losses due to the irreversibilities in the cycle and their effect on the machine performance.

Recent books had major contributions to the Stirling machine analysis [1-4]. Others presented optimization models of endoreversible and exoirreversible cycles [5-8]. The effects of the non linear form of the heat transfer law (i.e., radiative) related to the solar applications [9-12] were also investigated.

Our analysis of a Stirling engine irreversible cycle has followed several steps [13-17] aiming at the optimization of the engine performance, particularly of the heat transfer surface conductance or area distribution among the machine heat exchangers. The optimization criterion was the maximum engine power output. A synthesis of some results produced by the method is presented in this paper.

2. General Presentation of the Optimization Method

2.1. THE STIRLING ENGINE CYCLE WITH IRREVERSIBILITIES

The Stirling engine cycle is shown on T-S diagram in Fig. 1. The diagram is modified to include the effects of heat transfer through a temperature difference at the source and sink, machine thermal losses and for incomplete regeneration. The finite thermal capacities of the source and the sink are suggested by their continuos temperature variation. Additional heat from the external source is shown to be needed in the process, Q_{x-2}, due to incomplete regeneration. The H_i notation on the source temperature variation identifies the separation between the two heat exchangers.

A. Bejan and E. Mamut (eds.), Thermodynamic Optimization of Complex Energy Systems, 403–410.

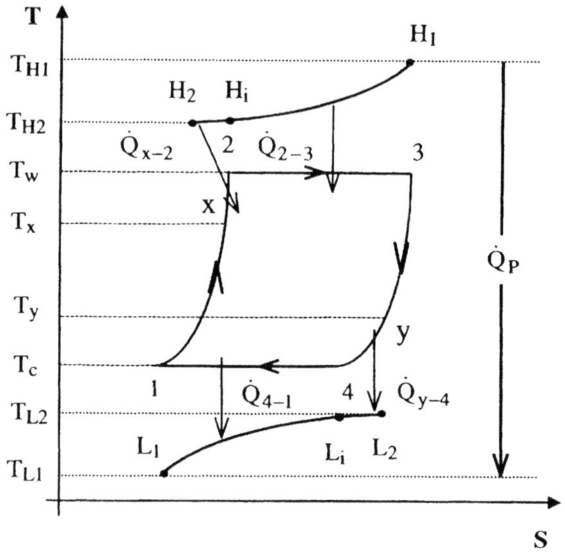

Figure 1. Stirling engine cycle with irreversibilities.

Similarly, the unregenerated heat is shown being rejected in the process, Q_{y-4}, and the L_i notation corresponds to the separation between the two heat exchangers at the sink. The heat losses of the engine are shown between the extreme temperature levels of the source and sink, T_{H1} and T_{L1}. Note that when $T_{S1} = T_{S2}$, the source and the sink become thermostats.

The heat transfer rate available at the source replaces in our model the source of constant temperature, usually considered. Thus, the source temperature can also be optimized.

2.2. MATHEMATICAL MODEL

The Stirling engine is analyzed using a mathematical model based on the First and Second Law of Thermodynamics applied to the cycle, and the energy balance at the source. By hypothesis, the regenerated heat corresponds to the temperature difference ($T_w - T_y$ or $T_x - T_c$). Thus, the heat regeneration factor introduced in the model is defined [18] as

$$\eta_R = \frac{T_x - T_c}{T_w - T_c} = \frac{T_w - T_y}{T_w - T_c} \tag{1}$$

Note that this definition is the same as the heat exchanger effectiveness ε.

Generally, the Stirling machine regenerator can be considered as a counterflow heat exchanger. In order to make the approach more general, a parallel-flow heat exchanger was also considered. Taking account of the fluid flow in the machine regenerator, it results that $C_{max} = C_{min}$. Hence, adjusting the effectiveness expressions of a counterflow and a parallel-flow heat exchanger [19] yields

$$\varepsilon_{R.cf} = 1 - \frac{1}{1 + NTU_R} \quad (2)$$

$$\varepsilon_{R.pf} = \frac{1 - \exp[-2NTU_R]}{2} \quad (3)$$

where NTU_R is the number of heat transfer units corresponding to the regenerator. Between the extreme temperature levels of the source and the sink, the heat losses are accounted for by an additional equivalent heat transfer conductance:

$$\dot{Q}_{loss} = \left(\bar{h}A\right)_{loss,eq} \cdot \left(T_{Hl} - T_{Ll}\right) \quad (4)$$

To make the results more general, the model was conceived to investigate the influence of the heat transfer law form on the engine performance. The two nonlinear forms of the heat transfer law at the source and sink are given by

$$\dot{Q}_S = \bar{h}A\left(T_S - T_f\right)^{n_j} \quad (5)$$

$$\dot{Q}_S = \bar{h}A\left(T_S^{n_i} - T_f^{n_i}\right) \quad (6)$$

where the values of n_i and n_j corresponding to the physical cases are: $n_j = 1/4$ for natural convection ; $n_j = 3$ for boiling ; $n_i = 1$ for Newton's law of cooling ; $n_i = 4$ for radiative heat transfer. We notice $\bar{h}$ - heat transfer coefficient, A - heat transfer surface area, T_S - temperature of the heat reservoirs, and T_f - working fluid temperature.

The resulting nonlinear system of equations is :

First Law of Thermodynamics $\quad \dot{W} = \dot{Q}_{2-3} - \left|\dot{Q}_{4-1}\right| \quad (7)$

Second Law of Thermodynamics $\quad \dfrac{\dot{Q}_{2-3}}{T_w} - \dfrac{\left|\dot{Q}_{4-1}\right|}{T_c} = 0 \quad (8)$

Energy balance at the source

$$\dot{m}c_v\left(T_w - T_c\right)\left(1 - \eta_R\right) + \dot{Q}_{2-3} + \left(\bar{h}A\right)_{loss,eq}\left(T_H - T_L\right) = \dot{\phi} \quad (9)$$

where $\dot{m}c_v$ is the flow capacity rate of the working fluid, and $\dot{\phi}$ is heat transfer rate available at the source. The mathematical modeling also includes other equations, especially for the heat transfer in the heat exchangers that permit to express the intermediate model variables. Hence, it results a reduced number of constraints and variables to be optimized.

In order to render the model more general and to facilitate the extension of the analysis, all calculations have been made in non-dimensional form of the equations. Thus, the temperatures were divided by the sink temperature, T_L, the heat transfer rates and the engine power output, by the product $\dot{m} c_v T_L$, respectively. The term "reduced" were added to each non-dimensional value.

The optimization analysis of the Stirling engine results in a system of nonlinear equations that are solved using the method of Lagrangian undetermined multipliers. The non-dimensional system of equations consists generally from an objective function and two constraints:

$$\dot{W}_r = \dot{Q}_{2-3r} - \left|\dot{Q}_{4-1r}\right| \tag{10}$$

$$\frac{\dot{Q}_{2-3r}}{\theta_w} - \frac{\left|\dot{Q}_{4-1r}\right|}{\theta_c} = 0 \tag{11}$$

$$(\theta_w - \theta_c)(1-\eta_R) + \dot{Q}_{2-3r} + \gamma\, NTU_{loss}(\theta_H - 1) = \dot{\phi}_r \tag{12}$$

where γ is the ratio of specific heats, c_p/c_v. The general variables of the model are the reduced temperatures : θ_H, θ_w, and θ_c. For source and sink of finite thermal capacity, instead of θ_H,, the outlet temperature of the source, θ_{H2},, is considered. Also, depending on the constraint, in this case one or two variables are added to the model. These variables are reduced heat transfer conductance or area.

The model general parameters are (1) the reduced heat transfer rate available at the source, (2) the NTU_{loss} corresponding to the heat losses of the engine. Also depending on the case, the NTU_T corresponding to the total heat transfer surface conductance or area of the engine heat exchangers and the ratio of heat transfer coefficients, $\overline{rhi}$, are added. The optimization method aims at indicating the optimum state and optimum distribution of the heat transfer surface conductance or area for maximum engine power output.

2.3. STUDIED CASES INVENTORY

The optimization models were developed for several cases characterized by

(i) source and sink of constant temperature and finite thermal capacity ;
(ii) linear and nonlinear form of the heat transfer law in the heat exchangers at the source and sink ;
(iii) regenerator conductance or area included or not in the fixed corresponding parameter of the engine.

By combining the above system configurations with the constraints of fixed (1) heat transfer surface conductance or area at the source and sink and (2) total heat transfer surface conductance or area of the engine, it results a significant ramification of the cases tree. Only some simple cases permitted analytical solution to the optimization problem, the others being solved numerically.

3. Results

The preceding optimization method of the Stirling engine provides the optimum state and distribution of the heat transfer surface conductance or area among the engine heat exchangers. Calculations have been made for illustration using helium as the working fluid and with the following variation range of fixed parameters : $\dot{\phi}_r$=0.1÷10, NTU_T =0÷5, NTU_{loss} =0÷5, $\overline{rhi}$ =0.1÷10. From the results we have obtained, the effect of the type of heat reservoirs and that of the heat transfer law form on the engine performance was selected for presentation in the paper.

3.1. EFFECT OF THE FINITE THERMAL CAPACITY OF THE HEAT RESERVOIRS COMPARATIVELY TO THERMOSTATS

The analysis of the effect of the finite thermal capacity of the source and the sink on the optimum state point, maximum power output of the engine and cycle efficiency is shown in Table 1. The temperatures are not substantially affected. It can be seen that all the temperatures increase for the finite thermal capacity, especially the source temperature. It is quite normal because it represents the inlet source temperature, θ_{HI}^*.

TABLE 1. State point variables, maximum engine power output and cycle thermal efficiency for the case with regenerator included in the model and total heat transfer area fixed.

	Counterflow Regenerator		Parallel-flow Regenerator	
	Thermostat CF	Finite Capacity CF	Thermostat PF	Finite Capacity PF
θ_c^*	1,170	1,240	1,150	1,220
θ_w^*	1,460	1,500	1,430	1,450
θ_H^*	1,580	1,830	1,530	1,800
$\lvert\dot{W}_r\rvert_{max}$	**0,065**	**0,030**	**0,060**	**0,030**
η_t^*	0,140	0,100	0,130	0,085

The maximum power output and the thermal efficiency of the cycle are mostly affected. It is registered a reduction of 40-50% of the power and of 25-30% of the cycle efficiency for the case of finite thermal capacity.

Other results of the calculations are shown plotted in Fig. 2. The heat transfer area is mostly concentrated in the second heat exchanger of the sink, the regenerator and the first heat exchanger of the source. The comparison of the distributions indicates that the effect of the finite thermal capacity consists of the reduction of the heat transfer area of both second heat exchangers of the source and sink, while the heat transfer area of the other engine heat exchangers increases. Also we note that the regenerator heat transfer area is important, mostly for the finite thermal capacity case.

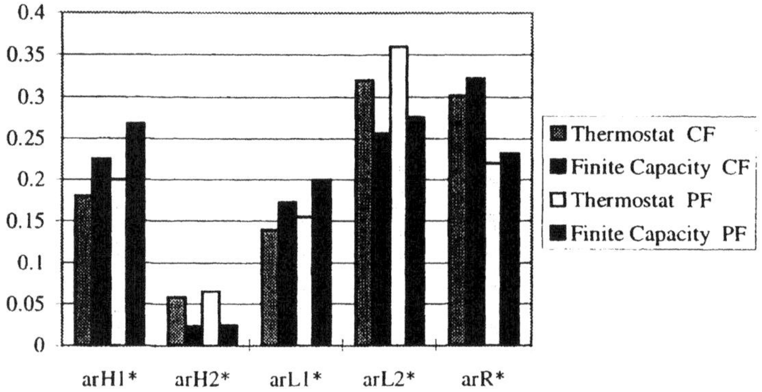

Figure 2. Comparison of the optimum distribution of the heat transfer area among the engine heat exchangers corresponding to the cases with constant or variable source and sink temperature.

3.2. EFFECT OF THE FORM OF THE HEAT TRANSFER LAW

The values corresponding to the optimum state point, maximum reduced power output and thermal efficiency of the cycle for symmetrical and non symmetrical combinations of the linear and radiative nonlinear form of the heat transfer law are presented in Table 2. One can se that the minimum temperature difference at the thermal reservoirs (θ_H^* -θ_w^*, θ_c^*-1) is registered for the radiative nonlinear form of the heat transfer law in

TABLE 2. Optimum state point, maximum power output and the corresponding thermal efficiency of the cycle for the total heat transfer area fixed and regenerator included in the model.

	Linear Law		Combination of Radiative Nonlinear Law - Linear Law					
	CF	PF	CF			PF		
	$n_H = 1$ $n_L = 1$	$n_H = 1$ $n_L = 1$	$n_H = 4$ $n_L = 1$	$n_H = 4$ $n_L = 4$	$n_H = 1$ $n_L = 4$	$n_H = 4$ $n_L = 1$	$n_H = 4$ $n_L = 4$	$n_H = 1$ $n_L = 4$
θ_c^*	1,170	1,150	**1,150**	**1,055**	1,070	**1,130**	**1,030**	1,050
θ_w^*	1,460	1,430	1,500	**1,450**	1,400	1,430	**1,360**	1,350
θ_H^*	1,580	1,530	1,520	**1,460**	**1,480**	1,450	**1,373**	**1,420**
$\lvert \dot{W}_r \rvert_{max}$	0,065	0,060	0,078	**0,104**	0,090	0,074	**0,093**	0,082
η_t^*	0,140	0,130	0,165	**0,200**	0,175	0,142	**0,165**	0,150

any combination. It means that the irreversibilities due to the heat transfer at temperature difference are minimum in these cases. The maximum power output and cycle efficiency of the cycle are attained in the case of symmetrical combination of the radiative nonlinear form of the heat transfer law and for the convective nonlinear form, that of the combination $n_H = 3$; $n_L = 1$. It is obvious that the engine performance in the case of symmetrical combination of the linear form of the heat transfer law usually adopted in thermal machine analysis are low. Generally, the engine performance decreases in this order: (1) combinations of radiative nonlinear - linear form, (2) symmetrical combination of linear form and (3) combinations of the linear - convective nonlinear form.

The heat transfer area distribution among the engine heat exchangers for three combinations of the convective nonlinear form with the linear form of the heat transfer law is shown in Fig. 3. This figure indicates that the heat transfer distribution changes substantially for the last combination (n_H = 1 ; n_L = 3) comparing with those corresponding to the others two, especially for the source heat exchangers. It appears that the heat transfer area must be greather for the heat exchangers with linear heat transfer law form. The same conclusion results for the combinations of the radiative nonlinear - linear form of the heat transfer law.

The regenerator heat transfer area is important comparatively to the area distribution in the other heat exchanger. Thus, for some combinations of the heat transfer law form, it may have the greatest value among the engine heat exchangers.

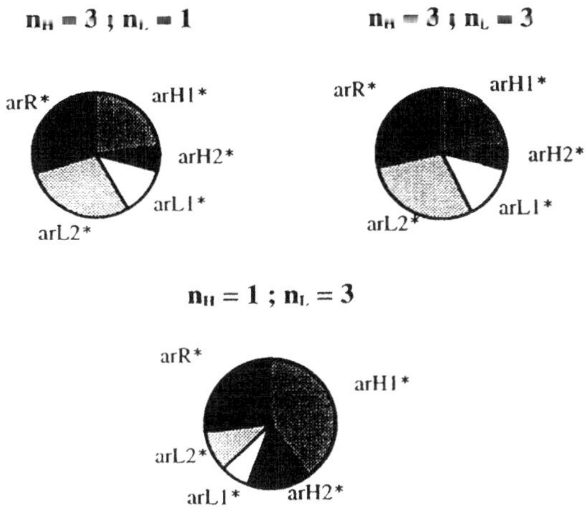

Figure 3. Heat transfer area distribution among the engine heat exchangers for three combinations of the convective nonlinear form with the linear form of the heat transfer law.

The regenerator type does not influence substantially the engine performance. Although the counterflow heat exchanger modeling the regenerator is preferred due to the higher power output and cycle efficiency performed.

4. Conclusion

A synthesis of the results of an optimization method of the Stirling engine with irreversibilities is presented that makes possible the prediction and comparison of the performance of solar Stirling engines operating under the same conditions [20]. It has shown how the state parameters and the distribution of the heat transfer area among the engine heat exchangers are affected by the finite thermal capacity of the heat reservoirs and the heat transfer law form in the heat exchangers.

5. References

1. Walker, G., Reader, G., Fauvel, O.R. and Bingham, E.R. (1994) *The Stirling Alternative*, Gordon and Breach Science Publishers, Amsterdam.
2. Organ, J.A. (1992) *Thermodynamics and Gas Dynamics of Stirling Cycle Machine*, Cambridge University Press, Cambridge.
3. Walker, G. (1983) *Cryocoolers - Part I : Fundamentals*, Plenum Press, New York.
4. Tew, R.C. (1986) Progress of Stirling cycle analysis and loss mechanism characterisation, *NASA TM - 88891*.
5. Feidt, M. (1996) *Thermodynamique et Optimisation Énergétique des Systèmes et Procédés*, TECH & DOC, Paris.
6. Radcenco, V. (1994) *Generalized Thermodynamics*, Tehnical Publisher, Bucharest.
7. Bejan, A. (1988) Theory of heat transfer - irreversible power plants, *Int. J. Heat Mass Transfer*, **31**, 6, 1211-1219.
8. Bejan, A. (1988) *Advanced Engineering Thermodynamics*, Wiley & Sons, New York.
9. De Vos, A. (1992) *Endoreversible Thermodynamics of Solar Energy Conversion*, Oxford Science Publishers, Oxford.
10. Gordon, J.M. (1990) Nonequilibrium thermodynamics for solar energy applications, *Advanced in Thermodynamics*, **4**, 95-120.
11. Chen, F. and Yan, Z. (1989) The effect of heat transfer law on performance of a two-heat source endoreversible cycle, *Journal Chem. Phys.*, **90**, 7, 3740-3743.
12. Blank, D. A., Davis, G.W. and Wu, C. (1994) Power optimization of an endoreversible Stirling cycle with regeneration, *Energy*, **19**, 1, 125-133.
13. Feidt, M., Costea, M. and Popescu, G. (1995) Optimal design of Stirling engine using a generalized heat transfer law model, *Workshop on Second Law Analysis of Energy Systems : towards the 21-st century*, Rome, Italy, 431-441.
14. Costea, M., Feidt, M. and Belabbas, B. (1995) Optimal distribution of the heat transfer conductances or areas for Stirling machines, *Proc of WAM of ASME*, San Francisco, USA, **35**, 35-44.
15. Feidt, M., Costea, M. and Belabbas, B. (1996) Influence of the regenerator on the optimal distribution of the heat transfer surface conductance in a Stirling engine, *Proc. of the Inter. Sympos. ECOS'96*, Stockholm, Sweden, 99-103.
16. Costea, M., Feidt, M. and Belabbas, B. (1996) Comparison of the optimal heat transfer surface area distribution in a Stirling engine with source and sink of finite thermal capacity, *Proc. of the Inter. Sympos. ESDA'96*, Montpellier, France, 207-214.
17. Popescu, G., Radcenco, V., Costea, M. and Feidt, M. (1996) Optimisation thermodynamique en temps fini du moteur de Stirling endo- et exo-irréversible, *Rev. Gén. Therm.*, **35**, 656-661.
18. Costea, M. (1997) Improvement of heat exchangers performance in view of the thermodynamic optimization of the Stirling machine ; Unsteady-state heat transfer in porous media, *Ph.D. Thesis*, P. U. Bucharest & U.H.P. Nancy I.
19. Incropera, F.P. and DeWitt, D.P. (1996) *Fundamentals of Heat and Mass Transfer*, Wiley&Sons, New York.
20 Stine, W. and Diver, R. (1994) A compendium of solar dish/Stirling technology, Sandia National Laboratories, Albuquerque, NM, USA, *Contract No. 67-3678*.

MINIMIZING LOSSES — TOOLS OF FINITE-TIME THERMODYNAMICS

B. ANDRESEN
Ørsted Laboratory, University of Copenhagen
Universitetsparken 5, DK-2100 Copenhagen Ø
Denmark

1. Introduction

Finite-time thermodynamics was 'invented' in 1975 by R. S. Berry, P. Salamon, and myself as a consequence of the first world oil crisis. It simply dawned on us that all the existing criteria of merit were based on reversible processes and therefore were totally unrealistic for most real processes. That made an evaluation of the potential for improvement, of minimizing the losses, for a given process quite difficult [1].

Finite-time thermodynamics is developed from a macroscopic point of view with heat conductances, friction coefficients, overall reaction rates, etc. rather than based on a microscopic knowledge of the processes involved. Consequently most of the ideas of traditional thermodynamics have been assimilated, e.g. the notions of thermodynamic potential (Sect. 2.1) and availability (Sect. 2.2). At the same time we have seen new concepts emerge, e.g. the non-equivalence of well-honored criteria of merit (Sect. 3.2), the importance of power as the objective, the generality of the endoreversible engine, and in particular thermodynamic length (Sects. 2.3 and 3.3). Several of these abstract concepts have been successfully applied to practical optimizations [2 – 9]. We will present some of these tools below, roughly in order of increasing complexity.

The model of Curzon and Ahlborn [10] has evolved into almost a classic paradigm of systems operating in finite time. This is a Carnot engine with the simple constraint that it be linked to its surroundings through *finite* heat conductances. Figure 1 illustrates the slightly more general endoreversible system with the triangle signifying any reversible engine. (The term endoreversible means 'internally reversible', i.e. all irreversibilities reside in the coupling of flows to the surroundings. In this case that means resistance to heat transfer and possibly friction.) It turns out that the results derived by Curzon and Ahlborn (and by Novikov [11] and Chambadal [12] for simpler systems) explicitly for an interior Carnot engine are equally valid for a general endoreversible system. The maximum efficiency of their engine is of course $\eta_C = 1 - T_L/T_H$, obtained at zero rate so that losses across the resistors vanish, but these authors showed that, when the system operates to produce *maximum power*, the efficiency of the engine is only

$$\eta_w = 1 - (T_L/T_H)^{1/2}. \qquad (1)$$

A. Bejan and E. Mamut (eds.), Thermodynamic Optimization of Complex Energy Systems, 411–420.

Besides the simplicity of the expression it is remarkable that it does not contain the value of the heat conductances.

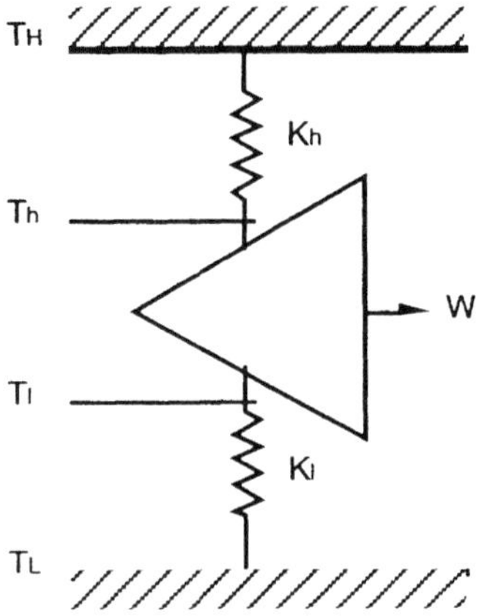

Figure 1. An endoreversible engine has all its losses associated with its coupling to the environment, there are no internal irreversibilities. This is illustrated here as resistances in the flows of heat to and from the working device indicated by a triangle. These unavoidable resistances cause the engine proper to work across a smaller temperature interval, $[T_h;T_l]$ than that between the reservoirs, $[T_H;T_L]$, one which depends on the rate of operation.

2. Performance Bound without Path

The smallest amount of information one can ask for concerning the performance of a system is a single number, e.g. the work or heat exchanged during the process, its efficiency, or any other figure of merit. In most cases this can be calculated without knowledge of the detailed path followed and is then computationally much simpler to obtain.

2.1. GENERALIZED POTENTIALS

In traditional thermodynamics potentials are used to describe the ability of a system to perform some kind of work under given constraints. These constraints are usually the constancy of some state variables like pressure, volume, temperature, entropy, chemical potential, particle number, etc. Under such conditions the decrease in thermodynamic potential P from state i to state f is equal to the amount of work that is produced when a reversible process carries out the transition, and hence is the upper bound to the amount of work produced by any other process,

$$W \leq W_{rev} = P_i - P_f. \tag{2}$$

In this section we will show that the constraints need not simply be the constancy of some state variable, and that the potentials may be generalized to contain constraints involving time [13]. The procedure will be a straight forward extension of the Legendre transformations [14] used in traditional thermodynamics [15, 16], and we will start with such an example.

In a reversible process heat and work can be expressed as inexact differentials, $dQ = TdS$, $dW = PdV$, i.e. they cannot by themselves be integrated, further constraints defining the integration path are required. Such a constraint could be that the process is isobaric, dP = 0. One can then add a suitable integrating zero-term, xdP to make dW an exact differential. The obvious choice is x = V,

$$dW = PdV = PdV + VdP = d(PV), \quad (3)$$

such that the isobaric work potential becomes $\mathcal{P} = PV$.

Now, the constraints need not be the constancy of one of the state variables. Consider a balloon with constant surface tension α. In equilibrium with an external pressure P_{ex} such a sphere of radius r has an internal pressure $P = P_{ex} + 2\alpha/r$, which can be rearranged into

$$(P - P_{ex})V^{1/3} = 2\alpha\,(4\pi/3)^{1/3}. \quad (4)$$

Since the right hand side of this equation is a constant, this means that $(P - P_{ex})V^{1/3}$ is an integral of motion for the fluid inside the balloon. We can then add a suitable amount of $d[(P - P_{ex})V^{1/3}]$ (equal to zero) to dW to make it exact,

$$dW = PdV = PdV + \frac{3}{2}\,V^{2/3}\,d[(P - P_{ex})V^{1/3}] = d[\frac{1}{2}\,V(3V - P_{ex})]. \quad (5)$$

Thus the work done by the coupled system, surface plus fluid, is given by the decrease in the potential $\mathcal{P} = \frac{1}{2}\,V(3V - P_{ex})$, regardless of path followed.

This procedure of adding a suitable amount of a 'zero-quantity' can be generalized to any differential constraint, even including time explicitly [13].

2.2. FINITE-TIME AVAILABILITY

One of the more powerful results in finite-time thermodynamics is the definition of a finite-time availability [17]. The traditional availability A of a system in contact with given surroundings is a state function with the quality that the decrease in its value in going from state i to state f is the maximum (and hence reversible) work that can be extracted during that process. The finite-time availability $\mathcal{A}$ retains this property and simply adds that the process is restricted to operate (go to completion) during time $\tau = t_f - t_i$. Then

$$\mathcal{A} = W_{max}(\tau) = \max\left[A(t_i) - A(t_f) - T_0 \int_{t_i}^{t_f} \dot{S}_{tot}\, dt \right], \quad (6)$$

where the last equality uses the Tolman-Fine form of the Second Law of thermodynamics [18].

The maximum search in Eq. (6) can either be constrained to exactly reach a given final state at time t_f (the initial state is always considered known), in which case ΔA is fixed, and the optimization becomes one of minimizing the entropy production, or also the final state may be included in the optimization, in which case $\mathcal{A}$ must be evaluated by optimal control. If the final state is specified, a solution may not exist if τ is too short, since only a certain set of states can be reached from a given initial state in time

τ. In addition, the finite-time availability does not necessarily have ΔA as its limit for very long times, because the system may contain internal relaxation processes which remain irreversible even for very slow operation. If there is a direct heat leak from the system reservoir to the surroundings, then a long process time may even reduce *A* to zero. On the other hand, losses are not always detrimental to the performance of a system if they open up new pathways — actually some processes depend on irreversibilities for their very existence (e.g. [19]).

2.3. THERMODYNAMIC LENGTH

For static purposes Weinhold [20, 21] defined a metric on the abstract space of equilibrium states of a system represented by all its extensive variables X_i as

$$\mathbf{M}_U = \left\{\frac{\partial^2 U}{\partial X_i \partial X_j}\right\}, \tag{7}$$

where U is the internal energy. Based on this metric Salamon and Berry [22] found a connection between the thermodynamic length along a process path and the (reversible) availability lost in the process. Specifically, if the system moves via states of local thermodynamic equilibrium from an initial equilibrium state i to a final equilibrium state f in time τ, then the dissipated availability $-\Delta A$ is bounded from below by the square of the distance (i.e. length of the shortest path) from i to f times ε/τ, where ε is a mean relaxation time of the system. If the process is endoreversible, the bound can be strengthened to

$$-\Delta A \geq L^2 \varepsilon/\tau, \tag{8}$$

where L is the length of the *traversed* path from i to f. Equality is achieved at constant thermodynamic speed $v = dL/dt$ corresponding to a temperature evolution given by [23]

$$\frac{dT}{dt} = - vT/\varepsilon(C)^{1/2}, \tag{9}$$

where C is the heat capacity of the system. For comparison, the bound from traditional thermodynamics is only $-\Delta A \geq 0$.

An analogous expression exists for the total entropy production during the process:

$$\Delta S^u \geq L^2 \varepsilon/\tau. \tag{10}$$

The length L is then calculated relative to the entropy metric

$$\mathbf{M}_S = -\left\{\frac{\partial^2 S}{\partial X_i \partial X_j}\right\}. \tag{11}$$

In statistical mechanics, where entropy takes the form $S(\{p_i\}) = -\sum_i p_i \ln p_i$, the metric $\mathbf{M}_S$ is particularly simple, being the diagonal matrix [24] $\mathbf{M}_S = -\{1/p_i\}$.

The same procedure of calculating metric bounds for dynamic systems has been applied to coding of messages [25] and to economics [26].

More recently [27] we have relaxed a number of the assumptions in the original work, primarily those restricting the system to be close to equilibrium at all times and the average form of the relaxation time ε. The more general bound replacing Eq. (10) then becomes

$$\Delta S^u \geq \frac{1}{\Xi}\left[\int_{\xi_i}^{\xi_f} \frac{1}{T\sqrt{C}} \left|\frac{dU}{d\xi}\right| \left(1 + \frac{\theta}{CT}\frac{dU}{d\xi} + \ldots\right)^{1/2} d\xi\right]^2 \tag{12}$$

with $\Xi = \xi_f - \xi_i$ being the total duration of the process in natural dimensionless time units, $d\xi = dt/\varepsilon(T)$, and where we have defined $\theta(T) = 1 + (T/2C)(\partial C/\partial T)$.

The equality (lower bound) in Eq. (12) is achieved when the integrand is a constant, i.e. when $dS^u/d\xi$ = constant. Consequently, constant rate of entropy production, when expressed in terms of natural time, is the path or operating strategy which produces the least overall entropy. One can express the optimal path in a form similar to Eq. (9):

$$\frac{dT}{dt}\left[1 + \frac{\theta(T)\,\varepsilon\,(dT/dt)}{T} + \ldots\right]^{1/2} = \text{constant} \times \frac{T}{\varepsilon(C)^{1/2}}\,. \tag{13}$$

The constant thermodynamic speed algorithm, Eq. (9), is thus the leading term of the general solution in an expansion about equilibrium behavior.

3. Optimal Path

3.1. OPTIMAL PATH CALCULATIONS

A knowledge of the maximum work that can be extracted during a given process, e.g. calculated by one of the procedures described in the previous section, may not by itself be sufficient. One may also want to know *how* this maximum work can be achieved, i.e. the time path of the thermodynamic variables of the system. The primary tool for obtaining this path is optimal control theory.

This is not the place to repeat the mechanics of optimal control calculations (see e.g. [28 – 30]). Let it suffice to point out that in order to set up the optimal control problem one must specify

- the controls, i.e. the variables that can be manipulated by the operator (they may be a volume, rate, voltage, heat conductance, etc.);
- limits on the controls and on the state variables, if any (in order to avoid unphysical situations such as negative temperatures and infinite speeds);
- the equations that govern the time evolution of the system (they will usually be differential equations describing heat transfer rates, chemical reaction rates, friction, and other loss mechanisms);

- the constraints that are imposed on the system (e.g. conserved quantities, the quantities held constant, or any requirements on reversibility. The constraints may either be differential, instantaneous i.e. algebraic, or integral i.e. not obeyed at each point but over the entire interval);
- the desired quantity to be maximized, called the objective function (usually expressed as an integral); and finally
- whether the duration of the process is fixed or part of the optimization.

Typical manipulation usually leads to a set of coupled, non-linear differential equations for which a qualitative analysis and a numerical solution are the only hope. Thus answering the more demanding question about the optimal time path rather than the standard question about maximum performance requires a considerably larger computational effort. On the other hand, once the time path is calculated, all other thermodynamic quantities may be calculated from it, much like the wave function is the basis of all information in quantum mechanics.

3.2. CRITERIA OF PERFORMANCE

Efficiency, the earliest criterion of performance for engines, measured how much water could be pumped out of a mine by burning a ton of coal. Other familiar criteria include effectiveness (efficiency relative to the Carnot efficiency), change of thermodynamic potential, and loss of availability, all of which are measures of work. Potentials for heat can also be defined but are less common. In addition total entropy production is indicative of inefficient operation [31]. The Curzon-Ahlborn analysis and many of our own analyses use a quite different criterion, that of *power*. This quantity is of course zero for any reversible system, and maximizing power forces us to deal with systems operating at finite rates. Other criteria of performance are the rate of entropy production and the rate of loss of availability. Under some circumstances, optimizing one of these quantities is equivalent to optimizing another. For example, minimizing the entropy production is equivalent to minimizing the loss of availability.

Salamon and Nitzan [32] have optimized the Curzon-Ahlborn engine for a number of these objective functions. Assuming the working fluid to be in contact with the hot reservoir for the period τ_1 and the cold reservoir for the period τ_2, the optimal time distributions are shown in Fig. 2. The diagonal $\tau = \tau_1 + \tau_2$ indicates fixed total cycle time, and only processes above the curve labeled 'zero power' produce positive power. It is quite obvious that different criteria of merit dictate different operating conditions for the process. Even when not knowing the precise objective function but only that it belongs to a specified class, one can sometimes say a good deal about the possible optimal behavior of the system. If one considers the Curzon-Ahlborn engine to be a model of a power plant which buys heat q (coal) at the unit price α and sells work w (electricity) at the unit price β, its net revenue is $\Pi = \beta w - \alpha q$. All solutions to the problem of maximizing this revenue are bounded on one side by the solutions to the maximum power problem (when α is significant compared to β) and on the other side by the solutions corresponding to minimum loss of availability (when coal and electricity are priced according to their availability contents). While this is a very simple example, this approach has far reaching possibilities for describing biological, ecological, and economic systems.

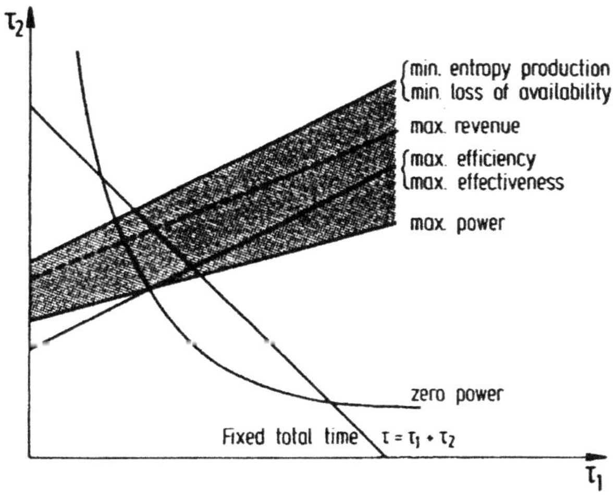

Figure 2. If an endoreversible engine (Fig. 1) spends time τ_1 in contact with the hot reservoir and τ_2 in contact with the cold reservoir, the optimal proportioning between τ_1 and τ_2 depends on what one chooses to optimize, as indicated. The locus of maximum revenue for a power producing system falls in the shaded area for any choice of prices, as described in the text. Only contact times above the hyperbola marked 'zero power' actually correspond to positive power production.

3.3. DESIGN OF DISTILLATION COLUMN BY EQUAL THERMODYNAMIC DISTANCE

In Sect. 2.3 we saw that the concept of thermodynamic length not only provided a lower bound on dissipation (Eqs. (8), (10), and (12)), it also predicted which path would achieve that bound, namely operation at constant thermodynamic speed. We will use that line of thought to improve the performance of a conventional distillation column.

The internal distribution of temperature and mole fractions of light and heavy components in a conventional distillation column where heat is added and withdrawn only in the reboiler and the condenser, respectively, is fixed exclusively by the laws of energy and mass conservation. In most columns that leads to an S-shaped curve of temperature versus plate number with most of the variation occurring near the end points of the column, connected with a flat stretch around the feed point (see Fig. 3). A qualitatively similar picture emerges for the mole fractions. This temperature profile implies that the major part of the entropy production in the distillation process occurs near the ends of the column and is thus not uniformly distributed. Even more importantly, if the number of plates is increased, essentially only the middle flat section is extended while the segments of rapid variation are unchanged. This means that dissipation does not approach zero as the number of plates goes to infinity.

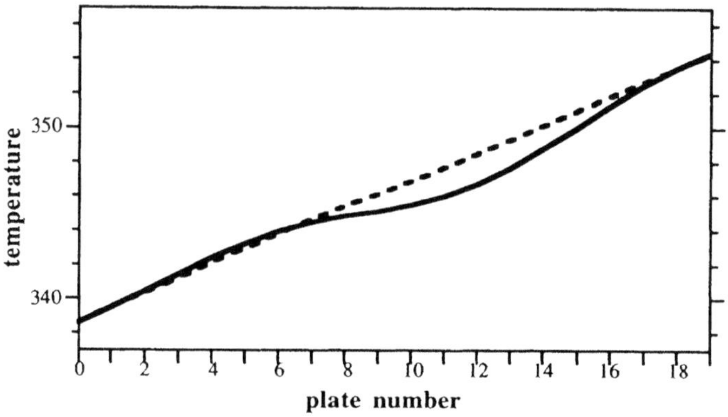

Figure 3. The temperature profile as a function of plate number, counted from the condenser, for conventional (solid) and equal-thermodynamic-distance (dashed) separation of methanol and isopropanol.

The principle of constant thermodynamic speed for continuous systems, or equivalently equal thermodynamic distance for discrete systems, aims at minimizing total dissipation in the column by distributing it evenly among the plates whatever their number [33]. In [34] a general quasistatic step process was optimized, i.e. a process composed of N discrete steps where the system equilibrates fully after each step. The standard description of a distillation column is exactly such a process where it is assumed that gas and liquid come to equilibrium at a particular temperature on each plate. Entropy is produced when the up- and down-moving flows encounter liquid on the next plate at slightly different temperature and composition.

The result of the optimization [34], for processes not too far from equilibrium, was that the minimum of the entropy production

$$\Delta S^u \geq L^2/2N \tag{14}$$

is achieved when the thermodynamic distance between each pair of plates is kept constant, i.e. they are equidistant in that measure. L is the total thermodynamic distance from one end of the column to the other. For any other distribution of the total length along the column dissipation is larger.

Using the metric Eq. (11), the thermodynamic distance $D_S{}^k$ from plate k–1 to plate k is equal to

$$D_S{}^k = (\Delta \mathbf{X}^k \, \mathbf{M}_S{}^k \, \Delta \mathbf{X}^k)^{1/2} \tag{15}$$

where $\Delta\mathbf{X}^k = \mathbf{X}^k - \mathbf{X}^{k-1}$ is the difference in extensities from plate k–1 to plate k. The total thermodynamic length of the column is thus $L_S = \sum_{k=1}^{N} D_S{}^k$.

Two consequences of this general result are immediate: dissipation (here entropy production) must be equally distributed along the column; and the total dissipation approaches zero as N, the number of plates, goes to infinity (Eq. (14)), i.e. the separation becomes reversible.

In binary distillation 8 extensive quantities are involved on each plate, besides the objective function entropy, namely enthalpy, volume, mole number of light component, and mole number of heavy component, each one for both gas and liquid, leading to an 8×8 metric matrix. Fortunately a number of relations allows to reduce the dimensionality of the problem dramatically. First of all, the two components are usually considered non-interacting, at once making $\mathbf{M}_S$ block-diagonal. Next, constant pressure in the column, energy and mass conservation, and the gas-liquid equilibrium equation reduce the problem to just one free variable which may conveniently be taken to be temperature. After a lot of algebra the plate-to-plate distance then becomes simply

$$D_S{}^k = [(C_r{}^k)^{1/2}/T^k]\,\Delta T \tag{16}$$

where C_r is an effective heat capacity involving all the constraints mentioned above in a complicated expression.

The computational procedure is to integrate Eq. (16) from the distillate temperature T_D to the reboiler temperature T_B (both of course given by the required product purities) to obtain the total thermodynamic column length L_S. The distance from one plate to the next must then be fixed at $D_S = L_S/N$ for optimal performance by adjusting the plate temperatures appropriately according to Eq. (24).

Obviously such freedom of adjustment does not exist in a conventional adiabatic column. Rather, it is necessary to allow individual heat exchange with each plate to maintain it at the desired temperature. This heat addition/removal is of course part of the energy balance used above. The result of the whole calculation is either a graph like Fig. 3 specifying the temperature of each plate in the column or a graph of the amount of heat added/removed at each plate. It should be emphasized that the total amount of heat used to perform a certain equal-thermodynamic-distance separation is only marginally different from that required by a conventional column, but a large part of it is used over a much smaller temperature difference than T_B to T_D leading to a correspondingly smaller entropy production.

4. Acknowledgments

Much of the work reported in this review has been done in collaboration with colleagues near and far. In particular I want to express my gratitude to profs. R. Stephen Berry, Peter Salamon, Jim Nulton, and Jeff Gordon.

References

1. Andresen, B. (1983) *Finite-Time Thermodynamics,* Physics Laboratory II, University of Copenhagen.
2. Hoffmann, K. H., Burzler, J., and Schubert, S. (1997) Endoreversible systems. *J. Non-Equil. Therm.* **22**, 311.
3. Hoffmann, K. H., Watowitch, S., and Berry, R. S. (1985) Optimal paths for thermodynamic systems: The ideal Diesel cycle, *J. Appl. Phys* **58**, 2125 – 2134.
4. Mozurkewich, M. and Berry, R. S. (1981) Finite-time thermodynamics: Engine performance improved by optimized piston motion, *Proc. Natl. Acad. Sci. USA* **78**, 1986.
5. Mozurkewich, M. and Berry, R. S. (1982) Optimal paths for thermodynamic systems: The ideal Otto cycle, *J. Appl. Phys* **53**, 34 – 42.
6. Gordon, J. M. and Zarmi, Y. (1989) Wind energy as a solar-driven heat engine: A thermodynamic approach, *Am. J. Phys.* **57**, 995 – 998.

7. Andresen, B. and Gordon, J. M. (1992) Optimal heating and cooling strategies for heat exchanger design, J. *Appl. Phys.* **71**, 76 –79.
8. Andresen, B. and Gordon, J. M. (1992) Optimal paths for minimizing entropy generation in a common class of finite-time heating and cooling processes, *Int. J. Heat and Fluid Flow* **13**, 294 – 298.
9. Gordon, J. M. and Ng, K. C. (1994) Thermodynamic modeling of reciprocating chillers, *J. Appl. Phys.* **57**, 2769 – 2774.
10. Curzon, F. L. and Ahlborn, B. (1975) Efficiency of a Carnot engine at maximum power output, *Am. J. Phys.* **43**, 22 – 24.
11. Novikov, I. I. (1957) The efficiency of atomic power stations (in Russian), *At. Enertg.* **3**, 409. English translation in (1958) *J. Nucl. Energy II* **7**, 125.
12. Chambadal, P. (1957) *Les centrales nucleaires*, Armand Colin, Paris., p. 41.
13. Salamon, P., Andresen, B., and Berry, R. S. (1977) Thermodynamics in finite time. II. Potentials for finite-time processes, *Phys. Rev. A* **15**, 2094 – 2102.
14. Hermann, R. (1973) *Geometry, Physics, and Systems*, Marcel Dekker, New York.
15. Callen, H. B. (1985) *Thermodynamics*, Wiley, New York.
16. Tisza, L. (1966) *Generalized Thermodynamics*, MIT Press, Cambridge, Massachusetts.
17. Andresen, B., Rubin, M. H., and Berry, R. S. (1983) Availability for finite-time processes: General theory and a model, *J. Chem. Phys.* **87**, 2704 – 2713.
18. Tolman, R. C. and Fine, P. C. (1948) On the irreversible production of entropy, *Rev. Mod. Phys.* **20**, 51 – 77.
19. Wheatley, J., Hofler, T., Swift, G. W., and Migliori, A. (1983) Experiments with an intrinsically irreversible acoustic heat engine, *Phys. Rev. Lett.* **50**, 499.
20. Weinhold, F. (1975) Metric geometry of equilibrium thermodynamics, *J. Chem. Phys.* **63**, 2479 – 2482.
21. Weinhold, F. (1978) Geometrical aspects of equilibrium thermodynamics, in D. Henderson and H. Eyring (eds.), *Theoretical Chemistry, Advances and Perspectives*, Vol. 3, Academic Press, New York, p. 15 – 54.
22. Salamon, P. and Berry, R. S. (1983) Thermodynamic length and dissipated availability, *Phys. Rev. Lett.* **51**, 1127 – 1130.
23. Nulton, J. and Salamon, P. (1988) Statistical mechanics and combinatorial optimization, *Phys. Rev. A* **37**, 1351 – 1356.
24. Feldmann, T., Andresen, B., Qi, A., and Salamon, P. (1985) Thermodynamic lengths and intrinsic time scales in molecular relaxation, *J. Chem. Phys.* **83**, 5849 – 5853.
25. Flick, J. D., Salamon, P., and Andresen, B. (1987) Metric bounds on losses in adaptive coding, *Info. Sci.* **42**, 239 – 253.
26. Salamon, P., Komlos, J., Andresen, B., and Nulton J. (1987) A geometric view of welfare gains with non-instantaneous adjustment, *Math. Soc. Sci.* **13**, 153 – 163.
27. Andresen, B. and Gordon, J. M. (1994) On constant thermodynamic speed for minimizing entropy production in thermodynamic processes and simulated annealing, *Phys. Rev. E.* **50**, 4346 – 4351.
28. Boltyanskii, V. G. (1971) *Mathematical Methods of Optimal Control*, Holt, Reinhart, and Winston, New York.
29. Tolle, H. (1975) *Optimization Methods*, Springer, New York.
30. Leondes, C. T. (ed.) (1964 and onward), *Advances in Control Systems*, Academic Press, New York [especially articles in the first volumes].
31. Salamon, P., Nitzan, A., Andresen, B., and Berry, R. S. (1980) Minimum entropy production and the optimization of heat engines, *Phys. Rev. A* **21**, 2115 – 2129.
32. Salamon, P. and Nitzan, A. (1981), Finite-time optimization of a Newton's law Carnot cycle, *J. Chem. Phys.* **74**, 3546 – 3560.
33. Salamon, P., Andresen, B., Nulton, J., and Siragusa, G. (1998), The horse-carrot theorem and geometric optimization of distillation (in preparation).
34. Nulton, J., Salamon, P., Andresen, B., and Anmin, Q. (1985) Quasistatic processes as step equilibrations, *J. Chem. Phys.* **83**, 334 – 338.

PHYSICS VERSUS ENGINEERING OF FINITE-TIME THERMODYNAMIC MODELS AND OPTIMIZATIONS

P. SALAMON
Department of Mathematical and Computer Sciences
San Diego State University
San Diego, CA 92182-7720, USA

1. Introduction

Finite-time thermodynamics [1] is an extension of traditional thermodynamics that seeks to characterize in-principle limits to the performance of thermodynamic processes given the constraint that such processes take place in a finite time. This constraint forces any process that involves transport through a finite conductance to produce entropy and thus leads to strengthened versions of the second law and provides positive lower bounds on the associated entropy production.

The present paper differentiates between the physics and the engineering perspectives regarding the value of such studies and gives some of the highlights of what can be gained from physical (as opposed to engineering) studies of such processes. The difference can be characterized as follows: physical studies seek in-principle limits to what is possible while engineering studies seek to characterize the best possible process design given current technology and current prices. Most of the papers at this conference describe the engineering perspective; the present paper describes the physics perspective.

2. Bounds and the Role of Optimization

One important reason for studying thermodynamic processes in finite time is to pursue the quest for understanding the limits of what can be achieved in such processes quite aside from any direct implications for engineering. The simplest analyses of this kind reveal how different sources of irreversibility individually influence the limits to power generation and entropy production. In such analyses there is no pretense of counting **all** sources of irreversibility -- one merely counts those sources one is focusing on and thereby obtains **bounds** on what can happen in real processes.

The underlying mathematical fact that allows this is a basic lemma of optimization theory: solving an optimization problem with fewer constraints (additional controls) bounds the value of the objective for the original problem. Analyzing the optimal performance of a system when we improve the control that we are able to exercise, can only improve the optimal value of our objective and thereby gives a bound on what is physically possible.

One important example of this lemma in finite-time thermodynamics is the theorem of Orlov and Berry [2] which is striking in its simplicity and power. They begin by writing down the full dynamical equations for the working fluid, the heat baths, and the power source coupled to a heat engine complete with partial differential equations for the

A. Bejan and E. Mamut (eds.), Thermodynamic Optimization of Complex Energy Systems, 421–424.

distributed subsystems. The heat transfer to the working fluid is proportional to the temperature difference across a surface. They complete the careful formulation of their control problem by specifying an objective function. They then switch to a problem in which the temperature at each position inside the working fluid and the heat exchangers is controllable. This discards the partial differential equation constraints describing the dynamics of the temperature in the regions *inside* the engine and heat exchangers, thereby changing these temperatures from state variables to control variables. The optimum in the simplified problem bounds the optimum in the real problem since the simplified problem results by effecting better controls. Ignoring a great number of the dynamical equations converts state variables to control variables and enlarges the set of allowed processes. Of course this must be done carefully respecting the laws of thermodynamics -- these are constraints that we must not ignore.

Physically it is easy to ensure that the laws of thermodynamics are not violated. We simply take the additional (fictitious) controls on our process to be reversible, i.e. effected by reversible processes using additional work and heat sources as needed. Thus we assume reversible control over all but a few of the degrees of freedom in our system. We then ask what is the least entropy production, or maximum power output, or the optimal value of various other possible objectives, when we control the rest of the universe in an ideal way -- namely reversibly. This focuses the analysis on the limitations to energy conversion inherent in that part of the system or device which we did **not** assume to be reversibly controllable. The approach is not limited to heat engines and the results bound the possible values of the objective in a more meaningful way than the bounds provided by classical thermodynamics. While the additional controls may be unrealistic from an engineering point of view, they do not violate physical laws and can help us extract information regarding the limitations to real processes imposed by specific modes of irreversibility which are the focus of a specific study. Since typical applications of the Orlov-Berry theorem require the controls to maintain uniform intensities inside a system, they can even be realized by real processes in which the transport coefficients inside a system are large compared to the transport coefficients across the boundary. This is usually referred to as endoreversibility which is often useful in simplifying the analysis of limitations to the behavior of thermodynamic systems[1]. Note that this is not always needed for our analysis but only when the dissipation inside a system is not part of the irreversible modes which are the focus of our study. Note further that the term "endo" is slightly misleading since this argument can be equally well applied to subsystems such as thermal reservoirs which form part of the "environment" and are thus more aptly labeled "exo".

A second theorem due to A. Tsirlin [4] serves to simplify finite-time thermodynamic studies even further. This theorem states that provided the constraints and objective of a problem depend only on net (or average) process quantities, i.e. on integrals over time, the optimal way to control the system is to take our controls to be piecewise constant in time. These two theorems make simple models interesting and able to give useful, rigorous bounds regarding what can be achieved during the control of real processes. The simplified problems often lead to solutions which tell us to run the process infinitely fast or infinitesimally slow. The interesting domain is in between: well posed problems which

---

[1] Prior to the Orlov-Berry theorem the argument required an assumption of the separability of the time scales for the transport processes within a system and across system boundaries [3].

ignore enough irreversibility to be tractable and yet count enough irreversibilities to keep the solution finite. The value of the objective obtained in the simpler optimization bounds the value of the objective in real processes in a more realistic way than a reversible bound, although it is still often very far from what can be achieved in reality.

The results must be used with great care however. Other features of the optimal solution obtained cannot be directly compared with similar features of real processes. A simple and familiar example is provided by the well known equation of Curzon-Ahlborn-Novikov (CAN) [5, 6] for the efficiency at maximum power

$$\eta = 1 - (T_c/T_h)^{1/2}. \tag{1}$$

The striking simplicity of this equation, as well as its similarity to the well known efficiency of Carnot, has brought it rapid acceptance. Nonetheless, as pointed out in [7], this efficiency is surpassed in real power plants! The explanation is that this equation does **not** represent the maximum efficiency but rather the efficiency at maximum power. The associated maximum power

$$P = \kappa \ (T_h^{1/2} - T_c^{1/2})^2 / 4 \tag{2}$$

must be more than the power delivered in any process which corresponds to the CAN calculation augmented by additional dynamical equations. This much is mathematical fact. Note that there is **no** implication regarding a comparison between the efficiency obtained at this power and efficiencies in real processes -- even when the real processes are also designed to maximize power, albeit with additional constraints!

3. Mathematical Structure

As the above discussion illustrates, understanding the physics is different from understanding the engineering. Physicists are not directly interested in how to design or operate a specific device or process but rather in understanding what limitations are set for the process by the thermodynamics laws. In mathematical physics, such understanding usually comes by studying the mathematical structure of the set of possibilities. This is not a problem that engineers have been concerned with, even when very similar optimizations were performed for very similar systems. It is a problem worth studying.

As an example of the type of insight that can be gained from an examination of the mathematical structure, we cite a result that has become known as the horse-carrot theorem [8, 9]. This theorem quantifies the minimum entropy production associated with bringing a system along a specific sequence of states, i.e. with making it traverse a certain path in its state space. The theorem concerns a Riemannian metric structure on this state space[2] and says that the minimum entropy produced in the process is the square of the length of the process divided by twice the number of relaxations. A recent application of the theorem says something of interest to engineering by specifying the optimal temperature profile for a distillation column [10].

[2] The Riemannian metric is given by the second derivative matrix of the entropy of the system as a function of the extensive variables.

4. Difficulties of Communication

There exist nontrivial barriers to communication between physicists and engineers working in this area. To make this point, I cite an anecdote regarding the use of the phrase "industrial example". In theorist-speak, this phrase means that *some* industry might care about it *some* day. In a referee report, the esteemed organizer of our conference, Adrian Bejan, dismissed geothermal power plants as "hardly an industrial example" since "its share of the global energy industry is meaningless."[3]

As a second example I consider the word "applied". In a mathematics department, this word is used to mean that the arguments and results relate to something outside the sphere of mathematics, even when this "use" is extremely esoteric, e.g. characterizing certain cross sections of a fiber bundle over a seventeen dimensional space-time. Note that this use is very far from what "applied" typically means in a physics department which is still dramatically different from what "applied" means in an engineering department.

Other examples of words that mean very different things to engineers, physicists, and mathematicians include "realistic," "general," "optimal," and even "understanding."

5. Conclusions

In our modern multicultural and multidisciplinary world it is important to adopt a wider perspective. While the type of physical studies outlined above may be of interest to engineering design only in its earliest stages, there is a long tradition of physics *ultimately* leading to principles important for engineering. It is important to accept the fact that our field is interdisciplinary and to try to move beyond the perspectives offered by our own, narrowly defined disciplines.

References

1. Sieniutycz, S. and Salamon, P. (1990) *Finite–Time Thermodynamics and Thermoeconomics*, Taylor and Francis, New York.
2. Orlov, V. and Berry, R. S. (1990) Power output from an irreversible heat engine with a nonuniform working fluid, *Physical Review* A **42**, 7230-7235.
3. Salamon, P., Nitzan, A., Andresen, B. and Berry, R. S. (1980) Minimum entropy generation and the optimization of heat engines, *Physical Review A* **21**, 2115-2129.
4. Tsirlin, A. M. (1986) *Optimal Control of Technological Processes*, Energoatomizdat, Moscow.
5. Novikov, I. I. (1958) The efficiency of nuclear power stations, *Journal of Nuclear Energy II* (U.S.S.R.) **7**, 125-128.
6. Curzon, F. L. and Ahlborn, B. (1975) Efficiency of a Carnot engine at maximum power output, *American Journal of Physics* **43**, 22-24.
7. Gyftopoulos, E. (1997) Fundamentals of analysis of processes, *Energy Conversion and Management* **38**, 1525-1533.
8. Salamon, P. and Berry, R. S. (1983) Thermodynamic length and dissipated availability, *Physical Review Letters* **51**, 1127 - 1130.
9. Nulton, J., Salamon, P., Andresen, B. and Anmin, Q. (1985) Quasistatic processes as step equilibrations, *Journal of Chemical Physics* **83**, 334-338.
10. Salamon, P. and Nulton, J.D. (1998), The geometry of separation processes: a horse-carrot theorem for steady flow systems, *Europhysics Letters* **42**, 571-576.

---

[3] We remark that in Iceland, the overwhelming majority of the energy industry is geothermal. Thus the perspectives offered by different cultural and ethnic affiliations also create barriers to communications.

OPTIMIZATION AND SIMULATION OF TIME DEPENDENT HEAT DRIVEN REFRIGERATORS WITH CONTINUOUS TEMPERATURE CONTROL

J.V.C. VARGAS
Department of Mechanical Engineering
Universidade Federal do Paraná, Curitiba, PR, 81531-990, Brazil

1. Introduction

Economic and environmental considerations have brought new interest to heat driven refrigerators. A large research effort has been brought to bear on refrigeration systems of this kind in recent years [1, 2]. These systems can make use of waste heat and renewable energy sources such as solar energy [2, 3]. In countries in which electricity and fossil fuels are available only intermittently or are very expensive, solar refrigeration may be an alternative for the preservation of foodstuffs and medical supplies [4], since a "reservoir of cold" can also be generated through the formation of subcooled ice for use when the sun is obscured. However, when uninterrupted precisely controlled cooling is required, heat driven systems powered by gas, electricity or oil are needed either as an alternative to solar thermal energy and waste heat or as a supplementary thermal energy source. The improvement and optimization of the design and control of heat driven refrigerators is a crucial issue regardless of the energy sources used. Heat driven refrigerators, with an effective control system, may well be able to combine economy with a low environmental impact.

To be effective, a theoretical analysis of a thermal system must be capable of capturing "realistic" features of the heat transfer processes, and the unit. Several studies have dealt with these issues by using the method of entropy generation minimization [1, 2, 5-7] (exergetic optimization), which pursues realistic models and takes into account the inherent irreversibility of heat, mass and fluid-flow, this is also known as the method of finite-time thermodynamics.

In this paper, a mathematical model is introduced to simulate the transient behavior of a heat driven refrigerator that uses a fuel burning heater as the high temperature source. External irreversibilities are accounted for. A generalized version of a power law control system, first introduced by Vargas and Parise [8], is then applied to control the fuel flow rate to the system in closed loop operation. The power law approach replaces the well known proportional-integral-derivative [9] combined actions. An experimental validation of the steady state behavior of the model is

A. Bejan and E. Mamut (eds.), Thermodynamic Optimization of Complex Energy Systems, 425–435.

conducted through direct comparison of the theoretical results with experimental data obtained from an existing absorption refrigeration unit [10].

The objectives of this study, therefore, were (1) to examine how the thermal conductance allocation should be optimized for maximum refrigeration rate and minimum "pull-down" time, (2) to investigate the fuel consumption of the system under "on-off" and closed loop operation, and (3) to present a model for the transient simulation of heat driven refrigerators for design and optimization purposes.

2. Mathematical Model

The block diagram of Fig. 1 shows the main features of the system with continuous temperature control (closed loop). From left to right: center, summer (circle), power law control action, net gain, fuel pump gain, absorption refrigerator and ambient temperature blocks; top, integration block; bottom, amplifier gain and thermal sensitivity blocks.

The sole difference between the two systems is in the nature of the control action. In the "on-off" system the fuel flow rate can only have one of two values: $\dot{M}$ or zero. In the system with "feed-back", the value of $\dot{M}$ varies continuously in a manner determined by the control system.

The general features of the absorption refrigerator - refrigerated space model are shown in Fig. 2, which are represented by elements in the block diagram shown in Fig. 1.

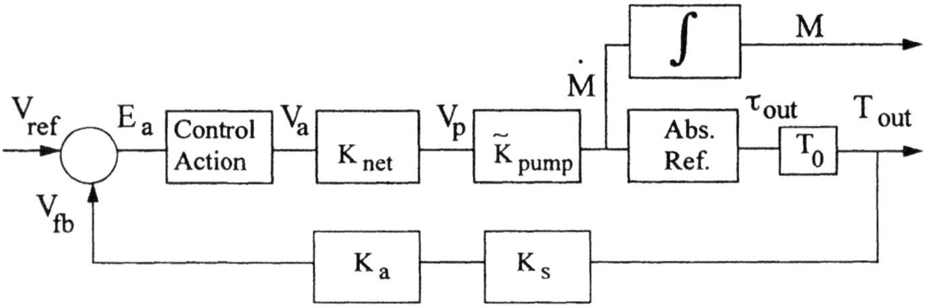

Figure 1. Closed loop block diagram.

The cycle requires negligible work input. The cycle is driven by the heat transfer rate $\dot{Q}_H$ received from the hot source temperature at T_H, which is determined by the operating temperature of the generator. The refrigeration load $\dot{Q}_L$ is removed from the refrigerated space, i.e., the evaporator drum shown in *Figure 3*, at T_L, and the heat transfer rate $\dot{Q}_0$ is rejected to the ambient, at T_0. The label reversible compartment means that no entropy is being generated between the heat exchangers temperatures seen by the working fluid, according to the endoreversible model.

The thermal inertia of the chilled two-fluid side of the evaporator is large enough such that the time to achieve the desired water level in the evaporator drum can

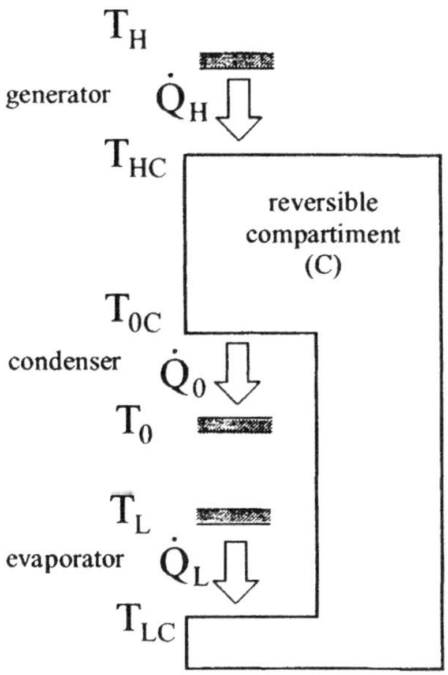

Figure 2. Absorption refrigerator model with three heat transfer irreversibilities.

be neglected when compared to the time evolution of the temperature inside the refrigerated space.This justifies the use of a quasi-steady model [1, 2]. Additionally, it is assumed that the water and trapped air in the evaporator drum are in thermal equilibrium, such that $T_w = T_{air} = T_L$.

Due to heat transfer, mixing and throttling, the refrigerator operation is irreversible. The endoreversible model considers the refrigerator to be internally reversible, such that only the external irreversibilities associated with the heat transfer interactions are taken into account, namely $\dot{Q}_H$, $\dot{Q}_L$ and $\dot{Q}_0$. Since the endoreversible transient model was developed based on previous steady-state models [1, 2], only the new equations introduced in the model for the present work are shown, for the sake of brevity.

The heat input rate, $\dot{Q}_H$ is proportional to the fuel mass flow rate, where, without loss of generality, we assume stoichiometric combustion, in an adiabatic chamber, and negligible heat loss to the ambient through the exhaust gases, as follows:

$$\dot{Q}_H = h_f \dot{m}_f \tag{1}$$

The refrigerated space model was based on an evaporator coil surrounded by a drum where chilling water circulates, as shown in Fig. 3. This configuration is part of an experimental study of a gas-fired absorption unit [10]. Therefore, the model allows for the validation of steady-state theoretical results against steady-state experimental data. The transient cooling of the refrigerated space is accounted for by writing the first-law of thermodynamics, as follows:

$$m_w c_w \frac{dT_w}{dt} + m_{air} c_{v,air} \frac{dT_{air}}{dt} = \dot{m}_w c_w \left(T_{in} - T_{out}\right) - \dot{Q}_L + q \tag{2}$$

where q accounts for the heat leakage from the ambient across the walls of the evaporator drum.

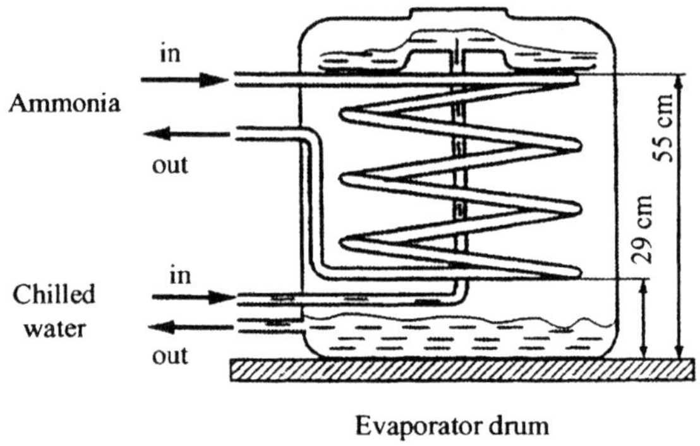

Figure 3. The refrigerated space of the experimental unit.

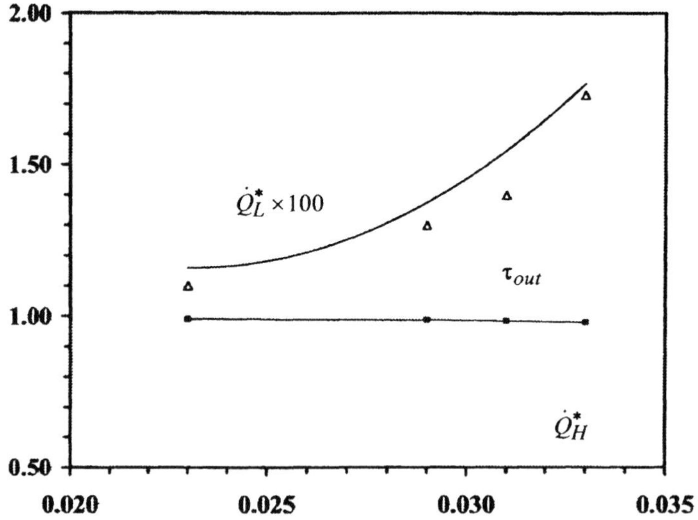

Figure 4. Comparison with experimental data (symbols) and theory (lines).

The factors $(UA)_H$, $(UA)_L$ and $(UA)_0$ represent the overall thermal conductances of the heat exchangers. The overall thermal conductance of the walls of the refrigerated space is given by $(UA)_W$. Since $(UA)_H$, $(UA)_L$ and $(UA)_0$ are not known, the following constraint is introduced at this stage:

$$UA = (UA)_H + (UA)_L + (UA)_0 \tag{3}$$

In addition, define the ratio w that accounts for the thermal conductance of the heat transfer area of the vessel walls,

$$w = (UA)_W \;/\; UA \tag{4}$$

The complete set of nondimensional equations for the model, including the absorption refrigerator steady state equations [1, 2], is

$$\dot{Q}_H^* = \dot{M}C \qquad \dot{Q}_H^* = y(\tau_H - \tau_{HC}) \tag{5,6}$$

$$\dot{Q}_0^* = (1-y-z)(\tau_{0C} - 1) \qquad \dot{Q}_L^* = z(\tau_L - \tau_{LC}) \tag{7,8}$$

$$\dot{Q}_H^* + \dot{Q}_L^* = \dot{Q}_0^* \qquad \frac{\dot{Q}_H^*}{\tau_{HC}} + \frac{\dot{Q}_L^*}{\tau_{LC}} = \frac{\dot{Q}_0^*}{\tau_{0C}} \tag{9,10}$$

$$\frac{d\tau_{out}}{d\theta} = 2 \cdot \frac{r(\tau_{in} - \tau_{out}) - \dot{Q}_L^* + w(1-\tau_L)}{\alpha + 1}; \quad \tau_L = \frac{\tau_{in} + \tau_{out}}{2} \tag{11,12}$$

where τ_L is the arithmetic mean of the chilled water inlet and outlet temperatures. The nondimensional groups have been defined as follows:

$$\tau_H = T_H / T_0 \;, \tau_L = T_L / T_0, \tau_{in} = T_{in} / T_0, \tau_{out} = T_{out} / T_0 \tag{13}$$

$$\tau_{HC} = T_{HC} / T_0, \tau_{LC} = T_{LC} / T_0, \tau_{0C} = T_{0C} / T_0 \tag{14}$$

$$\dot{Q}_1^* = \frac{\dot{Q}_1^*}{UAT_0}, \dot{Q}_H^* = \frac{\dot{Q}_H^*}{UAT_0}, \dot{Q}_0^* = \frac{\dot{Q}_0^*}{UAT_0}, \dot{Q}_L^* = \frac{\dot{Q}_L^*}{UAT_0} \tag{15}$$

$$\alpha = \frac{m_{air} c_{v,air}}{m_w c_w}, r = \frac{\dot{m}_w c_w}{UA} \tag{16}$$

$$C = \frac{h_f}{c_{p,air} T_0}, \theta = \frac{t}{m_w c_w / (UA)}, \dot{M} = \frac{\dot{m}_f}{UA / c_{p,air}} \tag{17}$$

The conductance allocation ratios are $y = (UA)_H / UA, z = (UA)_L / UA$.

According to the UA constraint in Eq. (3), the conductance allocated to the condenser is $(UA)_0 / UA = 1 - y - z$.

During off-cycle periods, the fuel mass flow rate drops to zero instantaneously and the water is kept circulating in the evaporator drum, thus, the model accounts for both on and off operation.

A complete treatment on these topics is available in the recent work of Vargas et al. [12]. Starting from $V_s = K_c^{(K_1 \Delta T_{control} + K_2)}$ where K_c, K_1 and K_2 are constants to be selected according to the system, the voltage input to the net gain block, V_a and, consequently, the voltage input to the fuel pump block is determined by $V_a = V_s$, if $V_s < V_{max}$, and $V_a = V_{max}$, if $V_s \geq V_{max}$.

3. Experimental Validation

The validation of the absorption refrigerator model presented in this study was performed by direct comparison of the model's steady state results with experimental data [10]. No available experimental data has so far been found in the literature with measurements of the transient behavior of an absorption unit.

The procedure consisted of evaluating the nondimensional parameters defined by the theoretical model for the particular unit tested by Horuz [10]. The unit was a *10 kW* gas-fired vapor absorption refrigerator water chiller using aqua-ammonia solution, with an evaporator configuration as shown in Fig. 3. The air cooled absorber and condenser heat exchangers account for the heat rejection of the cycle.

The unit uses a water-ethylene glycol solution, an antifreeze for applications in process cooling at lower temperatures. The composition of the solution used in the experiments was *30 %* by weight ethylene glycol and *70 %* by weight water, giving [11] $c_w = 3.77\ kJ/(kg.{}^\circ C)$ and $\rho_w = 1040\ kg/m^3$.

Through the available experimental data, thermal conductances of the heat exchangers were computed as $(UA)_L = 0.87\ kW/{}^\circ C$, $(UA)_0 = 0.89\ kW/{}^\circ C$ and $(UA)_H = 0.34\ kW/{}^\circ C$, therefore $(UA) = 2.1\ kW/{}^\circ C$. The other necessary data were $\dot{m}_w = 0.52\ kg/s$, $H_w = 0.29\ m$, $H_{air} = 0.26\ m$, $T_H = 130\ {}^\circ C$, $\dot{Q}_H = 23.2\ kW$ and $(UA)_W = 1.32\ W/{}^\circ C$. The resulting dimensionless parameters, initially kept fixed are $\tau_H = 1.4$, $y = 0.16$, $z = 0.41$, $w = 0.0007$, $r = 0.934$ and $\alpha = m_{air} c_{v,air} / m_w c_w = H_{air} \rho_{air} c_{v,air} / H_w \rho_w c_w = 1.64E-4$, and $\dot{Q}_H^* = 0.038$.

The experimental work of Horuz [10] analizes the effect of variations in several parameters on the performance of the unit. In this work, theoretical and experimental results are compared for the effect of the variation of chilled water inlet temperature, chilled water flow rate and heat input to the system.

The experimental measurements shown in Fig. 4 were obtained with $T_0 = 19\ {}^\circ C$. The water level in the evaporator drum was $H_w = 0.29\ m$. The experimental data are presented discretely with symbols (triangles and squares), whereas the model's results are presented with continuous lines.

In Fig. 4, it is seen that the increase in heat input provokes an increase in cooling capacity and a decrease in the water outlet temperature, in the tested heat input range, with $\tau_{in} = 0.990$ and $r = 0.934$. However, the decreasing rate of the water outlet temperature is considerably smaller than the increasing rate of heat input. Therefore, the decrease in water outlet temperature does not justify the increase in heat input. A transient analysis, such as the present model provides, to predict the "pull-down" capacity of the unity is mandatory for the designer, in order to achieve the project demands. The theoretical model predicts a plausible "pull-down" trend, but as yet no experimental data is available on which to base a comparison.

The analysis of Fig. 4 therefore validates the steady state results of the theoretical model against experimental data. Based on these observations, the next section uses the theoretical model to predict the "pull-down" capacity of an absorption unit, and to perform a simulation of its transient behavior during a specified time of operation.

4. Results and Discussion

The model uses a total thermal conductance constraint, UA, as shown in Eq. (3). Nevertheless, it is interesting to observe that, from a practical standpoint, for heat exchangers with the same geometrical operating characteristics, i.e., similar global heat transfer coefficients, the UA constraint turns into an area constraint. Therefore, for a given total thermal conductance inventory, the designer could access the optimal thermal conductance (or heat transfer area) allocation, such that (i) the cooling capacity is maximized, and (ii) the "pull-down" time is minimized, by the use of the model.

The first fundamental question, concerning the optimal fraction allocated to the condenser, has been answered in previous studies [1,2], where it was shown theoretically that $(UA)_0 / UA$ is insensitive to the external temperature levels (τ_H, τ_L) and equal to half of the total thermal conductance for maximum refrigeration rate. In general, the other half must be equally divided between generator and evaporator. To verify this trend, by means of a numerical example, Fig. 5 shows the effect of the variation of the total thermal conductance allocation on the steady state results for

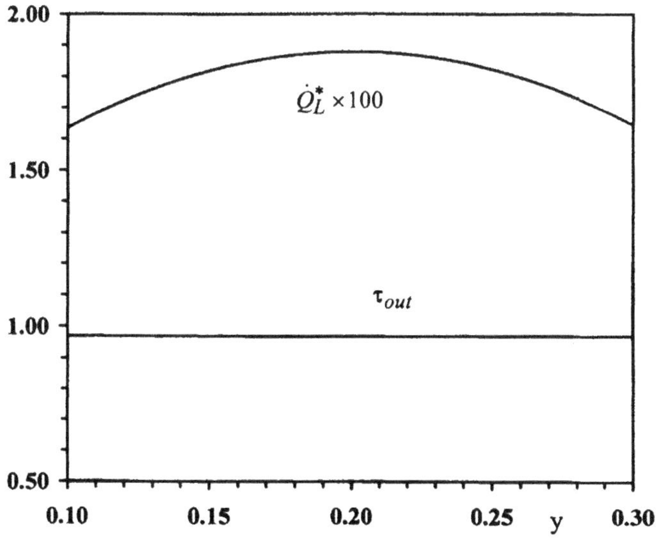

Figure 5. Maximum refrigeration rate.

$\dot{Q}_L^*$ and τ_{out}, using the experimental parameters of the tested unit. It has been further assumed that $(UA)_0 / UA = 0.5$, and once y is specified, z results, and vice-versa, the two collapsing in just one degree of freedom.

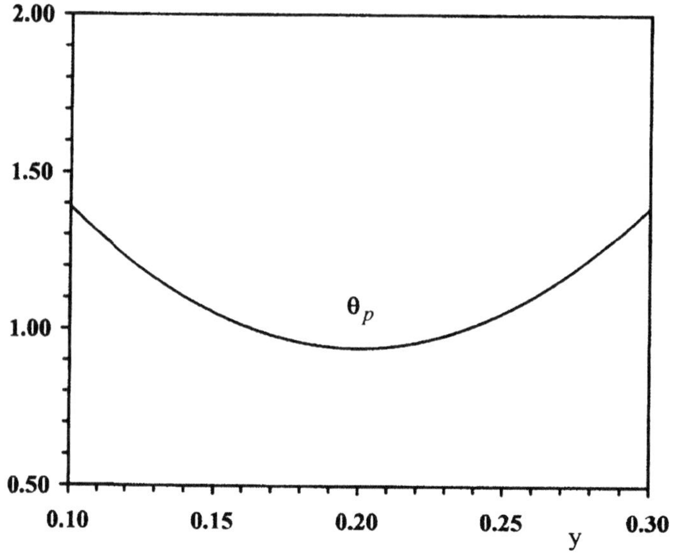

Figure 6. Minimum "pull-down" time.

The results show there is an optimal thermal conductance allocation to attain maximum refrigeration, i.e., $y_{opt} = z_{opt} \cong 0.25$. Therefore, by computing $\dot{Q}_{L,max}^* = 0.0188$ and $\tau_{L,min} = 0.967$ with the model, and comparing with $\dot{Q}_L^* = 0.017$ and $\tau_L = 0.970$, obtained experimentally, the existence of a better design for that unit may be inferred from the model, with a possible gain of *10.6 %* in refrigeration rate and *0.3 %* in chilled water outlet temperature.

The answer to the second fundamental question is given by the results shown in Fig. 6, where a desired chilled water outlet temperature, $\tau_{out,l} = 0.98$, was established. It verifies the existence of a minimum "pull-down" time, θ_p, when the thermal conductance is allocated in an optimal way ($y_{opt} = z_{opt} \cong 0.25$). Using the model to compute θ_p for the tested unit, it is found that $\theta_p = 1.2$, for $y = 0.16$ and $z = 0.41$, and $\theta_p = 1.0$ for $y_{opt} = z_{opt} = 0.25$, i.e., with the optimal thermal conductance distribution, a gain of 17 % is expected in the "pull-down" time. The importance of the transient analysis to identify the best design is thus stressed.

Next, using the optimal thermal conductance allocation, a simulation of the unit, operating with closed loop and "on-off" control, is performed for a specified period of time, $\theta_f = 4$. The following parameters were adopted in the simulation: $C = 0.15$, $r = 0.630$, $y = z = 0.25$, $w = 0.0007$, $\alpha = 1.64E-4$, $V_{max} = 4.8\,V$, $K_{net} = 25$, $K_2 = 0$, $K_a = 4000$, $K_s = 40\,\mu V/°C$, $K_c = 10\,V$, $K_1 = 10\,°C^{-1}$, $\tilde{K}_{pump} = 2.1E-3\,V^{-1}$, $T_0 = 17\,°C$, $T_{in} = 20\,°C$, $T_H = 130\,°C$, $T_{out,max} = 20\,°C$, $T_{out,min} = -10\,°C$ and $T_{out} = T_{in} = 20\,°C$, at $\theta = 0$ (initial condition). The temperature setpoint in closed loop was $T_{set} = 13\,°C$, and the "on-off" control higher- and lower-temperature were $T_{out,h} = 14\,°C$ and $T_{out,l} = 12\,°C$, respectively.

The comparison of fuel consumption between the "on-off" and closed loop systems, operating with the same parameters is depicted in Fig. 7. The results show that the closed loop system requires approximately half of the fuel the on-off system does, during the simulation time ($\theta_f = 4$).

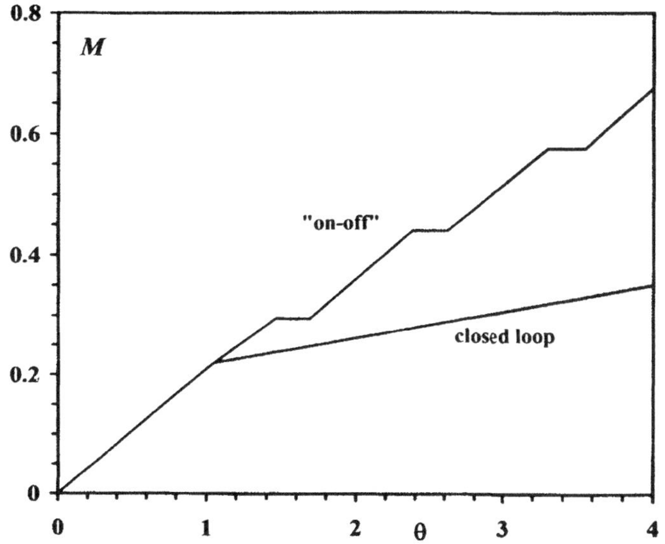

Figure 7. Fuel consumption comparison.

This reduction increases as the unit works longer, since the slope of the "on-off" fuel consumption curve is bigger than the closed loop one. The use of closed loop control in absorption refrigerators is therefore very promising. Since a simple control system was used and no additional cost is required with respect to the "on-off" system, the same electrical fuel pump can be used in both systems, with obvious economic advantages.

Next, the issue of how sensitive the closed loop control system is to changes in the chilled water inlet temperature is addressed. The total simulation time was increased to $\theta_f = 20$, and the water inlet temperature started as $20\,°C$ up to $\theta = 4$, then increased to $22\,°C$ up to $\theta = 10$, when it was set to $18\,°C$, for the remainder of the simulation period. The chilled water outlet temperature does not vary at all, in spite of the variations in the water inlet temperature, as it is shown by the results of Fig. 8.

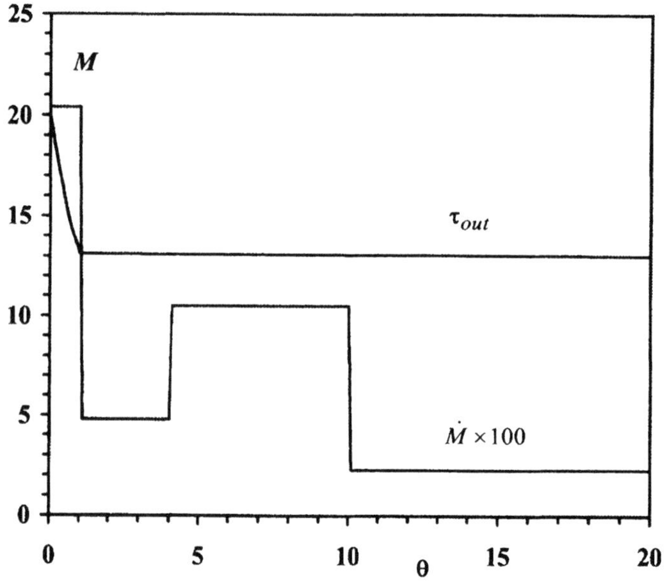

Figure 8. Closed loop transient simulation.

When the water inlet temperature increases, there is only a slight increase in the controlled temperature, which in this simulation was less than $0.05\ °C$. The fuel flow rate is automatically adjusted by the control action, according to the variations in the water inlet temperature, such that the water outlet temperature does not vary.

Finally, it is important to observe that there are three degrees of freedom in the control action that have to be selected appropriately. The controller parameters, $K_c = 10\,V$, $K_1 = 10\,°C^{-1}$ and $K_2 = 0$ were selected for this system by trial and error, such that $\Delta T_{control} = 0.05\ °C$ was the maximum deviation observed in the simulations performed, after stabilization at the temperature setpoint was achieved.

5. Conclusions

In this paper, a transient model of a heat driven refrigeration system, operating with continuous temperature control, was presented to investigate the behavior of the system under "on-off" and closed loop operation, for design and optimization purposes. The absorption refrigerator model was validated for steady state operation by direct comparison between theoretical results and experimental data, with good qualitative and quantitative agreement. Appropriate dimensionless groups were identified and the generalized results reported in charts using dimensionless variables.

The model was then used to identify the optimal allocation of a fixed thermal conductance inventory for minimum "pull-down" time combined with the maximum refrigeration rate. The most important conclusion of this second part of the study is that the gain in the reduction of the refrigerated space temperature is much smaller than the gain in the "pull-down" time. The importance of the transient analysis to identify the best design is thus clearly seen.

A simulation of the system operating with "on-off" and closed loop control was performed, for a specified period of time, in the third part of this study. The continuous temperature control eliminates undesirable oscillations in the controlled temperature, observed with the traditional "on-off" control. The most promising result is that a large reduction in fuel consumption, of about *50 %*, was obtained with the closed loop system, in comparison with the "on-off" control, even for a short simulation time ($\theta_f = 4$). Note that this result is plausible because, with the "on-off" control there are only two fuel flow rate settings, maximum (on) or zero (off) and the ontime ratio is about 70 to 80 %, whereas the closed loop system works at a well reduced fuel flow rate after the temperature setpoint is achieved. Since the same electrical fuel pump could be used in both systems, no changes in hardware are required, except for those required for the implementation of a simple control system. Hence, an obvious reduction, both in operational cost and in environmental impact, will be achieved, if a continuous temperature control is implemented on heat driven refrigerators. The response of the power law control system was tested by varying the thermal load during a longer period of simulation ($\theta_f = 20$), and no oscillations were observed on the controlled temperature, whereas fuel consumption was automatically kept to a minimum.

6. References

1. Bejan, A., Vargas, J. V. C. and Sokolov, M. (1995) Optimal allocation of a heat exchanger inventory in heat driven refrigerators, *International Journal of Heat and Mass Transfer* **38**, 2997-3005.
2. Vargas, J. V. C., Sokolov, M. and Bejan, A. (1996) Thermodynamic optimization of solar-driven refrigerators, *Journal of Solar Energy Engineering* **118**, 130-135.
3. Sokolov, M. and Hershgal, D. (1993) Optimal coupling and feasibility of a solar-powered year-round ejector air conditioner *Solar Energy* **50**, 507-516.
4. Worsfe-Schmidt, P. and Holm, F. (1989) Development field testing of a solar-powered block-ice plant, *Jutland Technological Institute Teknologiparken DK-8000 Aarhus C* Denmark FHO/VIM 1-17.
5. Bejan, A. (1989) Theory of heat transfer-irreversible refrigeration plants, *International Journal of Heat and Mass Transfer* **32**, 1631-1639.
6. Bejan, A. (1995) *Entropy Generation Minimization* CRC Press, Boca Raton, FL.
7. Bejan, A. (1988) Theory of heat transfer-irreversible power plants, *International Journal of Heat and Mass Transfer* **31**, 1211-1219.
8. Vargas, J. V. C. and Parise, J. A. R. (1995) Simulation in transient regime of a heat pump with closed loop and on-off control, *International Journal of Refrigeration* **18:4**, 235-243
9. Kuo, B. C. (1987) *Automatic Control Systems* Prentice-Hall International Editions.
10. Horuz, I. (1994) An experimental study of the use of vapor absorption refrigeration in road transport vehicles PhD thesis University of Strathclyde, Glasgow, UK.
11. Perry, R. H. and Chilton, C. H. (1974) *Chemical Engineering Handbook* International Student Edition, McGraw-Hill.
12. Vargas, J. V. C., Horuz, I., Callander, T. M. S., Fleming, J. S. and Parise, J. A. R (1998) Simulation of the transient response of heat driven refrigerators with continuous temperature control, *International Journal of Refrigeration*, in press.

INTELLIGENT COMPUTER AIDED DESIGN, ANALYSIS, OPTIMIZATION AND IMPROVEMENT OF THERMODYNAMIC SYSTEMS

C. WU
Department of Mechanical Engineering
U.S. Naval Academy
Annapolis, MD 21402, USA

1. Introduction

Research in Cognitive Science potentially offers tremendous benefits in education. Psychological studies of human learning yields insights as to what forms of instruction might be most beneficial for particular purposes and what mental models and misconceptions students might have. Artificial Intelligence (AI) research provides formal representations and reasoning techniques that can be used in new kinds of educational software which itself contains a deep understanding of the material being taught. These scientific advances, coupled with dramatic changes in computing technology provide new opportunities for exploring how intelligent software can be used in education.

A new kind of educational software called articulate virtual laboratories (AVL) has been implemented into the engineering curriculum at the US Naval Academy since 1996. An articulate virtual laboratory is designed to help students learn thermodynamics by scaffolding them in conceptual design tasks, using a simulated environment to allow them to create and analyze designs for complex thermodynamic systems and using software coaches to provide guidance.

2. Articulate Virtual Laboratory

The rapid increase in the power of microcomputers combined with rapidly falling prices, provides the opportunity to apply recent advances in artificial intelligence technology to new kinds of educational software. By exploiting advances in qualitative physics, educational software can use methods and concepts similar to those used by domain experts, including the tacit knowledge of the domain. By exploiting advances in reasoning technology, particularly truth-maintenance systems, educational software can be built that explains the principles it uses. The ability of qualitative physics to provide computer-assisted modeling capabilities combined with efficient reasoning techniques means that educational software can work with new problems in the domain, without reprogramming. Such generality will empower students and instructors by enabling them

A. Bejan and E. Mamut (eds.), Thermodynamic Optimization of Complex Energy Systems, 437–444.

to pose their own problems, and will thus provide a more motivating instructional context. By exploiting advances in analogical reasoning, we can use libraries of examples and previously-solved problems to provide a foundation for coaches that introduce students to appropriate problem-solving methods in a domain.

3. The Role of Design

Design is the core activity of engineering. The ability to develop useful artifacts by understanding how to apply fundamental principles to applications is a major reason why engineering education is so important in today's highly competitive environment. Providing design experience is clearly essential for training engineers. However, design activities also provide powerful motivation for learning fundamental physical principles. Design requires using knowledge in an integrated fashion rather than memorizing isolated facts. Therefore, design provides a meaningful context for learning physical principles. Getting students to think in design terms leads naturally to building a strong interest in understanding complex, real-world relationships.

The most important class of educational software for engineering education is oriented around conceptual design activities. In engineering, conceptual design is the early phase which identifies the basic features of the design. This includes specifying the physical processes and principles by which the design achieves its goals, as well as performance estimates and rough cost estimates.

Conceptual design is most appropriate for engineering education because it focuses on the fundamentals of a domain. On the other hand, detailed design is concerned with fleshing out the results of conceptual design, including figuring out the geometry of the system, detailed parts costs, operating and maintenance procedures for the resulting artifact, and all of the other issues which must be settled in order to build and operate the newly designed artifact. Engineering students need detailed design experience, of course. However, we believe that the largest pedagogical payoff will be in computer technology that simplifies broad-scale incorporation of conceptual design in engineering curricula. Conceptual design forces the designer to think directly about many of the major themes.

Despite the centrality of design in engineering, today's engineering curricula generally offer few design experiences. Some of the reasons come from the cost of, and sometimes danger in, building physical artifacts. Additional reasons that design exposure is limited include (1) the time-consuming nature of detailed design which strains already-overloaded curricula, and (2) the complexity of design which requires additional supervision to minimize unenlightening aspects of student explorations.

Virtual laboratories provide a partial solution to these dilemmas. A virtual laboratory is a software environment consisting of "parts" corresponding to physical parts, tools for assembling collections of these parts into systems, and facilities for analyzing and testing the systems. These software environments allow students to "build" their designs and try them out without expense or danger. Thus, these programs solve the cost and safety problems associated with design. However, they provide no coaching. They do not explain their results in qualitative terms that students can easily understand. They cannot teach students how to choose an appropriate model or how to build new models. They have no representation of the student's goals so they cannot assist the student in evaluating his or her design. Finally. they cannot help the student understand the connections between the simulation results and the fundamental principles used to generate those results. To overcome these limitations will require synthesizing artificial intelligence technology to develop articulate virtual laboratories, that is, intelligent learning environments with a deep understanding of engineering principles.

4. CyclePad

CyclePad is the conceptual CAD component of the engineering thermodynamics AVL developed by Professor K.D. Forbus [1] of Northwestern University and evaluated by the author [2]. CyclePad scaffolds the activities of students who are learning how to design and analyze thermodynamic cycles. An inventory shop of parts (e.g., compressors, turbines, pumps, heat exchangers, etc.) are combined into networks, potentially generating an unlimited set of designs of thermodynamic systems. CyclePad performances the operational characteristics and sensitivity analysis of the system.

The design of thermodynamic systems is a significant element of engineering education that requires a broad and deep understanding of the fundamentals of thermodynamics. Most introductory engineering thermodynamics textbooks devote several chapters to system analysis. In some advanced books, system analysis is the main concentration. Developing software architecture that scaffold students in conceptual design problems, allows professors to bring advanced concepts into the curriculum at an earlier time.

CyclePad is a conceptual CAD tool for visualizing and analyzing thermodynamic systems. Students construct their design by "wiring up" components and assigning numerical values for their parameters.

Due to the substantial amount of numerical calculation involved, students typically avoid exploring more than one or two alternatives and avoid performing advantageous trade-off studies. Since trade-off studies are one of the most effective means to building intuitions about thermodynamic systems, CyclePad's value lies in the ability of the students to incrementally derive the consequences of assumptions and automating the work, thus relieving the student of the tedium of numerical calculations. Another major stumbling block for students is learning how to model, that is, how to make the simplifying assumptions needed to derive reasonable answers. CyclePad helps by making the choices for modeling assumptions explicit in its representation and depiction of the design. Therefore, the students are encouraged to consider what reasonable assumptions might be applicable. One benefit of Articulate Virtual Laboratories is to decrease students' primary difficulties and intuitive but faulty conceptions within the domain, based on empirical evaluations of their performance using AVLs.

CyclePad' s knowledge base captures a substantial fraction of the knowledge in introductory thermodynamics text. The knowledge base consists of representations for the components of thermodynamic systems and qualitative representations of the physical processes inside them.

5. Using Articulate Virtual Laboratory in the Classroom

Undergraduate students in mechanical engineering at the U.S. Naval Academy are required to take two courses in thermodynamics. Since thermodynamics is such a problem-oriented course, students must learn to work cycles to understand thermodynamics. To advance and challenge the intellect, students must be able to calculate the problem in a reasonable amount of time by hand. Extraneous or more tedious work will only tend to frustrate and dissuade the students from contemplating further designs.

Normally, calculating system design is a lengthy, complex and tedious for a student. Analyzing cycles is more meaningful, time saving and fun for students using quick-change simulation software such as CyclePad in design. Students are able to change any parameter in the cycle and see the effect of the parameter on the performance of the cycle instantaneously. Design is an iterative learning process. By trial and error, students are able to choose various components in the design of power plants or choose different working fluids in the plants using CyclePad as a tool. They gained a tremendous amount of experience, confidence, and knowledge in a relative short time. Due to the quick iterative design abilities of CyclePad, instructors were able to assign more cycle analysis and design problems to their students.

6. Example: Improvement of a Brayton Cycle

Starting with the simplest possible Brayton cycle, improvements can be progressing through a number of modifications of the basic configuration. These modifications, each made at some additional components from the inventory shop of CyclePad, are typically evaluated for their contribution to performance criteria such as efficiency and power output. The modifications include the addition of a reheater, the addition of a regenerator, the addition of an inter-cooler, the addition of any combination of inter-coolers, reheaters and regenerators, etc. This gradual approach introduces to the students how Brayton cycles operate and how to improve plant performance.

An innovative Brayton cycle for potential naval shipboard power plant application is used as an example. The cycle consists of four compressors, one combustion chamber, two turbines, two coolers, one source and one sink as illustrated in Fig. 1.

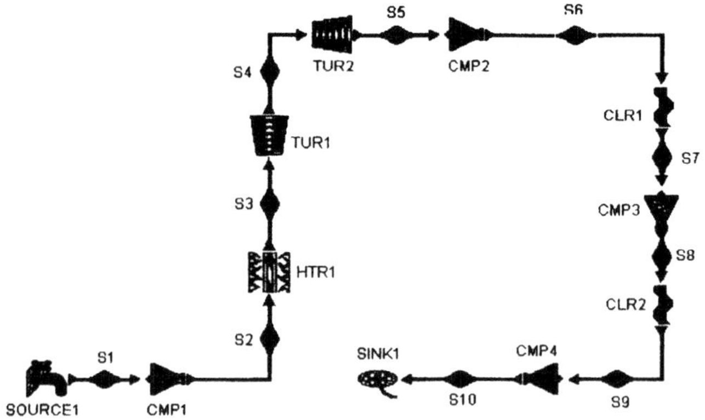

Figure 1. Example cycle.

In the cycle, air enters from the atmospheric source to an isentropic compressor at 293 K and 1 bar (state 1) and leaves at 8 bar (state 2); air enters the combustion chamber (heater) and leaves at 1373 K (state 3); air enters the high-pressure isentropic turbine and

leaves at 1 bar (state 4); air enters the low-pressure isentropic turbine and leaves at 0.04 bar (state 5); air enters the first-stage isentropic compressor and leaves at 0.2 bar (state 6); air enters the isobaric low-pressure inter-cooler and leaves at 293 K (state 7); air enters the second-stage isentropic compressor and leaves at 0.6 bar (state 8);
air enters the isobaric high-pressure inter-cooler and leaves at 293 K (state 9); and air enters the third-stage isentropic compressor and is discharged to the atmospheric sink at 1 bar (state 10). Once the cycle is built, the mode is changed to analysis.

The user can analyze a cycle the same way as traditionally done by defining each device (process) and inputting numerical values as shown in Fig. 2 and Fig. 3, respectively. The properties of the working fluid at each state, and the performance of each device and the cycle are quickly calculated by the software. The results of these calculation are displayed as shown in Fig. 4. The thermodynamic efficiency and the net power output of the cycle are 59.67% and 504.2 kW, respectively.

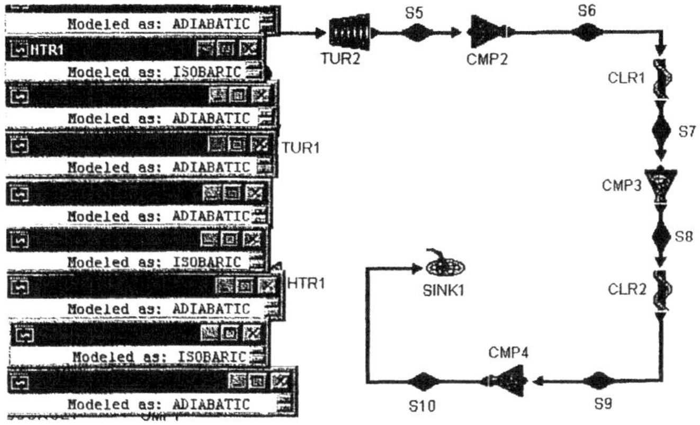

Figure 2. Processes of cycle.

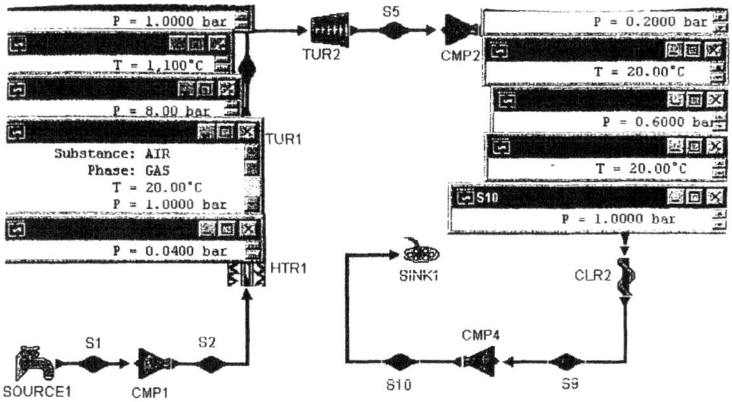

Figure 3. Input parameters of cycle

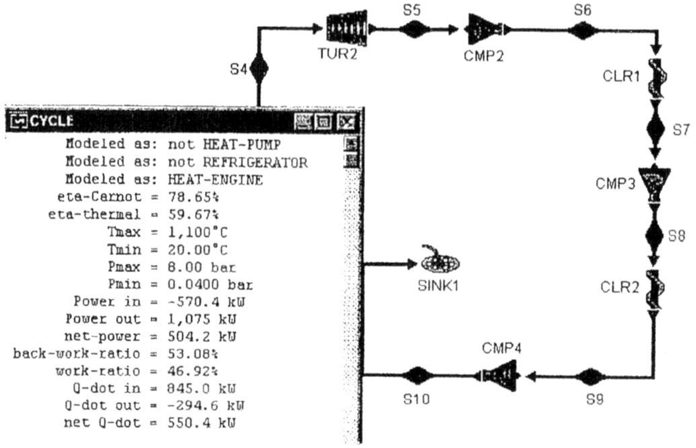

Figure 4. Results of cycle.

In order to find the effect of an input parameter to the performance of the combined cycle, one may do a sensitivity analysis. For example, one may choose the cycle efficiency (η) to be a dependent parameter, and the outlet pressure of the second-stage compressor (p_8) to be an independent parameter. Selecting a range for the pressure (p_8), a graphical (η-p_8) sensitivity analysis is immediately displayed as shown in Fig. 5. It is seen that the maximum cycle efficiency occurs at p_8 equals to 0.45 bar.

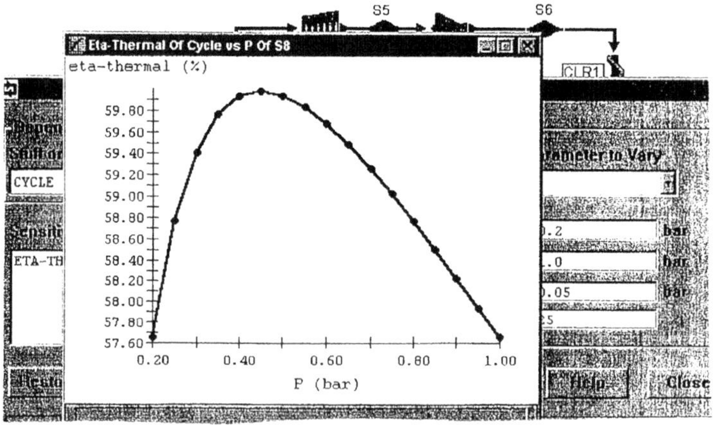

Figure 5. Graphical (η-p_8) sensitivity diagram.

Design of different arranged Brayton cycles can be easily performed by the usage of the computer software. Refinements of the three-stage compressor cycle could include more reheaters and more turbiness added, and more compressors and more inter-coolers added, etc.

7. Discussion

In the realm of thermodynamics, CyclePad is to a thermodynamist what a word processor is to a journalist. The benefits of using this software are numerous. The first is that significantly less time will be spent doing numerical analysis. As an engineer, this is much appreciated because computational work that would have taken hours before, can now be done in seconds. Second, as this paper has demonstrated, CyclePad is capable of analyzing thermodynamic systems with various working fluids. Third, due to its computer-assisted modeling capabilities, the software allows for individuals to immediately view the effects of varying input parameters, either through calculated results or in the form of graphs and diagrams, giving the user a greater appreciation of how a system actually works. More specifically, there is a feature that provides the user the opportunity to optimize a specified cycle parameter. This has been proven to be incredibly useful. Last, and most important, is the built-in coaching facility that provides definitions of terms and descriptions of calculations. CyclePad goes a step further by informing the user if a contradiction or an incompatibility exists between inputted parameters within a cycle and why.

When viewing the applicability of software of this nature on the drawing board, users could reap the benefits at all stages in an engineering career. For the young engineer just beginning the learning process, less time is spent doing iterations resulting in more time dedicated to reinforcing the fundamentals and gaining valuable experience. In the case of the seasoned engineer, who is well indoctrinated in the principles and has gained an "engineer's intuition", can augment his/her abilities by becoming more computer literate.

8. Conclusion

The intelligent computer software has been demonstrated to be a very effective tool in analyzing thermodynamic cycles. Any complicated cycle can be analyzed easily using CyclePad. The advantage of using CyclePad is not only does it provide quick and accurate cycle analysis predictions, but that it also forces the user to define each process in the cycle and to make reasonable input information to the cycle. Users using CyclePad in designing cycles could be freed from tedious table searching and hand calculations. They could spend more time doing optimization of operational parameters and selection of different combinations of working fluids of the thermodynamic systems.

Acknowledgement. This paper is a result of a research contract sponsored by ONR (Office of Naval Research) under the award number N0001498AF00002.

References

1. Forbus, K. (1994) Self-explanatory simulators: making computers partners in the modeling process, *Mathematics and Computers in Simulation* **36**, 91-97.
2. Wu, C. (1998) Using articulate virtual laboratories in teaching energy conversion at the U.S. Naval Academy, *Journal of Educational Technology Systems* **26**, 127-136.

A STUDY OF THE LARGE OSCILLATIONS OF THE THERMODYNAMIC PENDULLUM BY THE METHOD OF CUBICATION

G. STANESCU
Federal University of Paraná, Polytechnic Center, Mechanical Engineering Department, Box 19011, 81531 - 990 Curitiba, PR, Brazil

1. Introduction

Study of mechanically oscillating thermodynamic systems (thermodynamic pendullums or oscillators) has been neglected for a long time, partly because no exact analytic solution of the corresponding system of differential equations modeling the processes is known [1], and partly because of lack of practical interest. During the last two decades, space, industrial and domestic applications of free piston thermal machines (FPTMs) and progress in numerical methods have stimulated scientific interest in this field since the thermodynamic pendullum represents an appropriate physical model to study FPTMs functioning.

Study of small vibrations (perturbations) of thermodynamic oscillators [2 - 3] contributed to shape this new research area. It is very attractive because of simple analytic solutions, but did not provide general results. By considering the speed of processes to be infinitesimally small, Keller [2] determined the relaxation of small oscillations of a thermodynamic pendullum with friction. He used the classical thermodynamics laws, without conceptual difficulties generated by the finite non-vanishing speed of processes, and neglected simultaneously the inertia force of the moving piston mass. Due to the last hypothesis Keller's results are not suitable to model large oscillations and the FPTMs functioning. Brening [3] studied the dynamics of thermodynamic oscillators with thermal and mechanical attenuation. He developed a very laborious matrix method to determine approximately the characteristic frequencies of "pure thermal" and "pure mechanical" vibrating modes, corresponding to the limiting case of isothermal and adiabatic oscillator, respectively. Since it takes into account the heat exchange through the piston between the two gas masses, the Brening's method represents a better approach in studying FPTMs functioning and it makes possible to model the influence of heat flow into the regenerator on the FPTMs output.

FPTMs optimization to satisfy the "criteria of L^4" (low pollution, low cost, low loss, long life) emerged new problems related to the resonance of nonlinear oscillators [4] and to the existence and stability of limit cycles of motion [5]. The present paper presents a study of free and forced large vibrations of conservative and dissipative thermodynamic pendullums, based on the "cubication" of nonlinear oscillators method

A. Bejan and E. Mamut (eds.), Thermodynamic Optimization of Complex Energy Systems, 445–452.

of Yuste [5]. Similarly to the mathematical pendullum case [6], it is possible to calculate the period of large oscillations of the thermodynamic pendullum by using the elliptic integrals. We found that period of large oscillations depends on initial conditions (initial speed and position of the moving piston), contrarily to the small oscillations that develop during the same period [7].

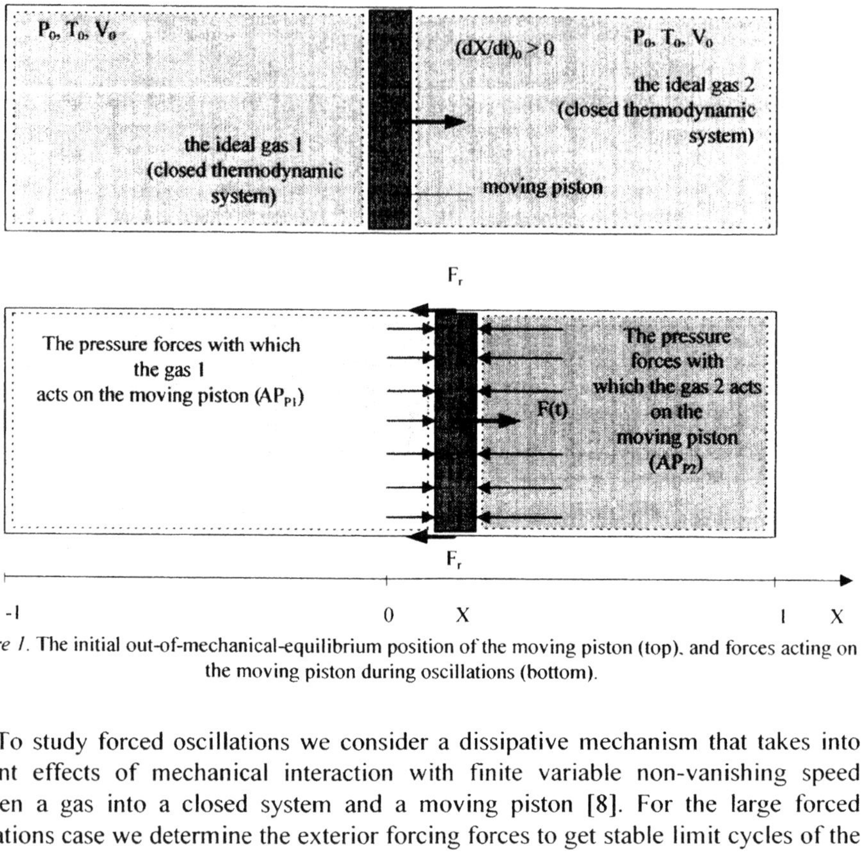

Figure 1. The initial out-of-mechanical-equilibrium position of the moving piston (top), and forces acting on the moving piston during oscillations (bottom).

To study forced oscillations we consider a dissipative mechanism that takes into account effects of mechanical interaction with finite variable non-vanishing speed between a gas into a closed system and a moving piston [8]. For the large forced oscillations case we determine the exterior forcing forces to get stable limit cycles of the movement of thermodynamic pendullum.

2. Physical Model

The thermodynamic pendullum or oscillator is modeled in here using a cylinder that contains a moving piston (Fig. 1). The piston is assumed free to move, with or without friction, and to provide perfect adiabatic insulation and to provide a seal between the gases in the cylinder. The two compartments within the cylinder, containing the same mass of an ideal gas, are each assumed to be closed system. The cylinder walls may be considered either adiabatic or non-adiabatic.

The piston motion is the result of forces, including externally imposed force function, continuously acting on it. The external forcing force may for instance be applied by a linear motor surrounding the cylinder. We assume that both external forcing force and frictional force depend on the instantaneous piston speed.

Thermodynamic processes occurring into gases during oscillations are assumed to be either adiabatic or isothermal. Initially, temperature of both gases is the same. For the sake of simplicity of equations, in Fig. 1 the zero point of the X-axis coincides with the initial out-of-mechanical-equilibrium position of the moving piston. The initial piston speed is assumed to be positive and non-zero. We consider a dissipating mechanism that takes into account losses generated by the finite speed of mechanical interaction between gases within the cylinder and the moving piston, as described in [8].

3. Mathematical Model

We consider the system in Fig. 1. The initial out-of-mechanical-equilibrium state is thus characterized by the same values of pressure P_0 and temperature T_0 of the ideal gas within the two compartments of the cylinder. For the sake of simplicity of equations we consider also that gases fill initially equal volumes. The heat produced by the mechanical friction between the piston and the cylinder does not return to the system, but is considered lost to the surroundings. Balancing the forces acting on the moving piston we can write the equation:

$$m\ddot{x} = P_1\left[1 - a\dot{x}/(3RT_1)^{1/2}\right]A - P_2\left[1 + a\dot{x}/(3RT_2)^{1/2}\right]A + F_r + F(t) \qquad (1)$$

where P_1, P_2 are the instantaneous pressures of gases, m is the mass of the moving piston, A is the cross section area of the piston, $F_r=-\mu^*\dot{x}$ is the friction force between the piston and the cylinder, and $F(t)=\eta^*\dot{x}$ is the exterior forcing force function, with coefficients $\mu^*>0$ and $\eta^*>0$.

To consider losses generated by the mechanical interaction with finite speed, we assume in Eq. (1) that the pressure with which a gas into a closed system acts on a moving piston is given by $P_P=P(1\pm aw/c)$ [8]. P is the interior instantaneous gas pressure, $w=\dot{x}$ is the instantaneous speed of the moving piston, $c=(3RT)^{1/2}$ is the average speed of molecules for the gas in the system, and $a=(3k)^{1/2}$ is a coefficient that depends on the nature of the gas, where $k=c_P/c_V$ is the ratio between the specific heats of gas at constant pressure and constant volume. The sign (+) corresponds to the case dV<0, and (-) sign to the case dV>0. The processes into gases are assumed to be reversible during oscillations, and therefore the instantaneous pressures and temperatures can be written

$$P_{1,2}(x) = P_0(1 \pm Ax/V_0)^{-n} \qquad T_{1,2}(x) = T_0(1 \pm Ax/V_0)^{1-n} \qquad (2a, b)$$

by using a polytropic function with a constant coefficient n. In Eqs. (2a) and (2b) x is the instantaneous position of the moving piston. We switch n=1 to study isothermal oscillations and n=k to study adiabatic oscillations.

Based on the Keller's idea, we calculate $P_1T_1^{-1/2}$ and $P_2T_2^{-1/2}$ in Eq. (1) and term into brackets in Eq. (2a) by using Maclaurin series expansions up to the fourth term

$$\left(1\pm\frac{Ax}{V_0}\right)^{-n}\cong 1\mp n\frac{Ax}{V_0}+\frac{n(n+1)}{2}\left(\frac{Ax}{V_0}\right)^2\mp\frac{n(n+1)(n+2)}{6}\left(\frac{Ax}{V_0}\right)^3 \quad (3a)$$

$$\left(1\pm\frac{Ax}{V_0}\right)^{-\frac{n+1}{2}}\cong 1\mp\frac{n+1}{2}\frac{Ax}{V_0}+\frac{(n+1)(n+3)}{8}\left(\frac{Ax}{V_0}\right)^2\mp\frac{(n+1)(n+3)(n+5)}{48}\left(\frac{Ax}{V_0}\right)^3 \quad (3b)$$

To get a higher precision we may consider, at least in principle, in the Maclaurin series expansions any large number of terms, but we will arrive to deal on more and more complicated differential equations. With $X = Ax / V_0$ we can rewrite Eq. (1)

$$\ddot{X}+c_1X+c_3X^3+\varepsilon(\alpha+\beta X^2)\dot{X}=0 \quad (4)$$

with coefficients $c_1=2nP_0A^2/(mV_0)$, $c_3=n(n+1)(n+2)P_0A^2/(3mV_0)$, $\varepsilon=aAP_0/[m(3RT_0)^{1/2}]$, $\alpha=2+(\mu^*-\eta^*)(3RT_0)^{1/2}/(aAP_0)$ and $\beta=(n+1)(n+3)/4$. We note in here that $-1 < X < 1$.

4. Method of Cubication

Solution of the autonomous non-linear oscillator in Eq. (4) can be approximated by the solution of associated "cubic oscillator"

$$\ddot{X}+c_1^*X+c_3^*X^3+\lambda^*\dot{X}=0 \quad (5)$$

as proved by Yuste [5]. Coefficients c_1^*, c_3^* and λ^* are determined using the principle of harmonic balance. With our notation the solution of Eq. (5) is given by

$$X(t) = X_m\,\mathrm{cn}(\omega t+\theta,\mu^2) = X_m\cos\Phi \quad (6a)$$

$$\Phi = \mathrm{am}(\omega t+\theta,\mu^2)\equiv \mathrm{am}(\psi,\mu^2) \quad (6b)$$

where X_m is the amplitude of oscillations, and

$$\mathrm{sn}(\psi,\mu^2) = \sin\Phi = \sin[\mathrm{am}(\psi,\mu^2)] \quad (7a)$$

$$\mathrm{cn}(\psi,\mu^2) = \cos\Phi = \cos[\mathrm{am}(\psi,\mu^2)] \quad (7b)$$

$$\mathrm{dn}(\psi,\mu^2) = \Delta\Phi = \{1-\mu^2\mathrm{sn}^2[\mathrm{am}(\psi,\mu^2)]\}^{1/2} \quad (7c)$$

are the Jacobi elliptic functions [9]. In defining the elliptic integral of the first kind ψ

$$\psi(\Phi,\mu^2)=\int_0^{\Phi}\frac{d\xi}{\sqrt{1-\mu^2\sin^2\xi}} \quad (8)$$

Φ is the amplitude and μ is the modulus.

In Eqs. (6a) and (6b) the frequency ω and the modulus μ can be written

$$\omega^2 = c_1^*+c_3^*X_m^2 \qquad \mu^2 = (c_3^*X_m^2/2)/(c_1^*+c_3^*X_m^2) \quad (9a, b)$$

For the "cubic" oscillator in Eq. (5), Yuste found $c_1^* = c_1$ and $c_3^* = c_3$. In our case coefficient λ^* is given by

$$\lambda^* = -\varepsilon[\,\alpha+\beta(T/Q)X_m^2\,] \quad (10)$$

where

$$Q=\int_0^{2\pi}\sin^2\Phi\Delta\Phi d\Phi \qquad T=\int_0^{2\pi}\sin^2\Phi\cos^2\Phi\Delta\Phi d\Phi \quad (11a, b)$$

The method of harmonic balance is especially useful to study the stability of limit cycles. In Eq. (5) we note the existence of such stable limit cycles for $\lambda^*= 0$. The trivial solution of the last equation corresponds to $a=\mu^*=\eta^*=0$. In this case we study the large oscillations of a conservative thermodynamic pendullum. When studying the limit cycles of the large forced oscillations with friction, the trivial solution of equation $\lambda^*=0$ corresponds to $a=0$ and $\mu^*=\eta^*$. Requiring $a\neq 0$, $\mu^*\neq 0$ and $\eta^*\neq 0$, we determine the non-trivial solution of equation $\lambda^*=0$ for large forced oscillations of a dissipative thermodynamic pendullum.

4.1. DISSIPATIVE THERMODYNAMIC PENDULLUM

For the thermodynamic pendullum in Fig. 1 we are interested in determining solutions of equation $\lambda^* = 0$, when $a \neq 0$, $\mu^* \neq 0$ and $\eta^* \neq 0$. To simplify the computation we set, without loss of generality, $P_0A^2 / (mV_0) = 1\ s^{-2}$. Coefficients in Eq. (5) become now $c_1^* = 2n$, $c_3^* = n(n+1)(n+2)/3$, $\varepsilon = aAP_0/[m(3RT_0)^{1/2}]$, $\alpha = 2 + (\mu^*-\eta^*)(3RT_0)^{1/2}/(aAP_0)$ and $\beta=(n+1)(n+3)/4$. There is no difficulty in obtaining numerical results when considering any other value of parameter $P_0A^2/(mV_0)$. To determine the limit cycles of oscillations of the dissipative thermodynamic pendullum we rewrite equation $\lambda^*=0$ in the equivalent form

$$X_m^2 = -\frac{\alpha}{\beta(n)}\frac{Q(\nu)}{T(\nu)} \tag{12a}$$

Eq. (12a) combined with the equality

$$X_m^2 = \frac{c_1^*(n)}{c_3^*(n)}\nu = \frac{6\nu}{(n+1)(n+2)} \tag{12b}$$

yield the following system of two algebraic equations

$$X_m^2 = -\frac{4\alpha}{(n+1)(n+3)}\frac{Q(\nu)}{T(\nu)} \qquad -\frac{3}{2}\frac{(n+3)}{(n+2)}\frac{\nu}{\alpha} = \frac{Q(\nu)}{T(\nu)} \tag{13a, b}$$

where $\nu = c_3^* X_m^2 / c_1^*$ is the non-linearity factor, $En = c_1^* X_m^2 (1+\nu/2)$ is the total energy of oscillation, and $V_m = (c_1^*)^2/(2c_3^*)$ is the maximum potential, as defined in [5].

It is apparent in Eqs. (11a) and (11b) that $Q(\nu)/T(\nu) > 0$, and therefore the system (13a), (13b) gets real solutions when

$$-\frac{3(n+3)}{2(n+2)}\frac{1}{\alpha} \leq -5 \qquad 0 < -\frac{4\alpha Q(\nu)/T(\nu)}{(n+1)(n+3)} < 1 \tag{14a, b}$$

or

$$-\frac{3(n+3)}{2(n+2)}\frac{1}{\alpha} \geq -1.329 \qquad 0 < -\frac{4\alpha Q(\nu)/T(\nu)}{(n+1)(n+3)} < 1 \tag{14a', b'}$$

If we arrange Eq. (13b) as $r\nu = Q(\nu)/T(\nu)$, the two limiting values of slope $r=-5$ and $r=-1.329$ correspond in Fig. 2 to the positions of line $r\nu$ when it is tangent to the $Q(\nu)/T(\nu)$ curve. The limit cycles can be written

$$X(t) = X_S\, \mathrm{cn}(\psi, \mu^2) \qquad \dot{X}(t) = -\omega\, X_S\, \mathrm{sn}(\psi, \mu^2)\, \mathrm{dn}(\psi, \mu^2) \tag{15a, b}$$

We label the solutions of system (12a) and (12b) by the subscript S, (X_S, ν_S). For $n=1$ or $n=k$ we study a hard oscillator ($c_1^* > 0$ and $c_3^* > 0$). This is the case when $0 \leq \mu^2 \leq 1/2$ and $0 \leq \nu < \infty$, with $\omega^2 = c_1^*\,(1+\nu_S)$ in terms of the non-linearity factor. The period of X(t) is $4K(\mu^2)$, where $K(\mu^2) = \psi(\tfrac{1}{2}\pi, \mu^2)$ is the complete elliptic integral of the first kind. The analytical expressions of Q and T functions are given by [5]

$$Q = (4 / 3\mu^2)[(2\mu^2 - 1)E(\mu^2) + (1 - \mu^2)K(\mu^2)] \tag{16a}$$

$$T = (4 / 15\mu^4)[2(\mu^4 + 1 - \mu^2)E(\mu^2) + (1 - \mu^2)(\mu^2 - 2)K(\mu^2)] \tag{16b}$$

where $\mu^2 = \nu_S / [2(1 + \nu_S)]$ and $E(\mu^2)$ is the complete elliptic integral of the second kind.

4.2. CONSERVATIVE THERMODYNAMIC PENDULLUM

Equation of large free oscillations ($\eta^* = 0$) of the conservative thermodynamic pendullum ($\mu^* = 0$) is written

$$\ddot{X} = -(c_1^* X + c_3^* X^3) \tag{17}$$

In here we neglect also the losses generated by the mechanical interaction with finite speed between gases in the cylinder and the moving piston (a=0). We consider the following initial conditions in Eq. (17): $X(0)=0$ and $\dot{X}(0)=\dot{X}_0>0$. Multiplying Eq. (17) by $2\dot{X}\,dt=2dX$ we have

$$d[(\dot{X})^2] = -(2c_1^* X + 2c_3^* X^3)dX \tag{18}$$

By integrating the left-hand side of the last equation between $\dot{X}_0$ and $\dot{X}$, and the right-hand side between 0 and X, we obtain

$$\dot{X}^2 - \dot{X}_0^2 = -(c_1^* X^2 + c_3^* X^4 / 2) \tag{19}$$

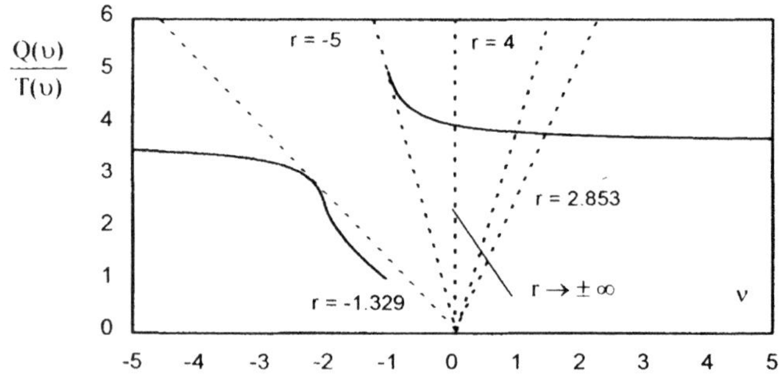

Figure 2. Solutions of Eq. (13b) for adiabatic oscillations (n = k = 1.4) plot the curve Q(v)/T(v) (heavy line) for $0 \le v \le 1.332$ ($r \ge 2.853$), and for isothermal oscillations (n=1) for $0 \le v \le 0.982$ ($r \ge 4$).

or, by deriving the instantaneous speed of the moving piston

$$dX/dt = \dot{X}_0\left[1 - X^2\left(c_1^* + c_3^* X^2/2\right)/\dot{X}_0^2\right]^{1/2} \tag{20}$$

We use the (+) sign in front of the square root since we considered before that $\dot{X}_0>0$, and therefore $dX/dt>0$ during the first quarter period. Separating the variables in Eq. (20) and integrating dt between 0 and t, we find

$$t = \frac{1}{\dot{X}_0}\int_0^X \frac{dy}{\sqrt{(1 - u^2 y^2)(1 + v^2 y^2)}} \tag{21}$$

with $u^2 = \frac{c_1^*}{2\dot{X}_0^2} + \left[\left(\frac{c_1^*}{2\dot{X}_0^2}\right)^2 + \frac{c_3^*}{2\dot{X}_0^2}\right]^{1/2}$ $\quad v^2 = -\frac{c_1^*}{2\dot{X}_0^2} + \left[\left(\frac{c_1^*}{2\dot{X}_0^2}\right)^2 + \frac{c_3^*}{2\dot{X}_0^2}\right]^{1/2}$ (22a, b)

Time t may be written as

$$t = \frac{1}{u\dot{X}_0} \int_0^{uX} \frac{d\xi}{\sqrt{(1-\xi^2)(1+w^2\xi^2)}} \tag{23}$$

when considering $\xi = uy$, and $w^2 = v^2 / u^2$. Now if we let $v^2 / (u^2 + v^2) = \mu^2$ and $\xi = \cos\zeta$, Eq. (23) is written

$$t = \frac{1}{\dot{X}_0\sqrt{u^2+v^2}} \int_{\pi/2}^{\Phi} \frac{-d\zeta}{\sqrt{1-\mu^2\sin^2\zeta}} = \frac{K(\mu^2) - \psi(\Phi,\mu^2)}{\dot{X}_0\sqrt{u^2+v^2}} \tag{24}$$

To derive X as an explicit function of time we rewrite the last equation as

$$\psi(\Phi, \mu^2) = K(\mu^2) - \dot{X}_0 t(u^2+v^2)^{1/2} \tag{25}$$

and, since μ^2 is a constant, we find

$$\Phi = \text{am}[K(\mu^2) - \dot{X}_0 t(u^2+v^2)^{1/2}] \tag{26}$$

Finally, we obtain

$$X(t) = X_m \cos\Phi = X_m \,\text{cn}\{[K(\mu^2) - \dot{X}_0 t(u^2+v^2)^{1/2}], \mu^2\} \tag{27}$$

with the amplitude X_m=1/u and the frequency ω

$$\omega = \dot{X}_0(u^2+v^2)^{1/2} \tag{28}$$

The instantaneous speed of the moving piston is given by

$$\dot{X}(t) = X_m\dot{X}_0(u^2+v^2)^{1/2}\sin\Phi\Delta\Phi \tag{29}$$

5. Numerical Results

Fig. 3 shows comparatively the analytic approximate limit cycle of oscillations of an adiabatic dissipative thermodynamic pendullum, given by Eq. (5), and the results of numerical integration of Eq. (4). We assumed the ideal gas, filling both compartments within the cylinder, to be diatomic (n=k=1.4), and chose α=-0.3 and ε=0.85. With these values Eq. (4) is written

$$\ddot{X} + 2.8X + 3.808X^3 + 0.85(2.64X^2 - 0.3)\dot{X} = 0 \tag{30}$$

The instantaneous position and speed of the moving piston are given by

$$X(t) = 0.647\ \text{cn}(2.129t, 0.191) \tag{31a}$$

$$\dot{X}(t) = -1.435\ \text{sn}(2.129t, 0.191)\times\text{dn}(2.129t, 0.191) \tag{31b}$$

with the frequency ω = 2.129 s^{-1}, the modulus μ = 0.437 and the period 4K(0.191)=6.62.

Solutions of systems (14a), (14b) and (14a′), (14b′) are $-0.675\leq\alpha<0$ for adiabatic oscillations, when n=k=1.4, and $-0.5\leq\alpha<0$ for isothermal oscillations, when n=1. The parameter α is indicative for the external forcing force F(t). It is apparent from the mathematical expression of η^* that F(t) depends also on the values of coefficient a, that accounts for dissipation. To get stable limit cycles of oscillations, the driving effort increases for large dissipation.

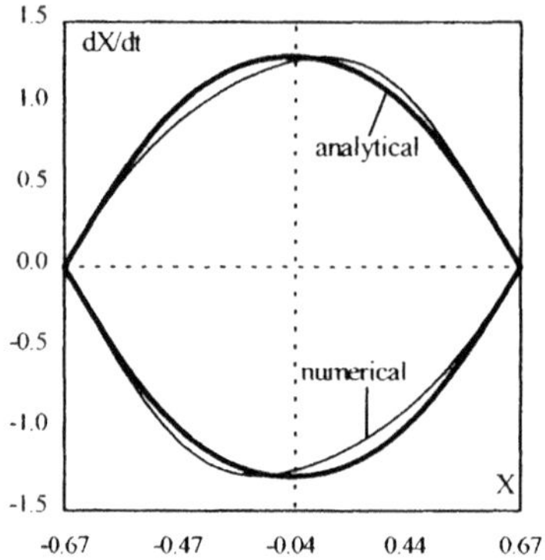

Figure 3. Numerical (thin line) and approximate analytical (heavy line) limit cycles of oscillations of the adiabatic dissipative thermodynamic pendullum Eq. (30).

6. Conclusions

The described method provides a tool to study free and forced non-linear oscillations of thermodynamic pendullums. By modeling actual FPTMs as thermodynamic pendullums it is possible to investigate, based on this method, important issues of their design, optimization and functioning, such as the existence and stability of limit cycles of oscillations.

For an external forcing force function depending on the instantaneous speed of the moving piston, it is also possible to determine its characteristics such that to obtain stable limit cycles. The method can be easily extended to study resonance in FPTMs functioning and to minimize the driving effort and maximize the FPTMs ouptut.

7. References

1. Minorsky, N. (1947) *Introduction to Non-Linear Mechanic*, Edwards Brothers, Inc., Ann Arbor, MI.
2. Keller, J. U. (1977) *Thermodynamik der Irreversiblen Prozesse*, Walter de Gruyter & Co., New York.
3. Brening, W. (1989) *Statistical Theory of Heat. Nonequilibrium Phenomena*, Springer-Verlag, Heidelberg.
4. Wargitsch, C. and Hübler, A. (1995) Resonances of non-linear oscillators, *Physical Review E* **51**, 1508-1519.
5. Yuste, B. S. (1992) Cubication of non-linear oscillators using the principle of harmonic balance, *Int. J. Non-Linear Mechanics* **27**, 347-356.
6. Salvadori, M. G. and Schwarz, R. J. (1956) *Differential Equations in Engineering Problems*, Englewood Cliffs, Prentice-Hall, Inc.
7. Petrescu, S., Stanescu, G., and Iordache, R. (1993) Study of reversible thermodynamicpendullum in the small oscillations hypothesis, *Energetica Review* **41**, 66-70, Bucharest.
8. Petrescu, S. and Harman, C. M. (1994) The connection between the first and second laws of thermodynamics for processes with finite speed. A direct method for approaching and optimization of irreversible processes, *Journal of Heat Transfer Society of Japan* **33**, 60-67.
9. Lösch, F. (1960) *Tables of Higher Functions*, McGraw-Hill Book Company, Inc., New York.
10. *Handbook of Tables for Mathematics* (1975), CRC Press, Inc., Cleveland, Ohio.

Author Index

Subject Index

GPSR Compliance

The European Union's (EU) General Product Safety Regulation (GPSR) is a set of rules that requires consumer products to be safe and our obligations to ensure this.

If you have any concerns about our products, you can contact us on ProductSafety@springernature.com

In case Publisher is established outside the EU, the EU authorized representative is:

Springer Nature Customer Service Center GmbH
Europaplatz 3
69115 Heidelberg, Germany

Batch number: 10755284

Printed by Printforce, the Netherlands